Features of this Text

Who will benefit from using this text?

This text can be used in Sophomore, Junior or Senior level courses in probability and stochastic processes. The mathematical exposition will appeal to students and practitioners in many areas. The examples, quizzes, and problems are typical of those encountered by practicing electrical and computer engineers. Professionals in the telecommunications and wireless industry will find it particularly useful.

What's New?

This text has been expanded with new introductory material:

- Over 160 new homework problems
- New chapters on *Sequential Trials, Derived Random Variables* and *Conditional Probability Models.*
- MATLAB examples and problems give students hands-on access to theory and applications. Every chapter includes guidance on how to use MATLAB to perform calculations and simulations relevant to the subject of the chapter.
- Advanced material online in *Signal Processing* and *Markov Chains* supplements.

Notable Features

The Friendly Approach

The friendly and accessible writing style gives students an intuitive feeling for the formal mathematics.

Quizzes and Homework Problems

An extensive collection of in-chapter quizzes provides checkpoints for readers to gauge their understanding. Hundreds of end-of-chapter problems are clearly marked as to their degree of difficulty from beginner to expert.

Student Companion Website www.wiley.com/college/yates

Available for download: All MATLAB m-files in the text, the *Quiz Solutions Manual*, a *Student Solutions Manual*, the *Signal Processing Supplement*, and the *Markov Chains Supplement*.

Instructor Support

Instructors can register for the Instructor Companion Site at www.wiley.com/college/yates

Probability and Stochastic Processes

A Friendly Introduction
for Electrical and Computer Engineers

Third Edition

International Student Version

Roy D. Yates
Rutgers, The State University of New Jersey

David J. Goodman
New York University

To Alissa, Brett, Daniel, Hannah, Leila, Milo, Theresa, Tony, and Zach

Preface

Welcome to the third edition

You are reading the international student version (ISV) of the third edition of our textbook. Although the fundamentals of probability and stochastic processes have not changed since we wrote the first edition, the world inside and outside universities is different now than it was in 1998. Outside of academia, applications of probability theory have expanded enormously in the past 16 years. Think of the 20 billion+ Web searches each month and the billions of daily computerized stock exchange transactions, each based on probability models, many of them devised by electrical and computer engineers.

Universities and secondary schools, recognizing the fundamental importance of probability theory to a wide range of subject areas, are offering courses in the subject to younger students than the ones who studied probability 16 years ago. At Rutgers, probability is now a required course for Electrical and Computer Engineering sophomores.

We have responded in several ways to these changes and to the suggestions of students and instructors who used the earlier editions. The first and second editions contain material found in postgraduate as well as advanced undergraduate courses. By contrast, the printed and e-book versions of this third edition focus on the needs of undergraduates studying probability for the first time.

The more advanced material in the earlier editions, covering random signal processing and Markov chains, is available at the companion website (`www.wiley.com/college/yates`). To promote intuition into the practical applications of the mathematics, we have expanded the number of examples and quizzes and homework problems to about 600, an increase of about 35 percent compared to the second edition. Many of the examples are mathematical exercises. Others are questions that are simple versions of the ones encountered by professionals working on practical applications.

In response to suggestions from international readers of the earlier editions, this third edition ISV is shorter than the United States third edition by about 35 pages. It contains less introductory material and fewer elementary examples. We have also omitted a few topics that are not central to the interests of electrical and computer and engineers. These topics are reliability analysis, confidence interval estimation, and random sums of random variables.

How the book is organized

Motivated by our teaching experience, we have rearranged the sequence in which we present the elementary material on probability models, counting methods, conditional probability models, and derived random variables. In this edition, the first chapter covers fundamentals, including axioms and probability of events, and the second chapter covers counting methods and sequential experiments. As before, we introduce discrete random variables and continuous random variables in separate chapters. The subject of Chapter 5 is multiple discrete and continuous random variables. The first and second editions present derived random variables and conditional random variables in the introductions to discrete and continuous random variables. In this third edition, derived random variables and conditional random variables appear in their own chapters, which cover both discrete and continuous random variables.

Chapter 8 introduces random vectors. It extends the material on multiple random variables in Chapter 5 and relies on principles of linear algebra to derive properties of random vectors that are useful in real-world data analysis and simulations. Chapter 12 on estimation relies on the properties of random vectors derived in Chapter 8. Chapters 9 through 12 cover subjects relevant to data analysis including Gaussian approximations based on the central limit theorem, estimates of model parameters, hypothesis testing, and estimation of random variables. Chapter 13 introduces stochastic processes in the context of the probability model that guides the entire book: an experiment consisting of a procedure and observations.

Each of the 92 sections of the 13 chapters ends with a quiz. By working on the quiz and checking the solution at the book's website, students will get quick feedback on how well they have grasped the material in each section.

We think that 60-80% (7 to 10 chapters) of the book would fit into a one semester undergraduate course for beginning students in probability. We anticipate that all courses will cover the first five chapters, and that instructors will select the remaining course content based on the needs of their students. The "roadmap" on page ix displays the thirteen chapter titles and suggests a few possible undergraduate syllabi.

The Signal Processing Supplement (SPS) and Markov Chains Supplement (MCS) are the final chapters of the third edition. They are now available at the book's website. They contain postgraduate-level material. We, and colleagues at other universities, have used these two chapters in graduate courses that move very quickly through the early chapters to review material already familiar to students and to fill in gaps in learning of diverse postgraduate populations.

What is distinctive about this book?

- The entire text adheres to a single model that begins with an experiment consisting of a procedure and observations.

- The mathematical logic is apparent to readers. Every fact is identified clearly as a definition, an axiom, or a theorem. There is an explanation, in simple English, of the intuition behind every concept when it first appears in the text.

- The mathematics of discrete random variables is introduced separately from the mathematics of continuous random variables.

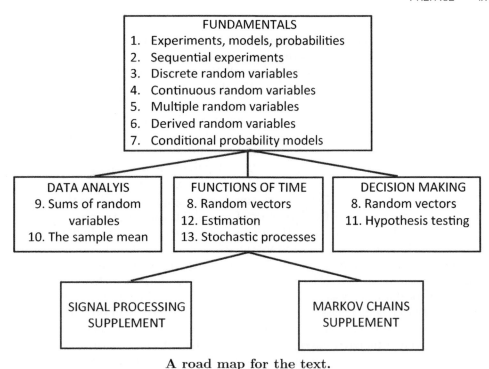

A road map for the text.

- Stochastic processes and statistical inference fit comfortably within the unifying model of the text.

- An abundance of exercises puts the theory to use. New ideas are augmented with detailed solutions of numerical examples.

- Each section begins with a brief statement of the important concepts introduced in the section and concludes with a simple quiz to help students gauge their grasp of the new material.

- Each problem at the end of a chapter is labeled with a reference to a section in the chapter and a degree of difficulty ranging from "easy" to "experts only." For example Problem 3.4.5 requires material from Section 3.4 but not from later sections. Each problem also has a label that reflects our estimate of degree of difficulty. Skiers will recognize the following symbols:

  ● Easy   ■ Moderate   ◆ Difficult   ◆◆ Experts Only

  Every ski area emphasizes that these designations are relative to the trails at that area. Similarly, the difficulty of our problems is relative to the other problems in this text.

- There is considerable support on the World Wide Web for students and instructors, including MATLAB programs and solutions to the quizzes and problems.

Further Reading

Libraries and bookstores contain an endless collection of textbooks at all levels covering the topics presented in this textbook. We know of two in comic book format [GS93, Pos01]. The reference list on page 453 is a brief sampling of books that can add breadth or depth to the material in this text. Most books on probability, statistics, stochastic processes, and random signal processing contain expositions of the basic principles of probability and random variables, covered in Chapters 1–5. In advanced texts, these expositions serve mainly to establish notation for more specialized topics. [LG11] and [Gub06] share our focus on electrical and computer engineering applications. [BT08], [Ros12] and [Dra67] and introduce the fundamentals of probability and random variables to a general audience of students with a calculus background. [KMT12] is a comprehensive graduate level textbook with a thorough presentation of fundamentals of probability, stochastic processes, and data analysis. It uses the basic theory to develop techniques including hidden Markov models, queuing theory, and machine learning used in many practical applications. [Bil12] is more advanced mathematically; it presents probability as a branch of measure theory. [MR10] and [SMM10] introduce probability theory in the context of data analysis. [Dav10] and [HL11] are beginners' introductions to MATLAB. [Ber98] is in a class by itself. It presents the concepts of probability from a historical perspective, focusing on the lives and contributions of mathematicians and others who stimulated major advances in probability and statistics and their application in various fields including psychology, economics, government policy, and risk management.

Acknowledgments

We are grateful for assistance and suggestions from many sources including our students at Rutgers and New York Universities, instructors who adopted the previous editions, reviewers, and the Wiley team.

At Wiley, we are pleased to acknowledge the encouragement and enthusiasm of our executive editor Daniel Sayre and the support of sponsoring editor Mary O'Sullivan, project editor Ellen Keohane, production editor Eugenia Lee, and cover designer Samantha Low.

We also convey special thanks to Ivan Seskar of WINLAB at Rutgers University for exercising his magic to make the WINLAB computers particularly hospitable to the electronic versions of the book and to the supporting material on the World Wide Web.

The organization and content of the second edition has benefited considerably from the input of many faculty colleagues including Alhussein Abouzeid at Rensselaer Polytechnic Institute, Krishna Arora at Florida State University, Frank Candocia at Florida International University, Robin Carr at Drexel University, Keith Chugg at USC, Charles Doering at University of Michigan, Roger Green at North Dakota State University, Witold Krzymien at University of Alberta, Edl Schamiloglu at University of New Mexico, Arthur David Snider at University of South Florida, Junshan Zhang at Arizona State University, and colleagues

Narayan Mandayam, Leo Razumov, Christopher Rose, Predrag Spasojević, and Wade Trappe at Rutgers.

Unique among our teaching assistants, Dave Famolari took the course as an undergraduate. Later as a teaching assistant, he did an excellent job writing homework solutions with a tutorial flavor. Other graduate students who provided valuable feedback and suggestions on the first edition include Ricki Abboudi, Zheng Cai, Pi-Chun Chen, Sorabh Gupta, Vahe Hagopian, Amar Mahboob, Ivana Maric, David Pandian, Mohammad Saquib, Sennur Ulukus, and Aylin Yener.

The first edition also benefited from reviews and suggestions conveyed to the publisher by D.L. Clark at California State Polytechnic University at Pomona, Mark Clements at Georgia Tech, Gustavo de Veciana at the University of Texas at Austin, Fred Fontaine at Cooper Union, Rob Frohne at Walla Walla College, Chris Genovese at Carnegie Mellon, Simon Haykin at McMaster, and Ratnesh Kumar at the University of Kentucky.

Finally, we acknowledge with respect and gratitude the inspiration and guidance of our teachers and mentors who conveyed to us when we were students the importance and elegance of probability theory. We cite in particular Robert Gallager and the late Alvin Drake of MIT and the late Colin Cherry of Imperial College of Science and Technology.

A Message to Students from the Authors

A lot of students find it hard to do well in this course. We think there are a few reasons for this difficulty. One reason is that some people find the concepts hard to use and understand. Many of them are successful in other courses but find the ideas of probability difficult to grasp. Usually these students recognize that learning probability theory is a struggle, and most of them work hard enough to do well. However, they find themselves putting in more effort than in other courses to achieve similar results.

Other people have the opposite problem. The work looks easy to them, and they understand everything they hear in class and read in the book. There are good reasons for assuming this is easy material. There are very few basic concepts to absorb. The terminology (like the word *probability*), in most cases, contains familiar words. With a few exceptions, the mathematical manipulations are not complex. You can go a long way solving problems with a four-function calculator.

For many people, this apparent simplicity is dangerously misleading because it is very tricky to apply the math to specific problems. A few of you will see things clearly enough to do everything right the first time. However, most people who do well in probability need to practice with a lot of examples to get comfortable with the work and to really understand what the subject is about. Students in this course end up like elementary school children who do well with multiplication tables and long division but bomb out on word problems. The hard part is figuring out what to do with the numbers, not actually doing it. Most of the work in this course is that way, and the only way to do well is to practice a lot. Taking the midterm and final are similar to running in a five-mile race. Most people can do it in a respectable time, provided they train for it. Some people look at the runners who do it and say, "I'm as strong as they are. I'll just go out there and join in." Without the training, most of them are exhausted and walking after a mile or two.

So, our advice to students is, if this looks really weird to you, keep working at it. You will probably catch on. If it looks really simple, don't get too complacent. It may be harder than you think. Get into the habit of doing the quizzes and problems, and if you don't answer all the quiz questions correctly, go over them until you understand each one.

We can't resist commenting on the role of probability and stochastic processes in our careers. The theoretical material covered in this book has helped both of us devise new communication techniques and improve the operation of practical systems. We hope you find the subject intrinsically interesting. If you master the basic ideas, you will have many opportunities to apply them in other courses and throughout your career.

We have worked hard to produce a text that will be useful to a large population of students and instructors. We welcome comments, criticism, and suggestions. Feel free to send us e-mail at *ryates@winlab.rutgers.edu* or *dgoodman@poly.edu*. In addition, the website `www.wiley.com/college/yates` provides a variety of supplemental materials, including the MATLAB code used to produce the examples in the text.

Roy D. Yates
Rutgers, The State University of New Jersey

David J. Goodman
New York University

February 19, 2014

Contents

1

Experiments, Models, and Probabilities

Getting Started with Probability

The title of this book is *Probability and Stochastic Processes*. We say and hear and read the word *probability* and its relatives (*possible, probable, probably*) in many contexts. Within the realm of applied mathematics, the meaning of *probability* is a question that has occupied mathematicians, philosophers, scientists, and social scientists for hundreds of years.

Everyone accepts that the probability of an event is a number between 0 and 1. Some people interpret probability as a physical property (like mass or volume or temperature) that can be measured. This is tempting when we talk about the probability that a coin flip will come up heads. This probability is closely related to the nature of the coin. Fiddling around with the coin can alter the probability of heads.

Another interpretation of probability relates to the knowledge that we have about something. We might assign a low probability to the truth of the statement, *It is raining now in Phoenix, Arizona*, because we know that Phoenix is in the desert. However, our knowledge changes if we learn that it was raining an hour ago in Phoenix. This knowledge would cause us to assign a higher probability to the truth of the statement, *It is raining now in Phoenix*.

Both views are useful when we apply probability theory to practical problems. Whichever view we take, we will rely on the abstract mathematics of probability, which consists of definitions, axioms, and inferences (theorems) that follow from the axioms. While the structure of the subject conforms to principles of pure logic, the terminology is not entirely abstract. Instead, it reflects the practical origins of probability theory, which was developed to describe phenomena that cannot be predicted with certainty. The point of view is different from the one we took when we started studying physics. There we said that if we do the same thing in the same way over and over again — send a space shuttle into orbit, for example —

the result will always be the same. To predict the result, we have to take account of all relevant facts.

The mathematics of probability begins when the situation is so complex that we just can't replicate everything important exactly, like when we fabricate and test an integrated circuit. In this case, repetitions of the same procedure yield different results. The situation is not totally chaotic, however. While each outcome may be unpredictable, there are consistent patterns to be observed when we repeat the procedure a large number of times. Understanding these patterns helps engineers establish test procedures to ensure that a factory meets quality objectives. In this repeatable procedure (making and testing a chip) with unpredictable outcomes (the quality of individual chips), the *probability* is a number between 0 and 1 that states the proportion of times we expect a certain thing to happen, such as the proportion of chips that pass a test.

As an introduction to probability and stochastic processes, this book serves three purposes:

- It introduces students to the logic of probability theory.

- It helps students develop intuition into how the theory relates to practical situations.

- It teaches students how to apply probability theory to solving engineering problems.

To exhibit the logic of the subject, we show clearly in the text three categories of theoretical material: definitions, axioms, and theorems. Definitions establish the logic of probability theory, and axioms are facts that we accept without proof. Theorems are consequences that follow logically from definitions and axioms. Each theorem has a proof that refers to definitions, axioms, and other theorems. Although there are dozens of definitions and theorems, there are only three axioms of probability theory. These three axioms are the foundation on which the entire subject rests. To meet our goal of presenting the logic of the subject, we could set out the material as dozens of definitions followed by three axioms followed by dozens of theorems. Each theorem would be accompanied by a complete proof.

While rigorous, this approach would completely fail to meet our second aim of conveying the intuition necessary to work on practical problems. To address this goal, we augment the purely mathematical material with a large number of examples of practical phenomena that can be analyzed by means of probability theory. We also interleave definitions and theorems, presenting some theorems with complete proofs, presenting others with partial proofs, and omitting some proofs altogether. We find that most engineering students study probability with the aim of using it to solve practical problems, and we cater mostly to this goal. We also encourage students to take an interest in the logic of the subject — it is very elegant — and we feel that the material presented is sufficient to enable these students to fill in the gaps we have left in the proofs.

Therefore, as you read this book you will find a progression of definitions, axioms, theorems, more definitions, and more theorems, all interleaved with examples and comments designed to contribute to your understanding of the theory. We also include brief quizzes that you should try to solve as you read the book. Each one will help you decide whether you have grasped the material presented just before the quiz. The problems at the end of each chapter give you more practice applying the material introduced in the chapter. They vary considerably in their level of difficulty. Some of them take you more deeply into the subject than the examples and quizzes do.

1.1   Applying Set Theory to Probability

> Probability is based on a repeatable experiment that consists of a procedure and observations. An *outcome* is an observation. An *event* is a set of outcomes.

The mathematical basis of probability is the theory of sets. We assume you have already encountered set theory and are familiar with such terms as *set, element, union, intersection* and *complement*. To work with sets mathematically it is necessary to define a *universal set*. This is the set of all things that we could possibly consider in a given context. We will use the letter S to denote the universal set. The *null set*, which is also important, may seem like it is not a set at all. By definition it has no elements. The notation for the null set is $\varnothing$. By definition $\varnothing$ is a subset of every set. For any set A, $\varnothing \subset A$.

In working with probability we will often refer to two important properties of collections of sets. Here are the definitions.

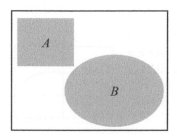

A collection of sets $A_1, \ldots, A_n$ is *mutually exclusive* if and only if

$$A_i \cap A_j = \varnothing, \qquad i \neq j. \tag{1.1}$$

The word *disjoint* is sometimes used as a synonym for mutually exclusive.

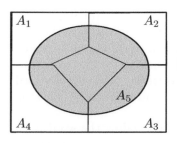

A collection of sets $A_1, \ldots, A_n$ is *collectively exhaustive* if and only if

$$A_1 \cup A_2 \cup \cdots \cup A_n = S. \tag{1.2}$$

In the definition of *collectively exhaustive*, we used the somewhat cumbersome notation $A_1 \cup A_2 \cup \cdots \cup A_n$ for the union of N sets. Just as $\sum_{i=1}^{n} x_i$ is a shorthand for $x_1 + x_2 + \cdots + x_n$, we will use a shorthand for unions and intersections of n sets:

$$\bigcup_{i=1}^{n} A_i = A_1 \cup A_2 \cup \cdots \cup A_n, \tag{1.3}$$

$$\bigcap_{i=1}^{n} A_i = A_1 \cap A_2 \cap \cdots \cap A_n. \tag{1.4}$$

We will see that collections of sets that are both mutually exclusive and collectively exhaustive are sufficiently useful to merit a definition.

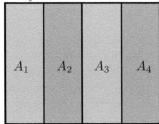

A collection of sets $A_1, \ldots, A_n$ is a *partition* if it is both mutually exclusive and collectively exhaustive.

══════**Example 1.1**══════

Phonesmart offers customers two kinds of smart phones, Apricot (A) and Banana (B). It is possible to buy a Banana phone with an optional external battery E. Apricot customers can buy a phone with an external battery (E) or an extra memory card (C) or both. Draw a Venn diagram that shows the relationship among the items A,B,C and E available to Phonesmart customers.

. . . .

Since each phone is either Apricot or Banana, A and B form a partition. Since the external battery E is available for both kinds of phones, E intersects both A and B. However, since the memory card C is available only to Apricot customers, $C \subset A$. A Venn diagram representing these facts is shown on the right.

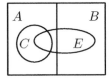

The mathematics we study is a branch of measure theory. Probability is a number that describes a set. The higher the number, the more probability there is. In this sense probability is like a quantity that measures a physical phenomenon; for example, a weight or a temperature. However, it is not necessary to think about probability in physical terms. We can do all the math abstractly, just as we defined sets and set operations in the previous paragraphs without any reference to physical phenomena.

Fortunately for engineers, the language of probability (including the word *probability* itself) makes us think of things that we experience. The basic model is a repeatable *experiment*. An experiment consists of a *procedure* and *observations*. There is uncertainty in what will be observed; otherwise, performing the experiment would be unnecessary. Some examples of experiments include

1. Flip a coin. Did it land with heads or tails facing up?

2. Walk to a bus stop. How long do you wait for the arrival of a bus?

3. Give a lecture. How many students are seated in the fourth row?

4. Transmit one of a collection of waveforms over a channel. What waveform arrives at the receiver?

5. Transmit one of a collection of waveforms over a channel. Which waveform does the receiver identify as the transmitted waveform?

For the most part, we will analyze *models* of actual physical experiments. We create models because real experiments generally are too complicated to analyze. For example, to describe *all* of the factors affecting your waiting time at a bus stop, you may consider

- The time of day. (Is it rush hour?)
- The speed of each car that passed by while you waited.
- The weight, horsepower, and gear ratios of each kind of bus used by the bus company.
- The psychological profile and work schedule of each bus driver. (Some drivers drive faster than others.)
- The status of all road construction within 100 miles of the bus stop.

It should be apparent that it would be difficult to analyze the effect of each of these factors on the likelihood that you will wait less than five minutes for a bus. Consequently, it is necessary to study a *model* of the experiment that captures the important part of the actual physical experiment. Since we will focus on the model of the experiment almost exclusively, we often will use the word *experiment* to refer to the model of an experiment.

Example 1.2

An experiment consists of the following procedure, observation, and model:

- Procedure: Monitor activity at a Phonesmart store.
- Observation: Observe which type of phone (Apricot or Banana) the next customer purchases.
- Model: Apricots and Bananas are equally likely. The result of each purchase is unrelated to the results of previous purchases.

As we have said, an experiment consists of both a procedure and observations. It is important to understand that two experiments with the same procedure but with different observations are different experiments. For example, consider these two experiments:

Example 1.3

Monitor the Phonesmart store until three customers purchase phones. Observe the sequence of Apricots and Bananas.

Example 1.4

Monitor the Phonesmart store until three customers purchase phones. Observe the number of Apricots.

These two experiments have the same procedure: monitor the Phonesmart store until three customers purchase phones. They are different experiments because they require different observations. We will describe models of experiments in terms of a set of possible experimental outcomes. In the context of probability, we give precise meaning to the word *outcome*.

Definition 1.1   Outcome

*An **outcome** of an experiment is any possible observation of that experiment.*

Implicit in the definition of an outcome is the notion that each outcome is distinguishable from every other outcome. As a result, we define the universal set of all possible outcomes. In probability terms, we call this universal set the *sample space*.

Definition 1.2   Sample Space

*The **sample space** of an experiment is the finest-grain, mutually exclusive, collectively exhaustive set of all possible outcomes.*

The *finest-grain* property simply means that all possible distinguishable outcomes are identified separately. The requirement that outcomes be mutually exclusive says that if one outcome occurs, then no other outcome also occurs. For the set of outcomes to be collectively exhaustive, every outcome of the experiment must be in the sample space.

Example 1.5

- The sample space in Example 1.2 is $S = \{a, b\}$ where a is the outcome "Apricot sold," and b is the outcome "Banana sold."
- The sample space in Example 1.3 is

$$S = \{aaa, aab, aba, abb, baa, bab, bba, bbb\} \tag{1.5}$$

- The sample space in Example 1.4 is $S = \{0, 1, 2, 3\}$.

In common speech, an event is something that occurs. In an experiment, we may say that an event occurs when a certain phenomenon is observed. To define an event mathematically, we must identify *all* outcomes for which the phenomenon is observed. That is, for each outcome, either the particular event occurs or it does not. In probability terms, we define an event in terms of the outcomes in the sample space.

Set Algebra	Probability
Set	Event
Universal set	Sample space
Element	Outcome

Table 1.1   The terminology of set theory and probability.

Definition 1.3 **Event**

*An **event** is a set of outcomes of an experiment.*

Table 1.1 relates the terminology of probability to set theory. All of this may seem so simple that it is boring. While this is true of the definitions themselves, applying them is a different matter. Defining the sample space and its outcomes are key elements of the solution of any probability problem. A probability problem arises from some practical situation that can be modeled as an experiment. To work on the problem, it is necessary to define the experiment carefully and then derive the sample space. Getting this right is a big step toward solving the problem.

Example 1.6

Observe the number of minutes a customer spends in the Phonesmart store. An outcome T is a nonnegative real number. The sample space is $S = \{T|T \geq 0\}$. The event "the customer stays longer than five minutes is $\{T|T > 5\}$.

Example 1.7

Monitor three customers in the Phonesmart store. Classify the behavior as buying (b) if a customer purchases a smartphone. Otherwise the behavior is no purchase (n). An outcome of the experiment is a sequence of three customer decisions. We can denote each outcome by a three-letter word such as bnb indicating that the first and third customers buy a phone and the second customer does not. We denote the event that customer i buys a phone by B_i and the event customer i does not buy a phone by N_i. The event $B_2 = \{nbn, nbb, bbn, bbb\}$. We can also express an outcome as an intersection of events B_i and N_j. For example the outcome $bnb = B_1 N_2 B_3$.

Quiz 1.1

Monitor three consecutive packets going through a Internet router. Based on the packet header, each packet can be classified as either video (v) if it was sent from a Youtube server or as ordinary data (d). Your observation is a sequence of three letters (each letter is either v or d). For example, two video packets followed by one data packet corresponds to vvd. Write the elements of the following sets:

$A_1 = \{\text{second packet is video}\},$ $B_1 = \{\text{second packet is data}\},$

$A_2 = \{\text{all packets are the same}\},$ $B_2 = \{\text{video and data alternate}\},$

$A_3 = \{\text{one or more video packets}\},$ $B_3 = \{\text{two or more data packets}\}.$

For each pair of events A_1 and B_1, A_2 and B_2, and so on, identify whether the pair of events is either mutually exclusive or collectively exhaustive or both.

1.2 Probability Axioms

> A probability model assigns a number between 0 and 1 to every event. The probability of the union of mutually exclusive events is the sum of the probabilities of the events in the union.

Thus far our model of an experiment consists of a procedure and observations. This leads to a set-theory representation with a sample space (universal set S), outcomes (s that are elements of S), and events (A that are sets of elements). To complete the model, we assign a probability $P[A]$ to every event, A, in the sample space. With respect to our physical idea of the experiment, the probability of an event is the proportion of the time that event is observed in a large number of runs of the experiment. This is the *relative frequency* notion of probability. Mathematically, this is expressed in the following axioms.

Definition 1.4   Axioms of Probability

A probability measure $P[\cdot]$ is a function that maps events in the sample space to real numbers such that

Axiom 1 *For any event A, $P[A] \geq 0$.*

Axiom 2 $P[S] = 1.$

Axiom 3 *For any countable collection $A_1, A_2, \ldots$ of mutually exclusive events*

$$P[A_1 \cup A_2 \cup \cdots] = P[A_1] + P[A_2] + \cdots.$$

We will build our entire theory of probability on these three axioms. Axioms 1 and 2 simply establish a probability as a number between 0 and 1. Axiom 3 states that the probability of the union of mutually exclusive events is the sum of the individual probabilities. We will use this axiom over and over in developing the theory of probability and in solving problems. In fact, it is really all we have to work with. Everything else follows from Axiom 3. To use Axiom 3 to solve a practical problem, we will learn in Section 1.4 to analyze a complicated event as the union of mutually exclusive events whose probabilities we can calculate. Then, we will add the probabilities of the mutually exclusive events to find the probability of the complicated event we are interested in.

A useful extension of Axiom 3 applies to the union of two mutually exclusive events.

Theorem 1.1

For mutually exclusive events A_1 and A_2,

$$\mathrm{P}\left[A_1 \cup A_2\right] = \mathrm{P}\left[A_1\right] + \mathrm{P}\left[A_2\right].$$

Although it may appear that Theorem 1.1 is a trivial special case of Axiom 3, this is not so. In fact, a simple proof of Theorem 1.1 may also use Axiom 2! If you are curious, Problem 1.2.13 gives the first steps toward a proof. It is a simple matter to extend Theorem 1.1 to any finite union of mutually exclusive sets.

Theorem 1.2

If $A = A_1 \cup A_2 \cup \cdots \cup A_m$ and $A_i \cap A_j = \varnothing$ for $i \neq j$, then

$$\mathrm{P}\left[A\right] = \sum_{i=1}^{m} \mathrm{P}\left[A_i\right].$$

In Chapter 10, we show that the probability measure established by the axioms corresponds to the idea of relative frequency. The correspondence refers to a sequential experiment consisting of n repetitions of the basic experiment. We refer to each repetition of the experiment as a *trial*. In these n trials, $N_A(n)$ is the number of times that event A occurs. The relative frequency of A is the fraction $N_A(n)/n$. Theorem 10.7 proves that $\lim_{n \to \infty} N_A(n)/n = \mathrm{P}[A]$.

Here we list some properties of probabilities that follow directly from the three axioms. While we do not supply the proofs, we suggest that students prove at least some of these theorems in order to gain experience working with the axioms.

Theorem 1.3

The probability measure $\mathrm{P}[\cdot]$ *satisfies*

(a) $\mathrm{P}[\varnothing] = 0$.

(b) $\mathrm{P}[A^c] = 1 - \mathrm{P}[A]$.

(c) *For any A and B (not necessarily mutually exclusive),*

$$\mathrm{P}\left[A \cup B\right] = \mathrm{P}\left[A\right] + \mathrm{P}\left[B\right] - \mathrm{P}\left[A \cap B\right].$$

(d) *If $A \subset B$, then $\mathrm{P}[A] \leq \mathrm{P}[B]$.*

Another consequence of the axioms can be expressed as the following theorem:

=======Theorem 1.4=======

The probability of an event $B = \{s_1, s_2, \ldots, s_m\}$ is the sum of the probabilities of the outcomes contained in the event:

$$P[B] = \sum_{i=1}^{m} P[\{s_i\}].$$

Proof Each outcome s_i is an event (a set) with the single element s_i. Since outcomes by definition are mutually exclusive, B can be expressed as the union of m mutually exclusive sets:

$$B = \{s_1\} \cup \{s_2\} \cup \cdots \cup \{s_m\} \tag{1.6}$$

with $\{s_i\} \cap \{s_j\} = \varnothing$ for $i \neq j$. Applying Theorem 1.2 with $B_i = \{s_i\}$ yields

$$P[B] = \sum_{i=1}^{m} P[\{s_i\}]. \tag{1.7}$$

Comments on Notation

We use the notation $P[\cdot]$ to indicate the probability of an event. The expression in the square brackets is an event. Within the context of one experiment, $P[A]$ can be viewed as a function that transforms event A to a number between 0 and 1.

Note that $\{s_i\}$ is the formal notation for a set with the single element s_i. For convenience, we will sometimes write $P[s_i]$ rather than the more complete $P[\{s_i\}]$ to denote the probability of this outcome.

We will also abbreviate the notation for the probability of the intersection of two events, $P[A \cap B]$. Sometimes we will write it as $P[A, B]$ and sometimes as $P[AB]$. Thus by definition, $P[A \cap B] = P[A, B] = P[AB]$.

Equally Likely Outcomes

A large number of experiments have a sample space $S = \{s_1, \ldots, s_n\}$ in which our knowledge of the practical situation leads us to believe that no one outcome is any more likely than any other. In these experiments we say that the n outcomes are *equally likely*. In such a case, the axioms of probability imply that every outcome has probability $1/n$.

=======Theorem 1.5=======

For an experiment with sample space $S = \{s_1, \ldots, s_n\}$ in which each outcome s_i is equally likely,

$$P[s_i] = 1/n \qquad 1 \leq i \leq n.$$

Proof Since all outcomes have equal probability, there exists p such that $P[s_i] = p$ for $i = 1, \ldots, n$. Theorem 1.4 implies

$$P[S] = P[s_1] + \cdots + P[s_n] = np. \tag{1.8}$$

Since Axiom 2 says $P[S] = 1$, $p = 1/n$.

Example 1.8

Roll a six-sided die in which all faces are equally likely. What is the probability of each outcome? Find the probabilities of the events: "Roll 4 or higher," "Roll an even number," and "Roll the square of an integer."

. . . .

The probability of each outcome is $P[i] = 1/6$ for $i = 1, 2, \ldots, 6$. The probabilities of the three events are

- $P[\text{Roll 4 or higher}] = P[4] + P[5] + P[6] = 1/2$.
- $P[\text{Roll an even number}] = P[2] + P[4] + P[6] = 1/2$.
- $P[\text{Roll the square of an integer}] = P[1] + P[4] = 1/3$.

Quiz 1.2

A student's test score T is an integer between 0 and 100 corresponding to the experimental outcomes $s_0, \ldots, s_{100}$. A score of 90 to 100 is an A, 80 to 89 is a B, 70 to 79 is a C, 60 to 69 is a D, and below 60 is a failing grade of F. If all scores between 51 and 100 are equally likely and a score of 50 or less never occurs, find the following probabilities:

(a) $P[\{s_{100}\}]$

(b) $P[A]$

(c) $P[F]$

(d) $P[T < 90]$

(e) $P[\text{a } C \text{ grade or better}]$

(f) $P[\text{student passes}]$

1.3 Conditional Probability

> Conditional probabilities correspond to a modified probability model that reflects partial information about the outcome of an experiment. The modified model has a smaller sample space than the original model.

As we suggested earlier, it is sometimes useful to interpret $P[A]$ as our knowledge of the occurrence of event A before an experiment takes place. If $P[A] \approx 1$, we have advance knowledge that A will almost certainly occur. $P[A] \approx 0$ reflects strong knowledge that A is unlikely to occur when the experiment takes place. With $P[A] \approx 1/2$, we have little knowledge about whether or not A will occur. Thus $P[A]$ reflects our knowledge of the occurrence of A *prior* to performing an experiment. Sometimes, we refer to $P[A]$ as the *a priori probability*, or the *prior probability*, of A.

In many practical situations, it is not possible to find out the precise outcome of an experiment. Rather than the outcome s_i, itself, we obtain information that the outcome is in the set B. That is, we learn that some event B has occurred, where B consists of several outcomes. Conditional probability describes our knowledge of A when we know that B has occurred but we still don't know the precise outcome. The notation for this new probability is $P[A|B]$. We read this as "the probability of A given B." Before going to the mathematical definition of conditional probability, we provide an example that gives an indication of how conditional probabilities can be used.

====**Example 1.9**====

Consider an experiment that consists of testing two integrated circuits (IC chips) that come from the same silicon wafer and observing in each case whether a chip is accepted (a) or rejected (r). The sample space of the experiment is $S = \{rr, ra, ar, aa\}$. Let B denote the event that the first chip tested is rejected. Mathematically, $B = \{rr, ra\}$. Similarly, let $A = \{rr, ar\}$ denote the event that the second chip is a failure.

The chips come from a high-quality production line. Therefore the prior probability $P[A]$ is very low. In advance, we are pretty certain that the second circuit will be accepted. However, some wafers become contaminated by dust, and these wafers have a high proportion of defective chips. When the first chip is a reject, the outcome of the experiment is in event B and $P[A|B]$, the probability that the second chip will also be rejected, is higher than the *a priori* probability $P[A]$ because of the likelihood that dust contaminated the entire wafer.

====**Definition 1.5**====**Conditional Probability**

The conditional probability of the event A given the occurrence of the event B is

$$P[A|B] = \frac{P[AB]}{P[B]}.$$

Conditional probability is defined only when $P[B] > 0$. In most experiments, $P[B] = 0$ means that it is certain that B never occurs. In this case, it is illogical to speak of the probability of A given that B occurs. Note that $P[A|B]$ is a respectable probability measure relative to a sample space that consists of all the outcomes in B. This means that $P[A|B]$ has properties corresponding to the three axioms of probability.

====Theorem 1.6====

A conditional probability measure $P[A|B]$ *has the following properties that correspond to the axioms of probability.*

Axiom 1: $P[A|B] \geq 0.$

Axiom 2: $P[B|B] = 1.$

Axiom 3: If $A = A_1 \cup A_2 \cup \cdots$ *with* $A_i \cap A_j = \emptyset$ *for* $i \neq j$, *then*

$$P[A|B] = P[A_1|B] + P[A_2|B] + \cdots$$

You should be able to prove these statements using Definition 1.5.

====Example 1.10====

With respect to Example 1.9, consider the *a priori* probability model

$$P[rr] = 0.01, \quad P[ra] = 0.01, \quad P[ar] = 0.01, \quad P[aa] = 0.97. \qquad (1.9)$$

Find the probability of $A =$ "second chip rejected" and $B =$ "first chip rejected." Also find the conditional probability that the second chip is a reject given that the first chip is a reject.

. . . .

We saw in Example 1.9 that A is the union of two mutually exclusive events (outcomes) rr and ar. Therefore, the a priori probability that the second chip is rejected is

$$P[A] = P[rr] + P[ar] = 0.02 \qquad (1.10)$$

This is also the a priori probability that the first chip is rejected:

$$P[B] = P[rr] + P[ra] = 0.02. \qquad (1.11)$$

The conditional probability of the second chip being rejected given that the first chip is rejected is, by definition, the ratio of $P[AB]$ to $P[B]$, where, in this example,

$$P[AB] = P[\text{both rejected}] = P[rr] = 0.01 \qquad (1.12)$$

Thus

$$P[A|B] = \frac{P[AB]}{P[B]} = 0.01/0.02 = 0.5. \qquad (1.13)$$

The information that the first chip is a reject drastically changes our state of knowledge about the second chip. We started with near certainty, $P[A] = 0.02$, that the second chip would not fail and ended with complete uncertainty about the quality of the second chip, $P[A|B] = 0.5$.

====Example 1.11====

Roll two fair four-sided dice. Let X_1 and X_2 denote the number of dots that appear on die 1 and die 2, respectively. Let A be the event $X_1 \geq 2$. What is $P[A]$? Let B

denote the event $X_2 > X_1$. What is $P[B]$? What is $P[A|B]$?

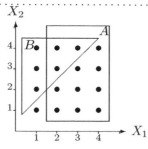

We begin by observing that the sample space has 16 elements corresponding to the four possible values of X_1 and the same four values of X_2. Since the dice are fair, the outcomes are equally likely, each with probability 1/16. We draw the sample space as a set of black circles in a two-dimensional diagram, in which the axes represent the events X_1 and X_2. Each outcome is a pair of values (X_1, X_2). The rectangle represents A. It contains 12 outcomes, each with probability 1/16.

To find $P[A]$, we add up the probabilities of outcomes in A, so $P[A] = 12/16 = 3/4$. The triangle represents B. It contains six outcomes. Therefore $P[B] = 6/16 = 3/8$. The event AB has three outcomes, $(2,3), (2,4), (3,4)$, so $P[AB] = 3/16$. From the definition of conditional probability, we write

$$P[A|B] = \frac{P[AB]}{P[B]} = \frac{1}{2}. \qquad (1.14)$$

We can also derive this fact from the diagram by restricting our attention to the six outcomes in B (the conditioning event) and noting that three of the six outcomes in B (one-half of the total) are also in A.

<hr />

Quiz 1.3

Monitor three consecutive packets going through an Internet router. Classify each one as either video (v) or data (d). Your observation is a sequence of three letters (each one is either v or d). For example, three video packets corresponds to vvv. The outcomes vvv and ddd each have probability 0.2 whereas each of the other outcomes vvd, vdv, vdd, dvv, dvd, and ddv has probability 0.1. Count the number of video packets N_V in the three packets you have observed. Describe in words and also calculate the following probabilities:

(a) $P[N_V = 2]$

(b) $P[N_V \geq 1]$

(c) $P[\{vvd\}|N_V = 2]$

(d) $P[\{ddv\}|N_V = 2]$

(e) $P[N_V = 2|N_V \geq 1]$

(f) $P[N_V \geq 1|N_V = 2]$

1.4   Partitions and the Law of Total Probability

A partition divides the sample space into mutually exclusive sets. The law of total probability expresses the probability of an event as the sum of the probabilities of outcomes that are in the separate sets of a partition.

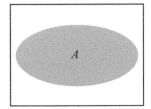 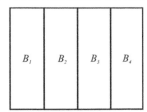 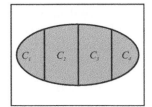

Figure 1.1 In this example of Theorem 1.7, the partition is $B = \{B_1, B_2, B_3, B_4\}$ and $C_i = A \cap B_i$ for $i = 1, \ldots, 4$. It should be apparent that $A = C_1 \cup C_2 \cup C_3 \cup C_4$.

Example 1.12

Flip four coins, a penny, a nickel, a dime, and a quarter. Examine the coins in order (penny, then nickel, then dime, then quarter) and observe whether each coin shows a head (h) or a tail (t). What is the sample space? How many elements are in the sample space?

The sample space consists of 16 four-letter words, with each letter either h or t. For example, the outcome $tthh$ refers to the penny and the nickel showing tails and the dime and quarter showing heads. There are 16 members of the sample space.

Example 1.13

Continuing Example 1.12, let $B_i = \{$outcomes with i heads$\}$. Each B_i is an event containing one or more outcomes. For example, $B_1 = \{ttth, ttht, thtt, httt\}$ contains four outcomes. The set $B = \{B_0, B_1, B_2, B_3, B_4\}$ is a partition. Its members are mutually exclusive and collectively exhaustive. It is not a sample space because it lacks the finest-grain property. Learning that an experiment produces an event B_1 tells you that one coin came up heads, but it doesn't tell you which coin it was.

The experiment in Example 1.12 and Example 1.13 refers to a "toy problem," one that is easily visualized but isn't something we would do in the course of our professional work. Mathematically, however, it is equivalent to many real engineering problems. For example, observe a pair of modems transmitting four bits from one computer to another. For each bit, observe whether the receiving modem detects the bit correctly (c) or makes an error (e). Or test four integrated circuits. For each one, observe whether the circuit is acceptable (a) or a reject (r).

In all of these examples, the sample space contains 16 four-letter words formed with an alphabet containing two letters. If we are interested only in the number of times one of the letters occurs, it is sufficient to refer only to the partition B, which does not contain all of the information about the experiment but does contain all of the information we need. The partition is simpler to deal with than the sample space because it has fewer members (there are five events in the partition and 16 outcomes in the sample space). The simplification is much more significant when the complexity of the experiment is higher. For example, in testing 20 circuits the sample space has $2^{20} = 1{,}048{,}576$ members, while the corresponding partition has only 21 members.

We observed in Section 1.2 that the entire theory of probability is based on a union of mutually exclusive events. The following theorem shows how to use a partition to represent an event as a union of mutually exclusive events.

======Theorem 1.7======

For a partition $B = \{B_1, B_2, \ldots\}$ and any event A in the sample space, let $C_i = A \cap B_i$. For $i \neq j$, the events C_i and C_j are mutually exclusive and

$$A = C_1 \cup C_2 \cup \cdots.$$

Figure 1.1 is a picture of Theorem 1.7.

======Example 1.14======

In the coin-tossing experiment of Example 1.12, let A equal the set of outcomes with less than three heads:

$$A = \{tttt, httt, thtt, ttht, ttth, hhtt, htht, htth, tthh, thth, thht\}. \tag{1.15}$$

From Example 1.13, let $B_i = \{\text{outcomes with } i \text{ heads}\}$. Since $\{B_0, \ldots, B_4\}$ is a partition, Theorem 1.7 states that

$$A = (A \cap B_0) \cup (A \cap B_1) \cup (A \cap B_2) \cup (A \cap B_3) \cup (A \cap B_4) \tag{1.16}$$

In this example, $B_i \subset A$, for $i = 0, 1, 2$. Therefore $A \cap B_i = B_i$ for $i = 0, 1, 2$. Also, for $i = 3$ and $i = 4$, $A \cap B_i = \varnothing$ so that $A = B_0 \cup B_1 \cup B_2$, a union of mutually exclusive sets. In words, this example states that the event "less than three heads" is the union of events "zero heads," "one head," and "two heads."

We advise you to make sure you understand Theorem 1.7 and Example 1.14. Many practical problems use the mathematical technique contained in the theorem. For example, find the probability that there are three or more bad circuits in a batch that comes from a fabrication machine.

The following theorem refers to a partition $\{B_1, B_2, \ldots, B_m\}$ and any event, A. It states that we can find the probability of A by adding the probabilities of the parts of A that are in the separate components of the event space.

======Theorem 1.8======

For any event A, and partition $\{B_1, B_2, \ldots, B_m\}$,

$$P[A] = \sum_{i=1}^{m} P[A \cap B_i].$$

Proof The proof follows directly from Theorem 1.7 and Theorem 1.2. In this case, the mutually exclusive sets are $C_i = \{A \cap B_i\}$.

Theorem 1.8 is often used when the sample space can be written in the form of a table. In this table, the rows and columns each represent a partition. This method is shown in the following example.

══════Example 1.15══════

A company has a model of email use. It classifies all emails as either long (l), if they are over 10 MB in size, or brief (b). It also observes whether the email is just text (t), has attached images (i), or has an attached video (v). This model implies an experiment in which the procedure is to monitor an email and the observation consists of the type of email, t, i, or v, and the length, l or b. The sample space has six outcomes: $S = \{lt, bt, li, bi, lv, bv\}$. In this problem, each email is classifed in two ways: by length and by type. Using L for the event that an email is long and B for the event that a email is brief, $\{L, B\}$ is a partition. Similarly, the text (T), image (I), and video (V) classification is a partition $\{T, I, V\}$. The sample space can be represented by a table in which the rows and columns are labeled by events and the intersection of each row and column event contains a single outcome. The corresponding table entry is the probability of that outcome. In this case, the table is

$$
\begin{array}{c|ccc}
 & T & I & V \\
\hline
L & 0.3 & 0.12 & 0.15 \\
B & 0.2 & 0.08 & 0.15
\end{array}
\qquad (1.17)
$$

For example, from the table we can read that the probability of a brief image email is $P[bi] = P[BI] = 0.08$. Note that $\{T, I, V\}$ is a partition corresponding to $\{B_1, B_2, B_3\}$ in Theorem 1.8. Thus we can apply Theorem 1.8 to find the probability of a long email:

$$
P[L] = P[LT] + P[LI] + P[LV] = 0.57. \qquad (1.18)
$$

Law of Total Probability

In many applications, we begin with information about conditional probabilities and use the law of total probability to calculate unconditional probabilities.

══════Theorem 1.9══════Law of Total Probability

For a partition $\{B_1, B_2, \ldots, B_m\}$ *with* $P[B_i] > 0$ *for all i,*

$$
P[A] = \sum_{i=1}^{m} P[A|B_i] P[B_i].
$$

Proof This follows from Theorem 1.8 and the identity $P[AB_i] = P[A|B_i] P[B_i]$, which is a direct consequence of the definition of conditional probability.

The usefulness of the result can be seen in the next example.

═══════Example 1.16═══════

A company has three machines B_1, B_2, and B_3 making 1 kΩ resistors. Resistors within 50 Ω of the nominal value are considered acceptable. It has been observed that 80% of the resistors produced by B_1 and 90% of the resistors produced by B_2 are acceptable. The percentage for machine B_3 is 60%. Each hour, machine B_1 produces 3000 resistors, B_2 produces 4000 resistors, and B_3 produces 3000 resistors. All of the resistors are mixed together at random in one bin and packed for shipment. What is the probability that the company ships an acceptable resistor?

....

Let $A = \{\text{resistor is acceptable}\}$. Using the resistor accuracy information to formulate a probability model, we write

$$P[A|B_1] = 0.8, \quad P[A|B_2] = 0.9, \quad P[A|B_3] = 0.6. \qquad (1.19)$$

The production figures state that $3000 + 4000 + 3000 = 10{,}000$ resistors per hour are produced. The fraction from machine B_1 is $P[B_1] = 3000/10{,}000 = 0.3$. Similarly, $P[B_2] = 0.4$ and $P[B_3] = 0.3$. Now it is a simple matter to apply the law of total probability to find the acceptable probability for all resistors shipped by the company:

$$
\begin{aligned}
P[A] &= P[A|B_1]P[B_1] + P[A|B_2]P[B_2] + P[A|B_3]P[B_3] \\
&= (0.8)(0.3) + (0.9)(0.4) + (0.6)(0.3) = 0.78. \qquad (1.20)
\end{aligned}
$$

For the whole factory, 78% of resistors are within 50 Ω of the nominal value.

Bayes' Theorem

When we have advance information about $P[A|B]$ and need to calculate $P[B|A]$, we refer to the following formula:

═══════Theorem 1.10═══════Bayes' theorem

$$P[B|A] = \frac{P[A|B]P[B]}{P[A]}.$$

Proof

$$P[B|A] = \frac{P[AB]}{P[A]} = \frac{P[A|B]P[B]}{P[A]}. \qquad (1.21)$$

Bayes' theorem is a simple consequence of the definition of conditional probability. It has a name because it is extremely useful for making inferences about phenomena that cannot be observed directly. Sometimes these inferences are described as "reasoning about causes when we observe effects." For example, let $\{B_1, \ldots, B_m\}$ be a partition that includes all possible states of something that interests us but that we cannot observe directly (for example, the machine that made a particular resistor). For each possible state, B_i, we know the prior probability $P[B_i]$ and $P[A|B_i]$, the probability that an event A occurs (the resistor meets a quality criterion) if B_i is the actual state. Now we observe the actual event (either the resistor passes or fails a test), and we ask about the thing we are interested in (the machines that might have produced the resistor). That is, we use Bayes' theorem to find $P[B_1|A], P[B_2|A], \ldots, P[B_m|A]$. In performing the calculations, we use the law of total probability to calculate the denominator in Theorem 1.10. Thus for state B_i,

$$P[B_i|A] = \frac{P[A|B_i]\,P[B_i]}{\sum_{i=1}^{m} P[A|B_i]\,P[B_i]}. \tag{1.22}$$

Example 1.17

In Example 1.16 about a shipment of resistors from the factory, we learned that:

- The probability that a resistor is from machine B_3 is $P[B_3] = 0.3$.

- The probability that a resistor is *acceptable* — i.e., within 50 Ω of the nominal value — is $P[A] = 0.78$.

- Given that a resistor is from machine B_3, the conditional probability that it is acceptable is $P[A|B_3] = 0.6$.

What is the probability that an acceptable resistor comes from machine B_3?

. . . .

Now we are given the event A that a resistor is within 50 Ω of the nominal value, and we need to find $P[B_3|A]$. Using Bayes' theorem, we have

$$P[B_3|A] = \frac{P[A|B_3]\,P[B_3]}{P[A]}. \tag{1.23}$$

Since all of the quantities we need are given in the problem description, our answer is

$$P[B_3|A] = (0.6)(0.3)/(0.78) = 0.23. \tag{1.24}$$

Similarly we obtain $P[B_1|A] = 0.31$ and $P[B_2|A] = 0.46$. Of all resistors within 50 Ω of the nominal value, only 23% come from machine B_3 (even though this machine produces 30% of all resistors). Machine B_1 produces 31% of the resistors that meet the 50 Ω criterion and machine B_2 produces 46% of them.

Quiz 1.4

Monitor customer behavior in the Phonesmart store. Classify the behavior as buying (B) if a customer purchases a smartphone. Otherwise the behavior is no purchase (N). Classify the time a customer is in the store as long (L) if the customer stays more than three minutes; otherwise classify the amount of time as rapid (R). Based on experience with many customers, we use the probability model $P[N] = 0.7$, $P[L] = 0.6$, $P[NL] = 0.35$. Find the following probabilities:

(a) $P[B \cup L]$ (b) $P[N \cup L]$

(c) $P[N \cup B]$ (d) $P[LR]$

1.5   Independence

Two events are independent if observing one event does not change the probability of observing the other event.

=======Definition 1.6=======Two Independent Events

*Events A and B are **independent** if and only if*

$$P[AB] = P[A] P[B].$$

When events A and B have nonzero probabilities, the following formulas are equivalent to the definition of independent events:

$$P[A|B] = P[A], \qquad P[B|A] = P[B]. \tag{1.25}$$

To interpret independence, consider probability as a description of our knowledge of the result of the experiment. $P[A]$ describes our prior knowledge (before the experiment is performed) that the outcome is included in event A. The fact that the outcome is in B is partial information about the experiment. $P[A|B]$ reflects our knowledge of A when we learn that B occurs. $P[A|B] = P[A]$ states that learning that B occurs does not change our information about A. It is in this sense that the events are independent.

Problem 1.5.11 asks the reader to prove that if A and B are independent, then A and B^c are also independent. The logic behind this conclusion is that if learning that event B occurs does not alter the probability of event A, then learning that B does not occur also should not alter the probability of A.

Keep in mind that **independent and mutually exclusive are *not* synonyms**. In some contexts these words can have similar meanings, but this is not the case in probability. Mutually exclusive events A and B have no outcomes in common and therefore $P[AB] = 0$. In most situations independent events are not mutually exclusive! Exceptions occur only when $P[A] = 0$ or $P[B] = 0$. When we have to calculate probabilities, knowledge that events A and B are *mutually exclusive* is very helpful. Axiom 3 enables us to *add* their probabilities to obtain the probability of the *union*. Knowledge that events C and D are *independent* is also very useful. Definition 1.6 enables us to *multiply* their probabilities to obtain the probability of the *intersection*.

=======Example 1.18=======

Suppose that for the experiment monitoring three purchasing decisions in Example 1.7, each outcome (a sequence of three decisions, each either buy or not buy) is equally likely. Are the events B_2 that the second customer purchases a phone and N_2 that the second customer does not purchase a phone independent? Are the events B_1 and B_2 independent?

. . . .

Each element of the sample space $S = \{bbb, bbn, bnb, bnn, nbb, nbn, nnb, nnn\}$ has probability $1/8$. Each of the events

$$B_2 = \{bbb, bbn, nbb, nbn\} \quad \text{and} \quad N_2 = \{bnb, bnn, nnb, nnn\} \tag{1.26}$$

contains four outcomes, so $P[B_2] = P[N_2] = 4/8$. However, $B_2 \cap N_2 = \varnothing$ and $P[B_2 N_2] = 0$. That is, B_2 and N_2 are mutually exclusive because the second customer cannot both purchase a phone and not purchase a phone. Since $P[B_2 N_2] \neq P[B_2] P[N_2]$, B_2 and N_2 are not independent. Learning whether or not the event B_2 (second customer buys a phone) occurs drastically affects our knowledge of whether or not the event N_2 (second customer does not buy a phone) occurs. Each of the events $B_1 = \{bnn, bnb, bbn, bbb\}$ and $B_2 = \{bbn, bbb, nbn, nbb\}$ has four outcomes, so $P[B_1] = P[B_2] = 4/8 = 1/2$. In this case, the intersection $B_1 \cap B_2 = \{bbn, bbb\}$ has probability $P[B_1 B_2] = 2/8 = 1/4$. Since $P[B_1 B_2] = P[B_1] P[B_2]$, events B_1 and B_2 are independent. Learning whether or not the event B_2 (second customer buys a phone) occurs does not affect our knowledge of whether or not the event B_1 (first customer buys a phone) occurs.

In this example we have analyzed a probability model to determine whether two events are independent. In many practical applications we reason in the opposite direction. Our knowledge of an experiment leads us to *assume* that certain pairs of events are independent. We then use this knowledge to build a probability model for the experiment.

Example 1.19

Integrated circuits undergo two tests. A mechanical test determines whether pins have the correct spacing, and an electrical test checks the relationship of outputs to inputs. We *assume* that electrical failures and mechanical failures occur independently. Our information about circuit production tells us that mechanical failures occur with probability 0.05 and electrical failures occur with probability 0.2. What is the probability model of an experiment that consists of testing an integrated circuit and observing the results of the mechanical and electrical tests?

. . . .

To build the probability model, we note that the sample space contains four outcomes:

$$S = \{(ma, ea), (ma, er), (mr, ea), (mr, er)\} \tag{1.27}$$

where m denotes mechanical, e denotes electrical, a denotes accept, and r denotes reject. Let M and E denote the events that the mechanical and electrical tests are acceptable. Our prior information tells us that $P[M^c] = 0.05$, and $P[E^c] = 0.2$. This implies $P[M] = 0.95$ and $P[E] = 0.8$. Using the independence assumption and

Definition 1.6, we obtain the probabilities of the four outcomes:

$$P[(ma, ea)] = P[ME] = P[M]P[E] = 0.95 \times 0.8 = 0.76, \tag{1.28}$$
$$P[(ma, er)] = P[ME^c] = P[M]P[E^c] = 0.95 \times 0.2 = 0.19, \tag{1.29}$$
$$P[(mr, ea)] = P[M^c E] = P[M^c]P[E] = 0.05 \times 0.8 = 0.04, \tag{1.30}$$
$$P[(mr, er)] = P[M^c E^c] = P[M^c]P[E^c] = 0.05 \times 0.2 = 0.01. \tag{1.31}$$

Thus far, we have considered independence as a property of a pair of events. Often we consider larger sets of independent events. For more than two events to be *independent*, the probability model has to meet a set of conditions. To define mutual independence, we begin with three sets.

Definition 1.7 Three Independent Events

A_1, A_2, and A_3 are **mutually independent** if and only if

(a) A_1 and A_2 are independent,

(b) A_2 and A_3 are independent,

(c) A_1 and A_3 are independent,

(d) $P[A_1 \cap A_2 \cap A_3] = P[A_1]P[A_2]P[A_3]$.

The final condition is a simple extension of Definition 1.6. The following example shows why this condition is insufficient to guarantee that "everything is independent of everything else," the idea at the heart of independence.

Example 1.20

In an experiment with equiprobable outcomes, the partition is $S = \{1, 2, 3, 4\}$. $P[s] = 1/4$ for all $s \in S$. Are the events $A_1 = \{1, 3, 4\}$, $A_2 = \{2, 3, 4\}$, and $A_3 = \varnothing$ mutually independent?

....

These three sets satisfy the final condition of Definition 1.7 because $A_1 \cap A_2 \cap A_3 = \varnothing$, and

$$P[A_1 \cap A_2 \cap A_3] = P[A_1]P[A_2]P[A_3] = 0. \tag{1.32}$$

However, A_1 and A_2 are not independent because, with all outcomes equiprobable,

$$P[A_1 \cap A_2] = P[\{3, 4\}] = 1/2 \neq P[A_1]P[A_2] = 3/4 \times 3/4. \tag{1.33}$$

Hence the three events are not mutually independent.

The definition of an arbitrary number of mutually independent events is an extension of Definition 1.7.

Definition 1.8 ════ **More than Two Independent Events**

If $n \geq 3$, the events $A_1, A_2, \ldots, A_n$ are mutually independent if and only if

(a) all collections of $n - 1$ events chosen from $A_1, A_2, \ldots A_n$ are mutually independent,

(b) $P[A_1 \cap A_2 \cap \cdots \cap A_n] = P[A_1] \, P[A_2] \cdots P[A_n]$.

This definition and Example 1.20 show us that when $n > 2$ it is a complex matter to determine whether or not n events are mutually independent. On the other hand, if we know that n events are mutually independent, it is a simple matter to determine the probability of the intersection of any subset of the n events. Just multiply the probabilities of the events in the subset.

Quiz 1.5

Monitor two consecutive packets going through a router. Classify each one as video (v) if it was sent from a Youtube server or as ordinary data (d) otherwise. Your observation is a sequence of two letters (either v or d). For example, two video packets corresponds to vv. The two packets are independent and the probability that any one of them is a video packet is 0.8. Denote the identity of packet i by C_i. If packet i is a video packet, then $C_i = v$; otherwise, $C_i = d$. Count the number N_V of video packets in the two packets you have observed. Determine whether the following pairs of events are independent:

(a) $\{N_V = 2\}$ and $\{N_V \geq 1\}$

(b) $\{N_V \geq 1\}$ and $\{C_1 = v\}$

(c) $\{C_2 = v\}$ and $\{C_1 = d\}$

(d) $\{C_2 = v\}$ and $\{N_V \text{ is even}\}$

1.6 MATLAB

The MATLAB programming environment can be used for studying probability models by performing numerical calculations, simulating experiments, and drawing graphs. Simulations make extensive use of the MATLAB random number generator `rand`. In addition to introducing aspects of probability theory, each chapter of this book concludes with a section that uses MATLAB to demonstrate with numerical examples the concepts presented in the chapter. All of the MATLAB programs in this book can be downloaded from the companion website. On the other hand, the MATLAB sections are not essential to understanding the theory. You can use this text to learn probability without using MATLAB.

Engineers studied and applied probability theory long before the invention of MATLAB. Nevertheless, MATLAB provides a convenient programming environment for solving probability problems and for building models of probabilistic systems. Versions of MATLAB, including a low-cost student edition, are available for most computer systems.

At the end of each chapter, we include a MATLAB section (like this one) that introduces ways that MATLAB can be applied to the concepts and problems of the chapter. We assume you already have some familiarity with the basics of running MATLAB. If you do not, we encourage you to investigate the built-in tutorial, books dedicated to MATLAB, and various Web resources.

MATLAB can be used two ways to study and apply probability theory. Like a sophisticated scientific calculator, it can perform complex numerical calculations and draw graphs. It can also simulate experiments with random outcomes. To simulate experiments, we need a source of randomness. MATLAB uses a computer algorithm, referred to as a *pseudorandom number generator*, to produce a sequence of numbers between 0 and 1. Unless someone knows the algorithm, it is impossible to examine some of the numbers in the sequence and thereby calculate others. The calculation of each random number is similar to an experiment in which all outcomes are equally likely and the sample space is all binary numbers of a certain length. (The length depends on the machine running MATLAB.) Each number is interpreted as a fraction, with a binary point preceding the bits in the binary number. To use the pseudorandom number generator to simulate an experiment that contains an event with probability p, we examine one number, r, produced by the MATLAB algorithm and say that the event occurs if $r < p$; otherwise it does not occur.

A MATLAB simulation of an experiment starts with `rand`: the random number generator `rand(m,n)` returns an $m \times n$ array of pseudorandom numbers. Similarly, `rand(n)` produces an $n \times n$ array and `rand(1)` is just a scalar random number. Each number produced by `rand(1)` is in the interval $(0, 1)$. Each time we use `rand`, we get new, seemingly unpredictable numbers. Suppose p is a number between 0 and 1. The comparison `rand(1) < p` produces a 1 if the random number is less than p; otherwise it produces a zero. Roughly speaking, the function `rand(1) < p` simulates a coin flip with $\mathrm{P}[\text{tail}] = p$.

Example 1.21

```
>> X=rand(1,4)
X =
   0.0879   0.9626   0.6627   0.2023
>> X<0.5
ans =
     1      0      0      1
```

Since `rand(1,4) < 0.5` compares four random numbers against 0.5, the result is a random sequence of zeros and ones that simulates a sequence of four flips of a fair coin. We associate the outcome 1 with {head} and 0 with {tail}.

MATLAB also has some convenient variations on `rand`. For example, `randi(k)` generates a random integer from the set $\{1, 2, \ldots, k\}$ and `randi(k,m,n)` generates an $m \times n$ array of such random integers.

Example 1.22

Use MATLAB to generate 12 random student test scores T as described in Quiz 1.2.

Since `randi(50,1,12)` generates 12 test scores from the set $\{1,\ldots,50\}$, we need only to add 50 to each score to obtain test scores in the range $\{51,\ldots,100\}$.

```
>> 50+randi(50,1,12)
ans =
    69    78    60    68    93    99    77    95    88    57    51    90
```

Finally, we note that MATLAB's random numbers are only seemingly unpredictable. In fact, MATLAB maintains a seed value that determines the subsequent "random" numbers that will be returned. This seed is controlled by the `rng` function; `s=rng` saves the current seed and `rng(s)` restores a previously saved seed. Initializing the random number generator with the same seed always generates the same sequence:

Example 1.23

```
>> s=rng;
>> 50+randi(50,1,12)
ans =
    89    76    80    80    72    92    58    56    77    78    59    58
>> rng(s);
>> 50+randi(50,1,12)
ans =
    89    76    80    80    72    92    58    56    77    78    59    58
```

When you run a simulation that uses `rand`, it normally doesn't matter how the `rng` seed is initialized. However, it can be instructive to use the same repeatable sequence of `rand` values when you are debugging your simulation.

Quiz 1.6

The number of characters in a tweet is equally likely to be any integer between 1 and 140. Simulate an experiment that generates 1000 tweets and counts the number of "long" tweets that have over 120 characters. Repeat this experiment 5 times.

Problems

Difficulty: ● Easy   ■ Moderate   ◆ Difficult   ◆◆ Experts Only

1.1.1● Ricardo's offers customers two kinds of pizza crust, Roman (R) and Neapolitan (N). All pizzas have cheese but not all pizzas have tomato sauce. Roman pizzas can have tomato sauce or they can be white (W); Neapolitan pizzas always have tomato sauce. It is possible to order a Roman pizza with mushrooms (M) added. A Neapolitan pizza can contain mushrooms or onions (O) or both, in addition to the tomato sauce and cheese. Draw a Venn diagram that shows the relationship among the ingredients N, M, O, T, and W in the menu of Ricardo's pizzeria.

1.1.2● A hypothetical wi-fi transmission can take place at any of three speeds depending on the condition of the radio channel between a laptop and an access point. The speeds are high (h) at 54 Mb/s, medium (m) at 11 Mb/s, and low (l) at 1 Mb/s. A user of the wi-fi connection can transmit a short signal corresponding to a mouse click (c), or a long signal corresponding to a tweet (t). Consider the experiment of monitoring wi-fi signals and observing the transmission speed and the length. An observation is a two-letter word, for example, a high-speed, mouse-click transmission is hm.

(a) What is the sample space?

(b) A_1 is the event "medium speed connection." What outcomes are in A_1?

(c) A_2 is the event "mouse click." What outcomes are in A_2?

(d) Let A_3 be the event "high speed connection or low speed connection." What are the outcomes in A_3?

(e) Are A_1, A_2, A_3 mutually exclusive?

(f) Are A_1, A_2, A_3 collectively exhaustive?

1.1.3● An integrated circuit factory has three machines X, Y, and Z. Test one integrated circuit produced by each machine. Either a circuit is acceptable (a) or it fails (f). An observation is a sequence of three test results corresponding to the circuits from machines X, Y, and Z, respectively. For example, aaf is the observation that the circuits from X and Y pass the test and the circuit from Z fails the test.

(a) What are the elements of the sample space of this experiment?

(b) What are the elements of the sets

$$Z_F = \{\text{circuit from } Z \text{ fails}\},$$
$$X_A = \{\text{circuit from } X \text{ is acceptable}\}.$$

(c) Are Z_F and X_A mutually exclusive?

(d) Are Z_F and X_A collectively exhaustive?

(e) What are the elements of the sets

$$C = \{\text{more than one circuit acceptable}\},$$
$$D = \{\text{at least two circuits fail}\}.$$

(f) Are C and D mutually exclusive?

(g) Are C and D collectively exhaustive?

1.1.4● Shuffle a deck of cards and turn over the first card. What is the sample space of this experiment? How many outcomes are in the event that the first card is a heart?

1.1.5● Find out the birthday (month and day but not year) of a randomly chosen person. What is the sample space of the experiment? How many outcomes are in the event that the person is born in July?

1.1.6● The sample space of an experiment consists of all undergraduates at a university. Give four examples of partitions.

1.1.7● The sample space of an experiment consists of the measured resistances of two resistors. Give four examples of partitions.

1.2.1● Find $\mathrm{P}[B]$ in each case:

(a) Events A and B are a partition and $\mathrm{P}[A] = 3\,\mathrm{P}[B]$.

(b) For events A and B, $\mathrm{P}[A \cup B] = \mathrm{P}[A]$ and $\mathrm{P}[A \cap B] = 0$.

(c) For events A and B, $\mathrm{P}[A \cup B] = \mathrm{P}[A] - \mathrm{P}[B]$.

1.2.2● You roll two fair six-sided dice; one die is red, the other is white. Let R_i be the event that the red die rolls i. Let W_j be the event that the white die rolls j.

(a) What is $\mathrm{P}[R_3 W_2]$?

(b) What is the $\mathrm{P}[S_5]$ that the sum of the two rolls is 5?

1.2.3● You roll two fair six-sided dice. Find the probability $\mathrm{P}[D_3]$ that the absolute value of the difference of the dice is 3.

1.2.4 Indicate whether each statement is true or false.

(a) If $P[A] = 2P[A^c]$, then $P[A] = 1/2$.

(b) For all A and B, $P[AB] \leq P[A]P[B]$.

(c) If $P[A] < P[B]$, then $P[AB] < P[B]$.

(d) If $P[A \cap B] = P[A]$, then $P[A] \geq P[B]$.

1.2.5 Computer programs are classified by the length of the source code and by the execution time. Programs with more than 150 lines in the source code are big (B). Programs with ≤ 150 lines are little (L). Fast programs (F) run in less than 0.1 seconds. Slow programs (W) require at least 0.1 seconds. Monitor a program executed by a computer. Observe the length of the source code and the run time. The probability model for this experiment contains the following information: $P[LF] = 0.5$, $P[BF] = 0.2$, and $P[BW] = 0.2$. What is the sample space of the experiment? Calculate the following probabilities: $P[W]$, $P[B]$, and $P[W \cup B]$.

1.2.6 There are two types of cellular phones, handheld phones (H) that you carry and mobile phones (M) that are mounted in vehicles. Phone calls can be classified by the traveling speed of the user as fast (F) or slow (W). Monitor a cellular phone call and observe the type of telephone and the speed of the user. The probability model for this experiment has the following information: $P[F] = 0.5$, $P[HF] = 0.2$, $P[MW] = 0.1$. What is the sample space of the experiment? Find the following probabilities $P[W]$, $P[MF]$, and $P[H]$.

1.2.7 Shuffle a deck of cards and turn over the first card. What is the probability that the first card is a heart?

1.2.8 You have a six-sided die that you roll once and observe the number of dots facing upwards. What is the sample space? What is the probability of each sample outcome? What is the probability of E, the event that the roll is even?

1.2.9 A student's score on a 10-point quiz is equally likely to be any integer between 0 and 10. What is the probability of an A, which requires the student to get a score of 9 or more? What is the probability the student gets an F by getting less than 4?

1.2.10 Use Theorem 1.3 to prove the following facts:

(a) $P[A \cup B] \geq P[A]$

(b) $P[A \cup B] \geq P[B]$

(c) $P[A \cap B] \leq P[A]$

(d) $P[A \cap B] \leq P[B]$

1.2.11 Use Theorem 1.3 to prove by induction the *union bound*: For any collection of events $A_1, \ldots, A_n$,

$$P[A_1 \cup A_2 \cup \cdots \cup A_n] \leq \sum_{i=1}^{n} P[A_i].$$

1.2.12 Using *only* the three axioms of probability, prove $P[\varnothing] = 0$.

1.2.13 Using the three axioms of probability and the fact that $P[\varnothing] = 0$, prove Theorem 1.2. Hint: Define $A_i = B_i$ for $i = 1, \ldots, m$ and $A_i = \varnothing$ for $i > m$.

1.2.14 For each fact stated in Theorem 1.3, determine which of the three axioms of probability are needed to prove the fact.

1.3.1 Mobile telephones perform *handoffs* as they move from cell to cell. During a call, a telephone either performs zero handoffs (H_0), one handoff (H_1), or more than one handoff (H_2). In addition, each call is either long (L), if it lasts more than three minutes, or brief (B). The following table describes the probabilities of the possible types of calls.

	H_0	H_1	H_2
L	0.1	0.1	0.2
B	0.4	0.1	0.1

(a) What is the probability that a brief call will have no handoffs?

(b) What is the probability that a call with one handoff will be long?

(c) What is the probability that a long call will have one or more handoffs?

1.3.2● You have a six-sided die that you roll once. Let R_i denote the event that the roll is i. Let G_j denote the event that the roll is greater than j. Let E denote the event that the roll of the die is even-numbered.

(a) What is $P[R_3|G_1]$, the conditional probability that 3 is rolled given that the roll is greater than 1?

(b) What is the conditional probability that 6 is rolled given that the roll is greater than 3?

(c) What is $P[G_3|E]$, the conditional probability that the roll is greater than 3 given that the roll is even?

(d) Given that the roll is greater than 3, what is the conditional probability that the roll is even?

1.3.3● You have a shuffled deck of three cards: 2, 3, and 4. You draw one card. Let C_i denote the event that card i is picked. Let E denote the event that the card chosen is a even-numbered card.

(a) What is $P[C_2|E]$, the probability that the 2 is picked given that an even-numbered card is chosen?

(b) What is the conditional probability that an even-numbered card is picked given that the 2 is picked?

1.3.4■ Phonesmart is having a sale on Bananas. If you buy one Banana at full price, you get a second at half price. When couples come in to buy a pair of phones, sales of Apricots and Bananas are equally likely. Moreover, given that the first phone sold is a Banana, the second phone is twice as likely to be a Banana rather than an Apricot. What is the probability that a couple buys a pair of Bananas?

1.3.5■ The basic rules of genetics were discovered in mid-1800s by Mendel, who found that each characteristic of a pea plant, such as whether the seeds were green or yellow,

is determined by two genes, one from each parent. In his pea plants, Mendel found that yellow seeds were a dominant trait over green seeds. A yy pea with two yellow genes has yellow seeds; a gg pea with two recessive genes has green seeds; a hybrid gy or yg pea has yellow seeds. In one of Mendel's experiments, he started with a parental generation in which half the pea plants were yy and half the plants were gg. The two groups were crossbred so that each pea plant in the first generation was gy. In the second generation, each pea plant was equally likely to inherit a y or a g gene from each first-generation parent. What is the probability $P[Y]$ that a randomly chosen pea plant in the second generation has yellow seeds?

1.3.6■ From Problem 1.3.5, what is the conditional probability of yy, that a pea plant has two dominant genes given the event Y that it has yellow seeds?

1.3.7■ You have a shuffled deck of three cards: 2, 3, and 4, and you deal out the three cards. Let E_i denote the event that ith card dealt is even numbered.

(a) What is $P[E_2|E_1]$, the probability the second card is even given that the first card is even?

(b) What is the conditional probability that the first two cards are even given that the third card is even?

(c) Let O_i represent the event that the ith card dealt is odd numbered. What is $P[E_2|O_1]$, the conditional probability that the second card is even given that the first card is odd?

(d) What is the conditional probability that the second card is odd given that the first card is odd?

1.3.8♦ Deer ticks can carry both Lyme disease and human granulocytic ehrlichiosis (HGE). In a study of ticks in the Midwest, it was found that 16% carried Lyme disease, 10% had HGE, and that 10% of the ticks that had either Lyme disease or HGE carried both diseases.

(a) What is the probability $P[LH]$ that a tick carries both Lyme disease (L) and HGE (H)?

(b) What is the conditional probability that a tick has HGE given that it has Lyme disease?

1.4.1● Given the model of handoffs and call lengths in Problem 1.3.1,

(a) What is the probability $P[H_0]$ that a phone makes no handoffs?

(b) What is the probability a call is brief?

(c) What is the probability a call is long or there are at least two handoffs?

1.4.2● For the telephone usage model of Example 1.15, let B_m denote the event that a call is billed for m minutes. To generate a phone bill, observe the duration of the call in integer minutes (rounding up). Charge for M minutes $M = 1, 2, 3, \ldots$ if the exact duration T is $M - 1 < t \leq M$. A more complete probability model shows that for $m = 1, 2, \ldots$ the probability of each event B_m is

$$P[B_m] = \alpha(1 - \alpha)^{m-1}$$

where $\alpha = 1 - (0.57)^{1/3} = 0.171$.

(a) Classify a call as long, L, if the call lasts more than three minutes. What is $P[L]$?

(b) What is the probability that a call will be billed for nine minutes or less?

1.4.3◆ Suppose a cellular telephone is equally likely to make zero handoffs (H_0), one handoff (H_1), or more than one handoff (H_2). Also, a caller is either on foot (F) with probability $5/12$ or in a vehicle (V).

(a) Given the preceding information, find three ways to fill in the following probability table:

	H_0	H_1	H_2
F			
V			

(b) Suppose we also learn that $1/4$ of all callers are on foot making calls with no handoffs and that $1/6$ of all callers are vehicle users making calls with a single handoff. Given these additional facts, find all possible ways to fill in the table of probabilities.

1.5.1● Is it possible for A and B to be independent events yet satisfy $A = B$?

1.5.2● Events A and B are equiprobable, mutually exclusive, and independent. What is $P[A]$?

1.5.3● At a Phonesmart store, each phone sold is twice as likely to be an Apricot as a Banana. Also each phone sale is independent of any other phone sale. If you monitor the sale of two phones, what is the probability that the two phones sold are the same?

1.5.4■ Use a Venn diagram in which the event areas are proportional to their probabilities to illustrate two events A and B that are independent.

1.5.5■ In an experiment, A and B are mutually exclusive events with probabilities $P[A] = 1/4$ and $P[B] = 1/8$.

(a) Find $P[A \cap B]$, $P[A \cup B]$, $P[A \cap B^c]$, and $P[A \cup B^c]$.

(b) Are A and B independent?

1.5.6■ In an experiment, C and D are independent events with probabilities $P[C] = 5/8$ and $P[D] = 3/8$.

(a) Determine the probabilities $P[C \cap D]$, $P[C \cap D^c]$, and $P[C^c \cap D^c]$.

(b) Are C^c and D^c independent?

1.5.7■ In an experiment, A and B are mutually exclusive events with probabilities $P[A \cup B] = 5/8$ and $P[A] = 3/8$.

(a) Find $P[B]$, $P[A \cap B^c]$, and $P[A \cup B^c]$.

(b) Are A and B independent?

1.5.8■ In an experiment, C, and D are independent events with probabilities $P[C \cap D] = 1/3$, and $P[C] = 1/2$.

(a) Find $P[D]$, $P[C \cap D^c]$, and $P[C^c \cap D^c]$.

(b) Find $P[C \cup D]$ and $P[C \cup D^c]$.

(c) Are C and D^c independent?

1.5.9■ In an experiment with equiprobable outcomes, the sample space is $S = \{1, 2, 3, 4\}$ and $P[s] = 1/4$ for all $s \in S$. Find three events in S that are pairwise independent but are not independent. (Note: Pairwise independent events meet the first three conditions of Definition 1.7).

1.5.10■ (Continuation of Problem 1.3.5) One of Mendel's most significant results was the conclusion that genes determining different characteristics are transmitted independently. In pea plants, Mendel found that round peas (r) are a dominant trait over wrinkled peas (w). Mendel crossbred a group of (rr, yy) peas with a group of (ww, gg) peas. In this notation, rr denotes a pea with two "round" genes and ww denotes a pea with two "wrinkled" genes. The first generation were either (rw, yg), (rw, gy), (wr, yg), or (wr, gy) plants with both hybrid shape and hybrid color. Breeding among the first generation yielded second-generation plants in which genes for each characteristic were equally likely to be either dominant or recessive. What is the probability $P[Y]$ that a second-generation pea plant has yellow seeds? What is the probability $P[R]$ that a second-generation plant has round peas? Are R and Y independent events? How many visibly different kinds of pea plants would Mendel observe in the second generation? What are the probabilities of each of these kinds?

1.5.11◆ For independent events A and B, prove that

(a) A and B^c are independent.

(b) A^c and B are independent.

(c) A^c and B^c are independent.

1.5.12◆ Use a Venn diagram in which the event areas are proportional to their probabilities to illustrate three events A, B, and C that are independent.

1.5.13◆ Use a Venn diagram in which event areas are in proportion to their probabilities to illustrate events A, B, and C that are pairwise independent but not independent.

1.6.1● Following Quiz 1.2, use MATLAB, but not the `randi` function, to generate a vector **T** of 200 independent test scores such that all scores between 51 and 100 are equally likely.

2

Sequential Experiments

Many applications of probability refer to sequential experiments in which the procedure consists of many actions performed in sequence, with an observation taken after each action. Each action in the procedure together with the outcome associated with it can be viewed as a separate experiment with its own probability model. In analyzing sequential experiments we refer to the separate experiments in the sequence as *subexperiments*.

2.1 Tree Diagrams

> Tree diagrams display the outcomes of the subexperiments in a sequential experiment. The labels of the branches are probabilities and conditional probabilities. The probability of an outcome of the entire experiment is the product of the probabilities of branches going from the root of the tree to a leaf.

Many experiments consist of a sequence of *subexperiments*. The procedure followed for each subexperiment may depend on the results of the previous subexperiments. We often find it useful to use a type of graph referred to as a *tree diagram* to represent the sequence of subexperiments. To do so, we assemble the outcomes of each subexperiment into sets in a partition. Starting at the root of the tree,[1] we represent each event in the partition of the first subexperiment as a branch and we label the branch with the probability of the event. Each branch leads to a node. The events in the partition of the second subexperiment appear as branches growing from every node at the end of the first subexperiment. The labels of the branches

---

[1] Unlike biological trees, which grow from the ground up, probabilities usually grow from left to right. Some of them have their roots on top and leaves on the bottom.

of the second subexperiment are the *conditional* probabilities of the events in the second subexperiment. We continue the procedure taking the remaining subexperiments in order. The nodes at the end of the final subexperiment are the leaves of the tree. Each leaf corresponds to an outcome of the entire sequential experiment. The probability of each outcome is the product of the probabilities and conditional probabilities on the path from the root to the leaf. We usually label each leaf with a name for the event and the probability of the event.

This is a complicated description of a simple procedure as we see in the following five examples.

Example 2.1

For the resistors of Example 1.16, we used A to denote the event that a randomly chosen resistor is "within 50 Ω of the nominal value." This could mean "acceptable." We use the notation N ("not acceptable") for the complement of A. The experiment of testing a resistor can be viewed as a two-step procedure. First we identify which machine (B_1, B_2, or B_3) produced the resistor. Second, we find out if the resistor is acceptable. Draw a tree for this sequential experiment. What is the probability of choosing a resistor from machine B_2 that is not acceptable?

....

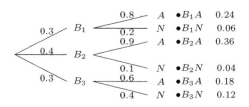

This two-step procedure is shown in the tree on the left. To use the tree to find the probability of the event B_2N, a nonacceptable resistor from machine B_2, we start at the left and find that the probability of reaching B_2 is $P[B_2] = 0.4$. We then move to the right to B_2N and multiply $P[B_2]$ by $P[N|B_2] = 0.1$ to obtain $P[B_2N] = (0.4)(0.1) = 0.04$.

We observe in this example a general property of all tree diagrams that represent sequential experiments. The probabilities on the branches leaving any node add up to 1. This is a consequence of the law of total probability and the property of conditional probabilities that corresponds to Axiom 3 (Theorem 1.6). Moreover, Axiom 2 implies that the probabilities of all of the leaves add up to 1.

Example 2.2

Traffic engineers have coordinated the timing of two traffic lights to encourage a run of green lights. In particular, the timing was designed so that with probability 0.8 a driver will find the second light to have the same color as the first. Assuming the first light is equally likely to be red or green, what is the probability $P[G_2]$ that the second light is green? Also, what is $P[W]$, the probability that you wait for at least one of the first two lights? Lastly, what is $P[G_1|R_2]$, the conditional probability of a green first light given a red second light?

....

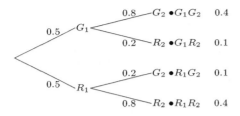

The tree for the two-light experiment is shown on the left. The probability that the second light is green is

$$P[G_2] = P[G_1G_2] + P[R_1G_2]$$
$$= 0.4 + 0.1 = 0.5. \qquad (2.1)$$

The event W that you wait for at least one light is the event that at least one light is red.

$$W = \{R_1G_2 \cup G_1R_2 \cup R_1R_2\}. \qquad (2.2)$$

The probability that you wait for at least one light is

$$P[W] = P[R_1G_2] + P[G_1R_2] + P[R_1R_2] = 0.1 + 0.1 + 0.4 = 0.6. \qquad (2.3)$$

An alternative way to the same answer is to observe that W is also the complement of the event that both lights are green. Thus,

$$P[W] = P[(G_1G_2)^c] = 1 - P[G_1G_2] = 0.6. \qquad (2.4)$$

To find $P[G_1|R_2]$, we need $P[R_2] = 1 - P[G_2] = 0.5$. Since $P[G_1R_2] = 0.1$, the conditional probability that you have a green first light given a red second light is

$$P[G_1|R_2] = \frac{P[G_1R_2]}{P[R_2]} = \frac{0.1}{0.5} = 0.2. \qquad (2.5)$$

The next example is the "Monty Hall" game, a famous problem with a solution that many regard as counterintuitive. Tree diagrams provide a clear explanation of the answer.

═══════**Example 2.3**═══════**Monty Hall**

In the Monty Hall game, a new car is hidden behind one of three closed doors while a goat is hidden behind each of the other two doors. Your goal is to select the door that hides the car. You make a preliminary selection and then a final selection. The game proceeds as follows:

- You select a door.

- The host, Monty Hall (who knows where the car is hidden), opens one of the two doors you didn't select to reveal a goat.

- Monty then asks you if you would like to switch your selection to the other unopened door.

- After you make your choice (either staying with your original door, or switching doors), Monty reveals the prize behind your chosen door.

To maximize your probability $P[C]$ of winning the car, *is switching to the other door either (a) a good idea, (b) a bad idea or (c) makes no difference?*

. . . .

To solve this problem, we will consider the "switch" and "do not switch" policies separately. That is, we will construct two different tree diagrams: The first describes what happens if you switch doors while the second describes what happens if you do not switch.

First we describe what is the same no matter what policy you follow. Suppose the doors are numbered 1, 2, and 3. Let H_i denote the event that the car is hidden behind door i. Also, let's assume you first choose door 1. (Whatever door you do choose, that door can be labeled door 1 and it would not change your probability of winning.) Now let R_i denote the event that Monty opens door i that hides a goat. If the car is behind door 1 Monty can choose to open door 2 or door 3 because both hide goats. He chooses door 2 or door 3 by flipping a fair coin. If the car is behind door 2, Monty opens door 3 and if the car is behind door 3, Monty opens door 2. Let C denote the event that you win the car and G the event that you win a goat. After Monty opens one of the doors, you decide whether to change your choice or stay with your choice of door 1. Finally, Monty opens the door of your final choice, either door 1 or the door you switched to.

The tree diagram in Figure 2.1(a) applies to the situation in which you change your choice. From this tree we learn that when the car is behind door 1 (event H_1) and Monty opens door 2 (event R_2), you switch to door 3 and then Monty opens door 3 to reveal a goat (event G). On the other hand, if the car is behind door 2, Monty reveals the goat behind door 3 and you switch to door 2 and win the car. Similarly, if the car is behind door 3, Monty reveals the goat behind door 2 and you switch to door 3 and win the car. For always switch, we see that

$$P[C] = P[H_2 R_3 C] + P[H_3 R_2 C] = 2/3. \qquad (2.6)$$

If you do not switch, the tree is shown in Figure 2.1(b). In this tree, when the car is behind door 1 (event H_1) and Monty opens door 2 (event R_2), you stay with door 1 and then Monty opens door 1 to reveal the car. On the other hand, if the car is behind door 2, Monty will open door 3 to reveal the goat. Since your final choice was door 1, Monty opens door 1 to reveal the goat. For do not switch,

$$P[C] = P[H_1 R_2 C] + P[H_1 R_3 C] = 1/3.$$

Thus switching is better; if you don't switch, you win the car only if you initially guessed the location of the car correctly, an event that occurs with probability 1/3. If you switch, you win the car when your initial guess was wrong, an event with probability 2/3.

Note that the two trees look largely the same because the key step where you make a choice is somewhat hidden because it is implied by the first two branches followed in the tree.

Quiz 2.1

In a cellular phone system, a mobile phone must be paged to receive a phone call. However, paging attempts don't always succeed because the mobile phone may not

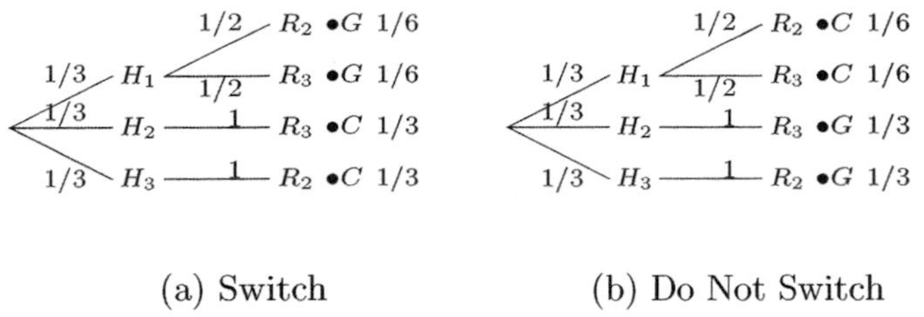

$1/2$ — R_2 •G 1/6

$1/3$ H_1 $1/2$ R_3 •G 1/6

$1/3$ H_2 —1— R_3 •C 1/3

$1/3$ H_3 —1— R_2 •C 1/3

$1/2$ — R_2 •C 1/6

$1/3$ H_1 $1/2$ R_3 •C 1/6

$1/3$ H_2 —1— R_3 •G 1/3

$1/3$ H_3 —1— R_2 •G 1/3

(a) Switch   (b) Do Not Switch

Figure 2.1   Tree Diagrams for Monty Hall

receive the paging signal clearly. Consequently, the system will page a phone up to three times before giving up. If the results of all paging attempts are independent and a single paging attempt succeeds with probability 0.8, sketch a probability tree for this experiment and find the probability $P[F]$ that the phone receives the paging signal clearly.

2.2   Counting Methods

In all applications of probability theory it is important to understand the sample space of an experiment. The methods in this section determine the number of outcomes in the sample space of a sequential experiment

Understanding the sample space is a key step in formulating and solving a probability problem. To begin, it is often useful to know the number of outcomes in the sample space. This number can be enormous as in the following simple example.

Example 2.4

Choose 7 cards at random from a deck of 52 different cards. Display the cards in the order in which you choose them. How many different sequences of cards are possible?

The procedure consists of seven subexperiments. In each subexperiment, the observation is the identity of one card. The first subexperiment has 52 possible outcomes corresponding to the 52 cards that could be drawn. For each outcome of the first subexperiment, the second subexperiment has 51 possible outcomes corresponding to the 51 remaining cards. Therefore there are 52×51 outcomes of the first two subexperiments. The total number of outcomes of the seven subexperiments is

$$52 \times 51 \times \cdots \times 46 = 674{,}274{,}182{,}400 \,. \tag{2.7}$$

Although many practical experiments are more complicated, the techniques for determining the size of a sample space all follow from the fundamental principle of counting in Theorem 2.1:

Theorem 2.1   Fundamental Principle of Counting

An experiment consists of two subexperiments. If one subexperiment has k outcomes and the other subexperiment has n outcomes, then the experiment has nk outcomes.

Example 2.5

There are two subexperiments. The first subexperiment is "Flip a coin and observe either heads H or tails T." The second subexperiment is "Roll a six-sided die and observe the number of spots." It has six outcomes, $1, 2, \ldots, 6$. The experiment, "Flip a coin and roll a die," has $2 \times 6 = 12$ outcomes:

$$(H, 1), \quad (H, 2), \quad (H, 3), \quad (H, 4), \quad (H, 5), \quad (H, 6),$$
$$(T, 1), \quad (T, 2), \quad (T, 3), \quad (T, 4), \quad (T, 5), \quad (T, 6).$$

Generally, if an experiment E has k subexperiments $E_1, \ldots, E_k$ where E_i has n_i outcomes, then E has $\prod_{i=1}^{k} n_i$ outcomes.

In Example 2.4, we chose an ordered sequence of seven objects out of a set of 52 *distinguishable objects*. In general, an ordered sequence of k distinguishable objects is called a *k-permutation*. We will use the notation $(n)_k$ to denote the number of possible k-permutations of n distinguishable objects. To find $(n)_k$, suppose we have n distinguishable objects, and the experiment is to choose a sequence of k of these objects. There are n choices for the first object, $n-1$ choices for the second object, etc. Therefore, the total number of possibilities is

$$(n)_k = n(n-1)(n-2)\cdots(n-k+1). \tag{2.8}$$

Multiplying the right side by $(n-k)!/(n-k)!$ yields our next theorem.

Theorem 2.2

The number of k-permutations of n distinguishable objects is

$$(n)_k = n(n-1)(n-2)\cdots(n-k+1) = \frac{n!}{(n-k)!}.$$

Sampling without Replacement

Sampling without replacement corresponds to a sequential experiment in which the sample space of each subexperiment depends on the outcomes of previous subexperiments. Choosing objects randomly from a collection is called *sampling*, and the chosen objects are known as a *sample*. A k-permutation is a type of sample obtained by specific rules for selecting objects from the collection. In particular, once we choose an object for a k-permutation, we remove the object from the collection and we cannot choose it again. Consequently, this procedure is called *sampling without replacement*.

Different outcomes in a k-permutation are distinguished by the order in which objects arrive in a sample. By contrast, in many practical problems, we are concerned only with the identity of the objects in a sample, not their order. For example, in many card games, only the set of cards received by a player is of interest. The order in which they arrive is irrelevant.

Example 2.6

Suppose there are four objects, A, B, C, and D, and we define an experiment in which the procedure is to choose two objects without replacement, arrange them in alphabetical order, and observe the result. In this case, to observe AD we could choose A first or D first or both A and D simultaneously. The possible outcomes of the experiment are AB, AC, AD, BC, BD, and CD.

In contrast to this example with six outcomes, the next example shows that the k-permutation corresponding to an experiment in which the observation is the sequence of two letters has $4!/2! = 12$ outcomes.

Example 2.7

Suppose there are four objects, A, B, C, and D, and we define an experiment in which the procedure is to choose two objects without replacement and observe the result. The 12 possible outcomes of the experiment are AB, AC, AD, BA, BC, BD, CA, CB, CD, DA, DB, and DC.

In Example 2.6, each outcome is a subset of the outcomes of a k-permutation. Each subset is called a k-*combination*. We want to find the number of k-combinations.

We use the notation $\binom{n}{k}$ to denote the number of k-combinations. The words for this number are "n choose k," the number of k-combinations of n objects. To find $\binom{n}{k}$, we perform the following two subexperiments to assemble a k-permutation of n distinguishable objects:

1. Choose a k-combination out of the n objects.

2. Choose a k-permutation of the k objects in the k-combination.

Theorem 2.2 tells us that the number of outcomes of the combined experiment is $(n)_k$. The first subexperiment has $\binom{n}{k}$ possible outcomes, the number we have to derive. By Theorem 2.2, the second experiment has $(k)_k = k!$ possible outcomes. Since there are $(n)_k$ possible outcomes of the combined experiment,

$$(n)_k = \binom{n}{k} \cdot k! \tag{2.9}$$

Rearranging the terms yields our next result.

Theorem 2.3

The number of ways to choose k objects out of n distinguishable objects is

$$\binom{n}{k} = \frac{(n)_k}{k!} = \frac{n!}{k!(n-k)!}.$$

We encounter $\binom{n}{k}$ in other mathematical studies. Sometimes it is called a *binomial coefficient* because it appears (as the coefficient of $x^k y^{n-k}$) in the expansion of the binomial $(x+y)^n$. In addition, we observe that

$$\binom{n}{k} = \binom{n}{n-k}. \tag{2.10}$$

The logic behind this identity is that choosing k out of n elements to be part of a subset is equivalent to choosing $n-k$ elements to be excluded from the subset.

In most contexts, $\binom{n}{k}$ is undefined except for integers n and k with $0 \le k \le n$. Here, we adopt the following definition that applies to all nonnegative integers n and all real numbers k:

Definition 2.1 *n* choose *k*

For an integer $n \ge 0$, we define

$$\binom{n}{k} = \begin{cases} \dfrac{n!}{k!(n-k)!} & k = 0, 1, \ldots, n, \\ 0 & \text{otherwise.} \end{cases}$$

This definition captures the intuition that given, say, $n = 33$ objects, there are zero ways of choosing $k = -5$ objects, zero ways of choosing $k = 8.7$ objects, and zero ways of choosing $k = 87$ objects. Although this extended definition may seem unnecessary, and perhaps even silly, it will make many formulas in later chapters more concise and easier for students to grasp.

Example 2.8

- The number of combinations of seven cards chosen from a deck of 52 cards is

$$\binom{52}{7} = \frac{52 \times 51 \times \cdots \times 46}{2 \times 3 \times \cdots \times 7} = 133{,}784{,}560, \tag{2.11}$$

which is the number of 7-combinations of 52 objects. By contrast, we found in Example 2.4 674,274,182,400 7-permutations of 52 objects. (The ratio is $7! = 5040$).

- There are 11 players on a basketball team. The starting lineup consists of five players. There are $\binom{11}{5} = 462$ possible starting lineups.

- There are $\binom{120}{60} \approx 10^{36}$. ways of dividing 120 students enrolled in a probability course into two sections with 60 students in each section.

- A baseball team has 15 field players and ten pitchers. Each field player can take any of the eight nonpitching positions. The starting lineup consists of one pitcher and eight field players. Therefore, the number of possible starting lineups is $N = \binom{10}{1}\binom{15}{8} = 64{,}350$. For each choice of starting lineup, the manager must submit to the umpire a batting order for the 9 starters. The number of possible batting orders is $N \times 9! = 23{,}351{,}328{,}000$ since there are N ways to choose the 9 starters, and for each choice of 9 starters, there are $9! = 362{,}880$ possible batting orders.

Example 2.9

There are four queens in a deck of 52 cards. You are given seven cards at random from the deck. What is the probability that you have no queens?

Consider an experiment in which the procedure is to select seven cards at random from a set of 52 cards and the observation is to determine if there are one or more queens in the selection. The sample space contains $H = \binom{52}{7}$ possible combinations of seven cards, each with probability $1/H$. There are $H_{NQ} = \binom{52-4}{7}$ combinations with no queens. The probability of receiving no queens is the ratio of the number of outcomes with no queens to the number of outcomes in the sample space. $H_{NQ}/H = 0.5504$.

Another way of analyzing this experiment is to consider it as a sequence of seven subexperiments. The first subexperiment consists of selecting a card at random and observing whether it is a queen. If it is a queen, an outcome with probability $4/52$ (because there are 52 outcomes in the sample space and four of them are in the event {queen}), stop looking for queens. Otherwise, with probability $48/52$, select another card from the remaining 51 cards and observe whether it is a queen. This outcome of this subexperiment has probability $4/51$. If the second card is not a queen, an outcome with probability $47/51$, continue until you select a queen or you have seven cards with no queen. Using Q_i and N_i to indicate a "Queen" or "No queen" on subexperiment i, the tree for this experiment is

The probability of the event N_7 that no queen is received in your seven cards is the product of the probabilities of the branches leading to N_7:

$$(48/52) \times (47/51) \cdots \times (42/46) = 0.5504. \tag{2.12}$$

Sampling with Replacement

Consider selecting an object from a collection of objects, replacing the selected object, and repeating the process several times, each time replacing the selected object before making another selection. We refer to this situation as *sampling with replacement*. Each selection is the procedure of a subexperiment. The subexperiments are referred to as *independent trials*. In this section we consider the number of possible outcomes that result from sampling with replacement. In the next section we derive probability models for for experiments that specify sampling with replacement.

=======Example 2.10=======

There are four queens in a deck of 52 cards. You are given seven cards at random from the deck. After receiving each card you return it to the deck and receive another card at random. Observe whether you have not received any queens among the seven cards you were given. What is the probability that you have received no queens?

. . . .

The sample space contains 52^7 outcomes. There are 48^7 outcomes with no queens. The ratio is $(48/52)^7 = 0.5710$, the probability of receiving no queens. If this experiment is considered as a sequence of seven subexperiments, the tree looks the same as the tree in Example 2.9, except that all the horizontal branches have probability $48/52$ and all the diagonal branches have probability $4/52$.

The result of Example 2.10 generalizes naturally when we want to choose with replacement a sample of n objects out of a collection of m distinguishable objects. The experiment consists of a sequence of n identical subexperiments with m outcomes in the sample space of each subexperiment. Hence there are m^n ways to choose with replacement a sample of n objects.

=======Theorem 2.4=======
Given m distinguishable objects, there are m^n ways to choose with replacement an ordered sample of n objects.

=======Example 2.11=======
There are $2^{10} = 1024$ binary sequences of length 10.

Sampling with replacement corresponds to performing n repetitions of an identical subexperiment. Using x_i to denote the outcome of the ith subexperiment, the result for n repetitions of the subexperiment is a sequence $x_1, \ldots, x_n$.

=======Example 2.12=======
A chip fabrication facility produces microprocessors. Each microprocessor is tested to determine whether it runs reliably at an acceptable clock speed. A subexperiment to test a microprocessor has sample space $S_{\text{sub}} = \{0, 1\}$ to indicate whether the test was a failure (0) or a success (1). For test i, we record $x_i = 0$ or $x_i = 1$ to indicate the result. In testing four microprocessors, the observation sequence, $x_1 x_2 x_3 x_4$, is one of 16 possible outcomes:

$$S = \left\{ \begin{array}{llllllll} 0000, & 0001, & 0010, & 0011, & 0100, & 0101, & 0110, & 0111, \\ 1000, & 1001, & 1010, & 1011, & 1100, & 1101, & 1110, & 1111 \end{array} \right\}.$$

Note that we can think of the observation sequence $x_1, \ldots, x_n$ as the result of sampling with replacement n times from a sample space S_{sub}. For sequences of identical subexperiments, we can express Theorem 2.4 as

Theorem 2.5

For n repetitions of a subexperiment with sample space $S_{sub} = \{s_0, \ldots, s_{m-1}\}$, the sample space S of the sequential experiment has m^n outcomes.

Example 2.13

There are ten students in a probability class; each earns a grade $s \in S_{\text{sub}} = \{A, B, C, F\}$. We use the notation x_i to denote the grade of the ith student. For example, the grades for the class could be

$$x_1 x_2 \cdots x_{10} = CBBACFBACF \tag{2.13}$$

The sample space S of possible sequences contains $4^{10} = 1{,}048{,}576$ outcomes.

In Example 2.13, repeating a subexperiment n times and recording the observation consists of constructing a word with n letters. In general, n repetitions of the same subexperiment consists of choosing symbols from the alphabet $\{s_0, \ldots, s_{m-1}\}$. In Example 2.12, $m = 2$ and we have a binary alphabet with symbols $s_0 = 0$ and $s_1 = 1$.

A more challenging problem than finding the number of possible combinations of m objects sampled with replacement from a set of n objects is to calculate the number of observation sequences such that each object appears a specified number of times. We start with the case in which each subexperiment is a trial with sample space $S_{\text{sub}} = \{0, 1\}$ indicating failure or success.

Example 2.14

For five subexperiments with sample space $S_{\text{sub}} = \{0, 1\}$, what is the number of observation sequences in which 0 appears $n_0 = 2$ times and 1 appears $n_1 = 3$ times?

The 10 five-letter words with 0 appearing twice and 1 appearing three times are:

$$\{00111, 01011, 01101, 01110, 10011, 10101, 10110, 11001, 11010, 11100\}.$$

Example 2.14 determines the number of outcomes in the sample space of an experiment with five subexperiments by listing all of the outcomes. Even in this simple example it is not a simple matter to determine all of the outcomes, and in most practical applications of probability there are far more then ten outcomes in the sample space of an experiment and listing them all is out of the question. On the other hand, the counting methods covered in this chapter provide formulas for quickly calculating the number of outcomes in a sample space.

In Example 2.14 each outcome corresponds to the position of three ones in a five-letter binary word. That is, each outcome is completely specified by choosing three positions that contain 1. There are $\binom{5}{3} = 10$ ways to choose three positions in a word. More generally, for length n binary words with n_1 1's, we choose $\binom{n}{n_1}$ slots to hold a 1.

---Theorem 2.6---

The number of observation sequences for n subexperiments with sample space $S = \{0, 1\}$ with 0 appearing n_0 times and 1 appearing $n_1 = n - n_0$ times is $\binom{n}{n_1}$.

Theorem 2.6 can be generalized to subexperiments with $m > 2$ elements in the sample space. For n trials of a subexperiment with sample space $S_{\text{sub}} = \{s_0, \ldots, s_{m-1}\}$, we want to find the number of outcomes in which s_0 appears n_0 times, s_1 appears n_1 times, and so on. Of course, there are no such outcomes unless $n_0 + \cdots + n_{m-1} = n$. The notation for the number of outcomes is

$$\binom{n}{n_0, \ldots, n_{m-1}}.$$

It is referred to as the *multinomial coefficient*. To derive a formula for the multinomial coefficient, we generalize the logic used in deriving the formula for the binomial coefficient. With n subexperiments, representing the observation sequence by n slots, we first choose n_0 positions in the observation sequence to hold s_0, then n_1 positions to hold s_1, and so on. The details can be found in the proof of the following theorem:

---Theorem 2.7---

For n repetitions of a subexperiment with sample space $S = \{s_0, \ldots, s_{m-1}\}$, the number of length $n = n_0 + \cdots + n_{m-1}$ observation sequences with s_i appearing n_i times is

$$\binom{n}{n_0, \ldots, n_{m-1}} = \frac{n!}{n_0! n_1! \cdots n_{m-1}!}.$$

Proof Let $M = \binom{n}{n_0, \ldots, n_{m-1}}$. Start with n empty slots and perform the following sequence of subexperiments:

Subexperiment	Procedure
0	Label n_0 slots as s_0.
1	Label n_1 slots as s_1.
$\vdots$	$\vdots$
$m-1$	Label the remaining n_{m-1} slots as s_{m-1}.

There are $\binom{n}{n_0}$ ways to perform subexperiment 0. After n_0 slots have been labeled, there are $\binom{n-n_0}{n_1}$ ways to perform subexperiment 1. After subexperiment $j - 1$, $n_0 + \cdots + n_{j-1}$ slots have already been filled, leaving $\binom{n-(n_0+\cdots+n_{j-1})}{n_j}$ ways to perform subexperiment j.

From the fundamental counting principle,

$$M = \binom{n}{n_0}\binom{n-n_0}{n_1}\binom{n-n_0-n_1}{n_2}\cdots\binom{n-n_0-\cdots-n_{m-2}}{n_{m-1}}$$

$$= \frac{n!}{(n-n_0)!n_0!}\frac{(n-n_0)!}{(n-n_0-n_1)!n_1!}\cdots\frac{(n-n_0-\cdots-n_{m-2})!}{(n-n_0-\cdots-n_{m-1})!n_{m-1}!}. \tag{2.14}$$

Canceling the common factors, we obtain the formula of the theorem.

Note that a binomial coefficient is the special case of the multinomial coefficient for an alphabet with $m = 2$ symbols. In particular, for $n = n_0 + n_1$,

$$\binom{n}{n_0, n_1} = \binom{n}{n_0} = \binom{n}{n_1}. \tag{2.15}$$

Lastly, in the same way that we extended the definition of the binomial coefficient, we will employ an extended definition for the multinomial coefficient.

=======Definition 2.2========Multinomial Coefficient

For an integer $n \geq 0$, we define

$$\binom{n}{n_0, \ldots, n_{m-1}} = \begin{cases} \dfrac{n!}{n_0!n_1!\cdots n_{m-1}!} & n_0 + \cdots + n_{m-1} = n; \\[2mm] & n_i \in \{0, 1, \ldots, n\}, i = 0, 1, \ldots, m-1, \\[2mm] 0 & otherwise. \end{cases}$$

=======Example 2.15=======

In Example 2.13, the professor uses a *curve* in determining student grades. When there are ten students in a probability class, the professor always issues two grades of A, three grades of B, three grades of C and two grades of F. How many different ways can the professor assign grades to the ten students?
. . . .
In Example 2.13, we determine that with four possible grades there are $4^{10} = 1{,}048{,}576$ ways of assigning grades to ten students. However, now we are limited to choosing $n_0 = 2$ students to receive an A, $n_1 = 3$ students to receive a B, $n_2 = 3$ students to receive a C and $n_3 = 4$ students to receive an F. The number of ways that fit the curve is the multinomial coefficient

$$\binom{n}{n_0, n_1, n_2, n_3} = \binom{10}{2, 3, 3, 2} = \frac{10!}{2!3!3!2!} = 25{,}200. \tag{2.16}$$

=======Quiz 2.2=======

Consider a binary code with 4 bits (0 or 1) in each code word. An example of a code word is 0110.

(a) How many different code words are there?

(b) How many code words have exactly two zeroes?

(c) How many code words begin with a zero?

(d) In a constant-ratio binary code, each code word has N bits. In every word, M of the N bits are 1 and the other $N - M$ bits are 0. How many different code words are in the code with $N = 8$ and $M = 3$?

2.3   Independent Trials

> Independent trials are identical subexperiments in a sequential experiment. The probability models of all the subexperiments are identical and independent of the outcomes of previous subexperiments. Sampling with replacement is one category of experiments with independent trials.

We now apply the counting methods of Section 2.2 to derive probability models for experiments consisting of independent repetitions of a subexperiment. We start with a simple subexperiment in which there are two outcomes: a success (1) occurs with probability p; otherwise, a failure (0) occurs with probability $1 - p$. The results of all trials of the subexperiment are mutually independent. An outcome of the complete experiment is a sequence of successes and failures denoted by a sequence of ones and zeroes. For example, $10101\ldots$ is an alternating sequence of successes and failures. Let E_{n_0,n_1} denote the event n_0 failures and n_1 successes in $n = n_0 + n_1$ trials. To find $P[E_{n_0,n_1}]$, we first consider an example.

=====**Example 2.16**=====
What is the probability $P[E_{2,3}]$ of two failures and three successes in five independent trials with success probability p.
. . . .
To find $P[E_{2,3}]$, we observe that the outcomes with three successes in five trials are 11100, 11010, 11001, 10110, 10101, 10011, 01110, 01101, 01011, and 00111. We note that the probability of each outcome is a product of five probabilities, each related to one subexperiment. In outcomes with three successes, three of the probabilities are p and the other two are $1 - p$. Therefore each outcome with three successes has probability $(1 - p)^2 p^3$.

From Theorem 2.6, we know that the number of such sequences is $\binom{5}{3}$. To find $P[E_{2,3}]$, we add up the probabilities associated with the 10 outcomes with 3 successes, yielding

$$P[E_{2,3}] = \binom{5}{3}(1 - p)^2 p^3. \tag{2.17}$$

In general, for $n = n_0 + n_1$ independent trials we observe that

- Each outcome with n_0 failures and n_1 successes has probability $(1 - p)^{n_0} p^{n_1}$.

- There are $\binom{n}{n_0} = \binom{n}{n_1}$ outcomes that have n_0 failures and n_1 successes.

Therefore the probability of n_1 successes in n independent trials is the sum of $\binom{n}{n_1}$ terms, each with probability $(1-p)^{n_0}p^{n_1} = (1-p)^{n-n_1}p^{n_1}$.

Theorem 2.8

The probability of n_0 failures and n_1 successes in $n = n_0 + n_1$ independent trials is

$$P[E_{n_0,n_1}] = \binom{n}{n_1}(1-p)^{n-n_1}p^{n_1} = \binom{n}{n_0}(1-p)^{n_0}p^{n-n_0}.$$

The second formula in this theorem is the result of multiplying the probability of n_0 failures in n trials by the number of outcomes with n_0 failures.

Example 2.17

In Example 1.16, we found that a randomly tested resistor was acceptable with probability $P[A] = 0.78$. If we randomly test 100 resistors, what is the probability of T_i, the event that i resistors test acceptable?

....

Testing each resistor is an independent trial with a success occurring when a resistor is acceptable. Thus for $0 \leq i \leq 100$,

$$P[T_i] = \binom{100}{i}(0.78)^i(1-0.78)^{100-i} \tag{2.18}$$

We note that our intuition says that since 78% of the resistors are acceptable, then in testing 100 resistors, the number acceptable should be near 78. However, $P[T_{78}] \approx 0.096$, which is fairly small. This shows that although we might expect the number acceptable to be close to 78, that does not mean that the probability of exactly 78 acceptable is high.

Now suppose we perform n independent repetitions of a subexperiment for which there are m possible outcomes for any subexperiment. That is, the sample space for each subexperiment is $(s_0, \ldots, s_{m-1})$ and every event in one subexperiment is independent of the events in all the other subexperiments. Therefore, in every subexperiment the probabilities of corresponding events are the same and we can use the notation $P[s_k] = p_k$ for all of the subexperiments.

An outcome of the experiment consists of a sequence of n subexperiment outcomes. In the probability tree of the experiment, each node has m branches and branch i has probability p_i. The probability of an outcome of the sequential experiment is just the product of the n branch probabilities on a path from the root of the tree to the leaf representing the outcome. For example, with $n = 5$, the outcome $s_2 s_0 s_3 s_2 s_4$ occurs with probability $p_2 p_0 p_3 p_2 p_4$. We want to find the probability of the event

$$E_{n_0,\ldots,n_{m-1}} = \{s_0 \text{ occurs } n_0 \text{ times}, \ldots, s_{m-1} \text{ occurs } n_{m-1} \text{ times}\} \tag{2.19}$$

Note that the notation $E_{n_0,\ldots,n_{m-1}}$ implies that the experiment consists of a sequence of $n = n_0 + \cdots + n_{m-1}$ trials.

To calculate $P[E_{n_0,\ldots,n_{m-1}}]$, we observe that the probability of the outcome

$$\underbrace{s_0 \cdots s_0}_{n_0 \text{ times}} \underbrace{s_1 \cdots s_1}_{n_1 \text{ times}} \cdots \underbrace{s_{m-1} \cdots s_{m-1}}_{n_{m-1} \text{ times}} \tag{2.20}$$

is

$$p_0^{n_0} p_1^{n_1} \cdots p_{m-1}^{n_{m-1}}. \tag{2.21}$$

Next, we observe that any other experimental outcome that is a reordering of the preceding sequence has the same probability because on each path through the tree to such an outcome there are n_i occurrences of s_i. As a result,

$$P\left[E_{n_0,\ldots,n_{m-1}}\right] = M p_1^{n_1} p_2^{n_2} \cdots p_r^{n_r} \tag{2.22}$$

where M, the number of such outcomes, is the multinomial coefficient $\binom{n}{n_0,\ldots,n_{m-1}}$ of Definition 2.2. Applying Theorem 2.7, we have the following theorem:

Theorem 2.9

A subexperiment has sample space $S_{sub} = \{s_0,\ldots,s_{m-1}\}$ *with* $P[s_i] = p_i$. *For* $n = n_0 + \cdots + n_{m-1}$ *independent trials, the probability of* n_i *occurences of* s_i, $i = 0,1,\ldots,m-1$, *is*

$$P\left[E_{n_0,\ldots,n_{m-1}}\right] = \binom{n}{n_0,\ldots,n_{m-1}} p_0^{n_0} \cdots p_{m-1}^{n_{m-1}}.$$

Example 2.18

A packet processed by an Internet router carries either audio information with probability $7/10$, video, with probability $2/10$, or text with probability $1/10$. Let $E_{a,v,t}$ denote the event that the router processes a audio packets, v video packets, and t text packets in a sequence of 100 packets. In this case,

$$P[E_{a,v,t}] = \binom{100}{a,v,t}\left(\frac{7}{10}\right)^a \left(\frac{2}{10}\right)^v \left(\frac{1}{10}\right)^t \tag{2.23}$$

Keep in mind that by the extended definition of the multinomial coefficient, $P[E_{a,v,t}]$ is nonzero only if $a + v + t = 100$ and a, v, and t are nonnegative integers.

Quiz 2.3

Data packets containing 100 bits are transmitted over a communication link. A transmitted bit is received in error (either a 0 sent is mistaken for a 1, or a 1 sent

```
Y =
  Columns 1 through 12
    47    52    48    46    54    48    47    48    59    44    49    48
  Columns 13 through 24
    42    52    40    40    47    48    48    48    53    49    45    61
  Columns 25 through 36
    60    59    49    47    49    45    48    51    48    53    52    53
  Columns 37 through 48
    56    54    60    53    52    51    58    47    50    48    44    49
  Columns 49 through 60
    50    46    52    50    51    51    57    50    49    56    44    56
```

Figure 2.2   The simulation output of 60 repeated experiments of 100 coin flips.

is mistaken for a 0) with probability $\epsilon = 0.01$, independent of the correctness of any other bit. The packet has been coded in such a way that if three or fewer bits are received in error, then those bits can be corrected. If more than three bits are received in error, then the packet is decoded with errors.

(a) Let $E_{k,100-k}$ denote the event that a received packet has k bits in error and $100 - k$ correctly decoded bits. What is $P[E_{k,100-k}]$ for $k = 0, 1, 2, 3$?

(b) Let C denote the event that a packet is decoded correctly. What is $P[C]$?

2.4   MATLAB

Two or three lines of MATLAB code are sufficient to simulate an arbitrary number of sequential trials.

We recall from Section 1.6 that rand(1,m)<p simulates m coin flips with $P[\text{heads}] = p$. Because MATLAB can simulate these coin flips much faster than we can actually flip coins, a few lines of MATLAB code can yield quick simulations of many experiments.

Example 2.19

Using MATLAB, perform 60 experiments. In each experiment, flip a coin 100 times and record the number of heads in a vector $\mathbf{Y}$ such that the jth element Y_j is the number of heads in subexperiment j.

```
>> X=rand(100,60)<0.5;
>> Y=sum(X,1)
```

The MATLAB code for this task appears on the left. The 100×60 matrix $\mathbf{X}$ has i,jth element X(i,j)=0 (tails) or X(i,j)=1 (heads) to indicate the result of flip i of subexperiment j. Since Y sums X across the first dimension, Y(j) is the number of heads in the jth subexperiment. Each Y(j) is between 0 and 100 and generally in the neighborhood of 50. The output of a sample run is shown in Figure 2.2.

===Example 2.20===

In testing a microprocessor all four grades, 1, 2, 3 and 4, have probability 0.25, independent of any other microprocessor. Simulate the testing of 100 microprocessors. Your output should be a 4×1 vector $\mathbf{X}$ such that X_i is the number of grade i microprocessors.

. . . .

```
%chiptest.m
G=ceil(4*rand(1,100));
T=1:4;
X=hist(G,T);
```

The first line generates a row vector G of random grades for 100 microprocessors. The possible test scores are in the vector T. Lastly, X=hist(G,T) returns a histogram vector X such that X(j) counts the number of elements G(i) that equal T(j).

Note that "help hist" will show the variety of ways that the hist function can be called. Morever, X=hist(G,T) does more than just count the number of elements of G that equal each element of T. In particular, hist(G,T) creates bins centered around each T(j) and counts the number of elements of G that fall into each bin.

Note that in MATLAB all variables are assumed to be matrices. In writing MATLAB code, X may be an $n \times m$ matrix, an $n \times 1$ column vector, a $1 \times m$ row vector, or a 1×1 scalar. In MATLAB, we write X(i,j) to index the i, jth element. By contrast, in this text, we vary the notation depending on whether we have a scalar X, or a vector or matrix $\mathbf{X}$. In addition, we use $X_{i,j}$ to denote the i, jth element. Thus, $\mathbf{X}$ and X (in a MATLAB code fragment) may both refer to the same variable.

===Quiz 2.4===

The flip of a thick coin yields heads with probability 0.4, tails with probability 0.5, or lands on its edge with probability 0.1. Simulate 100 thick coin flips. Your output should be a 3×1 vector $\mathbf{X}$ such that X_1, X_2, and X_3 are the number of occurrences of heads, tails, and edge.

Problems

Difficulty: ● Easy   ■ Moderate   ◆ Difficult   ◆◆ Experts Only

2.1.1● Suppose you flip a coin twice. On any flip, the coin comes up heads with probability 1/4. Use H_i and T_i to denote the result of flip i.

(a) What is the probability, $P[H_1|H_2]$, that the first flip is heads given that the second flip is heads?

(b) What is the probability that the first flip is heads and the second flip is tails?

2.1.2● For Example 2.2, suppose $P[G_1] = 1/2$, $P[G_2|G_1] = 3/4$, and $P[G_2|R_1] = 1/4$. Find $P[G_2]$, $P[G_2|G_1]$, and $P[G_1|G_2]$.

2.1.3● At the end of regulation time, a basketball team is trailing by one point and a player goes to the line for two free throws.

If the player makes exactly one free throw, the game goes into overtime. The probability that the first free throw is good is $1/2$. However, if the first attempt is good, the player relaxes and the second attempt is good with probability $3/4$. However, if the player misses the first attempt, the added pressure reduces the success probability to $1/4$. What is the probability that the game goes into overtime?

2.1.4● You have two biased coins. Coin A comes up heads with probability $1/4$. Coin B comes up heads with probability $3/4$. However, you are not sure which is which, so you choose a coin randomly and you flip it. If the flip is heads, you guess that the flipped coin is B; otherwise, you guess that the flipped coin is A. What is the probability $P[C]$ that your guess is correct?

2.1.5● Suppose you have two coins, one biased, one fair, but you don't know which coin is which. Coin 1 is biased. It comes up heads with probability $3/4$, while coin 2 comes up heads with probability $1/2$. Suppose you pick a coin at random and flip it. Let C_i denote the event that coin i is picked. Let H and T denote the possible outcomes of the flip. Given that the outcome of the flip is a head, what is $P[C_1|H]$, the probability that you picked the biased coin? Given that the outcome is a tail, what is the probability $P[C_1|T]$ that you picked the biased coin?

2.1.6■ Suppose that for the general population, 1 in 5000 people carries the human immunodeficiency virus (HIV). A test for the presence of HIV yields either a positive $(+)$ or negative $(-)$ response. Suppose the test gives the correct answer 99% of the time. What is $P[-|H]$, the conditional probability that a person tests negative given that the person does have the HIV virus? What is $P[H|+]$, the conditional probability that a randomly chosen person has the HIV virus given that the person tests positive?

2.1.7■ A machine produces photo detectors in pairs. Tests show that the first photo detector is acceptable with probability $3/5$.

When the first photo detector is acceptable, the second photo detector is acceptable with probability $4/5$. If the first photo detector is defective, the second photo detector is acceptable with probability $2/5$.

(a) Find the probability that exactly one photo detector of a pair is acceptable.

(b) Find the probability that both photo detectors in a pair are defective.

2.1.8■ You have two biased coins. Coin A comes up heads with probability $1/4$. Coin B comes up heads with probability $3/4$. However, you are not sure which is which so you flip each coin once, choosing the first coin randomly. Use H_i and T_i to denote the result of flip i. Let A_1 be the event that coin A was flipped first. Let B_1 be the event that coin B was flipped first. What is $P[H_1 H_2]$? Are H_1 and H_2 independent? Explain your answer.

2.1.9■ A particular birth defect of the heart is rare; a newborn infant will have the defect D with probability $P[D] = 10^{-4}$. In the general exam of a newborn, a particular heart arrhythmia A occurs with probability 0.99 in infants with the defect. However, the arrhythmia also appears with probability 0.1 in infants without the defect. When the arrhythmia is present, a lab test for the defect is performed. The result of the lab test is either positive (event T^+) or negative (event T^-). In a newborn with the defect, the lab test is positive with probability $p = 0.999$ independent from test to test. In a newborn without the defect, the lab test is negative with probability $p = 0.999$. If the arrhythmia is present and the test is positive, then heart surgery (event H) is performed.

(a) Given the arrythmia A is present, what is the probability the infant has the defect D?

(b) Given that an infant has the defect, what is the probability $P[H|D]$ that heart surgery is performed?

(c) Given that the infant does not have the defect, what is the probability

$q = P[H|D^c]$ that an unnecessary heart surgery is performed?

(d) Find the probability $P[H]$ that an infant has heart surgery performed for the arrythmia.

(e) Given that heart surgery is performed, what is the probability that the newborn does *not* have the defect?

2.1.10■ Suppose Dagwood (Blondie's husband) wants to eat a sandwich but needs to go on a diet. Dagwood decides to let the flip of a coin determine whether he eats. Using an unbiased coin, Dagwood will postpone the diet (and go directly to the refrigerator) if either (a) he flips heads on his first flip or (b) he flips tails on the first flip but then proceeds to get two heads out of the next three flips. Note that the first flip is *not* counted in the attempt to win two of three and that Dagwood never performs any unnecessary flips. Let H_i be the event that Dagwood flips heads on try i. Let T_i be the event that tails occurs on flip i.

(a) Draw the tree for this experiment. Label the probabilities of all outcomes.

(b) What are $P[H_3]$ and $P[T_3]$?

(c) Let D be the event that Dagwood must diet. What is $P[D]$? What is $P[H_1|D]$?

(d) Are H_3 and H_2 independent events?

2.1.11▦ The quality of each pair of photo detectors produced by the machine in Problem 2.1.7 is independent of the quality of every other pair of detectors.

(a) What is the probability of finding no good detectors in a collection of n pairs produced by the machine?

(b) How many pairs of detectors must the machine produce to reach a probability of 0.99 that there will be at least one acceptable photo detector?

2.1.12■ In Steven Strogatz's New York Times blog http://opinionator.blogs. nytimes.com/2010/04/25/chances-are/ ?ref=opinion, the following problem was posed to highlight the confusing character of conditional probabilities.

Before going on vacation for a week, you ask your spacey friend to water your ailing plant. Without water, the plant has a 90 percent chance of dying. Even with proper watering, it has a 20 percent chance of dying. And the probability that your friend will forget to water it is 30 percent. (a) What's the chance that your plant will survive the week? (b) If it's dead when you return, what's the chance that your friend forgot to water it? (c) If your friend forgot to water it, what's the chance it'll be dead when you return?

Solve parts (a), (b) and (c) of this problem.

2.1.13■ Each time a fisherman casts his line, a fish is caught with probability p, independent of whether a fish is caught on any other cast of the line. The fisherman will fish all day until a fish is caught and then he will quit and go home. Let C_i denote the event that on cast i the fisherman catches a fish. Draw the tree for this experiment and find $P[C_1]$, $P[C_2]$, and $P[C_n]$ as functions of p.

2.2.1◉ A Starburst candy package contains 12 individual candy pieces. Each piece is equally likely to be berry, orange, lemon, or cherry, independent of all other pieces.

(a) What is the probability that a Starburst package has only berry or cherry pieces and zero orange or lemon pieces?

(b) What is the probability that a Starburst package has no cherry pieces?

(c) What is the probability $P[F_1]$ that all twelve pieces of your Starburst are the same flavor?

2.2.2◉ Your Starburst candy has 12 pieces, three pieces of each of four flavors: berry, lemon, orange, and cherry, arranged in a random order in the pack. You draw the first three pieces from the pack.

(a) What is the probability they are all the same flavor?

(b) What is the probability they are all different flavors?

2.2.3 ▪ Your Starburst candy has 12 pieces, three pieces of each of four flavors: berry, lemon, orange, and cherry, arranged in a random order in the pack. You draw the first four pieces from the pack.

(a) What is the probability $P[F_1]$ they are all the same flavor?

(b) What is the probability $P[F_4]$ they are all different flavors?

(c) What is the probability $P[F_2]$ that your Starburst has exactly two pieces of each of two different flavors?

2.2.4 ● In a game of rummy, you are dealt a seven-card hand.

(a) What is the probability $P[R_7]$ that your hand has only red cards?

(b) What is the probability $P[F]$ that your hand has only face cards?

(c) What is the probability $P[R_7 F]$ that your hand has only red face cards? (The face cards are jack, queen, and king.)

2.2.5 ▪ In a game of poker, you are dealt a five-card hand.

(a) What is the probability $P[R_5]$ that your hand has only red cards?

(b) What is the probability of a "full house" with three-of-a-kind and two-of-a-kind?

2.2.6 ● Consider a binary code with 5 bits (0 or 1) in each code word. An example of a code word is 01010. How many different code words are there? How many code words have exactly three 0's?

2.2.7 ● Consider a language containing four letters: A, B, C, D. How many three-letter words can you form in this language? How many four-letter words can you form if each letter appears only once in each word?

2.2.8 ● On an American League baseball team with 15 field players and 10 pitchers, the manager selects a starting lineup with 8 field players, 1 pitcher, and 1 designated hitter. The lineup specifies the players for these positions and the positions in a batting order for the 8 field players and designated hitter. If the designated hitter must be chosen among all the field players, how many possible starting lineups are there?

2.2.9 ▪ Suppose that in Problem 2.2.8, the designated hitter can be chosen from among all the players. How many possible starting lineups are there?

2.2.10 ● At a casino, the only game is numberless roulette. On a spin of the wheel, the ball lands in a space with color red (r), green (g), or black (b). The wheel has 19 red spaces, 19 green spaces and 2 black spaces.

(a) In 40 spins of the wheel, find the probability of the event

$$A = \{19 \text{ reds}, 19 \text{ greens, and } 2 \text{ blacks}\}.$$

(b) In 40 spins of the wheel, find the probability of $G_{19} = \{19 \text{ greens}\}$.

(c) The only bets allowed are red and green. Given that you randomly choose to bet red or green, what is the probability p that your bet is a winner?

2.2.11 ▪ A basketball team has three pure centers, four pure forwards, four pure guards, and one swingman who can play either guard or forward. A pure position player can play only the designated position. If the coach must start a lineup with one center, two forwards, and two guards, how many possible lineups can the coach choose?

2.2.12 ◆ An instant lottery ticket consists of a collection of boxes covered with gray wax. For a subset of the boxes, the gray wax hides a special mark. If a player scratches off the correct number of the marked boxes (and no boxes without the mark), then that ticket is a winner. Design an instant lottery game in which a player scratches five boxes and the probability that a ticket is a winner is approximately 0.01.

2.3.1 ● Consider a binary code with 5 bits (0 or 1) in each code word. An example of a code word is 01010. In each code word, a bit is a zero with probability 0.8, independent of any other bit.

(a) What is the probability of the code word 00111?

(b) What is the probability that a code word contains exactly three ones?

2.3.2● The Boston Celtics have won 16 NBA championships over approximately 50 years. Thus it may seem reasonable to assume that in a given year the Celtics win the title with probability $p = 16/50 = 0.32$, independent of any other year. Given such a model, what would be the probability of the Celtics winning eight straight championships beginning in 1959? Also, what would be the probability of the Celtics winning the title in 10 out of 11 years, starting in 1959? Given your answers, do you trust this simple probability model?

2.3.3● Suppose each day that you drive to work a traffic light that you encounter is either green with probability 7/16, red with probability 7/16, or yellow with probability 1/8, independent of the status of the light on any other day. If over the course of five days, G, Y, and R denote the number of times the light is found to be green, yellow, or red, respectively, what is the probability that $P[G = 2, Y = 1, R = 2]$? Also, what is the probability $P[G = R]$?

2.3.4■ In a game between two equal teams, the home team wins with probability $p > 1/2$. In a best of three playoff series, a team with the home advantage has a game at home, followed by a game away, followed by a home game if necessary. The series is over as soon as one team wins two games. What is $P[H]$, the probability that the team with the home advantage wins the series? Is the home advantage increased by playing a three-game series rather than a one-game playoff? That is, is it true that $P[H] \geq p$ for all $p \geq 1/2$?

2.3.5♦ A collection of field goal kickers are divided into groups 1 and 2. Group i has $3i$ kickers. On any kick, a kicker from group i will kick a field goal with probability $1/(i+1)$, independent of the outcome of any other kicks.

(a) A kicker is selected at random from among all the kickers and attempts one field goal. Let K be the event that a field goal is kicked. Find $P[K]$.

(b) Two kickers are selected at random; K_j is the event that kicker j kicks a field goal. Are K_1 and K_2 independent?

(c) A kicker is selected at random and attempts 10 field goals. Let M be the number of misses. Find $P[M = 5]$.

2.4.1■ Build a MATLAB simulation of 50 trials of the experiment of Problem 2.1.5. Your output should be a pair of 50×1 vectors **C** and **H**. For the ith trial, H_i will record whether it was heads ($H_i = 1$) or tails ($H_i = 0$), and $C_i \in \{1, 2\}$ will record which coin was picked.

2.4.2■ Following Quiz 2.3, suppose the communication link has different error probabilities for transmitting 0 and 1. When a 1 is sent, it is received as a 0 with probability 0.01. When a 0 is sent, it is received as a 1 with probability 0.03. Each bit in a packet is still equally likely to be a 0 or 1. Packets have been coded such that if five or fewer bits are received in error, then the packet can be decoded. Simulate the transmission of 100 packets, each containing 100 bits. Count the number of packets decoded correctly.

2.4.3♦ Write a MATLAB function

```
N=countequal(G,T)
```

that duplicates the action of `hist(G,T)` in Example 2.20. Hint: Use `ndgrid`.

3

Discrete Random Variables

3.1 Definitions

> A random variable assigns numbers to outcomes in the sample
> space of an experiment.

Chapter 1 defines a probability model. It begins with a *physical* model of an experiment. An experiment consists of a procedure and observations. The set of all possible observations, S, is the sample space of the experiment. S is the beginning of the *mathematical* probability model. In addition to S, the mathematical model includes a rule for assigning numbers between 0 and 1 to sets A in S. Thus for every $A \subset S$, the model gives us a probability $\mathrm{P}[A]$, where $0 \le \mathrm{P}[A] \le 1$.

In this chapter and for most of the remainder of this book, we examine probability models that assign numbers to the outcomes in the sample space. When we observe one of these numbers, we refer to the observation as a *random variable*. In our notation, the name of a random variable is always a capital letter, for example, X. The set of possible values of X is the *range* of X. Since we often consider more than one random variable at a time, we denote the range of a random variable by the letter S with a subscript that is the name of the random variable. Thus S_X is the range of random variable X, S_Y is the range of random variable Y, and so forth. We use S_X to denote the range of X because the set of all possible values of X is analogous to S, the set of all possible outcomes of an experiment.

A probability model always begins with an experiment. Each random variable is related directly to this experiment. There are three types of relationships.

1. The random variable is the observation.

===Example 3.1===

The experiment is to attach a photo detector to an optical fiber and count the number of photons arriving in a one-microsecond time interval. Each observation is a random variable X. The range of X is $S_X = \{0, 1, 2, \ldots\}$. In this case, S_X, the range of X, and the sample space S are identical.

2. The random variable is a function of the observation.

──────Example 3.2──────

The experiment is to test six integrated circuits and after each test observe whether the circuit is accepted (a) or rejected (r). Each observation is a sequence of six letters where each letter is either a or r. For example, $s_8 = $ aaraaa. The sample space S consists of the 64 possible sequences. A random variable related to this experiment is N, the number of accepted circuits. For outcome s_8, $N = 5$ circuits are accepted. The range of N is $S_N = \{0, 1, \ldots, 6\}$.

3. The random variable is a function of another random variable.

──────Example 3.3──────

In Example 3.2, the net revenue R obtained for a batch of six integrated circuits is $5 for each circuit accepted minus $7 for each circuit rejected. (This is because for each bad circuit that goes out of the factory, it will cost the company $7 to deal with the customer's complaint and supply a good replacement circuit.) When N circuits are accepted, $6 - N$ circuits are rejected so that the net revenue R is related to N by the function

$$R = g(N) = 5N - 7(6 - N) = 12N - 42 \text{ dollars.} \tag{3.1}$$

Since $S_N = \{0, \ldots, 6\}$, the range of R is

$$S_R = \{-42, -30, -18, -6, 6, 18, 30\}. \tag{3.2}$$

The revenue associated with $s_8 = $ aaraaa and all other outcomes for which $N = 5$ is

$$g(5) = 12 \times 5 - 42 = 18 \text{ dollars} \tag{3.3}$$

If we have a probability model for the integrated circuit experiment in Example 3.2, we can use that probability model to obtain a probability model for the random variable. The remainder of this chapter will develop methods to characterize probability models for random variables. We observe that in the preceding examples, the value of a random variable can always be derived from the outcome of the underlying experiment. This is not a coincidence. The formal definition of a random variable reflects this fact.

Definition 3.1 ━━━ Random Variable

*A **random variable** consists of an experiment with a probability measure* P[·] *defined on a sample space S and a function that assigns a real number to each outcome in the sample space of the experiment.*

This definition acknowledges that a random variable is the result of an underlying experiment, but it also permits us to separate the experiment, in particular, the observations, from the process of assigning numbers to outcomes. As we saw in Example 3.1, the assignment may be implicit in the definition of the experiment, or it may require further analysis.

In some definitions of experiments, the procedures contain variable parameters. In these experiments, there can be values of the parameters for which it is impossible to perform the observations specified in the experiments. In these cases, the experiments do not produce random variables. We refer to experiments with parameter settings that do not produce random variables as *improper experiments*.

Example 3.4

The procedure of an experiment is to fire a rocket in a vertical direction from Earth's surface with initial velocity V km/h. The observation is T seconds, the time elapsed until the rocket returns to Earth. Under what conditions is the experiment improper?
....

At low velocities, V, the rocket will return to Earth at a random time T seconds that depends on atmospheric conditions and small details of the rocket's shape and weight. However, when $V > v^* \approx 40{,}000$ km/hr, the rocket will not return to Earth. Thus, the experiment is improper when $V > v^*$ because it is impossible to perform the specified observation.

On occasion, it is important to identify the random variable X by the function $X(s)$ that maps the sample outcome s to the corresponding value of the random variable X. As needed, we will write $\{X = x\}$ to emphasize that there is a set of sample points $s \in S$ for which $X(s) = x$. That is, we have adopted the shorthand notation

$$\{X = x\} = \{s \in S | X(s) = x\}. \tag{3.4}$$

Here are some more random variables:

- A, the number of students asleep in the next probability lecture;
- C, the number of texts you receive in the next hour;
- M, the number of minutes you wait until the next text arrives.

Random variables A and C are *discrete* random variables. The possible values of these random variables form a countable set. The underlying experiments have sample spaces that are discrete. The random variable M can be any nonnegative real number. It is a *continuous random variable*. Its experiment has a continuous sample space. In this chapter, we study the properties of discrete random variables. Chapter 4 covers continuous random variables.

====Definition 3.2====Discrete Random Variable

X is a **discrete** random variable if the range of X is a countable set

$$S_X = \{x_1, x_2, \ldots\}.$$

The defining characteristic of a discrete random variable is that the set of possible values can (in principle) be listed, even though the list may be infinitely long. Often, but not always, a discrete random variable takes on integer values. An exception is the random variable related to your probability grade. The experiment is to take this course and observe your grade. At Rutgers, the sample space is

$$S = \left\{F, D, C, C^+, B, B^+, A\right\}. \tag{3.5}$$

We use a function $G_1(\cdot)$ to map this sample space into a random variable. For example, $G_1(A) = 4$ and $G_1(F) = 0$. The table

Outcomes	F	D	C	C^+	B	B^+	A
G_1	0	1	2	2.5	3	3.5	4

is a concise description of the entire mapping.

G_1 is a discrete random variable with range $S_{G_1} = \{0, 1, 2, 2.5, 3, 3.5, 4\}$. Have you thought about why we transform letter grades to numerical values? We believe the principal reason is that it allows us to compute averages. This is also an important motivation for creating random variables by assigning numbers to the outcomes in a sample space. Unlike probability models defined on arbitrary sample spaces, random variables have *expected values*, which are closely related to averages of data sets. We introduce expected values formally in Section 3.5.

====Quiz 3.1====

A student takes two courses. In each course, the student will earn either a B or a C. To calculate a grade point average (GPA), a B is worth 3 points and a C is worth 2 points. The student's GPA G_2 is the sum of the points earned for each course divided by 2. Make a table of the sample space of the experiment and the corresponding values of the GPA, G_2.

3.2 Probability Mass Function

> The PMF of random variable X expresses the probability model of an experiment as a mathematical function. The function is the probability $P[X = x]$ for every number x.

Recall that the probability model of a discrete random variable assigns a number between 0 and 1 to each outcome in a sample space. When we have a discrete random variable X, we express the probability model as a probability mass function (PMF) $P_X(x)$. The argument of a PMF ranges over all real numbers.

═══════**Definition 3.3**═══════**Probability Mass Function (PMF)**

*The **probability mass function** (PMF) of the discrete random variable X is*

$$P_X(x) = \mathrm{P}[X = x]$$

Note that $X = x$ is an event consisting of all outcomes s of the underlying experiment for which $X(s) = x$. On the other hand, $P_X(x)$ is a function ranging over all real numbers x. For any value of x, the function $P_X(x)$ is the probability of the event $X = x$.

Observe our notation for a random variable and its PMF. We use an uppercase letter (X in the preceding definition) for the name of a random variable. We usually use the corresponding lowercase letter (x) to denote a possible value of the random variable. The notation for the PMF is the letter P with a subscript indicating the name of the random variable. Thus $P_R(r)$ is the notation for the PMF of random variable R. In these examples, r and x are dummy variables. The same random variables and PMFs could be denoted $P_R(u)$ and $P_X(u)$ or, indeed, $P_R(\cdot)$ and $P_X(\cdot)$.

We derive the PMF from the sample space, the probability model, and the rule that maps outcomes to values of the random variable. We then graph a PMF by marking on the horizontal axis each value with nonzero probability and drawing a vertical bar with length proportional to the probability.

═══════**Example 3.5**═══════

When the basketball player Wilt Chamberlain shot two free throws, each shot was equally likely either to be good (g) or bad (b). Each shot that was good was worth 1 point. What is the PMF of X, the number of points that he scored?

....

There are four outcomes of this experiment: gg, gb, bg, and bb. A simple tree diagram indicates that each outcome has probability $1/4$. The sample space and probabilities of the experiment and the corresponding values of X are given in the table:

Outcomes	bb	bg	gb	gg
$\mathrm{P}[\cdot]$	1/4	1/4	1/4	1/4
X	0	1	1	2

The random variable X has three possible values corresponding to three events:

$$\{X = 0\} = \{bb\}, \qquad \{X = 1\} = \{gb, bg\}, \qquad \{X = 2\} = \{gg\}. \qquad (3.6)$$

Since each outcome has probability $1/4$, these three events have probabilities

$$\mathrm{P}[X = 0] = 1/4, \qquad \mathrm{P}[X = 1] = 1/2, \qquad \mathrm{P}[X = 2] = 1/4. \qquad (3.7)$$

We can express the probabilities of these events in terms of the probability mass function

$$P_X(x) = \begin{cases} 1/4 & x = 0, \\ 1/2 & x = 1, \\ 1/4 & x = 2, \\ 0 & \text{otherwise.} \end{cases} \tag{3.8}$$

It is often useful or convenient to depict $P_X(x)$ in two other display formats: as a bar plot or as a table.

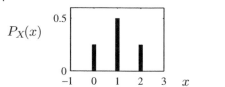

x	0	1	2
$P_X(x)$	1/4	1/2	1/4

Each PMF display format has its uses. The function definition (3.8) is best when $P_X(x)$ is given in terms of algebraic functions of x for various subsets of S_X. The bar plot is best for visualizing the probability masses. The table can be a convenient compact representation when the PMF is a long list of sample values and corresponding probabilities.

No matter how the $P_X(x)$ is formatted, the PMF of X states the value of $P_X(x)$ for every real number x. The first three lines of Equation (3.8) give the function for the values of X associated with nonzero probabilities: $x = 0$, $x = 1$ and $x = 2$. The final line is necessary to specify the function at all other numbers. Although it may look silly to see "$P_X(x) = 0$ otherwise" included in most formulas for a PMF, it is an essential part of the PMF. It is helpful to keep this part of the definition in mind when working with the PMF. However, in the bar plot and table representations of the PMF, it is understood that $P_X(x)$ is zero except for those values x explicitly shown.

The PMF contains all of our information about the random variable X. Because $P_X(x)$ is the probability of the event $\{X = x\}$, $P_X(x)$ has a number of important properties. The following theorem applies the three axioms of probability to discrete random variables.

Theorem 3.1

For a discrete random variable X with PMF $P_X(x)$ and range S_X:

(a) For any x, $P_X(x) \geq 0$.

(b) $\sum_{x \in S_X} P_X(x) = 1$.

(c) For any event $B \subset S_X$, the probability that X is in the set B is

$$P[B] = \sum_{x \in B} P_X(x).$$

Proof All three properties are consequences of the axioms of probability (Section 1.2).

First, $P_X(x) \geq 0$ since $P_X(x) = P[X = x]$. Next, we observe that every outcome $s \in S$ is associated with a number $x \in S_X$. Therefore, $P[x \in S_X] = \sum_{x \in S_X} P_X(x) = P[s \in S] = P[S] = 1$. Since the events $\{X = x\}$ and $\{X = y\}$ are mutually exclusive when $x \neq y$, B can be written as the union of mutually exclusive events $B = \bigcup_{x \in B}\{X = x\}$. Thus we can use Axiom 3 (if B is countably infinite) or Theorem 1.2 (if B is finite) to write

$$P[B] = \sum_{x \in B} P[X = x] = \sum_{x \in B} P_X(x). \tag{3.9}$$

Quiz 3.2

The random variable N has PMF

$$P_N(n) = \begin{cases} c/n & n = 1, 2, 3, \\ 0 & \text{otherwise.} \end{cases} \tag{3.10}$$

Find

(a) The value of the constant c

(b) $P[N = 1]$

(c) $P[N \geq 2]$

(d) $P[N > 3]$

3.3   Families of Discrete Random Variables

> In applications of probability, many experiments have similar probability mass functions. In a family of random variables, the PMFs of the random variables have the same mathematical form, differing only in the values of one or two parameters.

Thus far in our discussion of random variables we have described how each random variable is related to the outcomes of an experiment. We have also introduced the probability mass function, which contains the probability model of the experiment. In practical applications, certain families of random variables appear over and over again in many experiments. In each family, the probability mass functions of all the random variables have the same mathematical form. They differ only in the values of one or two parameters. This enables us to study in advance each family of random variables and later apply the knowledge we gain to specific practical applications.

In this section, we define six families of discrete random variables. There is one formula for the PMF of all the random variables in a family. Depending on the family, the PMF formula contains one or two parameters. By assigning numerical values to the parameters, we obtain a specific random variable. Our nomenclature for a family consists of the family name followed by one or two parameters in parentheses. For example, *binomial* (n, p) refers in general to the family of binomial random variables. *Binomial* $(7, 0.1)$ refers to the binomial random variable with parameters $n = 7$ and $p = 0.1$. Appendix A summarizes important properties of 17 families of random variables.

═══════Example 3.6═══════

Consider the following experiments:

- Flip a coin and let it land on a table. Observe whether the side facing up is heads or tails. Let X be the number of heads observed.

- Select a student at random and find out her telephone number. Let $X = 0$ if the last digit is even. Otherwise, let $X = 1$.

- Observe one bit transmitted by a modem that is downloading a file from the Internet. Let X be the value of the bit (0 or 1).

All three experiments lead to the probability mass function

$$P_X(x) = \begin{cases} 1/2 & x = 0, \\ 1/2 & x = 1, \\ 0 & \text{otherwise.} \end{cases} \tag{3.11}$$

Because all three experiments lead to the same probability mass function, they can all be analyzed the same way. The PMF in Example 3.6 is a member of the family of *Bernoulli* random variables.

═══════Definition 3.4═══════Bernoulli (p) Random Variable

X *is a* **Bernoulli** (p) *random variable if the PMF of* X *has the form*

$$P_X(x) = \begin{cases} 1 - p & x = 0, \\ p & x = 1, \\ 0 & \text{otherwise,} \end{cases}$$

where the parameter p *is in the range* $0 < p < 1$.

Many practical applications of probability produce sequential experiments with independent trials in which each subexperiment has two possible outcomes. A Bernoulli PMF represents the probability model for each subexperiment. We refer to subexperiments with two possible outcomes as *Bernoulli trials*.

In the following examples, we refer to tests of integrated circuits with two possible outcomes: accept (a) and reject (r). Each test in a sequence of tests is an independent trial with probability p of a reject. Depending on the observation, sequential experiments with Bernoulli trials have probability models represented by *Bernoulli, binomial, geometric,* and *Pascal* random variables. Other experiments produce *discrete uniform* random variables and *Poisson* random variables. These six families of random variables occur often in practical applications.

════Example 3.7════

Test one circuit and observe X, the number of rejects. What is $P_X(x)$ the PMF of random variable X?

. . . .

Because there are only two outcomes in the sample space, $X = 1$ with probability p and $X = 0$ with probability $1 - p$,

$$P_X(x) = \begin{cases} 1 - p & x = 0, \\ p & x = 1, \\ 0 & \text{otherwise.} \end{cases} \tag{3.12}$$

Therefore, the number of circuits rejected in one test is a Bernoulli (p) random variable.

════Example 3.8════

In a sequence of independent tests of integrated circuits, each circuit is rejected with probability p. Let Y equal the number of tests up to and including the first test that results in a reject. What is the PMF of Y?

. . . .

The procedure is to keep testing circuits until a reject appears. Using a to denote an accepted circuit and r to denote a reject, the tree is

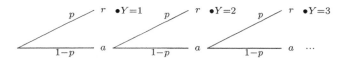

From the tree, we see that $P[Y = 1] = p$, $P[Y = 2] = p(1 - p)$, $P[Y = 3] = p(1 - p)^2$, and, in general, $P[Y = y] = p(1 - p)^{y-1}$. Therefore,

$$P_Y(y) = \begin{cases} p(1 - p)^{y-1} & y = 1, 2, \ldots \\ 0 & \text{otherwise.} \end{cases} \tag{3.13}$$

Y is referred to as a *geometric random variable* because the probabilities in the PMF constitute a geometric series.

In general, the number of Bernoulli trials that take place until the first observation of one of the two outcomes is a geometric random variable.

════Definition 3.5════Geometric (p) Random Variable

*X is a **geometric** (p) random variable if the PMF of X has the form*

$$P_X(x) = \begin{cases} p(1 - p)^{x-1} & x = 1, 2, \ldots \\ 0 & \text{otherwise.} \end{cases}$$

where the parameter p is in the range $0 < p < 1$.

======Example 3.9======

In a sequence of n independent tests of integrated circuits, each circuit is rejected with probability p. Let K equal the number of rejects in the n tests. Find the PMF $P_K(k)$.

. . . .

Adopting the vocabulary of Section 2.3, we call each discovery of a defective circuit a *success*, and each test is an independent trial with success probability p. The event $K = k$ corresponds to k successes in n trials. We refer to Theorem 2.8 to determine that the PMF of K is

$$P_K(k) = \binom{n}{k} p^k (1-p)^{n-k}. \tag{3.14}$$

K is an example of a *binomial random variable*.

We do not state the values of k for which $P_K(k) = 0$ in Equation (3.14) because $\binom{n}{k} = 0$ for $k \notin \{0, 1, \dots, n\}$.

======Definition 3.6======   **Binomial (n,p) Random Variable**

X *is a* **binomial** (n, p) *random variable if the PMF of X has the form*

$$P_X(x) = \binom{n}{x} p^x (1-p)^{n-x}$$

where $0 < p < 1$ and n is an integer such that $n \geq 1$.

Whenever we have a sequence of n independent Bernoulli trials each with success probability p, the number of successes is a binomial random variable. Note that a Bernoulli random variable is a binomial random variable with $n = 1$.

======Example 3.10======

Perform independent tests of integrated circuits in which each circuit is rejected with probability p. Observe L, the number of tests performed until there are k rejects. What is the PMF of L?

. . . .

For large values of k, it is not practical to draw the tree. In this case, $L = l$ if and only if there are $k - 1$ successes in the first $l - 1$ trials *and* there is a success on trial l so that

$$P[L = l] = P\left[\underbrace{k - 1 \text{ rejects in } l - 1 \text{ attempts}}_{A}, \underbrace{\text{reject on attempt } l}_{B}\right] \tag{3.15}$$

The events A and B are independent since the outcome of attempt l is not affected by the previous $l - 1$ attempts. Note that $P[A]$ is the binomial probability of $k - 1$ successes (i.e., rejects) in $l - 1$ trials so that

$$P[A] = \binom{l-1}{k-1} p^{k-1} (1-p)^{l-1-(k-1)} \tag{3.16}$$

Finally, since $P[B] = p$,

$$P_L(l) = P[A]P[B] = \binom{l-1}{k-1}p^k(1-p)^{l-k} \tag{3.17}$$

L is an example of a *Pascal* random variable.

<hr>

Definition 3.7 **Pascal** (k, p) **Random Variable**

X *is a **Pascal** (k, p) random variable if the PMF of X has the form*

$$P_X(x) = \binom{x-1}{k-1}p^k(1-p)^{x-k}$$

where $0 < p < 1$ and k is an integer such that $k \geq 1$.

<hr>

In general, the number of Bernoulli trials that take place until one of the two outcomes is observed k times is a Pascal random variable. For a Pascal (k, p) random variable X, $P_X(x)$ is nonzero only for $x = k, k+1, \ldots$. Definition 3.7 does not state the values of k for which $P_X(x) = 0$ because in Definition 3.6 we have $\binom{n}{x} = 0$ for $x \notin \{0, 1, \ldots, n\}$. Also note that the Pascal $(1, p)$ random variable is the geometric (p) random variable.

Example 3.11

In an experiment with equiprobable outcomes, the random variable N has the range $S_N = \{k, k+1, k+2, \cdots, l\}$, where k and l are integers with $k < l$. The range contains $l - k + 1$ numbers, each with probability $1/(l - k + 1)$. Therefore, the PMF of N is

$$P_N(n) = \begin{cases} 1/(l-k+1) & n = k, k+1, k+2, \ldots, l \\ 0 & \text{otherwise} \end{cases} \tag{3.18}$$

N is an example of a *discrete uniform* random variable.

<hr>

Definition 3.8 **Discrete Uniform** (k, l) **Random Variable**

X *is a **discrete uniform** (k, l) random variable if the PMF of X has the form*

$$P_X(x) = \begin{cases} 1/(l-k+1) & x = k, k+1, k+2, \ldots, l \\ 0 & \text{otherwise} \end{cases}$$

where the parameters k and l are integers such that $k < l$.

<hr>

To describe this discrete uniform random variable, we use the expression "X is uniformly distributed between k and l."

═══Example 3.12═══

Roll a fair die. The random variable N is the number of spots on the side facing up. Therefore, N is a discrete uniform $(1,6)$ random variable with PMF

$$P_N(n) = \begin{cases} 1/6 & n = 1, 2, 3, 4, 5, 6, \\ 0 & \text{otherwise.} \end{cases} \quad (3.19)$$

The probability model of a Poisson random variable describes phenomena that occur randomly in time. While the time of each occurrence is completely random, there is a known average number of occurrences per unit time. The Poisson model is used widely in many fields. For example, the arrival of information requests at a World Wide Web server, the initiation of telephone calls, and the emission of particles from a radioactive source are often modeled as Poisson random variables. We will return to Poisson random variables many times in this text. At this point, we consider only the basic properties.

═══Definition 3.9═══Poisson (α) Random Variable

X *is a **Poisson** (α) random variable if the PMF of X has the form*

$$P_X(x) = \begin{cases} \alpha^x e^{-\alpha}/x! & x = 0, 1, 2, \ldots, \\ 0 & otherwise, \end{cases}$$

where the parameter α is in the range $\alpha > 0$.

To describe a Poisson random variable, we will call the occurrence of the phenomenon of interest an *arrival*. A Poisson model often specifies an average rate, λ arrivals per second, and a time interval, T seconds. In this time interval, the number of arrivals X has a Poisson PMF with $\alpha = \lambda T$.

═══Example 3.13═══

The number of hits at a website in any time interval is a Poisson random variable. A particular site has on average $\lambda = 2$ hits per second. What is the probability that there are no hits in an interval of 0.25 seconds? What is the probability that there are no more than two hits in an interval of one second?

In an interval of 0.25 seconds, the number of hits H is a Poisson random variable with $\alpha = \lambda T = (2 \text{ hits/s}) \times (0.25 \text{ s}) = 0.5$ hits. The PMF of H is

$$P_H(h) = \begin{cases} 0.5^h e^{-0.5}/h! & h = 0, 1, 2, \ldots \\ 0 & \text{otherwise.} \end{cases}$$

The probability of no hits is

$$P[H = 0] = P_H(0) = (0.5)^0 e^{-0.5}/0! = 0.607. \tag{3.20}$$

In an interval of 1 second, $\alpha = \lambda T = (2\text{ hits/s}) \times (1\text{ s}) = 2$ hits. Letting J denote the number of hits in one second, the PMF of J is

$P_J(j)$

$$P_J(j) = \begin{cases} 2^j e^{-2}/j! & j = 0, 1, 2, \ldots \\ 0 & \text{otherwise.} \end{cases}$$

To find the probability of no more than two hits, we note that

$$\{J \le 2\} = \{J = 0\} \cup \{J = 1\} \cup \{J = 2\} \tag{3.21}$$

is the union of three mutually exclusive events. Therefore,

$$\begin{aligned} P[J \le 2] &= P[J = 0] + P[J = 1] + P[J = 2] \\ &= P_J(0) + P_J(1) + P_J(2) \\ &= e^{-2} + 2^1 e^{-2}/1! + 2^2 e^{-2}/2! = 0.677. \end{aligned} \tag{3.22}$$

Note that the units of λ and T have to be consistent. Instead of $\lambda = 0.5$ queries per second for $T = 10$ seconds, we could use $\lambda = 30$ queries per minute for the time interval $T = 1/6$ minutes to obtain the same $\alpha = 5$ queries, and therefore the same probability model.

In the following examples, we see that for a fixed rate λ, the shape of the Poisson PMF depends on the duration T over which arrivals are counted.

Example 3.14

Calls arrive at random times at a telephone switching office with an average of $\lambda = 0.25$ calls/second. The PMF of the number of calls that arrive in a $T = 2$-second interval is the Poisson (0.5) random variable with PMF

$P_J(j)$

$$P_J(j) = \begin{cases} (0.5)^j e^{-0.5}/j! & j = 0, 1, \ldots, \\ 0 & \text{otherwise.} \end{cases}$$

Note that we obtain the same PMF if we define the arrival rate as $\lambda = 60 \cdot 0.25 = 15$ calls per minute and derive the PMF of the number of calls that arrive in $2/60 = 1/30$ minutes.

====Quiz 3.3====

Each time a modem transmits one bit, the receiving modem analyzes the signal that arrives and decides whether the transmitted bit is 0 or 1. It makes an error with probability p, independent of whether any other bit is received correctly.

(a) If the transmission continues until the receiving modem makes its first error, what is the PMF of X, the number of bits transmitted?

(b) If $p = 0.1$, what is the probability that $X = 10$? What is the probability that $X \geq 10$?

(c) If the modem transmits 100 bits, what is the PMF of Y, the number of errors?

(d) If $p = 0.01$ and the modem transmits 100 bits, what is the probability of $Y = 2$ errors at the receiver? What is the probability that $Y \leq 2$?

(e) If the transmission continues until the receiving modem makes three errors, what is the PMF of Z, the number of bits transmitted?

(f) If $p = 0.25$, what is the probability of $Z = 12$ bits transmitted until the modem makes three errors?

3.4   Cumulative Distribution Function (CDF)

> Like the PMF, the CDF of random variable X expresses the probability model of an experiment as a mathematical function. The function is the probability $P[X \leq x]$ for every number x.

The PMF and CDF are closely related. Each can be obtained easily from the other.

====Definition 3.10==== ====Cumulative Distribution Function (CDF)====

*The **cumulative distribution function** (CDF) of random variable X is*

$$F_X(x) = P[X \leq x].$$

For any real number x, the CDF is the probability that the random variable X is no larger than x. All random variables have cumulative distribution functions, but only discrete random variables have probability mass functions. The notation convention for the CDF follows that of the PMF, except that we use the letter F with a subscript corresponding to the name of the random variable. Because $F_X(x)$ describes the probability of an event, the CDF has a number of properties.

====Theorem 3.2====

For any discrete random variable X with range $S_X = \{x_1, x_2, \ldots\}$ satisfying $x_1 \leq x_2 \leq \ldots$,

(a) $F_X(-\infty) = 0$ and $F_X(\infty) = 1$.

(b) For all $x' \geq x$, $F_X(x') \geq F_X(x)$.

(c) For $x_i \in S_X$ and ϵ, an arbitrarily small positive number,

$$F_X(x_i) - F_X(x_i - \epsilon) = P_X(x_i).$$

(d) $F_X(x) = F_X(x_i)$ for all x such that $x_i \leq x < x_{i+1}$.

Each property of Theorem 3.2 has an equivalent statement in words:

(a) Going from left to right on the x-axis, $F_X(x)$ starts at zero and ends at one.

(b) The CDF never decreases as it goes from left to right.

(c) For a discrete random variable X, there is a jump (discontinuity) at each value of $x_i \in S_X$. The height of the jump at x_i is $P_X(x_i)$.

(d) Between jumps, the graph of the CDF of the discrete random variable X is a horizontal line.

Another important consequence of the definition of the CDF is that the difference between the CDF evaluated at two points is the probability that the random variable takes on a value between these two points:

Theorem 3.3

For all $b \geq a$,

$$F_X(b) - F_X(a) = \mathrm{P}\left[a < X \leq b\right].$$

Proof To prove this theorem, express the event $E_{ab} = \{a < X \leq b\}$ as a part of a union of mutually exclusive events. Start with the event $E_b = \{X \leq b\}$. Note that E_b can be written as the union

$$E_b = \{X \leq b\} = \{X \leq a\} \cup \{a < X \leq b\} = E_a \cup E_{ab} \qquad (3.23)$$

Note also that E_a and E_{ab} are mutually exclusive so that $\mathrm{P}[E_b] = \mathrm{P}[E_a] + \mathrm{P}[E_{ab}]$. Since $\mathrm{P}[E_b] = F_X(b)$ and $\mathrm{P}[E_a] = F_X(a)$, we can write $F_X(b) = F_X(a) + \mathrm{P}[a < X \leq b]$. Therefore, $\mathrm{P}[a < X \leq b] = F_X(b) - F_X(a)$.

In working with the CDF, it is necessary to pay careful attention to the nature of inequalities, strict ($<$) or loose ($\leq$). The definition of the CDF contains a loose (less than or equal to) inequality, which means that the function is continuous from the right. To sketch a CDF of a discrete random variable, we draw a graph with the vertical value beginning at zero at the left end of the horizontal axis (negative numbers with large magnitude). It remains zero until x_1, the first value of x with nonzero probability. The graph jumps by an amount $P_X(x_i)$ at each x_i with nonzero probability. We draw the graph of the CDF as a staircase with jumps at each x_i with nonzero probability. The CDF is the upper value of every jump in the staircase.

Example 3.15

In Example 3.5, random variable X has PMF

$P_X(x)$

$$P_X(x) = \begin{cases} 1/4 & x = 0, \\ 1/2 & x = 1, \\ 1/4 & x = 2, \\ 0 & \text{otherwise.} \end{cases} \qquad (3.24)$$

Find and sketch the CDF of random variable X.

. . . .

Referring to the PMF $P_X(x)$, we derive the CDF of random variable X:

$F_X(x)$

$$F_X(x) = P[X \le x] = \begin{cases} 0 & x < 0, \\ 1/4 & 0 \le x < 1, \\ 3/4 & 1 \le x < 2 \\ 1 & x \ge 2. \end{cases}$$

Keep in mind that at the discontinuities $x = 0$, $x = 1$ and $x = 2$, the values of $F_X(x)$ are the upper values: $F_X(0) = 1/4$, $F_X(1) = 3/4$ and $F_X(2) = 1$. Math texts call this the *right hand limit* of $F_X(x)$.

Consider any finite random variable X with all elements of S_X between $x_{\min}$ and $x_{\max}$. For this random variable, the numerical specification of the CDF begins with

$$F_X(x) = 0, \qquad x < x_{\min},$$

and ends with

$$F_X(x) = 1, \qquad x \ge x_{\max}.$$

Like the statement "$P_X(x) = 0$ otherwise," the description of the CDF is incomplete without these two statements. The next example displays the CDF of an infinite discrete random variable.

Example 3.16

In Example 3.8, let the probability that a circuit is rejected equal $p = 1/4$. The PMF of Y, the number of tests up to and including the first reject, is the geometric $(1/4)$ random variable with PMF

$$P_Y(y) = \begin{cases} (1/4)(3/4)^{y-1} & y = 1, 2, \ldots \\ 0 & \text{otherwise.} \end{cases} \qquad (3.25)$$

What is the CDF of Y?

. . . .

Random variable Y has nonzero probabilities for all positive integers. For any integer

$n \geq 1$, the CDF is

$$F_Y(n) = \sum_{j=1}^{n} P_Y(j) = \sum_{j=1}^{n} \frac{1}{4} \left(\frac{3}{4}\right)^{j-1}. \tag{3.26}$$

Equation (3.26) is a geometric series. Familiarity with the geometric series is essential for calculating probabilities involving geometric random variables. Appendix B summarizes the most important facts. In particular, Math Fact B.4 implies $(1 - x)\sum_{j=1}^{n} x^{j-1} = 1 - x^n$. Substituting $x = 3/4$, we obtain

$$F_Y(n) = 1 - \left(\frac{3}{4}\right)^n. \tag{3.27}$$

The complete expression for the CDF of Y must show $F_Y(y)$ for all integer *and noninteger* values of y. For an integer-valued random variable Y, we can do this in a simple way using the *floor function* $\lfloor y \rfloor$, which is the largest integer less than or equal to y. In particular, if $n \leq y < n - 1$ for some integer n, then $\lfloor y \rfloor = n$ and

$$F_Y(y) = P[Y \leq y] = P[Y \leq n] = F_Y(n) = F_Y(\lfloor y \rfloor). \tag{3.28}$$

In terms of the floor function, we can express the CDF of Y as

$$F_Y(y) = \begin{cases} 0 & y < 1, \\ 1 - (3/4)^{\lfloor y \rfloor} & y \geq 1. \end{cases} \tag{3.29}$$

To find the probability that Y takes a value in the set $\{4, 5, 6, 7, 8\}$, we refer to Theorem 3.3 and compute

$$P[3 < Y \leq 8] = F_Y(8) - F_Y(3) = (3/4)^3 - (3/4)^8 = 0.322. \tag{3.30}$$

Quiz 3.4

Use the CDF $F_Y(y)$ to find the following probabilities:

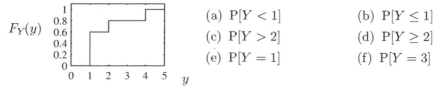

(a) $P[Y < 1]$ (b) $P[Y \leq 1]$

(c) $P[Y > 2]$ (d) $P[Y \geq 2]$

(e) $P[Y = 1]$ (f) $P[Y = 3]$

3.5   Averages and Expected Value

> An average is a number that describes a set of experimental observations. The expected value is a number that describes the probability model of an experiment.

The average value of a set of n numbers is a *statistic* of the the set of numbers. The average is a single number that describes the entire set. Statisticians work with several kinds of averages. The ones that are used the most are the *mean*, the *median*, and the *mode*.

The mean value of n numbers is the sum of the n numbers divided by n. An example is the mean value of the numerical grades of the students taking a midterm exam. The mean indicates the performance of the entire class. The median is another statistic that describes a set of numbers.

The median is a number in the middle of a data set. There is an equal number of data items below the median and above the median.

A third average is the mode of a set of numbers. The mode is the most common number in the set. There are as many or more numbers with that value than any other value. If there are two or more numbers with this property, the set of numbers is called *multimodal*.

====**Example 3.17**====

For one quiz, 10 students have the following grades (on a scale of 0 to 10):

$$9, 5, 10, 8, 4, 7, 5, 5, 8, 7 \qquad (3.31)$$

Find the mean, the median, and the mode.
. . . .
The sum of the ten grades is 68. The mean value is $68/10 = 6.8$. The median is 7, because there are four grades below 7 and four grades above 7. The mode is 5, because three students have a grade of 5, more than the number of students who received any other grade.

Example 3.17 and the preceding comments on averages apply to a set of numbers observed in a practical situation. The probability models of random variables characterize experiments with numerical outcomes, and in practical applications of probability, we assume that the probability models are related to the numbers observed in practice. Just as a statistic describes a set of numbers observed in practice, a *parameter* describes a probability model. Each parameter is a number that can be computed from the PMF or CDF of a random variable. When we use a probability model of a random variable to represent an application that results in a set of numbers, the *expected value* of the random variable corresponds to the mean value of the set of numbers. Expected values appear throughout the remainder of this textbook. Two notations for the expected value of random variable X are $E[X]$ and μ_X.

=====Definition 3.11=====Expected Value

The expected value of X is

$$E[X] = \mu_X = \sum_{x \in S_X} x P_X(x).$$

Expectation is a synonym for expected value. Sometimes the term *mean value* is also used as a synonym for expected value. We prefer to use mean value to refer to a *statistic* of a set of experimental data (the sum divided by the number of data items) to distinguish it from expected value, which is a *parameter* of a probability model. If you recall your studies of mechanics, the form of Definition 3.11 may look familiar. Think of point masses on a line with a mass of $P_X(x)$ kilograms at a distance of x meters from the origin. In this model, μ_X in Definition 3.11 is the center of mass. This is why $P_X(x)$ is called probability *mass* function.

=====Example 3.18=====

Random variable X in Example 3.5 has PMF

$$P_X(x) = \begin{cases} 1/4 & x = 0, \\ 1/2 & x = 1, \\ 1/4 & x = 2, \\ 0 & \text{otherwise.} \end{cases} \qquad (3.32)$$

What is $E[X]$?

. . . .

$$\begin{aligned} E[X] = \mu_X &= 0 \cdot P_X(0) + 1 \cdot P_X(1) + 2 \cdot P_X(2) \\ &= 0(1/4) + 1(1/2) + 2(1/4) = 1. \end{aligned} \qquad (3.33)$$

To understand how this definition of expected value corresponds to the notion of adding up a set of measurements, suppose we have an experiment that produces a random variable X and we perform n independent trials of this experiment. We denote the value that X takes on the ith trial by $x(i)$. We say that $x(1), \ldots, x(n)$ is a set of n sample values of X. We have, after n trials of the experiment, the sample average

$$m_n = \frac{1}{n} \sum_{i=1}^{n} x(i). \qquad (3.34)$$

Each $x(i)$ takes values in the set S_X. Out of the n trials, assume that each $x \in S_X$ occurs N_x times. Then the sum (3.34) becomes

$$m_n = \frac{1}{n} \sum_{x \in S_X} x N_x = \sum_{x \in S_X} x \frac{N_x}{n}. \qquad (3.35)$$

Recall our discussion in Section 1.2 of the relative frequency interpretation of probability. There we pointed out that if in n observations of an experiment, the event A occurs N_A times, we can interpret the probability of A as

$$P[A] = \lim_{n \to \infty} \frac{N_A}{n}. \tag{3.36}$$

N_A/n is the relative frequency of A. In the notation of random variables, we have the corresponding observation that

$$P_X(x) = \lim_{n \to \infty} \frac{N_x}{n}. \tag{3.37}$$

From Equation (3.35), this suggests that

$$\lim_{n \to \infty} m_n = \sum_{x \in S_X} x \left(\lim_{n \to \infty} \frac{N_x}{n} \right) = \sum_{x \in S_X} x P_X(x) = E[X]. \tag{3.38}$$

Equation (3.38) says that the definition of $E[X]$ corresponds to a model of doing the same experiment repeatedly. After each trial, add up all the observations to date and divide by the number of trials. We prove in Chapter 10 that the result approaches the expected value as the number of trials increases without limit. We can use Definition 3.11 to derive the expected value of each family of random variables defined in Section 3.3.

Theorem 3.4

The Bernoulli (p) random variable X has expected value $E[X] = p$.

Proof $E[X] = 0 \cdot P_X(0) + 1 P_X(1) = 0(1 - p) + 1(p) = p$.

Theorem 3.5

The geometric (p) random variable X has expected value $E[X] = 1/p$.

Proof Let $q = 1 - p$. The PMF of X becomes

$$P_X(x) = \begin{cases} pq^{x-1} & x = 1, 2, \ldots \\ 0 & \text{otherwise.} \end{cases} \tag{3.39}$$

The expected value $E[X]$ is the infinite sum

$$E[X] = \sum_{x=1}^{\infty} x P_X(x) = \sum_{x=1}^{\infty} x p q^{x-1} \tag{3.40}$$

Applying the identity of Math Fact B.7, we have

$$E[X] = p \sum_{x=1}^{\infty} x q^{x-1} = \frac{p}{q} \sum_{x=1}^{\infty} x q^x = \frac{p}{q} \frac{q}{1 - q^2} = \frac{p}{p^2} = \frac{1}{p}. \tag{3.41}$$

This result is intuitive if you recall the integrated circuit testing experiments and consider some numerical values. If the probability of rejecting an integrated circuit is $p = 1/5$, then on average, you have to perform $E[Y] = 1/p = 5$ tests until you observe the first reject. If $p = 1/10$, the average number of tests until the first reject is $E[Y] = 1/p = 10$.

═════Theorem 3.6═════

The Poisson (α) random variable in Definition 3.9 has expected value $E[X] = \alpha$.

Proof

$$E[X] = \sum_{x=0}^{\infty} x P_X(x) = \sum_{x=0}^{\infty} x \frac{\alpha^x}{x!} e^{-\alpha}. \tag{3.42}$$

We observe that $x/x! = 1/(x-1)!$ and also that the $x = 0$ term in the sum is zero. In addition, we substitute $\alpha^x = \alpha \cdot \alpha^{x-1}$ to factor α from the sum to obtain

$$E[X] = \alpha \sum_{x=1}^{\infty} \frac{\alpha^{x-1}}{(x-1)!} e^{-\alpha}. \tag{3.43}$$

Next we substitute $l = x - 1$, with the result

$$E[X] = \alpha \underbrace{\sum_{l=0}^{\infty} \frac{\alpha^l}{l!} e^{-\alpha}}_{1} = \alpha. \tag{3.44}$$

We can conclude that the sum in this formula equals 1 either by referring to the identity $e^\alpha = \sum_{l=0}^{\infty} \alpha^l/l!$ or by applying Theorem 3.1(b) to the fact that the sum is the sum of the PMF of a Poisson random variable L over all values in S_L and $P[S_L] = 1$.

In Section 3.3, we modeled the number of random arrivals in an interval of duration T by a Poisson random variable with parameter $\alpha = \lambda T$. We referred to λ as *the average rate* of arrivals with little justification. Theorem 3.6 provides the justification by showing that $\lambda = \alpha/T$ is the expected number of arrivals per unit time.

The next theorem provides, without derivations, the expected values of binomial, Pascal, and discrete uniform random variables.

======Theorem 3.7======

(a) For the binomial (n, p) random variable X of Definition 3.6,

$$\mathrm{E}[X] = np.$$

(b) For the Pascal (k, p) random variable X of Definition 3.7,

$$\mathrm{E}[X] = k/p.$$

(c) For the discrete uniform (k, l) random variable X of Definition 3.8,

$$\mathrm{E}[X] = (k + l)/2.$$

In the following theorem, we show that the Poisson PMF is a limiting case of a binomial PMF when the number of Bernoulli trials, n, grows without limit but the expected number of successes np remains constant at α, the expected value of the Poisson PMF. In the theorem, we let $\alpha = \lambda T$ and divide the T-second interval into n time slots each with duration T/n. In each slot, as n grows without limit and the duration, T/n, of each slot gets smaller and smaller we assume that there is either one arrival, with probability $p = \lambda T/n = \alpha/n$, or there is no arrival in the time slot, with probability $1 - p$.

======Theorem 3.8======

Perform n Bernoulli trials. In each trial, let the probability of success be α/n, where $\alpha > 0$ is a constant and $n > \alpha$. Let the random variable K_n be the number of successes in the n trials. As $n \to \infty$, $P_{K_n}(k)$ converges to the PMF of a Poisson (α) random variable.

Proof We first note that K_n is the binomial $(n, \alpha n)$ random variable with PMF

$$P_{K_n}(k) = \binom{n}{k} (\alpha/n)^k \left(1 - \frac{\alpha}{n}\right)^{n-k}. \tag{3.45}$$

For $k = 0, \ldots, n$, we can write

$$P_K(k) = \frac{n(n-1)\cdots(n-k+1)}{n^k} \frac{\alpha^k}{k!} \left(1 - \frac{\alpha}{n}\right)^{n-k}. \tag{3.46}$$

Notice that in the first fraction, there are k terms in the numerator. The denominator is n^k, also a product of k terms, all equal to n. Therefore, we can express this fraction as the product of k fractions, each of the form $(n - j)/n$. As $n \to \infty$, each of these fractions approaches 1. Hence,

$$\lim_{n \to \infty} \frac{n(n-1)\cdots(n-k+1)}{n^k} = 1. \tag{3.47}$$

Furthermore, we have

$$\left(1 - \frac{\alpha}{n}\right)^{n-k} = \frac{\left(1 - \frac{\alpha}{n}\right)^{n}}{\left(1 - \frac{\alpha}{n}\right)^{k}}. \tag{3.48}$$

As n grows without bound, the denominator approaches 1 and, in the numerator, we recognize the identity $\lim_{n\to\infty}(1 - \alpha/n)^{n} = e^{-\alpha}$. Putting these three limits together leads us to the result that for any integer $k \geq 0$,

$$\lim_{n\to\infty} P_{K_n}(k) = \begin{cases} \alpha^k e^{-\alpha}/k! & k = 0, 1, \ldots \\ 0 & \text{otherwise,} \end{cases} \tag{3.49}$$

which is the Poisson PMF.

Quiz 3.5

In a pay-as-you go cellphone plan, the cost of sending an SMS text message is 10 cents and the cost of receiving a text is 5 cents. For a certain subscriber, the probability of sending a text is $1/3$ and the probability of receiving a text is $2/3$. Let C equal the cost (in cents) of one text message and find

(a) The PMF $P_C(c)$

(b) The expected value $E[C]$

(c) The probability that four texts are received before a text is sent.

(d) The expected number of texts received before a text is sent.

3.6   Functions of a Random Variable

> A function $Y = g(X)$ of random variable X is another random variable. The PMF $P_Y(y)$ can be derived from $P_X(x)$ and $g(X)$.

In many practical situations, we observe sample values of a random variable and use these sample values to compute other quantities. One example that occurs frequently is an experiment in which the procedure is to monitor the data activity of a cellular telephone subscriber for a month and observe x the total number of megabytes sent and received. The telephone company refers to the price plan of the subscriber and calculates y dollars, the amount to be paid by the subscriber. If x is a sample value of a random variable X, Definition 3.1 implies that y is a sample value of a random variable Y. Because we obtain Y from another random variable, we refer to Y as a *derived random variable*.

Definition 3.12 ─────── Derived Random Variable

*Each sample value y of a **derived random variable** Y is a mathematical function $g(x)$ of a sample value x of another random variable X. We adopt the notation $Y = g(X)$ to describe the relationship of the two random variables.*

=====Example 3.19=====

A parcel shipping company offers a charging plan: \$1.00 for the first pound, \$0.90 for the second pound, etc., down to \$0.60 for the fifth pound, with rounding up for a fraction of a pound. For all packages between 6 and 10 pounds, the shipper will charge \$5.00 per package. (It will not accept shipments over 10 pounds.) Find a function $Y = g(X)$ for the charge in cents for sending one package.

· ·

When the package weight is an integer $X \in \{1, 2, \ldots, 10\}$ that specifies the number of pounds with rounding up for a fraction of a pound, the function

$$Y = g(X) = \begin{cases} 105X - 5X^2 & X = 1, 2, 3, 4, 5 \\ 500 & X = 6, 7, 8, 9, 10. \end{cases} \tag{3.50}$$

corresponds to the charging plan.

━━━━━━━

In this section we determine the probability model of a derived random variable from the probability model of the original random variable. We start with $P_X(x)$ and a function $Y = g(X)$. We use this information to obtain $P_Y(y)$.

Before we present the procedure for obtaining $P_Y(y)$, we alert students to the different nature of the functions $P_X(x)$ and $g(x)$. Although they are both functions with the argument x, they are entirely different. $P_X(x)$ describes the probability model of a random variable. It has the special structure prescribed in Theorem 3.1. On the other hand, $g(x)$ can be any function at all. When we combine $P_X(x)$ and $g(x)$ to derive the probability model for Y, we arrive at a PMF that also conforms to Theorem 3.1.

To describe Y in terms of our basic model of probability, we specify an experiment consisting of the following procedure and observation:

> `Sample value of` $Y = g(X)$
>
> `Perform an experiment and observe an outcome` s`.`
> `From` s`, find` x`, the corresponding value of random variable` X`.`
> `Observe` y `by calculating` $y = g(x)$`.`

This procedure maps each experimental outcome to a number, y, a sample value of a random variable, Y. To derive $P_Y(y)$ from $P_X(x)$ and $g(\cdot)$, we consider all of the possible values of x. For each $x \in S_X$, we compute $y = g(x)$. If $g(x)$ transforms different values of x into different values of y ($g(x_1) \neq g(x_2)$ if $x_1 \neq x_2$) we simply have

$$P_Y(y) = P[Y = g(x)] = P[X = x] = P_X(x). \tag{3.51}$$

The situation is a little more complicated when $g(x)$ transforms several values of x to the same y. For each $y \in S_Y$, we add the probabilities of all of the values $x \in S_x$ for which $g(x) = y$. Theorem 3.9 applies in general. It reduces to Equation (3.51) when $g(x)$ is a one-to-one transformation.

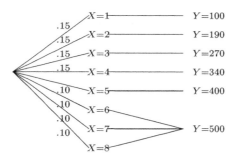

Figure 3.1 The derived random variable $Y = g(X)$ for Example 3.21.

Theorem 3.9

For a discrete random variable X, the PMF of $Y = g(X)$ is

$$P_Y(y) = \sum_{x:g(x)=y} P_X(x).$$

If we view $X = x$ as the outcome of an experiment, then Theorem 3.9 says that $P_Y(y)$ is the sum of the probabilities of all the outcomes $X = x$ for which $Y = y$.

Example 3.20

In Example 3.19, suppose all packages weigh 1, 2, 3, or 4 pounds with equal probability. Find the PMF and expected value of Y, the shipping charge for a package.
....

From the problem statement, the weight X has PMF

$$P_X(x) = \begin{cases} 1/4 & x = 1, 2, 3, 4, \\ 0 & \text{otherwise.} \end{cases} \tag{3.52}$$

The charge for a shipment, Y, has range $S_Y = \{100, 190, 270, 340\}$ corresponding to $S_X = \{1, \ldots, 4\}$. The experiment can be described by the following tree. Here each value of Y derives from a unique value of X. Hence, we can use Equation (3.51) to find $P_Y(y)$.

$$P_Y(y) = \begin{cases} 1/4 & y = 100, 190, 270, 340, \\ 0 & \text{otherwise.} \end{cases}$$

The expected shipping bill is

$$E[Y] = \frac{1}{4}(100 + 190 + 270 + 340)$$

$$= 225 \text{ cents.}$$

Example 3.21

Suppose the probability model for the weight in pounds X of a package in Example 3.19 is

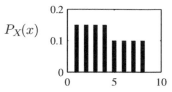

$$P_X(x) = \begin{cases} 0.15 & x = 1, 2, 3, 4, \\ 0.1 & x = 5, 6, 7, 8, \\ 0 & \text{otherwise.} \end{cases}$$

For the pricing plan given in Example 3.19, what is the PMF and expected value of Y, the cost of shipping a package?

....

Now we have three values of X, specifically $(6, 7, 8)$, transformed by $g(\cdot)$ into $Y = 500$. For this situation we need the more general view of the PMF of Y, given by Theorem 3.9. In particular, $y_6 = 500$, and we have to add the probabilities of the outcomes $X = 6$, $X = 7$, and $X = 8$ to find $P_Y(500)$. That is,

$$P_Y(500) = P_X(6) + P_X(7) + P_X(8) = 0.30. \tag{3.53}$$

The steps in the procedure are illustrated in the diagram of Figure 3.1. Applying Theorem 3.9, we have

$$P_Y(y) = \begin{cases} 0.15 & y = 100, 190, 270, 340, \\ 0.10 & y = 400, \\ 0.30 & y = 500, \\ 0 & \text{otherwise.} \end{cases}$$

For this probability model, the expected cost of shipping a package is

$$E[Y] = 0.15(100 + 190 + 270 + 340) + 0.10(400) + 0.30(500) = 325 \text{ cents.}$$

Quiz 3.6

Monitor three customers purchasing smartphones at the Phonesmart store and observe whether each buys an Apricot phone for \$450 or a Banana phone for \$300. The random variable N is the number of customers purchasing an Apricot phone. Assume N has PMF

$$P_N(n) = \begin{cases} 0.4 & n = 0, \\ 0.2 & n = 1, 2, 3, \\ 0 & \text{otherwise.} \end{cases} \tag{3.54}$$

M dollars is the amount of money paid by three customers.

(a) Express M as a function of N.   (b) Find $P_M(m)$ and $E[M]$.

3.7   Expected Value of a Derived Random Variable

> If $Y = g(X)$, $\mathrm{E}[Y]$ can be calculated from $P_X(x)$ and $g(X)$ without deriving $P_Y(y)$.

We encounter many situations in which we need to know only the expected value of a derived random variable rather than the entire probability model. Fortunately, to obtain this average, it is not necessary to compute the PMF or CDF of the new random variable. Instead, we can use the following property of expected values.

Theorem 3.10

Given a random variable X with PMF $P_X(x)$ and the derived random variable $Y = g(X)$, the expected value of Y is

$$\mathrm{E}[Y] = \mu_Y = \sum_{x \in S_X} g(x) P_X(x).$$

Proof From the definition of $\mathrm{E}[Y]$ and Theorem 3.9, we can write

$$\mathrm{E}[Y] = \sum_{y \in S_Y} y P_Y(y) = \sum_{y \in S_Y} y \sum_{x:g(x)=y} P_X(x) = \sum_{y \in S_Y} \sum_{x:g(x)=y} g(x) P_X(x), \qquad (3.55)$$

where the last double summation follows because $g(x) = y$ for each x in the inner sum. Since $g(x)$ transforms each possible outcome $x \in S_X$ to a value $y \in S_Y$, the preceding double summation can be written as a single sum over all possible values $x \in S_X$. That is,

$$\mathrm{E}[Y] = \sum_{x \in S_X} g(x) P_X(x). \qquad (3.56)$$

Example 3.22

In Example 3.20,

$$P_X(x) = \begin{cases} 1/4 & x = 1, 2, 3, 4, \\ 0 & \text{otherwise,} \end{cases} \quad \text{and} \quad g(X) = \begin{cases} 105X - 5X^2 & 1 \le X \le 5, \\ 500 & 6 \le X \le 10. \end{cases}$$
$$(3.57)$$

What is $\mathrm{E}[Y] = \mathrm{E}[g(X)]$?

. . . .

Applying Theorem 3.10 we have

$$\begin{aligned}
\mathrm{E}[Y] &= \sum_{x=1}^{4} P_X(x) g(x) \\
&= (1/4)[(105)(1) - (5)(1)^2] + (1/4)[(105)(2) - (5)(2)^2] \\
&\quad + (1/4)[(105)(3) - (5)(3)^2] + (1/4)[(105)(4) - (5)(4)^2] \\
&= (1/4)[100 + 190 + 270 + 340] = 225 \text{ cents.} \qquad (3.58)
\end{aligned}$$

This of course is the same answer obtained in Example 3.20 by first calculating $P_Y(y)$ and then applying Definition 3.11. As an exercise, you might want to compute $E[Y]$ in Example 3.21 directly from Theorem 3.10.

From this theorem we can derive some important properties of expected values. The first one has to do with the difference between a random variable and its expected value. When students learn their own grades on a midterm exam, they are quick to ask about the class average. Let's say one student has 73 and the class average is 80. She may be inclined to think of her grade as "seven points below average," rather than "73." In terms of a probability model, we would say that the random variable X points on the midterm has been transformed to the random variable

$$Y = g(X) = X - \mu_X \quad \text{points above average.} \tag{3.59}$$

The expected value of $X - \mu_X$ is zero, regardless of the probability model of X.

═══════Theorem 3.11═══════
For any random variable X,

$$E[X - \mu_X] = 0.$$

───────────

Proof Defining $g(X) = X - \mu_X$ and applying Theorem 3.10 yields

$$E[g(X)] = \sum_{x \in S_X} (x - \mu_X)P_X(x) = \sum_{x \in S_X} xP_X(x) - \mu_X \sum_{x \in S_X} P_X(x). \tag{3.60}$$

The first term on the right side is μ_X by definition. In the second term, $\sum_{x \in S_X} P_X(x) = 1$, so both terms on the right side are μ_X and the difference is zero.

───────────

Another property of the expected value of a function of a random variable applies to linear transformations.[1]

═══════Theorem 3.12═══════
For any random variable X,

$$E[aX + b] = a E[X] + b.$$

───────────

This follows directly from Definition 3.11 and Theorem 3.10. A linear transformation is essentially a scale change of a quantity, like a transformation from inches to centimeters or from degrees Fahrenheit to degrees Celsius. If we express the data (random variable X) in new units, the new average is just the old average transformed to the new units. (If the professor adds five points to everyone's grade, the average goes up by five points.)

───────────

[1]We call the transformation $aX + b$ linear although, strictly speaking, it should be called affine.

This is a rare example of a situation in which $E[g(X)] = g(E[X])$. *It is tempting, but usually wrong, to apply it to other transformations.* For example, if $Y = X^2$, it is usually the case that $E[Y] \neq (E[X])^2$. Expressing this in general terms, it is usually the case that $E[g(X)] \neq g(E[X])$.

Example 3.23

Recall from Examples 3.5 and 3.18 that X has PMF

$$P_X(x) = \begin{cases} 1/4 & x = 0, \\ 1/2 & x = 1, \\ 1/4 & x = 2, \\ 0 & \text{otherwise.} \end{cases} \tag{3.61}$$

What is the expected value of $V = g(X) = 4X + 7$?

. . . .

From Theorem 3.12,

$$E[V] = E[g(X)] = E[4X + 7] = 4E[X] + 7 = 4(1) + 7 = 11. \tag{3.62}$$

We can verify this result by applying Theorem 3.10:

$$\begin{aligned} E[V] &= g(0)P_X(0) + g(1)P_X(1) + g(2)P_X(2) \\ &= 7(1/4) + 11(1/2) + 15(1/4) = 11. \end{aligned} \tag{3.63}$$

Example 3.24

Continuing Example 3.23, let $W = h(X) = X^2$. What is $E[W]$?

. . . .

Theorem 3.10 gives

$$E[W] = \sum h(x)P_X(x) = (1/4)0^2 + (1/2)1^2 + (1/4)2^2 = 1.5. \tag{3.64}$$

Note that this is not the same as $h(E[W]) = (1)^2 = 1$.

Quiz 3.7

The number of memory chips M needed in a personal computer depends on how many application programs, A, the owner wants to run simultaneously. The number of chips M and the number of application programs A are described by

$$M = \begin{cases} 4 & \text{chips for 1 program,} \\ 4 & \text{chips for 2 programs,} \\ 6 & \text{chips for 3 programs,} \\ 8 & \text{chips for 4 programs,} \end{cases} \qquad P_A(a) = \begin{cases} 0.1(5 - a) & a = 1, 2, 3, 4, \\ 0 & \text{otherwise.} \end{cases} \tag{3.65}$$

(a) What is the expected number of programs $\mu_A = \mathrm{E}[A]$?

(b) Express M, the number of memory chips, as a function $M = g(A)$ of the number of application programs A.

(c) Find $\mathrm{E}[M] = \mathrm{E}[g(A)]$. Does $\mathrm{E}[M] = g(\mathrm{E}[A])$?

3.8 Variance and Standard Deviation

> The variance $\mathrm{Var}[X]$ measures the dispersion of sample values of X around the expected value $\mathrm{E}[X]$. When we view $\mathrm{E}[X]$ as an estimate of X, $\mathrm{Var}[X]$ is the mean square error.

In Section 3.5, we describe an average as a typical value of a random variable. It is one number that summarizes an entire probability model. After finding an average, someone who wants to look further into the probability model might ask, "How typical is the average?" or "What are the chances of observing an event far from the average?" In the example of the midterm exam, after you find out your score is 7 points above average, you are likely to ask, "How good is that? Is it near the top of the class or somewhere near the middle?" A measure of dispersion is an answer to these questions wrapped up in a single number. If this measure is small, observations are likely to be near the average. A high measure of dispersion suggests that it is not unusual to observe events that are far from the average.

The most important measures of dispersion are the standard deviation and its close relative, the variance. The variance of random variable X describes the difference between X and its expected value. This difference is the derived random variable, $Y = X - \mu_X$. Theorem 3.11 states that $\mu_Y = 0$, regardless of the probability model of X. Therefore μ_Y provides no information about the dispersion of X around μ_X. A useful measure of the likely difference between X and its expected value is the expected absolute value of the difference, $\mathrm{E}[|Y|]$. However, this parameter is not easy to work with mathematically in many situations, and it is not used often.

Instead we focus on $\mathrm{E}[Y^2] = \mathrm{E}[(X - \mu_X)^2]$, which is referred to as $\mathrm{Var}[X]$, the variance of X. The square root of the variance is σ_X, the standard deviation of X.

Definition 3.13 — Variance

*The **variance** of random variable X is*

$$\mathrm{Var}[X] = \mathrm{E}\left[(X - \mu_X)^2\right].$$

═══════**Definition 3.14**═══════**Standard Deviation**

*The **standard deviation** of random variable X is*

$$\sigma_X = \sqrt{\operatorname{Var}[X]}.$$

It is useful to take the square root of Var$[X]$ because σ_X has the same units (for example, exam points) as X. The units of the variance are squares of the units of the random variable (exam points squared). Thus σ_X can be compared directly with the expected value. Informally, we think of sample values within σ_X of the expected value, $x \in [\mu_X - \sigma_X, \mu_X + \sigma_X]$, as "typical" values of X and other values as "unusual." In many applications, about $2/3$ of the observations of a random variable are within one standard deviation of the expected value. Thus if the standard deviation of exam scores is 12 points, the student with a score of $+7$ with respect to the mean can think of herself in the middle of the class. If the standard deviation is 3 points, she is likely to be near the top.

The variance is also useful when you guess or predict the value of a random variable X. Suppose you are asked to make a prediction $\hat{x}$ before you perform an experiment and observe a sample value of X. The prediction $\hat{x}$ is also called a *blind estimate* of X since your prediction is an estimate of X without the benefit of any observation. Since you would like the prediction error $X - \hat{x}$ to be small, a popular approach is to choose $\hat{x}$ to minimize the expected square error

$$e = \operatorname{E}\left[(X - \hat{x})^2\right]. \tag{3.66}$$

Another name for e is the mean square error or MSE. With knowledge of the PMF $P_X(x)$, we can choose $\hat{x}$ to minimize the MSE.

═══════**Theorem 3.13**═══════

In the absence of observations, the minimum mean square error estimate of random variable X is

$$\hat{x} = \operatorname{E}[X].$$

Proof After substituting $\hat{X} = \hat{x}$, we expand the square in Equation (3.66) to write

$$e = \operatorname{E}\left[X^2\right] - 2\hat{x}\operatorname{E}[X] + \hat{x}^2. \tag{3.67}$$

To minimize e, we solve

$$\frac{de}{d\hat{x}} = -2\operatorname{E}[X] + 2\hat{x} = 0, \tag{3.68}$$

yielding $\hat{x} = \operatorname{E}[X]$.

When the estimate of X is $\hat{x} = \operatorname{E}[X]$, the MSE is

$$e^* = \operatorname{E}\left[(X - \operatorname{E}[X])^2\right] = \operatorname{Var}[X]. \tag{3.69}$$

Therefore, $E[X]$ is a best estimate of X and $\text{Var}[X]$ is the MSE associated with this best estimate.

Because $(X - \mu_X)^2$ is a function of X, $\text{Var}[X]$ can be computed according to Theorem 3.10.

$$\text{Var}[X] = \sigma_X^2 = \sum_{x \in S_X} (x - \mu_X)^2 P_X(x). \tag{3.70}$$

By expanding the square in this formula, we arrive at the most useful approach to computing the variance.

═══Theorem 3.14═══

$$\text{Var}[X] = E[X^2] - \mu_X^2 = E[X^2] - (E[X])^2.$$

Proof Expanding the square in (3.70), we have

$$\text{Var}[X] = \sum_{x \in S_X} x^2 P_X(x) - \sum_{x \in S_X} 2\mu_X x P_X(x) + \sum_{x \in S_X} \mu_X^2 P_X(x)$$

$$= E[X^2] - 2\mu_X \sum_{x \in S_X} x P_X(x) + \mu_X^2 \sum_{x \in S_X} P_X(x)$$

$$= E[X^2] - 2\mu_X^2 + \mu_X^2. \tag{3.71}$$

We note that $E[X]$ and $E[X^2]$ are examples of *moments* of the random variable X. $\text{Var}[X]$ is a *central moment* of X.

═══Definition 3.15═══Moments

For random variable X:

*(a) The nth **moment** is* $E[X^n]$.

*(b) The nth **central moment** is* $E[(X - \mu_X)^n]$.

Thus, $E[X]$ is the *first moment* of random variable X. Similarly, $E[X^2]$ is the *second moment*. Theorem 3.14 says that the variance of X is the second moment of X minus the square of the first moment.

Like the PMF and the CDF of a random variable, the set of moments of X is a complete probability model. We learn in Section 9.2 that the model based on moments can be expressed as a *moment generating function*.

===Example 3.25===

Continuing Examples 3.5, 3.18, and 3.23, we recall that X has PMF

$P_X(x)$

$$P_X(x) = \begin{cases} 1/4 & x = 0, \\ 1/2 & x = 1, \\ 1/4 & x = 2, \\ 0 & \text{otherwise}, \end{cases} \qquad (3.72)$$

and expected value $E[X] = 1$. What is the variance of X?

. . . .

In order of increasing simplicity, we present three ways to compute $\text{Var}[X]$.

- From Definition 3.13, define

$$W = (X - \mu_X)^2 = (X - 1)^2. \qquad (3.73)$$

We observe that $W = 0$ if and only if $X = 1$; otherwise, if $X = 0$ or $X = 2$, then $W = 1$. Thus $P[W = 0] = P_X(1) = 1/2$ and $P[W = 1] = P_X(0) + P_X(2) = 1/2$. The PMF of W is

$$P_W(w) = \begin{cases} 1/2 & w = 0, 1, \\ 0 & \text{otherwise}. \end{cases} \qquad (3.74)$$

Then

$$\text{Var}[X] = E[W] = (1/2)(0) + (1/2)(1) = 1/2. \qquad (3.75)$$

- Recall that Theorem 3.10 produces the same result without requiring the derivation of $P_W(w)$.

$$\begin{aligned} \text{Var}[X] &= E\left[(X - \mu_X)^2\right] \\ &= (0 - 1)^2 P_X(0) + (1 - 1)^2 P_X(1) + (2 - 1)^2 P_X(2) \\ &= 1/2. \end{aligned} \qquad (3.76)$$

- To apply Theorem 3.14, we find that

$$E\left[X^2\right] = 0^2 P_X(0) + 1^2 P_X(1) + 2^2 P_X(2) = 1.5. \qquad (3.77)$$

Thus Theorem 3.14 yields

$$\text{Var}[X] = E\left[X^2\right] - \mu_X^2 = 1.5 - 1^2 = 1/2. \qquad (3.78)$$

Note that $(X - \mu_X)^2 \geq 0$. Therefore, its expected value is also nonnegative. That is, for any random variable X

$$\text{Var}[X] \geq 0. \qquad (3.79)$$

The following theorem is related to Theorem 3.12

===Theorem 3.15===

$$\text{Var}\,[aX + b] = a^2\,\text{Var}\,[X].$$

Proof We let $Y = aX + b$ and apply Theorem 3.14. We first expand the second moment to obtain

$$\text{E}\,[Y^2] = \text{E}\,[a^2 X^2 + 2abX + b^2] = a^2\,\text{E}\,[X^2] + 2ab\mu_X + b^2. \tag{3.80}$$

Expanding the right side of Theorem 3.12 yields

$$\mu_Y^2 = a^2 \mu_X^2 + 2ab\mu_x + b^2. \tag{3.81}$$

Because $\text{Var}[Y] = \text{E}[Y^2] - \mu_Y^2$, Equations (3.80) and (3.81) imply that

$$\text{Var}\,[Y] = a^2\,\text{E}\,[X^2] - a^2\mu_X^2 = a^2(\text{E}\,[X^2] - \mu_X^2) = a^2\,\text{Var}\,[X]. \tag{3.82}$$

If we let $a = 0$ in this theorem, we have $\text{Var}[b] = 0$ because there is no dispersion around the expected value of a constant. If we let $a = 1$, we have $\text{Var}[X + b] = \text{Var}[X]$ because shifting a random variable by a constant does not change the dispersion of outcomes around the expected value.

===Example 3.26===
A printer automatically prints an initial cover page that precedes the regular printing of an X page document. Using this printer, the number of printed pages is $Y = X + 1$. Express the expected value and variance of Y as functions of $\text{E}[X]$ and $\text{Var}[X]$.
....
The expected number of transmitted pages is $\text{E}[Y] = \text{E}[X] + 1$. The variance of the number of pages sent is $\text{Var}[Y] = \text{Var}[X]$.

If we let $b = 0$ in Theorem 3.12, we have $\text{Var}[aX] = a^2\,\text{Var}[X]$ and $\sigma_{aX} = a\sigma_X$. Multiplying a random variable by a constant is equivalent to a scale change in the units of measurement of the random variable.

===Example 3.27===
The amplitude V (volts) of a sinusoidal signal is a random variable with PMF

$$P_V(v) = \begin{cases} 1/7 & v = -3, -2, \dots, 3, \\ 0 & \text{otherwise.} \end{cases}$$

A new voltmeter records the amplitude U in millivolts. Find the variance and standard deviation of U.

....

Note that $U = 1000V$. To use Theorem 3.15, we first find the variance of V. The expected value of the amplitude is

$$\mu_V = 1/7[-3 + (-2) + (-1) + 0 + 1 + 2 + 3] = 0 \text{ volts}. \tag{3.83}$$

The second moment is

$$\mathrm{E}\left[V^2\right] = 1/7[(-3)^2 + (-2)^2 + (-1)^2 + 0^2 + 1^2 + 2^2 + 3^2] = 4 \text{ volts}^2. \tag{3.84}$$

Therefore the variance is $\mathrm{Var}[V] = \mathrm{E}[V^2] - \mu_V^2 = 4 \text{ volts}^2$. By Theorem 3.15,

$$\mathrm{Var}\,[U] = 1000^2\,\mathrm{Var}[V] = 4{,}000{,}000 \text{ millivolts}^2, \tag{3.85}$$

and thus $\sigma_U = 2000$ millivolts.

The following theorem states the variances of the families of random variables defined in Section 3.3.

Theorem 3.16

(a) If X is Bernoulli (p), then

$$\mathrm{Var}[X] = p(1 - p).$$

(b) If X is geometric (p), then

$$\mathrm{Var}[X] = (1 - p)/p^2.$$

(c) If X is binomial (n, p), then

$$\mathrm{Var}[X] = np(1 - p).$$

(d) If X is Pascal (k, p), then

$$\mathrm{Var}[X] = k(1 - p)/p^2.$$

(e) If X is Poisson (α), then

$$\mathrm{Var}[X] = \alpha.$$

(f) If X is discrete uniform (k, l),

$$\mathrm{Var}[X] = (l - k)(l - k + 2)/12.$$

Quiz 3.8

In an experiment with three customers entering the Phonesmart store, the observation is N, the number of phones purchased. The PMF of N is

$$P_N(n) = \begin{cases} (4 - n)/10 & n = 0, 1, 2, 3 \\ 0 & \text{otherwise.} \end{cases} \tag{3.86}$$

Find

(a) The expected value $\mathrm{E}[N]$

(b) The second moment $\mathrm{E}[N^2]$

(c) The variance $\mathrm{Var}[N]$

(d) The standard deviation σ_N

3.9 MATLAB

> MATLAB programs calculate values of functions including PMFs and CDFs. Other MATLAB functions simulate experiments by generating random sample values of random variables.

This section presents two types of MATLAB programs based on random variables with arbitrary probability models and random variables in the families presented in Section 3.3. We start by calculating probabilities for any finite random variable with arbitrary PMF $P_X(x)$. We then compute PMFs and CDFs for the families of random variables introduced in Section 3.3. Based on the calculation of the CDF, we then develop a method for generating random sample values. Generating a random sample simulates performing an experiment that conforms to the probability model of a specific random variable. In subsequent chapters, we will see that MATLAB functions that generate random samples are building blocks for the simulation of more-complex systems. The MATLAB functions described in this section can be downloaded from the companion website.

PMFs and CDFs

For the most part, the PMF and CDF functions are straightforward. We start with a finite discrete random variable X defined by the set of sample values $S_X = \{s_1, \ldots, s_n\}$ and corresponding probabilities $p_i = P_X(s_i) = P[X = s_i]$. In MATLAB, we represent the range S_X by the column vector $\mathbf{s} = \begin{bmatrix} s_1 & \cdots & s_n \end{bmatrix}'$ and the corresponding probabilities by the vector $\mathbf{p} = \begin{bmatrix} p_1 & \cdots & p_n \end{bmatrix}'$.[2] The function y=finitepmf(sx,px,x) generates the probabilities of the elements of the m-dimensional vector $\mathbf{x} = \begin{bmatrix} x_1 & \cdots & x_m \end{bmatrix}'$. The output is $\mathbf{y} = \begin{bmatrix} y_1 & \cdots & y_m \end{bmatrix}'$ where $y_i = P_X(x_i)$. That is, for each requested x_i, finitepmf returns the value $P_X(x_i)$. If x_i is not in the sample space of X, $y_i = 0$.

═══ **Example 3.28** ═══

In Example 3.21, the random variable X, the weight of a package, has PMF

$$P_X(x) = \begin{cases} 0.15 & x = 1, 2, 3, 4, \\ 0.1 & x = 5, 6, 7, 8, \\ 0 & \text{otherwise.} \end{cases} \tag{3.87}$$

Write a MATLAB function that calculates $P_X(x)$. Calculate the probability of an x_i pound package for $x_1 = 2$, $x_2 = 2.5$, and $x_3 = 6$.

. . . .

The MATLAB function shipweightpmf(x) implements $P_X(x)$. We can then use shipweightpmf to calculate the desired probabilities:

---

[2] Although column vectors are supposed to appear as columns, we generally write a column vector $\mathbf{x}$ in the form of a transposed row vector $\begin{bmatrix} x_1 & \cdots & x_m \end{bmatrix}'$ to save space.

```
function y=shipweightpmf(x)
s=(1:8)';
p=[0.15*ones(4,1); 0.1*ones(4,1)];
y=finitepmf(s,p,x);
```

```
>> shipweightpmf([2 2.5 6])'
ans =
    0.1500  0  0.1000
```

We also can use MATLAB to calculate a PMF in a family of random variables by specifying the parameters of the PMF to be calculated. Although a PMF $P_X(x)$ is a scalar function of one variable, the nature of MATLAB makes it desirable to perform MATLAB PMF calculations with vector inputs and vector outputs. If y=xpmf(x) calculates $P_X(x)$, then for a vector input x, we produce a vector output y such that y(i)=xpmf(x(i)). That is, for vector input x, the output vector y is defined by $y_i = P_X(x_i)$.

Example 3.29

Write a MATLAB function geometricpmf(p,x) to calculate, for the sample values in vector x, $P_X(x)$ for a geometric (p) random variable.

. . . .

```
function pmf=geometricpmf(p,x)
%geometric(p) rv X
%out: pmf(i)=Prob[X=x(i)]
x=x(:);
pmf= p*((1-p).^(x-1));
pmf= (x>0).*(x==floor(x)).*pmf;
```

In geometricpmf.m, the last line ensures that values $x_i \notin S_X$ are assigned zero probability. Because x=x(:) reshapes x to be a column vector, the output pmf is always a column vector.

Example 3.30

Write a MATLAB function that calculates the Poisson (α) PMF.

. . . .

For an integer x, we could calculate $P_X(x)$ by the direct calculation

```
px= ((alpha^x)*exp(-alpha*x))/factorial(x)
```

This will yield the right answer as long as the argument x for the factorial function is not too large. In MATLAB version 6, factorial(171) causes an overflow. In addition, for $a > 1$, calculating the ratio $a^x/x!$ for large x can cause numerical problems because both a^x and $x!$ will be very large numbers, possibly with a small quotient. Another shortcoming of the direct calculation is apparent if you want to calculate $P_X(x)$ for the set of possible values $\mathbf{x} = [0, 1, \ldots, n]$. Calculating factorials is a lot of work for a computer and the direct approach fails to exploit the fact that if we have already calculated $(x-1)!$, we can easily compute $x! = x \cdot (x-1)!$. A more efficient calculation makes use of the observation

$$P_X(x) = \frac{a^x e^{-a}}{x!} = \frac{a}{x} P_X(x-1). \tag{3.88}$$

The poissonpmf.m function uses Equation (3.88) to calculate $P_X(x)$. Even this code is not perfect because MATLAB has limited range.

```
function pmf=poissonpmf(alpha,x)
%output: pmf(i)=P[X=x(i)]
x=x(:); k=(1:max(x))';
ip=[1;((alpha*ones(size(k)))./k)];
pb=exp(-alpha)*cumprod(ip);
  %pb= [P(X=0)...P(X=n)]
pmf=pb(x+1); %pb(1)=P[X=0]
pmf=(x>=0).*(x==floor(x)).*pmf;
  %pmf(i)=0 for zero-prob x(i)
```

In MATLAB, exp(-alpha) returns zero for alpha > 745.13. For these large values of alpha,

$$poissonpmf(alpha,x)$$

returns zero for all x. Problem 3.9.9 outlines a solution that is used in the version of poissonpmf.m on the companion website.

For the Poisson CDF, there is no simple way to avoid summing the PMF. The following example shows an implementation of the Poisson CDF. The code for a CDF tends to be more complicated than that for a PMF because if x is not an integer, $F_X(x)$ may still be nonzero. Other CDFs are easily developed following the same approach.

Example 3.31

Write a MATLAB function that calculates the CDF of a Poisson random variable.

```
function cdf=poissoncdf(alpha,x)
%output cdf(i)=Prob[X<=x(i)]
x=floor(x(:));
sx=0:max(x);
cdf=cumsum(poissonpmf(alpha,sx));
  %cdf from 0 to max(x)
okx=(x>=0);%x(i)<0 -> cdf=0
x=(okx.*x);%set negative x(i)=0
cdf= okx.*cdf(x+1);
  %cdf=0 for x(i)<0
```

Here we present the MATLAB code for the Poisson CDF. Since the sample values of a Poisson random variable X are integers, we observe that $F_X(x) = F_X(\lfloor x \rfloor)$ where $\lfloor x \rfloor$, equivalent to the MATLAB floor(x) function, denotes the largest integer less than or equal to x.

Example 3.32

In Example 3.13 a website has on average $\lambda = 2$ hits per second. What is the probability of no more than 130 hits in one minute? What is the probability of more than 110 hits in one minute?

Let M equal the number of hits in one minute (60 seconds). Note that M is a Poisson (α) random variable with $\alpha = 2 \times 60 = 120$ hits. The PMF of M is

$$P_M(m) = \begin{cases} (120)^m e^{-120}/m! & m = 0, 1, 2, \ldots \\ 0 & \text{otherwise.} \end{cases} \tag{3.89}$$

```
>> poissoncdf(120,130)
ans =
    0.8315
>> 1-poissoncdf(120,110)
ans =
    0.8061
```

The MATLAB solution shown on the left executes the following math calculations:

$$P[M \le 130] = \sum_{m=0}^{130} P_M(m), \qquad (3.90)$$

$$P[M > 110] = 1 - P[M \le 110]$$

$$= 1 - \sum_{m=0}^{110} P_M(m). \qquad (3.91)$$

Generating Random Samples

The programs described thus far in this section perform the familiar task of calculating a function of a single variable. Here, the functions are PMFs and CDFs. As described in Section 2.4, MATLAB can also be used to simulate experiments. In this section we present MATLAB programs that generate data conforming to families of discrete random variables. When many samples are generated by these programs, the relative frequency of data in an event in the sample space converges to the probability of the event. As in Chapter 2, we use `rand()` as a source of randomness. Let $R = $ `rand(1)`. Recall that `rand(1)` simulates an experiment that is equally likely to produce any real number in the interval $[0, 1]$. We will learn in Chapter 4 that to express this idea in mathematics, we say that for any interval $[a, b] \subset [0, 1]$,

$$P[a < R \le b] = b - a. \qquad (3.92)$$

For example, $P[0.4 < R \le 0.53] = 0.13$. Now suppose we wish to generate samples of discrete random variable K with $S_K = \{0, 1, \ldots\}$. Since $0 \le F_K(k-1) \le F_K(k) \le 1$, for all k, we observe that

$$P[F_K(k-1) < R \le F_K(k)] = F_K(k) - F_K(k-1) = P_K(k) \qquad (3.93)$$

This fact leads to the following approach (as shown in pseudocode) to using `rand()` to produce a sample of random variable K:

Random Sample of random variable K

Generate $R = $ `rand(1)`
Find $k^* \in S_K$ such that $F_K(k^*-1) < R \le F_K(k^*)$
Set $K = k^*$

A MATLAB function that uses `rand()` in this way simulates an experiment that produces samples of random variable K. Generally, this implies that before we can produce a sample of random variable K, we need to generate the CDF of K. We can reuse the work of this computation by defining our MATLAB functions such as

MATLAB **Functions**

PMF	CDF	Random Samples
finitepmf(sx,p,x)	finitecdf(sx,p,x)	finiterv(sx,p,m)
bernoullipmf(p,x)	bernoullicdf(p,x)	bernoullirv(p,m)
binomialpmf(n,p,x)	binomialcdf(n,p,x)	binomialrv(n,p,m)
geometricpmf(p,x)	geometriccdf(p,x)	geometricrv(p,m)
pascalpmf(k,p,x)	pascalcdf(k,p,x)	pascalrv(k,p,m)
poissonpmf(alpha,x)	poissoncdf(alpha,x)	poissonrv(alpha,m)
duniformpmf(k,l,x)	duniformcdf(k,l,x)	duniformrv(k,l,m)

Table 3.1   MATLAB functions for discrete random variables.

geometricrv(p,m) to generate m sample values each time. We now present the details associated with generating binomial random variables.

═══**Example 3.33**═══

Write a function that generates m samples of a binomial (n,p) random variable.

```
function x=binomialrv(n,p,m)
% m binomial(n,p) samples

r=rand(m,1);
cdf=binomialcdf(n,p,0:n);
x=count(cdf,r);
```

For vectors x and y, c=count(x,y) returns a vector c such that c(i) is the number of elements of x that are less than or equal to $y(i)$. In terms of our earlier pseudocode, $k^* = $ count(cdf,r). If count(cdf,r) $= 0$, then $r \leq P_X(0)$ and $k^* = 0$.

Generating binomial random variables is easy because the range is simply $\{0, \ldots, n\}$ and the minimum value is zero. The MATLAB code for geometricrv, poissonrv, and pascalrv is slightly more complicated because we need to generate enough terms of the CDF to ensure that we find k^*.

Table 3.1 contains a collection of functions for an arbitrary probability model and the six families of random variables introduced in Section 3.3. As in Example 3.28, the functions in the first row can be used for any discrete random variable X with a finite sample space. The argument **s** is the vector of sample values s_i of X, and **p** is the corresponding vector of probabilities $P[s_i]$ of those sample values. For PMF and CDF calculations, **x** is the vector of numbers for which the calculation is to be performed. In the function finiteserv, m is the number of random samples returned by the function. Each of the final six rows of the table contains for one family the pmf function for calculating values of the PMF, the cdf function for calculating values of the CDF, and the rv function for generating random samples. In each function description, x denotes a column vector $\mathbf{x} = \begin{bmatrix} x_1 & \cdots & x_m \end{bmatrix}'$. The pmf function output is a vector **y** such that $y_i = P_X(x_i)$. The cdf function output is a vector **y** such that $y_i = F_X(x_i)$. The rv function output is a vector $\mathbf{X} = \begin{bmatrix} X_1 & \cdots & X_m \end{bmatrix}'$ such that each X_i is a sample value of the random variable X. If $m = 1$, then the output is a single sample value of random variable X.

We present an additional example, partly because it demonstrates some useful MATLAB functions, and also because it shows how to generate the relative frequencies of random samples.

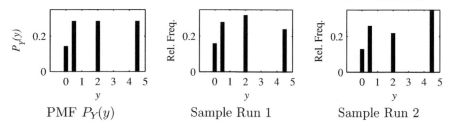

PMF $P_Y(y)$ Sample Run 1   Sample Run 2

Figure 3.2 The PMF of Y and the relative frequencies found in two sample runs of `voltpower(100)`. Note that in each run, the relative frequencies are close to (but not exactly equal to) the corresponding PMF.

Example 3.34

The amplitude V (volts) of a sinusoidal signal is a discrete uniform $(-3, 3)$ random variable. Let $Y = V^2/2$ watts denote the power of the transmitted signal. Simulate $n = 100$ trials of the experiment producing the power measurement Y. Plot the relative frequency of each $y \in S_Y$.

....

```
function voltpower(n)
v=duniformrv(-3,3,n);
y=(v.^2)/2;
yrange=0:max(y);
yfreq=(hist(y,yrange)/n)';
pmfplot(yrange,yfreq);
```

In `voltpower.m`, we calculate $Y = V^2/2$ for each of n samples of the voltage V. The function `unique(y)` assembles a vector of the unique values of `y`. As in Example 2.20, the function `hist(y,yrange)` produces a vector with jth element equal to the number of occurrences of `yrange(j)` in the vector `y`. The function `pmfplot.m` is a utility for producing PMF bar plots in the style of this text. Figure 3.2 shows the exact PMF $P_Y(y)$, which Problem 3.6.3 asks you to derive, along with the results of two runs of `voltpower(100)`.

Derived Random Variables

MATLAB can also calculate PMFs and CDFs of derived random variables. For this section, we assume X is a finite random variable with sample space $S_X = \{x_1, \ldots, x_n\}$ such that $P_X(x_i) = p_i$. We represent the properties of X by the vectors $\mathbf{s}_X = \begin{bmatrix} x_1 & \cdots & x_n \end{bmatrix}'$ and $\mathbf{p}_X = \begin{bmatrix} p_1 & \cdots & p_n \end{bmatrix}'$. In MATLAB notation, `sx` and `px` represent the vectors $\mathbf{s}_X$ and $\mathbf{p}_X$.

For derived random variables, we exploit a feature of `finitepmf(sx,px,x)` that allows the elements of `sx` to be repeated. Essentially, we use (`sx,px`), or equivalently $(\mathbf{s}_X, \mathbf{p}_X)$, to represent a random variable X described by the following experimental procedure:

```
Finite sample space

Roll an n-sided die such that side i has probability pi.
If side j appears, set X = xj.
```

A consequence of this approach is that if $x_2 = 3$ and $x_5 = 3$, then the probability of observing $X = 3$ is $P_X(3) = p_2 + p_5$.

Example 3.35

```
>> sx=[1 3 5 7 3];
>> px=[0.1 0.2 0.2 0.3 0.2];
>> pmfx=finitepmf(sx,px,1:7);
>> pmfx'
ans =
    0.10 0 0.40 0 0.20 0 0.30
```

finitepmf() accounts for multiple occurrences of a sample value. In the example on the left,

$$pmfx(3)=px(2)+px(5)=0.4.$$

It may seem unnecessary and perhaps even bizarre to allow these repeated values. However, we see in the next example that it is quite convenient for derived random variables $Y = g(X)$ with the property that $g(x_i)$ is the same for multiple x_i.

Example 3.36

Recall that in Example 3.21 the weight in pounds X of a package and the cost $Y = g(X)$ of shipping a package were described by

$$P_X(x) = \begin{cases} 0.15 & x = 1, 2, 3, 4, \\ 0.1 & x = 5, 6, 7, 8, \\ 0 & \text{otherwise,} \end{cases} \qquad Y = \begin{cases} 105X - 5X^2 & 1 \le X \le 5, \\ 500 & 6 \le X \le 10. \end{cases}$$

Write a function $y=\text{shipcostrv(m)}$ that outputs m samples of the shipping cost Y.

```
function y=shipcostrv(m)
sx=(1:8)';
px=[0.15*ones(4,1); ...
      0.1*ones(4,1)];
gx=(sx<=5).* ...
   (105*sx-5*(sx.^2))...
   + ((sx>5).*500);
y=finiterv(gx,px,m);
```

The vector gx is the mapping $g(x)$ for each $x \in S_X$. In gx, the element 500 appears three times, corresponding to $x = 6$, $x = 7$, and $x = 8$. The function y=finiterv(gx,px,m)) produces m samples of the shipping cost Y.

```
>> shipcostrv(9)'
ans =
    270 190 500 270 500 190 190 100 500
```

Quiz 3.9

In Section 3.5, it was argued that the average

$$m_n = \frac{1}{n}\sum_{i=1}^{n} x(i) \tag{3.94}$$

of samples $x(1), x(2), \ldots, x(n)$ of a random variable X will converge to $\mathrm{E}[X]$ as n becomes large. For a discrete uniform $(0, 10)$ random variable X, use MATLAB to examine this convergence.

(a) For 100 sample values of X, plot the sequence $m_1, m_2, \ldots, m_{100}$. Repeat this experiment five times, plotting all five m_n curves on common axes.

(b) Repeat part (a) for 1000 sample values of X.

Problems

Difficulty: ● Easy   ▦ Moderate   ◆ Difficult   ◆◆ Experts Only

3.2.1● The random variable N has PMF

$$P_N(n) = \begin{cases} c(1/2)^n & n = 0, 1, 2, \\ 0 & \text{otherwise.} \end{cases}$$

(a) What is the value of the constant c?

(b) What is $P[N \le 1]$?

3.2.2● The random variable V has PMF

$$P_V(v) = \begin{cases} cv^2 & v = 1, 2, 3, 4, \\ 0 & \text{otherwise.} \end{cases}$$

(a) Find the value of the constant c.

(b) Find $P[V \in \{u^2 | u = 1, 2, 3, \cdots\}]$.

(c) Find the probability that V is even.

(d) Find $P[V > 2]$.

3.2.3● The random variable X has PMF

$$P_X(x) = \begin{cases} c/x & x = 2, 4, 8, \\ 0 & \text{otherwise.} \end{cases}$$

(a) What is the value of the constant c?

(b) What is $P[X = 4]$?

(c) What is $P[X < 4]$?

(d) What is $P[3 \le X \le 9]$?

3.2.4● In each at-bat in a baseball game, mighty Casey swings at every pitch. The result is either a home run (with probability $q = 0.05$) or a strike. Of course, three strikes and Casey is out.

(a) What is the probability $P[H]$ that Casey hits a home run?

(b) For one at-bat, what is the PMF of N, the number of times Casey swings his bat?

3.2.5▦ A tablet computer transmits a file over a wi-fi link to an access point. Depending on the size of the file, it is transmitted as N packets where N has PMF

$$P_N(n) = \begin{cases} c/n & n = 1, 2, 3, \\ 0 & \text{otherwise.} \end{cases}$$

(a) Find the constant c.

(b) What is the probability that N is odd?

(c) Each packet is received correctly with probability p, and the file is received correctly if all N packets are received correctly. Find $P[C]$, the probability that the file is received correctly.

3.2.6▦ In college basketball, when a player is fouled while not in the act of shooting and the opposing team is "in the penalty," the player is awarded a "1 and 1." In the 1 and 1, the player is awarded one free throw, and if that free throw goes in the player is awarded a second free throw. Find the PMF of Y, the number of points scored in a 1 and 1 given that any free throw goes in with probability p, independent of any other free throw.

3.2.7▦ You roll a 6-sided die repeatedly. Starting with roll $i = 1$, let R_i denote the result of roll i. If $R_i > i$, then you will roll again; otherwise you stop. Let N denote the number of rolls.

(a) What is $P[N > 3]$?

(b) Find the PMF of N.

3.2.8■ You are manager of a ticket agency that sells concert tickets. You assume that people will call three times in an attempt to buy tickets and then give up. You want to make sure that you are able to serve at least 95% of the people who want tickets. Let p be the probability that a caller gets through to your ticket agency. What is the minimum value of p necessary to meet your goal?

3.2.9■ In the ticket agency of Problem 3.2.8, each telephone ticket agent is available to receive a call with probability 0.2. If all agents are busy when someone calls, the caller hears a busy signal. What is the minimum number of agents that you have to hire to meet your goal of serving 95% of the customers who want tickets?

3.2.10■ Suppose when a baseball player gets a hit, a single is twice as likely as a double, which is twice as likely as a triple, which is twice as likely as a home run. Also, the player's batting average, i.e., the probability the player gets a hit, is 0.300. Let B denote the number of bases touched safely during an at-bat. For example, $B = 0$ when the player makes an out, $B = 1$ on a single, and so on. What is the PMF of B?

3.2.11◆ When someone presses SEND on a cellular phone, the phone attempts to set up a call by transmitting a SETUP message to a nearby base station. The phone waits for a response, and if none arrives within 0.5 seconds it tries again. If it doesn't get a response after $n = 6$ tries, the phone stops transmitting messages and generates a busy signal.

(a) Draw a tree diagram that describes the call setup procedure.

(b) If all transmissions are independent and the probability is p that a SETUP message will get through, what is the PMF of K, the number of messages transmitted in a call attempt?

(c) What is the probability that the phone will generate a busy signal?

(d) As manager of a cellular phone system, you want the probability of a busy sig-

nal to be less than 0.02. If $p = 0.9$, what is the minimum value of n necessary to achieve your goal?

3.3.1● In a package of M&Ms, Y, the number of yellow M&Ms, is uniformly distributed between 5 and 15.

(a) What is the PMF of Y?

(b) What is $P[Y < 10]$?

(c) What is $P[Y > 12]$?

(d) What is $P[8 \leq Y \leq 12]$?

3.3.2● In a bag of 25 M&Ms, each piece is equally likely to be red, green, orange, blue, or brown, independent of the color of any other piece. Find the the PMF of R, the number of red pieces. What is the probability a bag has no red M&Ms?

3.3.3● When a conventional paging system transmits a message, the probability that the message will be received by the pager it is sent to is p. To be confident that a message is received at least once, a system transmits the message n times.

(a) Assuming all transmissions are independent, what is the PMF of K, the number of times the pager receives the same message?

(b) Assume $p = 0.8$. What is the minimum value of n that produces a probability of 0.95 of receiving the message at least once?

3.3.4● You roll a pair of fair dice until you roll "doubles" (i.e., both dice are the same). What is the expected number, $E[N]$, of rolls?

3.3.5● When you go fishing, you attach m hooks to your line. Each time you cast your line, each hook will be swallowed by a fish with probability h, independent of whether any other hook is swallowed. What is the PMF of K, the number of fish that are hooked on a single cast of the line?

3.3.6● Any time a child throws a Frisbee, the child's dog catches the Frisbee with probability p, independent of whether the Frisbee is caught on any previous throw.

When the dog catches the Frisbee, it runs away with the Frisbee, never to be seen again. The child continues to throw the Frisbee until the dog catches it. Let X denote the number of times the Frisbee is thrown.

(a) What is the PMF $P_X(x)$?

(b) If $p = 0.2$, what is the probability that the child will throw the Frisbee more than four times?

3.3.7● When a two-way paging system transmits a message, the probability that the message will be received by the pager it is sent to is p. When the pager receives the message, it transmits an acknowledgment signal (ACK) to the paging system. If the paging system does not receive the ACK, it sends the message again.

(a) What is the PMF of N, the number of times the system sends the same message?

(b) The paging company wants to limit the number of times it has to send the same message. It has a goal of $P[N \leq 3] \geq 0.95$. What is the minimum value of p necessary to achieve the goal?

3.3.8●

(a) Starting on day 1, you buy one lottery ticket each day. Each ticket is a winner with probability 0.1. Find the PMF of K, the number of tickets you buy up to and including your fifth winning ticket.

(b) L is the number of flips of a fair coin up to and including the 33rd occurrence of tails. What is the PMF of L?

(c) Starting on day 1, you buy one lottery ticket each day. Each ticket is a winner with probability 0.01. Let M equal the number of tickets you buy up to and including your first winning ticket. What is the PMF of M?

3.3.9● The number of bytes B in an HTML file is the geometric $(2.5 \cdot 10^{-5})$ random variable. What is the probability $P[B > 500,000]$ that a file has over 500,000 bytes?

3.3.10● The number of buses that arrive at a bus stop in T minutes is a Poisson random variable B with expected value $T/5$.

(a) What is the PMF of B, the number of buses that arrive in T minutes?

(b) What is the probability that in a two-minute interval, three buses will arrive?

(c) What is the probability of no buses arriving in a 10-minute interval?

(d) How much time should you allow so that with probability 0.99 at least one bus arrives?

3.3.11● In a wireless automatic meter-reading system, a base station sends out a wake-up signal to nearby electric meters. On hearing the wake-up signal, a meter transmits a message indicating the electric usage. Each message is repeated eight times.

(a) If a single transmission of a message is successful with probability p, what is the PMF of N, the number of successful message transmissions?

(b) I is an indicator random variable such that $I = 1$ if at least one message is transmitted successfully; otherwise $I = 0$. Find the PMF of I.

3.3.12● A Zipf $(n, \alpha = 1)$ random variable X has PMF

$$P_X(x) = \begin{cases} c(n)/x & x = 1, 2, \ldots, n, \\ 0 & \text{otherwise.} \end{cases}$$

The constant $c(n)$ is set so that $\sum_{x=1}^{n} P_X(x) = 1$. Calculate $c(n)$ for $n = 1, 2, \ldots, 6$.

3.3.13▪ In a bag of 64 "holiday season" M&Ms, each M&M is equally likely to be red or green, independent of any other M&M in the bag.

(a) If you randomly grab four M&Ms, what is the probability $P[E]$ that you grab an equal number of red and green M&Ms?

(b) What is the PMF of G, the number of green M&Ms in the bag?

(c) You begin eating randomly chosen M&Ms one by one. Let R equal the number of red M&Ms you eat before you eat your first green M&M. What is the PMF of R?

3.3.14▨ A radio station gives a pair of concert tickets to the sixth caller who knows the birthday of the performer. For each person who calls, the probability is 0.75 of knowing the performer's birthday. All calls are independent.

(a) What is the PMF of L, the number of calls necessary to find the winner?

(b) What is the probability of finding the winner on the tenth call?

(c) What is the probability that the station will need nine or more calls to find a winner?

3.3.15▨ In a packet voice communications system, a source transmits packets containing digitized speech to a receiver. Because transmission errors occasionally occur, an acknowledgment (ACK) or a negative acknowledgment (NAK) is transmitted back to the source to indicate the status of each received packet. When the transmitter gets a NAK, the packet is retransmitted. Voice packets are delay sensitive, and a packet can be transmitted a maximum of d times. If a packet transmission is an independent Bernoulli trial with success probability p, what is the PMF of T, the number of times a packet is transmitted?

3.3.16◆ At Newark airport, your jet joins a line as the tenth jet waiting for takeoff. At Newark, takeoffs and landings are synchronized to the minute. In each one-minute interval, an arriving jet lands with probability $p = 2/3$, independent of an arriving jet in any other minute. Such an arriving jet blocks any waiting jet from taking off in that one-minute interval. However, if there is no arrival, then the waiting jet at the head of the line takes off. Each take-off requires exactly one minute.

(a) Let L_1 denote the number of jets that land before the jet at the front of the line takes off. Find the PMF $P_{L_1}(l)$.

(b) Let W denote the number of minutes you wait until your jet takes off. Find $P[W = 10]$. (Note that if no jets land for ten minutes, then one waiting jet will take off each minute and $W = 10$.)

(c) What is the PMF of W?

3.3.17◆ Suppose each day (starting on day 1) you buy one lottery ticket with probability 1/2; otherwise, you buy no tickets. A ticket is a winner with probability p independent of the outcome of all other tickets. Let N_i be the event that on day i you do *not* buy a ticket. Let W_i be the event that on day i, you buy a winning ticket. Let L_i be the event that on day i you buy a losing ticket.

(a) What are $P[W_{33}]$, $P[L_{87}]$, and $P[N_{99}]$?

(b) Let K be the number of the day on which you buy your first lottery ticket. Find the PMF $P_K(k)$.

(c) Find the PMF of R, the number of losing lottery tickets you have purchased in m days.

(d) Let D be the number of the day on which you buy your jth losing ticket. What is $P_D(d)$? Hint: If you buy your jth losing ticket on day d, how many losers did you have after $d - 1$ days?

3.3.18◆ The Sixers and the Celtics play a best out of five playoff series. The series ends as soon as one of the teams has won three games. Assume that either team is equally likely to win any game independently of any other game played. Find

(a) The PMF $P_N(n)$ for the total number N of games played in the series;

(b) The PMF $P_W(w)$ for the number W of Celtics wins in the series;

(c) The PMF $P_L(l)$ for the number L of Celtics losses in the series.

3.3.19◆ For a binomial random variable K representing the number of successes in n trials, $\sum_{k=0}^{n} P_K(k) = 1$. Use this fact to prove the binomial theorem for any $a > 0$ and $b > 0$. That is, show that

$$(a+b)^n = \sum_{k=0}^{n} \binom{n}{k} a^k b^{n-k}.$$

3.4.1 Discrete random variable Y has the CDF $F_Y(y)$ as shown:

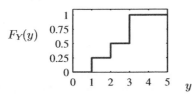

Use the CDF to find the following probabilities:

(a) $P[Y < 1]$ and $P[Y \leq 1]$

(b) $P[Y > 2]$ and $P[Y \geq 2]$

(c) $P[Y = 3]$ and $P[Y > 3]$

(d) $P_Y(y)$

3.4.2 The random variable X has CDF

$$F_X(x) = \begin{cases} 0 & x < -1, \\ 0.2 & -1 \leq x < 0, \\ 0.7 & 0 \leq x < 1, \\ 1 & x \geq 1. \end{cases}$$

(a) Draw a graph of the CDF.

(b) Write $P_X(x)$, the PMF of X. Be sure to write the value of $P_X(x)$ for all x from $-\infty$ to ∞.

3.4.3 Following Example 3.16, show that a geometric (p) random variable K has CDF

$$F_K(k) = \begin{cases} 0 & k < 1, \\ 1 - (1-p)^{\lfloor k \rfloor} & k \geq 1. \end{cases}$$

3.4.4 At the One Top Pizza Shop, a pizza sold has mushrooms with probability $p = 2/3$. On a day in which 100 pizzas are sold, let N equal the number of pizzas sold before the first pizza with mushrooms is sold. What is the PMF of N? What is the CDF of N?

3.4.5 In Problem 3.2.10, find and sketch the CDF of B, the number of bases touched safely during an at-bat.

3.4.6 In Problem 3.2.6, find and sketch the CDF of Y, the number of points scored in a 1 and 1 for $p = 1/4$, $p = 1/2$, and $p = 3/4$.

3.4.7 In Problem 3.2.11, find and sketch the CDF of N, the number of attempts made by the cellular phone for $p = 1/2$.

3.5.1 It costs 20 cents to receive a photo and 30 cents to send a photo from a cellphone. C is the cost of one photo (either sent or received). The probability of receiving a photo is 0.6. The probability sending a photo is 0.4.

(a) Find $P_C(c)$, the PMF of C.

(b) What is $E[C]$, the expected value of C?

3.5.2

(a) The number of trains J that arrive at the station in time t minutes is a Poisson random variable with $E[J] = t$. Find t such that $P[J > 0] = 0.9$.

(b) The number of buses K that arrive at the station in one hour is a Poisson random variable with $E[K] = 10$. Find $P[K = 10]$.

(c) In a 1 ms interval, the number of hits L on a Web server is a Poisson random variable with expected value $E[L] = 2$ hits. What is $P[L \leq 1]$?

3.5.3 You simultaneously flip a pair of fair coins. Your friend gives you one dollar if both coins come up heads. You repeat this ten times and your friend gives you X dollars. Find $E[X]$, the expected number of dollars you receive. What is the probability that you do "worse than average"?

3.5.4 A packet received by your smartphone is error-free with probability 0.95, independent of any other packet.

(a) Out of 10 packets received, let X equal the number of packets received with errors. What is the PMF of X?

(b) In one hour, your smartphone receives 12,000 packets. Let X equal the number of packets received with errors. What is $E[X]$?

3.5.5● Find the expected value of the random variable Y in Problem 3.4.1.

3.5.6● Find the expected value of the random variable X in Problem 3.4.2.

3.5.7● Use Definition 3.11 to calculate the expected value of a binomial $(4, 1/2)$ random variable X.

3.5.8● X is the discrete uniform $(1, 5)$ random variable.

(a) What is $P[X = E[X]]$?

(b) What is $P[X > E[X]]$?

3.5.9● K is the geometric $(1/11)$ random variable.

(a) What is $P[K = E[K]]$?

(b) What is $P[K > E[K]]$

(c) What is $P[K < E[K]]$?

3.5.10● At a casino, people line up to pay $20 each to be a contestant in the following game: The contestant flips a fair coin repeatedly. If she flips heads 20 times in a row, she walks away with $R = 20$ million dollars; otherwise she walks away with $R = 0$ dollars.

(a) Find the PMF of R, the reward earned by the contestant.

(b) The casino counts "losing contestants" who fail to win the 20 million dollar prize. Let L equal the number of losing contestants before the first winning contestant. What is the PMF of L?

(c) Why does the casino offer this game?

3.5.11■ Find $P[K < E[K]]$ when

(a) K is geometric $(1/3)$.

(b) K is binomial $(6, 1/2)$.

(c) K is Poisson (3).

(d) K is discrete uniform $(0, 6)$.

3.5.12■ Suppose you go to a casino with exactly $63. At this casino, the only game is roulette and the only bets allowed are red and green. The payoff for a winning bet is the amount of the bet. In addition, the wheel is fair so that $P[\text{red}] = P[\text{green}] = 1/2$. You have the following strategy: First, you bet $1. If you win the bet, you quit and leave the casino with $64. If you lose, you then bet $2. If you win, you quit and go home. If you lose, you bet $4. In fact, whenever you lose, you double your bet until either you win a bet or you lose all of your money. However, as soon as you win a bet, you quit and go home. Let Y equal the amount of money that you take home. Find $P_Y(y)$ and $E[Y]$. Would you like to play this game every day?

3.5.13◆ In a TV game show, there are three identical-looking suitcases. The first suitcase has 3 dollars, the second has 30 dollars and the third has 300 dollars. You start the game by randomly choosing a suitcase. *Between the two unchosen suitcases,* the game show host opens the suitcase with more money. The host then asks you if you want to keep your suitcase or switch to the other remaining suitcase. After you make your decision, you open your suitcase and keep the D dollars inside. Should you switch suitcases? To answer this question, solve the following subproblems and use the following notation:

• C_i is the event that you first choose the suitcase with i dollars.

• O_i denotes the event that the host opens a suitcase with i dollars.

In addition, you may wish to go back and review the Monty Hall problem in Example 2.3.

(a) Suppose you never switch; you always stick with your original choice. Use a tree diagram to find the PMF $P_D(d)$ and expected value $E[D]$.

(b) Suppose you always switch. Use a tree diagram to find the PMF $P_D(d)$ and expected value $E[D]$.

(c) Perhaps your rule for switching should depend on how many dollars are in the suitcase that the host opens? What is the optimal strategy to maximize $E[D]$? Hint: Consider making a random decision; if the host opens a suitcase with i dollars, let α_i denote the probability that you switch.

3.5.14 You are a contestant on a TV game show; there are four identical-looking suitcases containing $100, $200, $400, and $800. You start the game by randomly choosing a suitcase. *Among the three unchosen suitcases*, the game show host opens the suitcase that holds the median amount of money. (For example, if the unopened suitcases contain $100, $400 and $800, the host opens the $400 suitcase.) The host then asks you if you want to keep your suitcase or switch one of the other remaining suitcases. For your analysis, use the following notation for events:

- C_i is the event that you choose a suitcase with i dollars.
- O_i denotes the event that the host opens a suitcase with i dollars.
- R is the reward in dollars that you keep.

(a) You refuse the host's offer and open the suitcase you first chose. Find the PMF of R and the expected value $E[R]$.

(b) You always switch and randomly choose one of the two remaining suitcases with equal probability. You receive the R dollars in this chosen suitcase. Sketch a tree diagram for this experiment, and find the PMF and expected value of R.

(c) Can you do better than either always switching or always staying with your original choice? Explain.

3.5.15 You are a contestant on a TV game show; there are four identical-looking suitcases containing $200, $400, $800, and $1600. You start the game by randomly choosing a suitcase. *Among the three unchosen suitcases*, the game show host opens

the suitcase that holds the least money. The host then asks you if you want to keep your suitcase or switch one of the other remaining suitcases. For the following analysis, use the following notation for events:

- C_i is the event that you choose a suitcase with i dollars.
- O_i denotes the event that the host opens a suitcase with i dollars.
- R is the reward in dollars that you keep.

(a) You refuse the host's offer and open the suitcase you first chose. Find the PMF of R and the expected value $E[R]$.

(b) You switch and randomly choose one of the two remaining suitcases. You receive the R dollars in this chosen suitcase. Sketch a tree diagram for this experiment, and find the PMF and expected value of R.

3.5.16 Let binomial random variable X_n denote the number of successes in n Bernoulli trials with success probability p. Prove that $E[X_n] = np$. Hint: Use the fact that $\sum_{x=0}^{n-1} P_{X_{n-1}}(x) = 1$.

3.5.17 Prove that if X is a nonnegative integer-valued random variable, then

$$E[X] = \sum_{k=0}^{\infty} P[X > k].$$

3.5.18 At the gym, a weightlifter can bench press a maximum of 100 kg. For a mass of m kg, $(1 \leq m \leq 100)$, the maximum number of repetitions she can complete is R, a geometric random variable with expected value $100/m$.

(a) In terms of the mass m, what is the PMF of R?

(b) When she performs one repetition, she lifts the m kg mass a height $h = 5/9.8$ meters and thus does work $w = mgh = 5m$ Joules. For R repetitions, she does $W = 5mR$ Joules of work. What is the expected work $E[W]$ that she will complete?

(c) A friend offers to pay her 1000 dollars if she can perform 1000 Joules of weightlifting work. What mass m in the range $1 \leq m \leq 100$ should she use to maximize her probability of winning the money? For the best choice of m, what is the probability that she wins the money?

3.5.19◆ At the gym, a weightlifter can bench press a maximum of 200 kg. For a mass of m kg, $(1 \leq m \leq 200)$, the maximum number of repetitions she can complete is R, a geometric random variable with expected value $200/m$.

(a) In terms of the mass m, what is the PMF of R?

(b) When she performs one repetition, she lifts the m kg mass a height $h = 4/9.8$ meters and thus does work $w = mgh = 4m$ Joules. For R repetitions, she does $W = 4mR$ Joules of work. What is the expected work $E[W]$ that she will complete?

(c) A friend offers to pay her 1000 dollars if she can perform 1000 Joules of weightlifting work. What mass m in the range $1 \leq m \leq 200$ should she use to maximize her probability of winning the money?

3.6.1● Given the random variable Y in Problem 3.4.1, let $U = g(Y) = Y^2$.

(a) Find $P_U(u)$.

(b) Find $F_U(u)$.

(c) Find $E[U]$.

3.6.2● Given the random variable X in Problem 3.4.2, let $V = g(X) = |X|$.

(a) Find $P_V(v)$.

(b) Find $F_V(v)$.

(c) Find $E[V]$.

3.6.3● The amplitude V (volts) of a sinusoidal signal is a discrete uniform $(-3, 3)$ random variable. Let $Y = V^2/2$ watts denote the power of the transmitted signal. Find $P_Y(y)$.

3.6.4■ A source transmits data packets to a receiver over a radio link. The receiver uses error detection to identify packets that have been corrupted by radio noise. When a packet is received error free, the receiver sends an acknowledgment (ACK) back to the source. When the receiver gets a packet with errors, a negative acknowledgment (NAK) message is sent back to the source. Each time the source receives a NAK, the packet is retransmitted. We assume that each packet transmission is independently corrupted by errors with probability q.

(a) Find the PMF of X, the number of times that a packet is transmitted by the source.

(b) Suppose each packet takes 1 millisecond to transmit and that the source waits an additional millisecond to receive the acknowledgment message (ACK or NAK) before retransmitting. Let T equal the time required until the packet is successfully received. What is the relationship between T and X? What is the PMF of T?

3.6.5■ Suppose that a cellular phone costs $20 per month with 30 minutes of use included and that each additional minute of use costs $0.50. If the number of minutes you use in a month is a geometric random variable M with expected value of $E[M] = 1/p = 30$ minutes, what is the PMF of C, the cost of the phone for one month?

3.6.6■ A professor tries to count the number of students attending lecture. For each student in the audience, the professor either counts the student properly (with probability p) or overlooks (and does not count) the student with probability $1 - p$. The exact number of attending students is 70.

(a) The number of students counted by the professor is a random variable N. What is the PMF of N?

(b) Let $U = 70 - N$ denote the number of uncounted students. What is the PMF of U?

(c) What is the probability that the under-count U is 2 or more?

(d) For what value of p does $E[U] = 2$?

3.6.7 A forgetful professor tries to count the M&Ms in a package; however, the professor often loses his place and double counts an M&M. For each M&M in the package, the professor counts the M&M and then, with probability p counts the M&M again. The exact number of M&Ms in the pack is 20.

(a) Find the PMF of R, the number of double-counted M&Ms.

(b) Find the PMF of N, the number of M&Ms counted by the professor.

3.7.1 Starting on day $n = 1$, you buy one lottery ticket each day. Each ticket costs 1 dollar and is independently a winner that can be cashed for 5 dollars with probability 0.1; otherwise the ticket is worthless Let X_n equal your net profit after n days. What is $E[X_n]$?

3.7.2 For random variable T in Quiz 3.6, first find the expected value $E[T]$ using Theorem 3.10. Next, find $E[T]$ using Definition 3.11.

3.7.3 In a certain lottery game, the chance of getting a winning ticket is exactly one in a thousand. Suppose a person buys one ticket each day (except on the leap year day February 29) over a period of fifty years. What is the expected number $E[T]$ of winning tickets in fifty years? If each winning ticket is worth $1000, what is the expected amount $E[R]$ collected on these winning tickets? Lastly, if each ticket costs $2, what is your expected net profit $E[Q]$?

3.7.4 Suppose an NBA basketball player shooting an uncontested 2-point shot will make the basket with probability 0.6. However, if you foul the shooter, the shot will be missed, but two free throws will be awarded. Each free throw is an independent Bernoulli trial with success probability p. Based on the expected number of points the shooter will score, for what values of p may it be desirable to foul the shooter?

3.7.5 It can take up to four days after you call for service to get your computer repaired. The computer company charges for repairs according to how long you have to wait. The number of days D until the service technician arrives and the service charge C, in dollars, are described by

d	1	2	3	4
$P_D(d)$	0.2	0.4	0.3	0.1

and

$$C = \begin{cases} 90 & \text{for 1-day service,} \\ 70 & \text{for 2-day service,} \\ 40 & \text{for 3-day service,} \\ 40 & \text{for 4-day service.} \end{cases}$$

(a) What is the expected waiting time $\mu_D = E[D]$?

(b) What is the expected deviation $E[D - \mu_D]$?

(c) Express C as a function of D.

(d) What is the expected value $E[C]$?

3.7.6 True or False: For any random variable X, $E[1/X] = 1/E[X]$.

3.7.7 For the cellular phone in Problem 3.6.5, express the monthly cost C as a function of M, the number of minutes used. What is the expected monthly cost $E[C]$?

3.7.8 A new cellular phone billing plan costs $15 per month plus $1 for each minute of use. If the number of minutes you use the phone in a month is a geometric random variable with expected value $1/p$, what is the expected monthly cost $E[C]$ of the phone? For what values of p is this billing plan preferable to the billing plan of Problem 3.6.5 and Problem 3.7.7?

3.7.9 A particular circuit works if all 10 of its component devices work. Each circuit is tested before leaving the factory. Each working circuit can be sold for k dollars, but each nonworking circuit is worthless and must be thrown away. Each circuit can be built with either ordinary devices or ultra-reliable devices. An ordinary device has a failure probability of $q = 0.1$ and costs \$1. An ultrareliable device has a failure probability of $q/2$ but costs \$3. Assuming device failures are independent, should you build your circuit with ordinary devices or ultra-reliable devices in order to maximize your expected profit $E[R]$? Keep in mind that your answer will depend on k.

3.7.10♦♦ In the New Jersey state lottery, each \$1 ticket has six randomly marked numbers out of $1, \ldots, 46$. A ticket is a winner if the six marked numbers match six numbers drawn at random at the end of a week. For each ticket sold, 50 cents is added to the pot for the winners. If there are k winning tickets, the pot is divided equally among the k winners. Suppose you bought a winning ticket in a week in which $2n$ tickets are sold and the pot is n dollars.

(a) What is the probability q that a random ticket will be a winner?

(b) Find the PMF of K_n, the number of other (besides your own) winning tickets.

(c) What is the expected value of W_n, the prize for your winning ticket?

3.7.11♦♦ If there is no winner for the lottery described in Problem 3.7.10, then the pot is carried over to the next week. Suppose that in a given week, an r dollar pot is carried over from the previous week and $2n$ tickets sold. Answer the following questions.

(a) What is the probability q that a random ticket will be a winner?

(b) If you own one of the $2n$ tickets sold, what is the expected value of V, the value (i.e., the amount you win) of that ticket? Is it ever possible that $E[V] > 1$?

(c) Suppose that in the instant before the ticket sales are stopped, you are given the opportunity to buy one of each possible ticket. For what values (if any) of n and r should you do it?

3.8.1● In an experiment to monitor two packets, the PMF of N, the number of video packets, is

n	0	1	2
$P_N(n)$	0.2	0.7	0.1

Find $E[N]$, $E[N^2]$, $\text{Var}[N]$, and σ_N.

3.8.2● Find the variance of the random variable Y in Problem 3.4.1.

3.8.3● Find the variance of the random variable X in Problem 3.4.2.

3.8.4■ Let X have the binomial PMF

$$P_X(x) = \binom{4}{x}(1/2)^4.$$

(a) Find the standard deviation of X.

(b) What is $P[\mu_X - \sigma_X \leq X \leq \mu_X + \sigma_X]$, the probability that X is within one standard deviation of the expected value?

3.8.5■ X is the binomial $(5, 0.5)$ random variable.

(a) Find the standard deviation of X.

(b) Find $P[\mu_X - \sigma_X \leq X \leq \mu_X + \sigma_X]$, the probability that X is within one standard deviation of the expected value.

3.8.6■ Show that the variance of $Y = aX + b$ is $\text{Var}[Y] = a^2 \text{Var}[X]$.

3.8.7■ Given a random variable X with expected value μ_X and variance σ_X^2, find the expected value and variance of

$$Y = \frac{X - \mu_X}{\sigma_X}.$$

3.8.8■ In real-time packet data transmission, the time between successfully received packets is called the *interarrival time*, and randomness in packet interarrival times is called *jitter*. Jitter is undesirable. One measure of jitter is the standard deviation of the packet interarrival time. From Problem 3.6.4, calculate the jitter σ_T. How large must the successful transmission probability q be to ensure that the jitter is less than 2 milliseconds?

3.8.9◆ Random variable K has a Poisson (α) distribution. Derive the properties $\mathrm{E}[K] = \mathrm{Var}[K] = \alpha$. Hint: $\mathrm{E}[K^2] = \mathrm{E}[K(K-1)] + \mathrm{E}[K]$.

3.8.10● For the delay D in Problem 3.7.5, what is the standard deviation σ_D of the waiting time?

3.9.1● Let X be the binomial $(100, 1/2)$ random variable. Let E_2 denote the event that X is a perfect square. Calculate $\mathrm{P}[E_2]$.

3.9.2● Write a MATLAB function `x=shipweight8(m)` that produces m random sample values of the package weight X with PMF given in Example 3.21.

3.9.3● Use the `unique` function to write a MATLAB script `shipcostpmf.m` that outputs the pair of vectors `sy` and `py` representing the PMF $P_Y(y)$ of the shipping cost Y in Example 3.21.

3.9.4● For $m = 10$, $m = 100$, and $m = 1000$, use MATLAB to find the average cost of sending m packages using the model of Example 3.21. Your program input should have the number of trials m as the input. The output should be $\overline{Y} = \frac{1}{m}\sum_{i=1}^{m} Y_i$, where Y_i is the cost of the ith package. As m becomes large, what do you observe?

3.9.5■ The Zipf $(n, \alpha = 1)$ random variable X introduced in Problem 3.3.12 is often used to model the "popularity" of a collection of n objects. For example, a Web server can deliver one of n Web pages. The pages are numbered such that the page 1 is the most requested page, page 2 is the second most requested page, and so on. If page k is requested, then $X = k$.

To reduce external network traffic, an ISP gateway caches copies of the k most popular pages. Calculate, as a function of n for $1 \leq n \leq 1000$, how large k must be to ensure that the cache can deliver a page with probability 0.75.

3.9.6■ Generate n independent samples of the Poisson (5) random variable Y. For each $y \in S_Y$, let $n(y)$ denote the number of times that y was observed. Thus $\sum_{y \in S_Y} n(y) = n$ and the relative frequency of y is $R(y) = n(y)/n$. Compare the relative frequency of y against $P_Y(y)$ by plotting $R(y)$ and $P_Y(y)$ on the same graph as functions of y for $n = 100$, $n = 1000$ and $n = 10,000$. How large should n be to have reasonable agreement?

3.9.7■ Test the convergence of Theorem 3.8. For $\alpha = 10$, plot the PMF of K_n for $(n, p) = (10, 1)$, $(n, p) = (100, 0.1)$, and $(n, p) = (1000, 0.01)$ and compare each result with the Poisson (α) PMF.

3.9.8■ Use the result of Problem 3.4.3 and the Random Sample Algorithm on Page 91 to write a MATLAB function `k=geometricrv(p,m)` that generates m samples of a geometric (p) random variable.

3.9.9◆ Find n^*, the smallest value of n for which the function `poissonpmf(n,n)` shown in Example 3.30 reports an error. What is the source of the error? Write a function `bigpoissonpmf(alpha,n)` that calculates `poissonpmf(n,n)` for values of n much larger than n^*. Hint: For a Poisson (α) random variable K,

$$P_K(k) = \exp\left(-\alpha + k\ln(\alpha) - \sum_{j=1}^{k}\ln(j)\right).$$

4

Continuous Random Variables

4.1 Continuous Sample Space

> A random variable X is *continuous* if the range S_X consists of one or more intervals. For each $x \in S_X$, $P[X = x] = 0$.

Until now, we have studied discrete random variables. By definition, the range of a discrete random variable is a countable set of numbers. This chapter analyzes random variables that range over continuous sets of numbers. A continuous set of numbers, sometimes referred to as an *interval*, contains all of the real numbers between two limits. Many experiments lead to random variables with a range that is a continuous interval. Examples include measuring T, the arrival time of a particle ($S_T = \{t | 0 \leq t < \infty\}$); measuring V, the voltage across a resistor ($S_V = \{v| - \infty < v < \infty\}$); and measuring the phase angle A of a sinusoidal radio wave ($S_A = \{a | 0 \leq a < 2\pi\}$). We will call T, V, and A *continuous random variables*, although we will defer a formal definition until Section 4.2.

Consistent with the axioms of probability, we assign numbers between zero and one to all events (sets of elements) in the sample space. A distinguishing feature of the models of continuous random variables is that the probability of each individual outcome is zero! To understand this intuitively, consider an experiment in which the observation is the arrival time of the professor at a class. Assume this professor always arrives between 8:55 and 9:05. We model the arrival time as a random variable T minutes relative to 9:00 o'clock. Therefore, $S_T = \{t| - 5 \leq t \leq 5\}$. Think about predicting the professor's arrival time. The more precise the prediction, the lower the chance it will be correct. For example, you might guess the interval $-1 \leq T \leq 1$ minute (8:59 to 9:01). Your probability of being correct is higher than if you guess $-0.5 \leq T \leq 0.5$ minute (8:59:30 to 9:00:30). As your prediction becomes more and more precise, the probability that it will be correct gets smaller and smaller. The chance that the professor will arrive within a femtosecond of 9:00

is microscopically small (on the order of 10^{-15}), and the probability of a precise 9:00 arrival is zero.

One way to think about continuous random variables is that the *amount of probability* in an interval gets smaller and smaller as the interval shrinks. This is like the mass in a continuous volume. Even though any finite volume has some mass, there is no mass at a single point. In physics, we analyze this situation by referring to densities of matter. Similarly, we refer to *probability density functions* to describe probabilities related to continuous random variables. The next section introduces these ideas formally by describing an experiment in which the sample space contains all numbers between zero and one.

In many practical applications of probability, we encounter uniform random variables. The range of a uniform random variable is an interval with finite limits. The probability model of a uniform random variable states that any two intervals of equal size within the range have equal probability. To introduce many concepts of continuous random variables, we will refer frequently to a uniform random variable with limits 0 and 1. Most computer languages include a random number generator. In MATLAB, this is the **rand** function introduced in Chapter 1. These random number generators produce a sequence of pseudorandom numbers that approximate the properties of outcomes of repeated trials of an experiment with a probability model that is a continuous uniform random variable.

In the following example, we examine this random variable by defining an experiment in which the procedure is to spin a pointer in a circle of circumference one meter. This model is very similar to the model of the phase angle of the signal that arrives at the radio receiver of a cellular telephone. Instead of a pointer with stopping points that can be anywhere between 0 and 1 meter, the phase angle can have any value between 0 and 2π radians. By referring to the spinning pointer in the examples in this chapter, we arrive at mathematical expressions that illustrate the main properties of continuous random variables. The formulas that arise from analyzing phase angles in communications engineering models have factors of 2π that do not appear in the examples in this chapter. Example 4.1 defines the sample space of the pointer experiment and demonstrates that all outcomes have probability zero.

Example 4.1

Suppose we have a wheel of circumference one meter and we mark a point on the perimeter at the top of the wheel. In the center of the wheel is a radial pointer that we spin. After spinning the pointer, we measure the distance, X meters, around the circumference of the wheel going clockwise from the marked point to the pointer position as shown in Figure 4.1. Clearly, $0 \le X < 1$. Also, it is reasonable to believe that if the spin is hard enough, the pointer is just as likely to arrive at any part of the circle as at any other. For a given x, what is the probability $P[X = x]$?

. . . .

This problem is surprisingly difficult. However, given that we have developed methods for discrete random variables in Chapter 3, a reasonable approach is to find a discrete approximation to X. As shown on the right side of Figure 4.1, we can mark the perimeter with n equal-length arcs numbered 1 to n and let Y denote the number

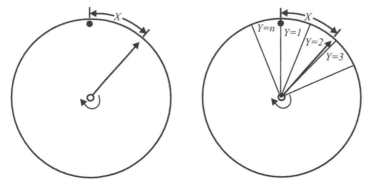

Figure 4.1   The random pointer on disk of circumference 1.

of the arc in which the pointer stops. Y is a discrete random variable with range $S_Y = \{1, 2, \ldots, n\}$. Since all parts of the wheel are equally likely, all arcs have the same probability. Thus the PMF of Y is

$$P_Y(y) = \begin{cases} 1/n & y = 1, 2, \ldots, n, \\ 0 & \text{otherwise.} \end{cases} \tag{4.1}$$

From the wheel on the right side of Figure 4.1, we can deduce that if $X = x$, then $Y = \lceil nx \rceil$, where the notation $\lceil a \rceil$ is defined as the smallest integer greater than or equal to a. Note that the event $\{X = x\} \subset \{Y = \lceil nx \rceil\}$, which implies that

$$P\left[X = x\right] \leq P\left[Y = \lceil nx \rceil\right] = \frac{1}{n}. \tag{4.2}$$

We observe this is true no matter how finely we divide up the wheel. To find $P[X = x]$, we consider larger and larger values of n. As n increases, the arcs on the circle decrease in size, approaching a single point. The probability of the pointer arriving in any particular arc decreases until we have in the limit,

$$P\left[X = x\right] \leq \lim_{n \to \infty} P\left[Y = \lceil nx \rceil\right] = \lim_{n \to \infty} \frac{1}{n} = 0. \tag{4.3}$$

This demonstrates that $P[X = x] \leq 0$. The first axiom of probability states that $P[X = x] \geq 0$. Therefore, $P[X = x] = 0$. This is true regardless of the outcome, x. It follows that every outcome has probability zero.

Just as in the discussion of the professor arriving in class, similar reasoning can be applied to other experiments to show that for any continuous random variable, the probability of any individual outcome is zero. This is a fundamentally different situation than the one we encountered in our study of discrete random variables. Clearly a probability mass function defined in terms of probabilities of individual outcomes has no meaning in this context. For a continuous random variable, the interesting probabilities apply to intervals.

4.2   The Cumulative Distribution Function

The CDF $F_X(x)$ is a probability model for any random variable. The CDF $F_X(x)$ is a continuous function if and only if X is a continuous random variable.

Example 4.1 shows that when X is a continuous random variable, $P[X = x] = 0$ for $x \in S_X$. This implies that when X is continuous, it is impossible to define a probability mass function $P_X(x)$. On the other hand, we will see that the cumulative distribution function, $F_X(x)$ in Definition 3.10, is a very useful probability model for a continuous random variable. We repeat the definition here.

=====Definition 4.1=====Cumulative Distribution Function (CDF)
*The **cumulative distribution function** (CDF) of random variable X is*

$$F_X(x) = P[X \leq x].$$

The key properties of the CDF, described in Theorem 3.2 and Theorem 3.3, apply to *all* random variables. Graphs of all cumulative distribution functions start at zero on the left and end at one on the right. All are nondecreasing, and, most importantly, the probability that the random variable is in an interval is the difference in the CDF evaluated at the ends of the interval.

=====Theorem 4.1=====
For any random variable X,
(a) $F_X(-\infty) = 0$ *(b)* $F_X(\infty) = 1$
(c) $P[x_1 < X \leq x_2] = F_X(x_2) - F_X(x_1)$

Although these properties apply to any CDF, there is one important difference between the CDF of a discrete random variable and the CDF of a continuous random variable. Recall that for a discrete random variable X, $F_X(x)$ has zero slope everywhere except at values of x with nonzero probability. At these points, the function has a discontinuity in the form of a jump of magnitude $P_X(x)$. By contrast, the defining property of a continuous random variable X is that $F_X(x)$ is a continuous function of X.

=====Definition 4.2=====Continuous Random Variable
X *is a **continuous random variable** if the CDF $F_X(x)$ is a continuous function.*

=====Example 4.2=====
In the wheel-spinning experiment of Example 4.1, find the CDF of X.

We begin by observing that any outcome $x \in S_X = [0, 1)$. This implies that $F_X(x) = 0$ for $x < 0$, and $F_X(x) = 1$ for $x \geq 1$. To find the CDF for x between 0 and 1 we consider the event $\{X \leq x\}$, with x growing from 0 to 1. Each event corresponds to an arc on the circle in Figure 4.1. The arc is small when $x \approx 0$ and it includes nearly the whole circle when $x \approx 1$. $F_X(x) = P[X \leq x]$ is the probability that the pointer stops somewhere in the arc. This probability grows from 0 to 1 as the arc increases to include the whole circle. Given our assumption that the pointer has no preferred stopping places, it is reasonable to expect the probability to grow in proportion to the fraction of the circle occupied by the arc $\{X \leq x\}$. This fraction is simply x. To be more formal, we can refer to Figure 4.1 and note that with the circle divided into n arcs,

$$\{Y \leq \lceil nx \rceil - 1\} \subset \{X \leq x\} \subset \{Y \leq \lceil nx \rceil\}. \tag{4.4}$$

Therefore, the probabilities of the three events are related by

$$F_Y(\lceil nx \rceil - 1) \leq F_X(x) \leq F_Y(\lceil nx \rceil). \tag{4.5}$$

Note that Y is a discrete random variable with CDF

$$F_Y(y) = \begin{cases} 0 & y < 0, \\ k/n & (k-1)/n < y \leq k/n, k = 1, 2, \ldots, n, \\ 1 & y > 1. \end{cases} \tag{4.6}$$

Thus for $x \in [0, 1)$ and for all n, we have

$$\frac{\lceil nx \rceil - 1}{n} \leq F_X(x) \leq \frac{\lceil nx \rceil}{n}. \tag{4.7}$$

In Problem 4.2.3, we ask the reader to verify that $\lim_{n\to\infty} \lceil nx \rceil / n = x$. This implies that as $n \to \infty$, both fractions approach x. The CDF of X is

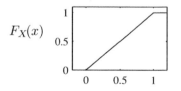

$$F_X(x) = \begin{cases} 0 & x < 0, \\ x & 0 \leq x < 1, \\ 1 & x \geq 1. \end{cases} \tag{4.8}$$

Quiz 4.2

The cumulative distribution function of the random variable Y is

$$F_Y(y) = \begin{cases} 0 & y < 0, \\ y/4 & 0 \leq y \leq 4, \\ 1 & y > 4. \end{cases} \tag{4.9}$$

Sketch the CDF of Y and calculate the following probabilities:

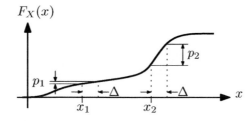

Figure 4.2   The graph of an arbitrary CDF $F_X(x)$.

(a) $P[Y \leq -1]$ (b) $P[Y \leq 1]$

(c) $P[2 < Y \leq 3]$ (d) $P[Y > 1.5]$

4.3   Probability Density Function

> Like the CDF, the PDF $f_X(x)$ is a probability model for a continuous random variable X. $f_X(x)$ is the derivative of the CDF. It is proportional to the probability that X is close to x.

The slope of the CDF contains the most interesting information about a continuous random variable. The slope at any point x indicates the probability that X is *near* x. To understand this intuitively, consider the graph of a CDF $F_X(x)$ given in Figure 4.2. Theorem 4.1(c) states that the probability that X is in the interval of width Δ to the right of x_1 is

$$p_1 = P\left[x_1 < X \leq x_1 + \Delta\right] = F_X(x_1 + \Delta) - F_X(x_1). \tag{4.10}$$

Note in Figure 4.2 that this is less than the probability of the interval of width Δ to the right of x_2,

$$p_2 = P\left[x_2 < X \leq x_2 + \Delta\right] = F_X(x_2 + \Delta) - F_X(x_2). \tag{4.11}$$

The comparison makes sense because both intervals have the same length. If we reduce Δ to focus our attention on outcomes nearer and nearer to x_1 and x_2, both probabilities get smaller. However, their relative values still depend on the average slope of $F_X(x)$ at the two points. This is apparent if we rewrite Equation (4.10) in the form

$$P\left[x_1 < X \leq x_1 + \Delta\right] = \frac{F_X(x_1 + \Delta) - F_X(x_1)}{\Delta}\Delta. \tag{4.12}$$

Here the fraction on the right side is the average slope, and Equation (4.12) states that the probability that a random variable is in an interval near x_1 is the average

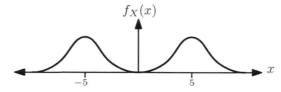

Figure 4.3   The PDF of the modem receiver voltage X.

slope over the interval times the length of the interval. By definition, the limit of the average slope as $\Delta \to 0$ is the derivative of $F_X(x)$ evaluated at x_1.

We conclude from the discussion leading to Equation (4.12) that the slope of the CDF in a region near any number x is an indicator of the probability of observing the random variable X near x. Just as the amount of matter in a small volume is the density of the matter times the size of volume, the amount of probability in a small region is the slope of the CDF times the size of the region. This leads to the term *probability density*, defined as the slope of the CDF.

---**Definition 4.3**---**Probability Density Function (PDF)**
*The **probability density function** (PDF) of a continuous random variable X is*

$$f_X(x) = \frac{dF_X(x)}{dx}.$$

This definition displays the conventional notation for a PDF. The name of the function is a lowercase f with a subscript that is the name of the random variable. As with the PMF and the CDF, the argument is a dummy variable: $f_X(x)$, $f_X(u)$, and $f_X(\cdot)$ are all the same PDF.

The PDF is a complete probability model of a continuous random variable. While there are other functions that also provide complete models (the CDF and the moment generating function that we study in Chapter 9), the PDF is the most useful. One reason for this is that the graph of the PDF provides a good indication of the likely values of observations.

---**Example 4.3**---
Figure 4.3 depicts the PDF of a random variable X that describes the voltage at the receiver in a modem. What are probable values of X?

....

Note that there are two places where the PDF has high values and that it is low elsewhere. The PDF indicates that the random variable is likely to be near -5 V (corresponding to the symbol 0 transmitted) and near $+5$ V (corresponding to a 1 transmitted). Values far from ± 5 V (due to strong distortion) are possible but much less likely.

Another reason the PDF is the most useful probability model is that it plays a key role in calculating the expected value of a continuous random variable, the subject of the next section. Important properties of the PDF follow directly from Definition 4.3 and the properties of the CDF.

═══Theorem 4.2═══

For a continuous random variable X with PDF $f_X(x)$,

(a) $f_X(x) \geq 0$ for all x,   *(b) $F_X(x) = \int_{-\infty}^{x} f_X(u)\, du$,*

(c) $\int_{-\infty}^{\infty} f_X(x)\, dx = 1$.

Proof The first statement is true because $F_X(x)$ is a nondecreasing function of x and therefore its derivative, $f_X(x)$, is nonnegative. The second fact follows directly from the definition of $f_X(x)$ and the fact that $F_X(-\infty) = 0$. The third statement follows from the second one and Theorem 4.1(b).

Given these properties of the PDF, we can prove the next theorem, which relates the PDF to the probabilities of events.

═══Theorem 4.3═══

$$P\left[x_1 < X \leq x_2\right] = \int_{x_1}^{x_2} f_X(x)\, dx.$$

Proof From Theorem 4.1(c) and Theorem 4.2(b),

$$P\left[x_1 < X \leq x_2\right] = F_X(x_2) - F_X(x_1)$$

$$= \int_{-\infty}^{x_2} f_X(x)\, dx - \int_{-\infty}^{x_1} f_X(x)\, dx = \int_{x_1}^{x_2} f_X(x)\, dx. \tag{4.13}$$

Theorem 4.3 states that the probability of observing X in an interval is the area under the PDF graph between the two end points of the interval. This property of the PDF is depicted in Figure 4.4. Theorem 4.2(c) states that the area under the entire PDF graph is one. Note that the value of the PDF can be any nonnegative number. It is not a probability and need not be between zero and one. To gain further insight into the PDF, it is instructive to reconsider Equation (4.12). For very small values of Δ, the right side of Equation (4.12) approximately equals $f_X(x_1)\Delta$. When Δ becomes the infinitesimal dx, we have

$$P\left[x < X \leq x + dx\right] = f_X(x)\, dx. \tag{4.14}$$

Equation (4.14) is useful because it permits us to interpret the integral of Theorem 4.3 as the limiting case of a sum of probabilities of events $\{x < X \leq x + dx\}$.

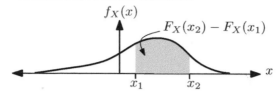

Figure 4.4   The PDF and CDF of X.

=======**Example 4.4**=======

For the experiment in Examples 4.1 and 4.2, find the PDF of X and the probability of the event $\{1/4 < X \le 3/4\}$.

. . . .

Taking the derivative of the CDF in Equation (4.8), $f_X(x) = 0$ when $x < 0$ or $x \ge 1$. For x between 0 and 1 we have $f_X(x) = dF_X(x)/dx = 1$. Thus the PDF of X is

$$f_X(x) = \begin{cases} 1 & 0 \le x < 1, \\ 0 & \text{otherwise.} \end{cases} \qquad (4.15)$$

The fact that the PDF is constant over the range of possible values of X reflects the fact that the pointer has no favorite stopping places on the circumference of the circle. To find the probability that X is between $1/4$ and $3/4$, we can use either Theorem 4.1 or Theorem 4.3. Thus

$$\mathrm{P}\left[1/4 < X \le 3/4\right] = F_X(3/4) - F_X(1/4) = 1/2, \qquad (4.16)$$

and equivalently,

$$\mathrm{P}\left[1/4 < X \le 3/4\right] = \int_{1/4}^{3/4} f_X(x)\,dx = \int_{1/4}^{3/4} dx = 1/2. \qquad (4.17)$$

When the PDF and CDF are both known, it is easier to use the CDF to find the probability of an interval. However, in many cases we begin with the PDF, in which case it is usually easiest to use Theorem 4.3 directly. The alternative is to find the CDF explicitly by means of Theorem 4.2(b) and then to use Theorem 4.1.

=======**Example 4.5**=======

Consider an experiment that consists of spinning the pointer in Example 4.1 three times and observing Y meters, the maximum value of X in the three spins. In Example 8.3, we show that the CDF of Y is

$$F_Y(y) = \begin{cases} 0 & y < 0, \\ y^3 & 0 \le y \le 1, \\ 1 & y > 1. \end{cases} \qquad (4.18)$$

Find the PDF of Y and the probability that Y is between $1/4$ and $3/4$.
....

We apply Definition 4.3 to the CDF $F_Y(y)$. When $F_Y(y)$ is piecewise differentiable, we take the derivative of each piece:

$$f_Y(y) = \frac{dF_Y(y)}{dy} = \begin{cases} 3y^2 & 0 < y \le 1, \\ 0 & \text{otherwise.} \end{cases} \qquad (4.19)$$

Note that the PDF has values between 0 and 3. Its integral between any pair of numbers is less than or equal to 1. The graph of $f_Y(y)$ shows that there is a higher probability of finding Y at the right side of the range of possible values than at the left side. This reflects the fact that the maximum of three spins produces higher numbers than individual spins. Either Theorem 4.1 or Theorem 4.3 can be used to calculate the probability of observing Y between $1/4$ and $3/4$:

$$P[1/4 < Y \le 3/4] = F_Y(3/4) - F_Y(1/4) = (3/4)^3 - (1/4)^3 = 13/32, \qquad (4.20)$$

and equivalently,

$$P[1/4 < Y \le 3/4] = \int_{1/4}^{3/4} f_Y(y)\, dy = \int_{1/4}^{3/4} 3y^2\, dy = 13/32. \qquad (4.21)$$

Note that this probability is less than $1/2$, which is the probability of $1/4 < X \le 3/4$ calculated in Example 4.4 for one spin of the pointer.

When we work with continuous random variables, it is usually not necessary to be precise about specifying whether or not a range of numbers includes the endpoints. This is because individual numbers have probability zero. In Example 4.2, there are four different events defined by the words X *is between 1/4 and 3/4*:

$$A = \{1/4 < X < 3/4\}, \qquad B = \{1/4 < X \le 3/4\},$$
$$C = \{1/4 \le X < 3/4\}, \qquad D = \{1/4 \le X \le 3/4\}.$$

While they are all different events, they all have the same probability because they differ only in whether they include $\{X = 1/4\}$, $\{X = 3/4\}$, or both. Since these two events have zero probability, their inclusion or exclusion does not affect the probability of the range of numbers. This is quite different from the situation we encounter with discrete random variables. For example, suppose random variable X has PMF

$$P_X(x) = \begin{cases} 1/6 & x = 1/4, x = 1/2, \\ 2/3 & x = 3/4, \\ 0 & \text{otherwise.} \end{cases} \qquad (4.22)$$

For this random variable X, the probabilities of the four sets are

$$P[A] = 1/6, \quad P[B] = 5/6, \quad P[C] = 1/3, \quad P[D] = 1.$$

So we see that the nature of an inequality in the definition of an event does not affect the probability when we examine continuous random variables. With discrete random variables, it is critically important to examine the inequality carefully.

If we compare other characteristics of discrete and continuous random variables, we find that with discrete random variables, many facts are expressed as sums. With continuous random variables, the corresponding facts are expressed as integrals. For example, when X is discrete,

$$P[B] = \sum_{x \in B} P_X(x). \qquad \text{(Theorem 3.1(c))}$$

When X is continuous and $B = [x_1, x_2]$,

$$P[x_1 < X \le x_2] = \int_{x_1}^{x_2} f_X(x) \, dx. \qquad \text{(Theorem 4.3)}$$

═══════Quiz 4.3═══════

Random variable X has probability density function

$$f_X(x) = \begin{cases} cxe^{-x/2} & x \ge 0, \\ 0 & \text{otherwise.} \end{cases} \qquad (4.23)$$

Sketch the PDF and find the following:

(a) the constant c

(b) the CDF $F_X(x)$

(c) $P[0 \le X \le 4]$

(d) $P[-2 \le X \le 2]$

4.4   Expected Values

> Like the expected value of a discrete random variable, the expected value, $E[X]$, of a continuous random variable X is a typical value of X. It is an important property of the probability model of X.

The primary reason that random variables are useful is that they permit us to compute averages. For a discrete random variable Y, the expected value,

$$E[Y] = \sum_{y_i \in S_Y} y_i P_Y(y_i), \qquad (4.24)$$

is a sum of the possible values y_i, each multiplied by its probability. For a continuous random variable X, this definition is inadequate because all possible values of X have probability zero. However, we can develop a definition for the expected value of the continuous random variable X by examining a discrete approximation of X. For a small Δ, let

$$Y = \Delta \left\lfloor \frac{X}{\Delta} \right\rfloor, \tag{4.25}$$

where the notation $\lfloor a \rfloor$ denotes the largest integer less than or equal to a. Y is an approximation to X in that $Y = k\Delta$ if and only if $k\Delta \leq X < k\Delta + \Delta$. Since the range of Y is $S_Y = \{\ldots, -\Delta, 0, \Delta, 2\Delta, \ldots\}$, the expected value is

$$\mathrm{E}\,[Y] = \sum_{k=-\infty}^{\infty} k\Delta\,\mathrm{P}\,[Y = k\Delta] = \sum_{k=-\infty}^{\infty} k\Delta\,\mathrm{P}\,[k\Delta \leq X < k\Delta + \Delta]. \tag{4.26}$$

As Δ approaches zero and the intervals under consideration grow smaller, Y more closely approximates X. Furthermore, $\mathrm{P}[k\Delta \leq X < k\Delta + \Delta]$ approaches $f_X(k\Delta)\Delta$ so that for small Δ,

$$\mathrm{E}\,[X] \approx \sum_{k=-\infty}^{\infty} k\Delta f_X\,(k\Delta)\,\Delta. \tag{4.27}$$

In the limit as Δ goes to zero, the sum converges to the integral in Definition 4.4.

Definition 4.4 Expected Value

*The **expected value** of a continuous random variable X is*

$$\mathrm{E}\,[X] = \int_{-\infty}^{\infty} x f_X\,(x)\;dx.$$

When we consider Y, the discrete approximation of X, the intuition developed in Section 3.5 suggests that $\mathrm{E}[Y]$ is what we will observe if we add up a very large number n of independent observations of Y and divide by n. This same intuition holds for the continuous random variable X. As $n \to \infty$, the average of n independent samples of X will approach $\mathrm{E}[X]$. In probability theory, this observation is known as the *Law of Large Numbers*, Theorem 10.6.

Example 4.6

In Example 4.4, we found that the stopping point X of the spinning wheel experiment was a uniform random variable with PDF

$$f_X\,(x) = \begin{cases} 1 & 0 \leq x < 1, \\ 0 & \text{otherwise.} \end{cases} \tag{4.28}$$

Find the expected stopping point $\mathrm{E}[X]$ of the pointer.

. . . .

$$\mathrm{E}\,[X] = \int_{-\infty}^{\infty} x f_X\,(x)\;dx = \int_{0}^{1} x\,dx = 1/2 \text{ meter.} \tag{4.29}$$

With no preferred stopping points on the circle, the average stopping point of the pointer is exactly halfway around the circle.

Corresponding to functions of discrete random variables described in Section 3.6, we have functions $g(X)$ of a continuous random variable X. A function of a continuous random variable is also a random variable; however, this random variable is not necessarily continuous!

Example 4.7

Let X be a uniform random variable with PDF

$$f_X(x) = \begin{cases} 1 & 0 \leq x < 1, \\ 0 & \text{otherwise.} \end{cases} \tag{4.30}$$

Let $W = g(X) = 0$ if $X \leq 1/2$, and $W = g(X) = 1$ if $X > 1/2$. W is a discrete random variable with range $S_W = \{0, 1\}$.

Regardless of the nature of the random variable $W = g(X)$, its expected value can be calculated by an integral that is analogous to the sum in Theorem 3.10 for discrete random variables.

Theorem 4.4

The expected value of a function, $g(X)$, of random variable X is

$$E[g(X)] = \int_{-\infty}^{\infty} g(x) f_X(x)\ dx.$$

Many of the properties of expected values of discrete random variables also apply to continuous random variables. Definition 3.13 and Theorems 3.11, 3.12, 3.14, and 3.15 apply to all random variables. All of these relationships are written in terms of expected values in the following theorem, where we use both notations for expected value, $E[X]$ and μ_X, to make the expressions clear and concise.

Theorem 4.5

For any random variable X,
(a) $E[X - \mu_X] = 0$,   (b) $E[aX + b] = a\,E[X] + b$,
(c) $\text{Var}[X] = E[X^2] - \mu_X^2$,   (d) $\text{Var}[aX + b] = a^2\,\text{Var}[X]$.

The method of calculating expected values depends on the type of random variable, discrete or continuous. Theorem 4.4 states that $E[X^2]$, the mean square value of X, and $\text{Var}[X]$ are the integrals

$$E[X^2] = \int_{-\infty}^{\infty} x^2 f_X(x)\ dx, \qquad \text{Var}[X] = \int_{-\infty}^{\infty} (x - \mu_X)^2 f_X(x)\ dx. \tag{4.31}$$

Our interpretation of expected values of discrete random variables carries over to continuous random variables. First, $E[X]$ represents a typical value of X, and the variance describes the dispersion of outcomes relative to the expected value. Second, $E[X]$ is a best guess for X in the sense that it minimizes the mean square error (MSE) and $Var[X]$ is the MSE associated with the guess. Furthermore, if we view the PDF $f_X(x)$ as the density of a mass distributed on a line, then $E[X]$ is the center of mass.

Example 4.8

Find the variance and standard deviation of the pointer position in Example 4.1.

To compute $Var[X]$, we use Theorem 4.5(c): $Var[X] = E[X^2] - \mu_X^2$. We calculate $E[X^2]$ directly from Theorem 4.4 with $g(X) = X^2$:

$$E\left[X^2\right] = \int_{-\infty}^{\infty} x^2 f_X(x)\,dx = \int_0^1 x^2\,dx = 1/3 \text{ m}^2. \tag{4.32}$$

In Example 4.6, we have $E[X] = 1/2$. Thus $Var[X] = 1/3 - (1/2)^2 = 1/12$, and the standard deviation is $\sigma_X = \sqrt{Var[X]} = 1/\sqrt{12} = 0.289$ meters.

Quiz 4.4

The probability density function of the random variable Y is

$$f_Y(y) = \begin{cases} 3y^2/2 & -1 \le y \le 1, \\ 0 & \text{otherwise.} \end{cases} \tag{4.33}$$

Sketch the PDF and find the following:

(a) the expected value $E[Y]$

(b) the second moment $E[Y^2]$

(c) the variance $Var[Y]$

(d) the standard deviation σ_Y

4.5   Families of Continuous Random Variables

> The families of continuous uniform random variables, exponential random variables, and Erlang random variables are related to the families of discrete uniform random variables, geometric random variables, and Pascal random variables, respectively.

Section 3.3 introduces several families of discrete random variables that arise in a wide variety of practical applications. In this section, we introduce three important families of continuous random variables: uniform, exponential, and Erlang. We devote all of Section 4.6 to Gaussian random variables. Like the families of discrete random variables, the PDFs of the members of each family all have the same mathematical form. They differ only in the values of one or two parameters. We have already encountered an example of a continuous *uniform random variable* in the wheel-spinning experiment. The general definition is

═══════Definition 4.5═══════Uniform Random Variable
X is a uniform (a, b) random variable if the PDF of X is

$$f_X(x) = \begin{cases} 1/(b-a) & a \le x < b, \\ 0 & otherwise, \end{cases}$$

where the two parameters are $b > a$.

Expressions that are synonymous with *X is a uniform random variable* are *X is uniformly distributed* and *X has a uniform distribution*.

If X is a uniform random variable there is an equal probability of finding an outcome x in any interval of length $\Delta < b - a$ within $S_X = [a, b)$. We can use Theorem 4.2(b), Theorem 4.4, and Theorem 4.5 to derive the following properties of a uniform random variable.

═══════Theorem 4.6═══════
If X is a uniform (a, b) random variable,

- *The CDF of X is*
$$F_X(x) = \begin{cases} 0 & x \le a, \\ (x-a)/(b-a) & a < x \le b, \\ 1 & x > b. \end{cases}$$

- *The expected value of X is* $\mathrm{E}[X] = (b+a)/2.$
- *The variance of X is* $\mathrm{Var}[X] = (b-a)^2/12.$

═══════Example 4.9═══════
The phase angle, Θ, of the signal at the input to a modem is uniformly distributed between 0 and 2π radians. What are the PDF, CDF, expected value, and variance of Θ?
....
From the problem statement, we identify the parameters of the uniform (a, b) random variable as $a = 0$ and $b = 2\pi$. Therefore the PDF and CDF of Θ are

$$f_\Theta(\theta) = \begin{cases} 1/(2\pi) & 0 \le \theta < 2\pi, \\ 0 & otherwise, \end{cases} \qquad F_\Theta(\theta) = \begin{cases} 0 & \theta \le 0, \\ \theta/(2\pi) & 0 < x \le 2\pi, \\ 1 & x > 2\pi. \end{cases} \qquad (4.34)$$

The expected value is $\mathrm{E}[\Theta] = b/2 = \pi$ radians, and the variance is $\mathrm{Var}[\Theta] = (2\pi)^2/12 = \pi^2/3$ rad^2.

The relationship between the family of discrete uniform random variables and the family of continuous uniform random variables is fairly direct. The following theorem expresses the relationship formally.

Theorem 4.7

Let X be a uniform (a, b) random variable, where a and b are both integers. Let $K = \lceil X \rceil$. Then K is a discrete uniform $(a + 1, b)$ random variable.

Proof Recall that for any x, $\lceil x \rceil$ is the smallest integer greater than or equal to x. It follows that the event $\{K = k\} = \{k - 1 < x \le k\}$. Therefore,

$$\mathrm{P}\left[K = k\right] = P_K(k) = \int_{k-1}^{k} P_X(x)\, dx = \begin{cases} 1/(b - a) & k = a + 1, a + 2, \ldots, b, \\ 0 & \text{otherwise.} \end{cases} \tag{4.35}$$

This expression for $P_K(k)$ conforms to Definition 3.8 of a discrete uniform $(a + 1, b)$ PMF.

The continuous relatives of the family of geometric random variables, Definition 3.5, are the members of the family of *exponential random variables*.

Definition 4.6   Exponential Random Variable

*X is an **exponential** (λ) **random variable** if the PDF of X is*

$$f_X(x) = \begin{cases} \lambda e^{-\lambda x} & x \ge 0, \\ 0 & \text{otherwise,} \end{cases}$$

where the parameter $\lambda > 0$.

Example 4.10

The duration T of a telephone call is often modeled as an exponential (λ) random variable.. If $\lambda = 1/3$, what is $\mathrm{E}[T]$, the expected duration of a telephone call? What are the variance and standard deviation of T? What is the probability that a call duration is within ± 1 standard deviation of the expected call duration?
....

From Definition 4.6, T has PDF

$$f_T(t) = \begin{cases} (1/3)e^{-t/3} & t \ge 0, \\ 0 & \text{otherwise.} \end{cases} \tag{4.36}$$

With $f_T(t)$, we use Definition 4.4 to calculate the expected duration of a call:

$$\mathrm{E}\left[T\right] = \int_{-\infty}^{\infty} t f_T(t)\, dt = \int_{0}^{\infty} t \frac{1}{3} e^{-t/3}\, dt. \tag{4.37}$$

Integration by parts (Appendix B, Math Fact B.10) yields

$$\mathrm{E}\left[T\right] = -t e^{-t/3} \Big|_{0}^{\infty} + \int_{0}^{\infty} e^{-t/3}\, dt = 3 \text{ minutes.} \tag{4.38}$$

To calculate the variance, we begin with the second moment of T:

$$\mathrm{E}\left[T^2\right] = \int_{-\infty}^{\infty} t^2 f_T(t)\, dt = \int_0^{\infty} t^2 \frac{1}{3} e^{-t/3}\, dt. \tag{4.39}$$

Again integrating by parts, we have

$$\mathrm{E}\left[T^2\right] = -t^2 e^{-t/3}\Big|_0^{\infty} + \int_0^{\infty} (2t) e^{-t/3}\, dt = 2\int_0^{\infty} t e^{-t/3}\, dt. \tag{4.40}$$

With the knowledge that $\mathrm{E}[T] = 3$, we observe that $\int_0^{\infty} t e^{-t/3}\, dt = 3\,\mathrm{E}[T] = 9$. Thus $\mathrm{E}[T^2] = 6\,\mathrm{E}[T] = 18$ and

$$\mathrm{Var}\,[T] = \mathrm{E}\left[T^2\right] - (\mathrm{E}\,[T])^2 = 18 - 3^2 = 9 \text{ minutes}^2. \tag{4.41}$$

The standard deviation is $\sigma_T = \sqrt{\mathrm{Var}[T]} = 3$ minutes. The probability that the call duration is within 1 standard deviation of the expected value is

$$\mathrm{P}\left[0 \le T \le 6\right] = F_T(6) - F_T(0) = 1 - e^{-2} = 0.865 \tag{4.42}$$

To derive general expressions for the CDF, the expected value, and the variance of an exponential random variable, we apply Theorem 4.2(b), Theorem 4.4, and Theorem 4.5 to the exponential PDF in Definition 4.6.

Theorem 4.8

If X is an exponential (λ) random variable,

- *The CDF of X is*
$$F_X(x) = \begin{cases} 1 - e^{-\lambda x} & x \ge 0, \\ 0 & otherwise. \end{cases}$$

- *The expected value of X is* $\mathrm{E}\,[X] = 1/\lambda.$

- *The variance of X is* $\mathrm{Var}\,[X] = 1/\lambda^2.$

The following theorem shows the relationship between the family of exponential random variables and the family of geometric random variables.

Theorem 4.9

If X is an exponential (λ) random variable, then $K = \lceil X \rceil$ is a geometric (p) random variable with $p = 1 - e^{-\lambda}$.

Proof As in the Theorem 4.7 proof, the definition of K implies $P_K(k) = \mathrm{P}[k - 1 < X \le k]$. Referring to the CDF of X in Theorem 4.8, we observe

$$\begin{aligned} P_K(k) &= F_x(k) - F_x(k - 1) \\ &= \begin{cases} e^{-\lambda(k-1)} - e^{-\lambda k} & k = 1, 2, \ldots \\ 0 & \text{otherwise}, \end{cases} = \begin{cases} (e^{-\lambda})^{k-1}(1 - e^{-\lambda}) & k = 1, 2, \ldots \\ 0 & \text{otherwise}. \end{cases} \end{aligned} \tag{4.43}$$

If we let $p = 1 - e^{-\lambda}$, we have

$$P_K(k) = \begin{cases} p(1-p)^{k-1} & k = 1, 2, \ldots \\ 0 & \text{otherwise,} \end{cases} \tag{4.44}$$

which conforms to Definition 3.5 of a geometric (p) random variable with $p = 1 - e^{-\lambda}$.

Example 4.11

Phone company A charges \$0.15 per minute for telephone calls. For any fraction of a minute at the end of a call, they charge for a full minute. Phone Company B also charges \$0.15 per minute. However, Phone Company B calculates its charge based on the exact duration of a call. If T, the duration of a call in minutes, is an exponential $(\lambda = 1/3)$ random variable, what are the expected revenues per call $E[R_A]$ and $E[R_B]$ for companies A and B?

....

Because T is an exponential random variable, we have in Theorem 4.8 (and in Example 4.10) $E[T] = 1/\lambda = 3$ minutes per call. Therefore, for phone company B, which charges for the exact duration of a call,

$$E[R_B] = 0.15\, E[T] = \$0.45 \text{ per call.} \tag{4.45}$$

Company A, by contrast, collects \$0.15$\lceil T \rceil$ for a call of duration T minutes. Theorem 4.9 states that $K = \lceil T \rceil$ is a geometric random variable with parameter $p = 1 - e^{-1/3}$. Therefore, the expected revenue for Company A is

$$E[R_A] = 0.15\, E[K] = 0.15/p = (0.15)(3.53) = \$0.529 \text{ per call.} \tag{4.46}$$

In Theorem 9.9, we show that the sum of a set of independent identically distributed exponential random variables is an *Erlang* random variable.

Definition 4.7   **Erlang Random Variable**

X *is an* **Erlang** (n, λ) *random variable if the PDF of X is*

$$f_X(x) = \begin{cases} \dfrac{\lambda^n x^{n-1} e^{-\lambda x}}{(n-1)!} & x \geq 0, \\ 0 & \textit{otherwise,} \end{cases}$$

where the parameter $\lambda > 0$, and the parameter $n \geq 1$ is an integer.

The parameter n is often called the *order* of an Erlang random variable. Problem 4.5.16 outlines a procedure to verify that the integral of the Erlang PDF over all x is 1. The Erlang $(n = 1, \lambda)$ random variable is identical to the exponential (λ) random variable. Just as the exponential (λ) random variable is related to the geometric $(1 - e^{-\lambda})$ random variable, the Erlang (n, λ) continuous random variable is related to the Pascal $(n, 1 - e^{-\lambda})$ discrete random variable.

════════**Theorem 4.10**════════

If X is an Erlang (n, λ) random variable, then

(a) $\mathrm{E}[X] = \dfrac{n}{\lambda}$,   (b) $\mathrm{Var}[X] = \dfrac{n}{\lambda^2}$.

By comparing Theorem 4.8 and Theorem 4.10, we see for X, an Erlang (n, λ) random variable, and Y, an exponential (λ) random variable, that $\mathrm{E}[X] = n\,\mathrm{E}[Y]$ and $\mathrm{Var}[X] = n\,\mathrm{Var}[Y]$. In the following theorem, we can also connect Erlang and Poisson random variables.

════════**Theorem 4.11**════════

Let K_α denote a Poisson (α) random variable. For any $x > 0$, the CDF of an Erlang (n, λ) random variable X satisfies

$$F_X(x) = 1 - F_{K_{\lambda x}}(n - 1) = \begin{cases} 1 - \sum_{k=0}^{n-1} \frac{(\lambda x)^k e^{-\lambda x}}{k!}, & x \geq 0, \\ 0 & \text{otherwise.} \end{cases}$$

Problem 4.5.18 outlines a proof of Theorem 4.11. Theorem 4.11 states that the probability that the Erlang (n, λ) random variable is $\leq x$ is the probability that the Poisson (λx) random variable is $\geq n$ because the sum in Theorem 4.11 is the CDF of the Poisson (λx) random variable evaluated at $n - 1$.

The mathematical relationships between the geometric, Pascal, exponential, Erlang, and Poisson random variables derive from the widely-used *Poisson process* model for arrivals of customers to a service facility. Formal definitions and theorems for the Poisson process appear in Section 13.4. The arriving customers can be, for example, shoppers at the Phonesmart store, packets at an Internet router, or requests to a Web server. In this model, the number of customers that arrive in a τ-minute time period is a Poisson $(\lambda \tau)$ random variable. Under continuous monitoring, the time that we wait for one arrival is an exponential (λ) random variable and the time we wait for n arrivals is an Erlang (n, λ) random variable. On the other hand, when we monitor arrivals in discrete one-minute intervals, the number of intervals we wait until we observe a nonempty interval (with one or more arrivals) is a geometric $(p = 1 - e^{-\lambda})$ random variable and the number of intervals we wait for n nonempty intervals is a Pascal (n, p) random variable.

════════**Quiz 4.5**════════

Continuous random variable X has $\mathrm{E}[X] = 3$ and $\mathrm{Var}[X] = 9$. Find the PDF, $f_X(x)$, if

(a) X is an exponential random variable,

(b) X is a continuous uniform random variable.

(c) X is an Erlang random variable.

4.6   Gaussian Random Variables

> The family of Gaussian random variables appears in more practical
> applications than any other family. The graph of a Gaussian PDF
> is a bell-shaped curve.

Bell-shaped curves appear in many applications of probability theory. The probability models in these applications are members of the family of *Gaussian random
variables*. Chapter 9 contains a mathematical explanation for the prevalence of
Gaussian random variables in models of practical phenomena. Because they occur
so frequently in practice, Gaussian random variables are sometimes referred to as
normal random variables.

=====Definition 4.8=====Gaussian Random Variable

X is a Gaussian (μ, σ) random variable if the PDF of X is

$$f_X(x) = \frac{1}{\sqrt{2\pi\sigma^2}} e^{-(x-\mu)^2/2\sigma^2},$$

where the parameter μ can be any real number and the parameter $\sigma > 0$.

Many statistics texts use the notation X *is* $N[\mu, \sigma^2]$ as shorthand for *X is a Gaussian (μ, σ) random variable*. In this notation, the N denotes *normal*. The graph
of $f_X(x)$ has a bell shape, where the center of the bell is $x = \mu$ and σ reflects the
width of the bell. If σ is small, the bell is narrow, with a high, pointy peak. If σ is
large, the bell is wide, with a low, flat peak. (The height of the peak is $1/\sigma\sqrt{2\pi}$.)
Figure 4.5 contains two examples of Gaussian PDFs with $\mu = 2$. In Figure 4.5(a),
$\sigma = 0.5$, and in Figure 4.5(b), $\sigma = 2$. Of course, the area under any Gaussian PDF
is $\int_{-\infty}^{\infty} f_X(x)\,dx = 1$. Furthermore, the parameters of the PDF are the expected
value of X and the standard deviation of X.

=====Theorem 4.12=====

If X is a Gaussian (μ, σ) random variable,

$$\mathrm{E}\,[X] = \mu \qquad \mathrm{Var}\,[X] = \sigma^2.$$

The proof of Theorem 4.12, as well as the proof that the area under a Gaussian
PDF is 1, employs integration by parts and other calculus techniques. We leave
them as an exercise for the reader in Problem 4.6.13.

It is impossible to express the integral of a Gaussian PDF between noninfinite
limits as a function that appears on most scientific calculators. Instead, we usually
find integrals of the Gaussian PDF by referring to tables, such as Table 4.1 (p. 129),
that have been obtained by numerical integration. To learn how to use this table,
we introduce the following important property of Gaussian random variables.

=====Theorem 4.13=====

If X is Gaussian (μ, σ), $Y = aX + b$ is Gaussian $(a\mu + b, a\sigma)$.

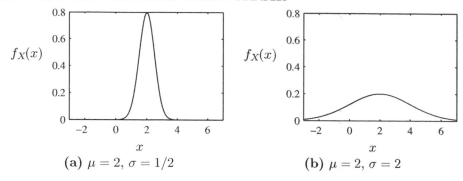

Figure 4.5   Two examples of a Gaussian random variable X with expected value μ and standard deviation σ.

The theorem states that any linear transformation of a Gaussian random variable produces another Gaussian random variable. This theorem allows us to relate the properties of an arbitrary Gaussian random variable to the properties of a specific random variable.

═══════Definition 4.9═══════Standard Normal Random Variable

*The **standard normal random variable** Z is the Gaussian $(0,1)$ random variable.*

Theorem 4.12 indicates that $E[Z] = 0$ and $\text{Var}[Z] = 1$. The tables that we use to find integrals of Gaussian PDFs contain values of $F_Z(z)$, the CDF of Z. We introduce the special notation $\Phi(z)$ for this function.

═══════Definition 4.10═══════Standard Normal CDF

The CDF of the standard normal random variable Z is

$$\Phi(z) = \frac{1}{\sqrt{2\pi}} \int_{-\infty}^{z} e^{-u^2/2}\, du.$$

Given a table of values of $\Phi(z)$, we use the following theorem to find probabilities of a Gaussian random variable with parameters μ and σ.

═══════Theorem 4.14═══════

If X is a Gaussian (μ, σ) random variable, the CDF of X is

$$F_X(x) = \Phi\left(\frac{x - \mu}{\sigma}\right).$$

The probability that X is in the interval $(a, b]$ is

$$P\left[a < X \le b\right] = \Phi\left(\frac{b - \mu}{\sigma}\right) - \Phi\left(\frac{a - \mu}{\sigma}\right).$$

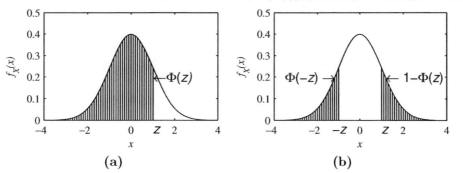

Figure 4.6 Symmetry properties of the Gaussian $(0, 1)$ PDF.

In using this theorem, we transform values of a Gaussian random variable, X, to equivalent values of the standard normal random variable, Z. For a sample value x of the random variable X, the corresponding sample value of Z is

$$z = \frac{x - \mu}{\sigma} \qquad (4.47)$$

Note that z is dimensionless. It represents x as a number of standard deviations relative to the expected value of X. Table 4.1 presents $\Phi(z)$ for $0 \leq z \leq 2.99$. People working with probability and statistics spend a lot of time referring to tables like Table 4.1. It seems strange to us that $\Phi(z)$ isn't included in every scientific calculator. For many people, it is far more useful than many of the functions included in ordinary scientific calculators.

Example 4.12

Suppose your score on a test is $x = 46$, a sample value of the Gaussian $(61, 10)$ random variable. Express your test score as a sample value of the standard normal random variable, Z.

. . . .

Equation (4.47) indicates that $z = (46 - 61)/10 = -1.5$. Therefore your score is 1.5 *standard deviations less than the expected value*.

To find probabilities of Gaussian random variables, we use the values of $\Phi(z)$ presented in Table 4.1. Note that this table contains entries only for $z \geq 0$. For negative values of z, we apply the following property of $\Phi(z)$.

Theorem 4.15

$$\Phi(-z) = 1 - \Phi(z).$$

Figure 4.6 displays the symmetry properties of $\Phi(z)$. Both graphs contain the standard normal PDF. In Figure 4.6(a), the shaded area under the PDF is $\Phi(z)$. Since the area under the PDF equals 1, the unshaded area under the PDF is $1-\Phi(z)$. In Figure 4.6(b), the shaded area on the right is $1 - \Phi(z)$ and the shaded area on the left is $\Phi(-z)$. This graph demonstrates that $\Phi(-z) = 1 - \Phi(z)$.

Example 4.13

If X is a Gaussian ($\mu = 61, \sigma = 10$) random variable, what is P$[51 < X \leq 71]$?

Applying Equation (4.47), $Z = (X - 61)/10$ and

$$\{51 < X \leq 71\} = \left\{-1 \leq \frac{X - 61}{10} \leq 1\right\} = \{-1 < Z \leq 1\}. \tag{4.48}$$

The probability of this event is

$$
\begin{aligned}
\mathrm{P}\left[-1 < Z \leq 1\right] &= \Phi(1) - \Phi(-1) \\
&= \Phi(1) - [1 - \Phi(1)] = 2\Phi(1) - 1 = 0.683. \tag{4.49}
\end{aligned}
$$

The solution to Example 4.13 reflects the fact that in an experiment with a Gaussian probability model, 68.3% of the outcomes are within ± 1 standard deviation of the expected value. Similarly, about 95% ($2\Phi(2) - 1$) of the outcomes are within two standard deviations of the expected value.

Tables of $\Phi(z)$ are useful for obtaining numerical values of integrals of a Gaussian PDF over intervals near the expected value. Regions farther than three standard deviations from the expected value (corresponding to $|z| \geq 3$) are in the *tails* of the PDF. When $|z| > 3$, $\Phi(z)$ is very close to one; for example, $\Phi(3) = 0.9987$ and $\Phi(4) = 0.9999768$. The properties of $\Phi(z)$ for extreme values of z are apparent in the *standard normal complementary CDF*.

Definition 4.11   Standard Normal Complementary CDF

*The **standard normal complementary CDF** is*

$$Q(z) = \mathrm{P}\left[Z > z\right] = \frac{1}{\sqrt{2\pi}} \int_z^\infty e^{-u^2/2} \, du = 1 - \Phi(z).$$

Although we may regard both $\Phi(3) = 0.9987$ and $\Phi(4) = 0.9999768$ as being very close to one, we see in Table 4.2 that $Q(3) = 1.35 \cdot 10^{-3}$ is almost two orders of magnitude larger than $Q(4) = 3.17 \cdot 10^{-5}$.

Example 4.14

In an optical fiber transmission system, the probability of a bit error is $Q(\sqrt{\gamma/2})$, where γ is the signal-to-noise ratio. What is the minimum value of γ that produces a bit error rate not exceeding 10^{-6}?

Referring to Table 4.1, we find that $Q(z) < 10^{-6}$ when $z \geq 4.75$. Therefore, if

z	$\Phi(z)$	z	$\Phi(z)$	z	$\Phi(z)$	z	$\Phi(z)$	z	$\Phi(z)$	z	$\Phi(z)$
0.00	0.5000	0.50	0.6915	1.00	0.8413	1.50	0.9332	2.00	0.97725	2.50	0.99379
0.01	0.5040	0.51	0.6950	1.01	0.8438	1.51	0.9345	2.01	0.97778	2.51	0.99396
0.02	0.5080	0.52	0.6985	1.02	0.8461	1.52	0.9357	2.02	0.97831	2.52	0.99413
0.03	0.5120	0.53	0.7019	1.03	0.8485	1.53	0.9370	2.03	0.97882	2.53	0.99430
0.04	0.5160	0.54	0.7054	1.04	0.8508	1.54	0.9382	2.04	0.97932	2.54	0.99446
0.05	0.5199	0.55	0.7088	1.05	0.8531	1.55	0.9394	2.05	0.97982	2.55	0.99461
0.06	0.5239	0.56	0.7123	1.06	0.8554	1.56	0.9406	2.06	0.98030	2.56	0.99477
0.07	0.5279	0.57	0.7157	1.07	0.8577	1.57	0.9418	2.07	0.98077	2.57	0.99492
0.08	0.5319	0.58	0.7190	1.08	0.8599	1.58	0.9429	2.08	0.98124	2.58	0.99506
0.09	0.5359	0.59	0.7224	1.09	0.8621	1.59	0.9441	2.09	0.98169	2.59	0.99520
0.10	0.5398	0.60	0.7257	1.10	0.8643	1.60	0.9452	2.10	0.98214	2.60	0.99534
0.11	0.5438	0.61	0.7291	1.11	0.8665	1.61	0.9463	2.11	0.98257	2.61	0.99547
0.12	0.5478	0.62	0.7324	1.12	0.8686	1.62	0.9474	2.12	0.98300	2.62	0.99560
0.13	0.5517	0.63	0.7357	1.13	0.8708	1.63	0.9484	2.13	0.98341	2.63	0.99573
0.14	0.5557	0.64	0.7389	1.14	0.8729	1.64	0.9495	2.14	0.98382	2.64	0.99585
0.15	0.5596	0.65	0.7422	1.15	0.8749	1.65	0.9505	2.15	0.98422	2.65	0.99598
0.16	0.5636	0.66	0.7454	1.16	0.8770	1.66	0.9515	2.16	0.98461	2.66	0.99609
0.17	0.5675	0.67	0.7486	1.17	0.8790	1.67	0.9525	2.17	0.98500	2.67	0.99621
0.18	0.5714	0.68	0.7517	1.18	0.8810	1.68	0.9535	2.18	0.98537	2.68	0.99632
0.19	0.5753	0.69	0.7549	1.19	0.8830	1.69	0.9545	2.19	0.98574	2.69	0.99643
0.20	0.5793	0.70	0.7580	1.20	0.8849	1.70	0.9554	2.20	0.98610	2.70	0.99653
0.21	0.5832	0.71	0.7611	1.21	0.8869	1.71	0.9564	2.21	0.98645	2.71	0.99664
0.22	0.5871	0.72	0.7642	1.22	0.8888	1.72	0.9573	2.22	0.98679	2.72	0.99674
0.23	0.5910	0.73	0.7673	1.23	0.8907	1.73	0.9582	2.23	0.98713	2.73	0.99683
0.24	0.5948	0.74	0.7704	1.24	0.8925	1.74	0.9591	2.24	0.98745	2.74	0.99693
0.25	0.5987	0.75	0.7734	1.25	0.8944	1.75	0.9599	2.25	0.98778	2.75	0.99702
0.26	0.6026	0.76	0.7764	1.26	0.8962	1.76	0.9608	2.26	0.98809	2.76	0.99711
0.27	0.6064	0.77	0.7794	1.27	0.8980	1.77	0.9616	2.27	0.98840	2.77	0.99720
0.28	0.6103	0.78	0.7823	1.28	0.8997	1.78	0.9625	2.28	0.98870	2.78	0.99728
0.29	0.6141	0.79	0.7852	1.29	0.9015	1.79	0.9633	2.29	0.98899	2.79	0.99736
0.30	0.6179	0.80	0.7881	1.30	0.9032	1.80	0.9641	2.30	0.98928	2.80	0.99744
0.31	0.6217	0.81	0.7910	1.31	0.9049	1.81	0.9649	2.31	0.98956	2.81	0.99752
0.32	0.6255	0.82	0.7939	1.32	0.9066	1.82	0.9656	2.32	0.98983	2.82	0.99760
0.33	0.6293	0.83	0.7967	1.33	0.9082	1.83	0.9664	2.33	0.99010	2.83	0.99767
0.34	0.6331	0.84	0.7995	1.34	0.9099	1.84	0.9671	2.34	0.99036	2.84	0.99774
0.35	0.6368	0.85	0.8023	1.35	0.9115	1.85	0.9678	2.35	0.99061	2.85	0.99781
0.36	0.6406	0.86	0.8051	1.36	0.9131	1.86	0.9686	2.36	0.99086	2.86	0.99788
0.37	0.6443	0.87	0.8078	1.37	0.9147	1.87	0.9693	2.37	0.99111	2.87	0.99795
0.38	0.6480	0.88	0.8106	1.38	0.9162	1.88	0.9699	2.38	0.99134	2.88	0.99801
0.39	0.6517	0.89	0.8133	1.39	0.9177	1.89	0.9706	2.39	0.99158	2.89	0.99807
0.40	0.6554	0.90	0.8159	1.40	0.9192	1.90	0.9713	2.40	0.99180	2.90	0.99813
0.41	0.6591	0.91	0.8186	1.41	0.9207	1.91	0.9719	2.41	0.99202	2.91	0.99819
0.42	0.6628	0.92	0.8212	1.42	0.9222	1.92	0.9726	2.42	0.99224	2.92	0.99825
0.43	0.6664	0.93	0.8238	1.43	0.9236	1.93	0.9732	2.43	0.99245	2.93	0.99831
0.44	0.6700	0.94	0.8264	1.44	0.9251	1.94	0.9738	2.44	0.99266	2.94	0.99836
0.45	0.6736	0.95	0.8289	1.45	0.9265	1.95	0.9744	2.45	0.99286	2.95	0.99841
0.46	0.6772	0.96	0.8315	1.46	0.9279	1.96	0.9750	2.46	0.99305	2.96	0.99846
0.47	0.6808	0.97	0.8340	1.47	0.9292	1.97	0.9756	2.47	0.99324	2.97	0.99851
0.48	0.6844	0.98	0.8365	1.48	0.9306	1.98	0.9761	2.48	0.99343	2.98	0.99856
0.49	0.6879	0.99	0.8389	1.49	0.9319	1.99	0.9767	2.49	0.99361	2.99	0.99861

Table 4.1   The standard normal CDF $\Phi(y)$.

z	$Q(z)$	z	$Q(z)$	z	$Q(z)$	z	$Q(z)$	z	$Q(z)$
3.00	$1.35 \cdot 10^{-3}$	3.40	$3.37 \cdot 10^{-4}$	3.80	$7.23 \cdot 10^{-5}$	4.20	$1.33 \cdot 10^{-5}$	4.60	$2.11 \cdot 10^{-6}$
3.01	$1.31 \cdot 10^{-3}$	3.41	$3.25 \cdot 10^{-4}$	3.81	$6.95 \cdot 10^{-5}$	4.21	$1.28 \cdot 10^{-5}$	4.61	$2.01 \cdot 10^{-6}$
3.02	$1.26 \cdot 10^{-3}$	3.42	$3.13 \cdot 10^{-4}$	3.82	$6.67 \cdot 10^{-5}$	4.22	$1.22 \cdot 10^{-5}$	4.62	$1.92 \cdot 10^{-6}$
3.03	$1.22 \cdot 10^{-3}$	3.43	$3.02 \cdot 10^{-4}$	3.83	$6.41 \cdot 10^{-5}$	4.23	$1.17 \cdot 10^{-5}$	4.63	$1.83 \cdot 10^{-6}$
3.04	$1.18 \cdot 10^{-3}$	3.44	$2.91 \cdot 10^{-4}$	3.84	$6.15 \cdot 10^{-5}$	4.24	$1.12 \cdot 10^{-5}$	4.64	$1.74 \cdot 10^{-6}$
3.05	$1.14 \cdot 10^{-3}$	3.45	$2.80 \cdot 10^{-4}$	3.85	$5.91 \cdot 10^{-5}$	4.25	$1.07 \cdot 10^{-5}$	4.65	$1.66 \cdot 10^{-6}$
3.06	$1.11 \cdot 10^{-3}$	3.46	$2.70 \cdot 10^{-4}$	3.86	$5.67 \cdot 10^{-5}$	4.26	$1.02 \cdot 10^{-5}$	4.66	$1.58 \cdot 10^{-6}$
3.07	$1.07 \cdot 10^{-3}$	3.47	$2.60 \cdot 10^{-4}$	3.87	$5.44 \cdot 10^{-5}$	4.27	$9.77 \cdot 10^{-6}$	4.67	$1.51 \cdot 10^{-6}$
3.08	$1.04 \cdot 10^{-3}$	3.48	$2.51 \cdot 10^{-4}$	3.88	$5.22 \cdot 10^{-5}$	4.28	$9.34 \cdot 10^{-6}$	4.68	$1.43 \cdot 10^{-6}$
3.09	$1.00 \cdot 10^{-3}$	3.49	$2.42 \cdot 10^{-4}$	3.89	$5.01 \cdot 10^{-5}$	4.29	$8.93 \cdot 10^{-6}$	4.69	$1.37 \cdot 10^{-6}$
3.10	$9.68 \cdot 10^{-4}$	3.50	$2.33 \cdot 10^{-4}$	3.90	$4.81 \cdot 10^{-5}$	4.30	$8.54 \cdot 10^{-6}$	4.70	$1.30 \cdot 10^{-6}$
3.11	$9.35 \cdot 10^{-4}$	3.51	$2.24 \cdot 10^{-4}$	3.91	$4.61 \cdot 10^{-5}$	4.31	$8.16 \cdot 10^{-6}$	4.71	$1.24 \cdot 10^{-6}$
3.12	$9.04 \cdot 10^{-4}$	3.52	$2.16 \cdot 10^{-4}$	3.92	$4.43 \cdot 10^{-5}$	4.32	$7.80 \cdot 10^{-6}$	4.72	$1.18 \cdot 10^{-6}$
3.13	$8.74 \cdot 10^{-4}$	3.53	$2.08 \cdot 10^{-4}$	3.93	$4.25 \cdot 10^{-5}$	4.33	$7.46 \cdot 10^{-6}$	4.73	$1.12 \cdot 10^{-6}$
3.14	$8.45 \cdot 10^{-4}$	3.54	$2.00 \cdot 10^{-4}$	3.94	$4.07 \cdot 10^{-5}$	4.34	$7.12 \cdot 10^{-6}$	4.74	$1.07 \cdot 10^{-6}$
3.15	$8.16 \cdot 10^{-4}$	3.55	$1.93 \cdot 10^{-4}$	3.95	$3.91 \cdot 10^{-5}$	4.35	$6.81 \cdot 10^{-6}$	4.75	$1.02 \cdot 10^{-6}$
3.16	$7.89 \cdot 10^{-4}$	3.56	$1.85 \cdot 10^{-4}$	3.96	$3.75 \cdot 10^{-5}$	4.36	$6.50 \cdot 10^{-6}$	4.76	$9.68 \cdot 10^{-7}$
3.17	$7.62 \cdot 10^{-4}$	3.57	$1.78 \cdot 10^{-4}$	3.97	$3.59 \cdot 10^{-5}$	4.37	$6.21 \cdot 10^{-6}$	4.77	$9.21 \cdot 10^{-7}$
3.18	$7.36 \cdot 10^{-4}$	3.58	$1.72 \cdot 10^{-4}$	3.98	$3.45 \cdot 10^{-5}$	4.38	$5.93 \cdot 10^{-6}$	4.78	$8.76 \cdot 10^{-7}$
3.19	$7.11 \cdot 10^{-4}$	3.59	$1.65 \cdot 10^{-4}$	3.99	$3.30 \cdot 10^{-5}$	4.39	$5.67 \cdot 10^{-6}$	4.79	$8.34 \cdot 10^{-7}$
3.20	$6.87 \cdot 10^{-4}$	3.60	$1.59 \cdot 10^{-4}$	4.00	$3.17 \cdot 10^{-5}$	4.40	$5.41 \cdot 10^{-6}$	4.80	$7.93 \cdot 10^{-7}$
3.21	$6.64 \cdot 10^{-4}$	3.61	$1.53 \cdot 10^{-4}$	4.01	$3.04 \cdot 10^{-5}$	4.41	$5.17 \cdot 10^{-6}$	4.81	$7.55 \cdot 10^{-7}$
3.22	$6.41 \cdot 10^{-4}$	3.62	$1.47 \cdot 10^{-4}$	4.02	$2.91 \cdot 10^{-5}$	4.42	$4.94 \cdot 10^{-6}$	4.82	$7.18 \cdot 10^{-7}$
3.23	$6.19 \cdot 10^{-4}$	3.63	$1.42 \cdot 10^{-4}$	4.03	$2.79 \cdot 10^{-5}$	4.43	$4.71 \cdot 10^{-6}$	4.83	$6.83 \cdot 10^{-7}$
3.24	$5.98 \cdot 10^{-4}$	3.64	$1.36 \cdot 10^{-4}$	4.04	$2.67 \cdot 10^{-5}$	4.44	$4.50 \cdot 10^{-6}$	4.84	$6.49 \cdot 10^{-7}$
3.25	$5.77 \cdot 10^{-4}$	3.65	$1.31 \cdot 10^{-4}$	4.05	$2.56 \cdot 10^{-5}$	4.45	$4.29 \cdot 10^{-6}$	4.85	$6.17 \cdot 10^{-7}$
3.26	$5.57 \cdot 10^{-4}$	3.66	$1.26 \cdot 10^{-4}$	4.06	$2.45 \cdot 10^{-5}$	4.46	$4.10 \cdot 10^{-6}$	4.86	$5.87 \cdot 10^{-7}$
3.27	$5.38 \cdot 10^{-4}$	3.67	$1.21 \cdot 10^{-4}$	4.07	$2.35 \cdot 10^{-5}$	4.47	$3.91 \cdot 10^{-6}$	4.87	$5.58 \cdot 10^{-7}$
3.28	$5.19 \cdot 10^{-4}$	3.68	$1.17 \cdot 10^{-4}$	4.08	$2.25 \cdot 10^{-5}$	4.48	$3.73 \cdot 10^{-6}$	4.88	$5.30 \cdot 10^{-7}$
3.29	$5.01 \cdot 10^{-4}$	3.69	$1.12 \cdot 10^{-4}$	4.09	$2.16 \cdot 10^{-5}$	4.49	$3.56 \cdot 10^{-6}$	4.89	$5.04 \cdot 10^{-7}$
3.30	$4.83 \cdot 10^{-4}$	3.70	$1.08 \cdot 10^{-4}$	4.10	$2.07 \cdot 10^{-5}$	4.50	$3.40 \cdot 10^{-6}$	4.90	$4.79 \cdot 10^{-7}$
3.31	$4.66 \cdot 10^{-4}$	3.71	$1.04 \cdot 10^{-4}$	4.11	$1.98 \cdot 10^{-5}$	4.51	$3.24 \cdot 10^{-6}$	4.91	$4.55 \cdot 10^{-7}$
3.32	$4.50 \cdot 10^{-4}$	3.72	$9.96 \cdot 10^{-5}$	4.12	$1.89 \cdot 10^{-5}$	4.52	$3.09 \cdot 10^{-6}$	4.92	$4.33 \cdot 10^{-7}$
3.33	$4.34 \cdot 10^{-4}$	3.73	$9.57 \cdot 10^{-5}$	4.13	$1.81 \cdot 10^{-5}$	4.53	$2.95 \cdot 10^{-6}$	4.93	$4.11 \cdot 10^{-7}$
3.34	$4.19 \cdot 10^{-4}$	3.74	$9.20 \cdot 10^{-5}$	4.14	$1.74 \cdot 10^{-5}$	4.54	$2.81 \cdot 10^{-6}$	4.94	$3.91 \cdot 10^{-7}$
3.35	$4.04 \cdot 10^{-4}$	3.75	$8.84 \cdot 10^{-5}$	4.15	$1.66 \cdot 10^{-5}$	4.55	$2.68 \cdot 10^{-6}$	4.95	$3.71 \cdot 10^{-7}$
3.36	$3.90 \cdot 10^{-4}$	3.76	$8.50 \cdot 10^{-5}$	4.16	$1.59 \cdot 10^{-5}$	4.56	$2.56 \cdot 10^{-6}$	4.96	$3.52 \cdot 10^{-7}$
3.37	$3.76 \cdot 10^{-4}$	3.77	$8.16 \cdot 10^{-5}$	4.17	$1.52 \cdot 10^{-5}$	4.57	$2.44 \cdot 10^{-6}$	4.97	$3.35 \cdot 10^{-7}$
3.38	$3.62 \cdot 10^{-4}$	3.78	$7.84 \cdot 10^{-5}$	4.18	$1.46 \cdot 10^{-5}$	4.58	$2.32 \cdot 10^{-6}$	4.98	$3.18 \cdot 10^{-7}$
3.39	$3.49 \cdot 10^{-4}$	3.79	$7.53 \cdot 10^{-5}$	4.19	$1.39 \cdot 10^{-5}$	4.59	$2.22 \cdot 10^{-6}$	4.99	$3.02 \cdot 10^{-7}$

Table 4.2 The standard normal complementary CDF $Q(z)$.

$\sqrt{\gamma/2} \geq 4.75$, or $\gamma \geq 45$, the probability of error is less than 10^{-6}. Although $10^{(-6)}$ seems a very small number, most practical optical fiber transmission systems have considerably lower binary error rates.

Keep in mind that $Q(z)$ is the probability that a Gaussian random variable exceeds its expected value by more than z standard deviations. We can observe from Table 4.2, $Q(3) = 0.0013$. This means that the probability that a Gaussian random variable is more than three standard deviations above its expected value is approximately one in a thousand. In conversation we refer to the event $\{X - \mu_X > 3\sigma_X\}$ as a *three-sigma event*. It is unlikely to occur. Table 4.2 indicates that the probability of a 5σ event is on the order of 10^{-7}.

Quiz 4.6

X is the Gaussian $(0, 1)$ random variable and Y is the Gaussian $(0, 2)$ random variable. Sketch the PDFs $f_X(x)$ and $f_Y(y)$ on the same axes and find:

(a) $P[-1 < X \leq 1]$, (b) $P[-1 < Y \leq 1]$,

(c) $P[X > 3.5]$, (d) $P[Y > 3.5]$.

4.7 Delta Functions, Mixed Random Variables

> X is a *mixed* random variable if S_X has at least one sample value with nonzero probability (like a discrete random variable) and also has sample values that cover an interval (like a continuous random variable). The PDF of a mixed random variable contains finite nonzero values and delta functions multiplied by probabilities.

Thus far, our analysis of continuous random variables parallels our analysis of discrete random variables in Chapter 2. Because of the different nature of discrete and continuous random variables, we represent the probability model of a discrete random variable as a PMF and we represent the probability model of a continuous random variable as a PDF. These functions are important because they enable us to calculate probabilities of events and parameters of probability models (such as the expected value and the variance). Calculations containing a PMF involve sums. The corresponding calculations for a PDF contain integrals.

In this section, we introduce the unit impulse function $\delta(x)$ as a mathematical tool that unites the analyses of discrete and continuous random variables. The unit impulse, often called the *delta function*, allows us to use the same formulas to describe calculations with both types of random variables. It does not alter the calculations, it just provides a new notation for describing them. This is especially convenient when we refer to a *mixed random variable*, which has properties of both continuous and discrete random variables.

The delta function is not completely respectable mathematically because it is zero everywhere except at one point, and there it is infinite. Thus at its most interesting point it has no numerical value at all. While $\delta(x)$ is somewhat disreputable,

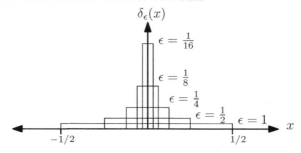

Figure 4.7 As $\epsilon \to 0$, $d_\epsilon(x)$ approaches the delta function $\delta(x)$. For each ϵ, the area under the curve of $d_\epsilon(x)$ equals 1.

it is extremely useful. There are various definitions of the delta function. All of them share the key property presented in Theorem 4.16. Here is the definition adopted in this book.

═══════**Definition 4.12**═══════**Unit Impulse (Delta) Function**

Let

$$d_\epsilon(x) = \begin{cases} 1/\epsilon & -\epsilon/2 \leq x \leq \epsilon/2, \\ 0 & otherwise. \end{cases}$$

*The **unit impulse function** is*

$$\delta(x) = \lim_{\epsilon \to 0} d_\epsilon(x).$$

The mathematical problem with Definition 4.12 is that $d_\epsilon(x)$ has no limit at $x = 0$. As indicated in Figure 4.7, $d_\epsilon(0)$ just gets bigger and bigger as $\epsilon \to 0$. Although this makes Definition 4.12 somewhat unsatisfactory, the useful properties of the delta function are readily demonstrated when $\delta(x)$ is approximated by $d_\epsilon(x)$ for very small ϵ. We now present some properties of the delta function. We state these properties as theorems even though they are not theorems in the usual sense of this text because we cannot prove them. Instead of theorem proofs, we refer to $d_\epsilon(x)$ for small values of ϵ to indicate why the properties hold.

Although $d_\epsilon(0)$ blows up as $\epsilon \to 0$, the area under $d_\epsilon(x)$ is the integral

$$\int_{-\infty}^{\infty} d_\epsilon(x)\, dx = \int_{-\epsilon/2}^{\epsilon/2} \frac{1}{\epsilon}\, dx = 1. \tag{4.50}$$

That is, the area under $d_\epsilon(x)$ is always 1, no matter how small the value of ϵ. We conclude that the area under $\delta(x)$ is also 1:

$$\int_{-\infty}^{\infty} \delta(x)\, dx = 1. \tag{4.51}$$

This result is a special case of the following property of the delta function.

Theorem 4.16

For any continuous function $g(x)$,

$$\int_{-\infty}^{\infty} g(x)\delta(x - x_0)\, dx = g(x_0).$$

Theorem 4.16 is often called the *sifting property* of the delta function. We can see that Equation (4.51) is a special case of the sifting property for $g(x) = 1$ and $x_0 = 0$. To understand Theorem 4.16, consider the integral

$$\int_{-\infty}^{\infty} g(x)d_\epsilon(x - x_0)\, dx = \frac{1}{\epsilon}\int_{x_0-\epsilon/2}^{x_0+\epsilon/2} g(x)\, dx. \tag{4.52}$$

On the right side, we have the average value of $g(x)$ over the interval $[x_0 - \epsilon/2, x_0 + \epsilon/2]$. As $\epsilon \to 0$, this average value must converge to $g(x_0)$.

The delta function has a close connection to the unit step function.

Definition 4.13   Unit Step Function

*The **unit step function** is*

$$u(x) = \begin{cases} 0 & x < 0, \\ 1 & x \geq 0. \end{cases}$$

Theorem 4.17

$$\int_{-\infty}^{x} \delta(v)\, dv = u(x).$$

To understand Theorem 4.17, we observe that for any $x > 0$, we can choose $\epsilon \leq 2x$ so that

$$\int_{-\infty}^{-x} d_\epsilon(v)\, dv = 0, \qquad \int_{-\infty}^{x} d_\epsilon(v)\, dv = 1. \tag{4.53}$$

Thus for any $x \neq 0$, in the limit as $\epsilon \to 0$, $\int_{-\infty}^{x} d_\epsilon(v)\, dv = u(x)$. Note that we have not yet considered $x = 0$. In fact, it is not completely clear what the value of $\int_{-\infty}^{0} \delta(v)\, dv$ should be. Reasonable arguments can be made for 0, 1/2, or 1. We have adopted the convention that $\int_{-\infty}^{0} \delta(x)\, dx = 1$. We will see that this is a particularly convenient choice when we reexamine discrete random variables.

Theorem 4.17 allows us to write

$$\delta(x) = \frac{du(x)}{dx}.$$
(4.54)

Equation (4.54) embodies a certain kind of consistency in its inconsistency. That is, $\delta(x)$ does not really exist at $x = 0$. Similarly, the derivative of $u(x)$ does not really exist at $x = 0$. However, Equation (4.54) allows us to use $\delta(x)$ to define a generalized PDF that applies to discrete random variables as well as to continuous random variables.

Consider the CDF of a discrete random variable, X. Recall that it is constant everywhere except at points $x_i \in S_X$, where it has jumps of height $P_X(x_i)$. Using the definition of the unit step function, we can write the CDF of X as

$$F_X(x) = \sum_{x_i \in S_X} P_X(x_i)\, u(x - x_i).$$
(4.55)

From Definition 4.3, we take the derivative of $F_X(x)$ to find the PDF $f_X(x)$. Referring to Equation (4.54), the PDF of the discrete random variable X is

$$f_X(x) = \sum_{x_i \in S_X} P_X(x_i)\, \delta(x - x_i).$$
(4.56)

When the PDF includes delta functions of the form $\delta(x - x_i)$, we say there is an impulse at x_i. When we graph a PDF $f_X(x)$ that contains an impulse at x_i, we draw a vertical arrow labeled by the constant that multiplies the impulse. We draw each arrow representing an impulse at the same height because the PDF is always infinite at each such point. For example, the graph of $f_X(x)$ from Equation (4.56) is

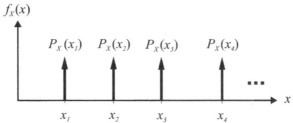

Using delta functions in the PDF, we can apply the formulas in this chapter to all random variables. In the case of discrete random variables, these formulas are equivalent to the ones presented in Chapter 3. For example, if X is a discrete random variable, Definition 4.4 becomes

$$E[X] = \int_{-\infty}^{\infty} x \sum_{x_i \in S_X} P_X(x_i)\, \delta(x - x_i)\, dx.$$
(4.57)

By writing the integral of the sum as a sum of integrals and using the sifting property of the delta function,

$$E[X] = \sum_{x_i \in S_X} \int_{-\infty}^{\infty} x P_X(x_i)\, \delta(x - x_i)\, dx = \sum_{x_i \in S_X} x_i P_X(x_i),$$
(4.58)

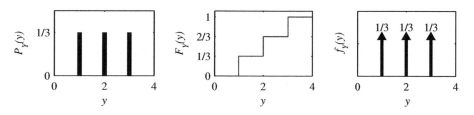

Figure 4.8 The PMF, CDF, and PDF of the discrete random variable Y.

which is Definition 3.11.

Example 4.15

Suppose Y takes on the values $1, 2, 3$ with equal probability. The PMF and the corresponding CDF of Y are

$$P_Y(y) = \begin{cases} 1/3 & y = 1, 2, 3, \\ 0 & \text{otherwise,} \end{cases} \qquad F_Y(y) = \begin{cases} 0 & y < 1, \\ 1/3 & 1 \le y < 2, \\ 2/3 & 2 \le y < 3, \\ 1 & y \ge 3. \end{cases} \tag{4.59}$$

Using the unit step function $u(y)$, we can write $F_Y(y)$ more compactly as

$$F_Y(y) = \frac{1}{3}u(y-1) + \frac{1}{3}u(y-2) + \frac{1}{3}u(y-3). \tag{4.60}$$

The PDF of Y is

$$f_Y(y) = \frac{dF_Y(y)}{dy} = \frac{1}{3}\delta(y-1) + \frac{1}{3}\delta(y-2) + \frac{1}{3}\delta(y-3). \tag{4.61}$$

We see that the discrete random variable Y can be represented graphically either by a PMF $P_Y(y)$ with bars at $y = 1, 2, 3$, by a CDF with jumps at $y = 1, 2, 3$, or by a PDF $f_Y(y)$ with impulses at $y = 1, 2, 3$. These three representations are shown in Figure 4.8. The expected value of Y can be calculated either by summing over the PMF $P_Y(y)$ or integrating over the PDF $f_Y(y)$. Using the PDF, we have

$$\begin{aligned} \mathrm{E}[Y] &= \int_{-\infty}^{\infty} y f_Y(y)\, dy \\ &= \int_{-\infty}^{\infty} \frac{y}{3}\delta(y-1)\, dy + \int_{-\infty}^{\infty} \frac{y}{3}\delta(y-2)\, dy + \int_{-\infty}^{\infty} \frac{y}{3}\delta(y-3)\, dy \\ &= 1/3 + 2/3 + 1 = 2. \end{aligned} \tag{4.62}$$

When $F_X(x)$ has a discontinuity at x, we use $F_X(x^+)$ and $F_X(x^-)$ to denote the upper and lower limits at x. That is,

$$F_X\left(x^-\right) = \lim_{h \to 0^+} F_X(x-h), \qquad F_X\left(x^+\right) = \lim_{h \to 0^+} F_X(x+h). \qquad (4.63)$$

Using this notation, we can say that if the CDF $F_X(x)$ has a jump at x_0, then $f_X(x)$ has an impulse at x_0 weighted by the height of the discontinuity $F_X(x_0^+) - F_X(x_0^-)$.

Example 4.16

For the random variable Y of Example 4.15,

$$F_Y\left(2^-\right) = 1/3, \qquad F_Y\left(2^+\right) = 2/3. \qquad (4.64)$$

Theorem 4.18

For a random variable X, we have the following equivalent statements:
(a) $\mathrm{P}[X = x_0] = q$ *(b)* $P_X(x_0) = q$
(c) $F_X(x_0^+) - F_X(x_0^-) = q$ *(d)* $f_X(x_0) = q\delta(0)$

In Example 4.15, we saw that $f_Y(y)$ consists of a series of impulses. The value of $f_Y(y)$ is either 0 or ∞. By contrast, the PDF of a continuous random variable has nonzero, finite values over intervals of x. In the next example, we encounter a random variable that has continuous parts and impulses.

Definition 4.14   Mixed Random Variable

X is a **mixed** random variable if and only if $f_X(x)$ contains both impulses and nonzero, finite values.

Example 4.17

Observe someone dialing a telephone and record the duration of the call. In a simple model of the experiment, $1/3$ of the calls never begin either because no one answers or the line is busy. The duration of these calls is 0 minutes. Otherwise, with probability $2/3$, a call duration is uniformly distributed between 0 and 3 minutes. Let Y denote the call duration. Find the CDF $F_Y(y)$, the PDF $f_Y(y)$, and the expected value $\mathrm{E}[Y]$.
....

Let A denote the event that the phone was answered. $\mathrm{P}[A] = 2/3$ and $\mathrm{P}[A^c] = 1/3$. Since $Y \geq 0$, we know that for $y < 0$, $F_Y(y) = 0$. Similarly, we know that for $y > 3$, $F_Y(y) = 1$. For $0 \leq y \leq 3$, we apply the law of total probability to write

$$F_Y(y) = \mathrm{P}\left[Y \leq y\right] = \mathrm{P}\left[Y \leq y | A^c\right] \mathrm{P}\left[A^c\right] + \mathrm{P}\left[Y \leq y | A\right] \mathrm{P}\left[A\right]. \qquad (4.65)$$

When A^c occurs, $Y = 0$, so that for $0 \leq y \leq 3$, $P[Y \leq y|A^c] = 1$. When A occurs, the call duration is uniformly distributed over $[0, 3]$, so that for $0 \leq y \leq 3$, $P[Y \leq y|A] = y/3$. So, for $0 \leq y \leq 3$,

$$F_Y(y) = (1/3)(1) + (2/3)(y/3) = 1/3 + 2y/9. \qquad (4.66)$$

The complete CDF of Y is

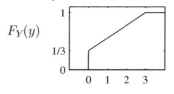

$$F_Y(y) = \begin{cases} 0 & y < 0, \\ 1/3 + 2y/9 & 0 \leq y < 3, \\ 1 & y \geq 3. \end{cases}$$

Consequently, the corresponding PDF $f_Y(y)$ is

$$f_Y(y) = \begin{cases} \delta(y)/3 + 2/9 & 0 \leq y \leq 3, \\ 0 & \text{otherwise.} \end{cases}$$

For the mixed random variable Y, it is easiest to calculate $E[Y]$ using the PDF:

$$E[Y] = \int_{-\infty}^{\infty} y\frac{1}{3}\delta(y)\, dy + \int_0^3 \frac{2}{9}y\, dy = 0 + \frac{2}{9}\frac{y^2}{2}\bigg|_0^3 = 1 \text{ minute.} \qquad (4.67)$$

In Example 4.17, we see that with probability $1/3$, Y resembles a discrete random variable; otherwise, Y behaves like a continuous random variable. This behavior is reflected in the impulse in the PDF of Y. In many practical applications of probability, mixed random variables arise as functions of continuous random variables. Electronic circuits perform many of these functions. Example 6.8 in Section 6.3 gives one example.

Before going any further, we review what we have learned about random variables. For any random variable X,

- X always has a CDF $F_X(x) = P[X \leq x]$.

- If $F_X(x)$ is piecewise flat with discontinuous jumps, then X is discrete.

- If $F_X(x)$ is a continuous function, then X is continuous.

- If $F_X(x)$ is a piecewise continuous function with discontinuities, then X is mixed.

- When X is discrete or mixed, the PDF $f_X(x)$ contains one or more delta functions.

Quiz 4.7

The cumulative distribution function of random variable X is

$$F_X(x) = \begin{cases} 0 & x < -1, \\ (x+1)/4 & -1 \le x < 1, \\ 1 & x \ge 1. \end{cases} \tag{4.68}$$

Sketch the CDF and find the following:

(a) $P[X \le 1]$

(b) $P[X < 1]$

(c) $P[X = 1]$

(d) the PDF $f_X(x)$

4.8 MATLAB

> Built-in MATLAB functions, either alone or with additional code, can be used to calculate PDFs and CDFs of several random variable families. The **rand** and **randn** functions simulate experiments that generate sample values of continuous uniform $(0, 1)$ random variables and Gaussian $(0, 1)$ random variables, respectively.

Probability Functions

Table 4.3 describes MATLAB functions related to four families of continuous random variables introduced in this chapter: uniform, exponential, Erlang, and Gaussian. The functions calculate directly the CDFs and PDFs of uniform and exponential random variables.

```
function F=erlangcdf(n,lambda,x)
F=1.0-poissoncdf(lambda*x,n-1);
```

For Erlang and Gaussian random variables, the PDFs can be calculated directly but the CDFs require numerical integration. For Erlang random variables, `erlangcdf` uses Theorem 4.11. For the Gaussian CDF, we use the built-in MATLAB error function

$$\text{erf}(x) = \frac{2}{\sqrt{\pi}} \int_0^x e^{-u^2} \, du. \tag{4.69}$$

It is related to the Gaussian CDF by

$$\Phi(x) = \frac{1}{2} + \frac{1}{2}\text{erf}\left(\frac{x}{\sqrt{2}}\right), \tag{4.70}$$

which is how we implement the MATLAB function `phi(x)`. In each function description in Table 4.3, x denotes a vector $\mathbf{x} = \begin{bmatrix} x_1 & \cdots & x_m \end{bmatrix}'$. The **pdf** function output is a vector $\mathbf{y}$ such that $y_i = f_X(x_i)$. The **cdf** function output is a vector $\mathbf{y}$

Random Variable	MATLAB Function	Function Output
X Uniform (a,b)	y=uniformpdf(a,b,x)	$y_i = f_X(x_i)$
	y=uniformcdf(a,b,x)	$y_i = F_X(x_i)$
	x=uniformrv(a,b,m)	$\mathbf{X} = \begin{bmatrix} X_1 & \cdots & X_m \end{bmatrix}'$
X Exponential (λ)	y=exponentialpdf(lambda,x)	$y_i = f_X(x_i)$
	y=exponentialcdf(lambda,x)	$y_i = F_X(x_i)$
	x=exponentialrv(lambda,m)	$\mathbf{X} = \begin{bmatrix} X_1 & \cdots & X_m \end{bmatrix}'$
X Erlang (n,λ)	y=erlangpdf(n,lambda,x)	$y_i = f_X(x_i)$
	y=erlangcdf(n,lambda,x)	$y_i = F_X(x_i)$
	x=erlangrv(n,lambda,m)	$\mathbf{X} = \begin{bmatrix} X_1 & \cdots & X_m \end{bmatrix}'$
X Gaussian (μ,σ^2)	y=gausspdf(mu,sigma,x)	$y_i = f_X(x_i)$
	y=gausscdf(mu,sigma,x)	$y_i = F_X(x_i)$
	x=gaussrv(mu,sigma,m)	$\mathbf{X} = \begin{bmatrix} X_1 & \cdots & X_m \end{bmatrix}'$

Table 4.3 MATLAB functions for continuous random variables.

such that $y_i = F_X(x_i)$. The rv function output is a vector $\mathbf{X} = \begin{bmatrix} X_1 & \cdots & X_m \end{bmatrix}'$ such that each X_i is a sample value of the random variable X. If $m = 1$, then the output is a single sample value of random variable X.

Random Samples

Now that we have introduced continuous random variables, we can say that the built-in function y=rand(m,n) is MATLAB's approximation to a uniform $(0,1)$ random variable. It is an approximation for two reasons. First, rand produces pseudorandom numbers; the numbers seem random but are actually the output of a deterministic algorithm. Second, rand produces a double precision floating point number, represented in the computer by 64 bits. Thus MATLAB distinguishes no more than 2^{64} unique double precision floating point numbers. By comparision, there are uncountably infinite real numbers in $[0,1)$. Even though rand is not random and does not have a continuous range, we can for all practical purposes use it as a source of independent sample values of the uniform $(0,1)$ random variable.

We have already employed the rand function to generate random samples of uniform $(0,1)$ random variables. Conveniently, MATLAB also includes the built-in function randn to generate random samples of standard normal random variables.

```
function x=gaussrv(mu,sigma,m)
x=mu +(sigma*randn(m,1));
```

Thus gaussrv generates Gaussian (μ,σ) random variables by stretching and shifting standard normal random variables. For other continuous random variables, we use a technique described in Theorem 6.5 that transforms a uniform $(0,1)$ random variable U into other types of random variables. This is explained in the MATLAB section of Chapter 6.

====Quiz 4.8====

Use the MATLAB function `uniformrv` and the result of Theorem 4.7 to write a MATLAB function `y=duniform(a,b,m)` that produces m samples of a discrete uniform (a, b) random variable.

Problems

Difficulty: ● Easy   ▨ Moderate   ◆ Difficult   ◆◆ Experts Only

4.2.1● The cumulative distribution function of random variable X is

$$F_X(x) = \begin{cases} 0 & x < -1, \\ (x+1)/2 & -1 \le x < 1, \\ 1 & x \ge 1. \end{cases}$$

(a) What is $P[X > 1/2]$?

(b) What is $P[-1/2 < X \le 3/4]$?

(c) What is $P[|X| \le 1/2]$?

(d) What is the value of a such that $P[X \le a] = 0.8$?

4.2.2● The CDF of the continuous random variable V is

$$F_V(v) = \begin{cases} 0 & v < -5, \\ c(v+5)^2 & -5 \le v < 7, \\ 1 & v \ge 7. \end{cases}$$

(a) What is c?

(b) What is $P[V > 4]$?

(c) What is $P[-3 < V \le 0]$?

(d) What is the value of a such that $P[V > a] = 2/3$?

4.2.3● In this problem, we verify that $\lim_{n \to \infty} \lceil nx \rceil / n = x$.

(a) Verify that $nx \le \lceil nx \rceil \le nx + 1$.

(b) Use part (a) to show

$$\lim_{n \to \infty} \lceil nx \rceil / n = x.$$

(c) Use a similar argument to show that $\lim_{n \to \infty} \lfloor nx \rfloor / n = x$.

4.2.4● The CDF of random variable W is

$$F_W(w) = \begin{cases} 0 & w < -5, \\ \frac{w+5}{8} & -5 \le w < -3, \\ \frac{1}{4} & -3 \le w < 3, \\ \frac{1}{4} + \frac{3(w-3)}{8} & 3 \le w < 5, \\ 1 & w \ge 5. \end{cases}$$

(a) What is $P[W \le 4]$?

(b) What is $P[-2 < W \le 2]$?

(c) What is $P[W > 0]$?

(d) What is the value of a such that $P[W \le a] = 1/2$?

4.3.1● The random variable X has probability density function

$$f_X(x) = \begin{cases} cx & 0 \le x \le 2, \\ 0 & \text{otherwise.} \end{cases}$$

Use the PDF to find

(a) the constant c,

(b) $P[0 \le X \le 1]$,

(c) $P[-1/2 \le X \le 1/2]$,

(d) the CDF $F_X(x)$.

4.3.2● The cumulative distribution function of random variable X is

$$F_X(x) = \begin{cases} 0 & x < -1, \\ (x+1)/2 & -1 \le x < 1, \\ 1 & x \ge 1. \end{cases}$$

Find the PDF $f_X(x)$ of X.

4.3.3● Find the PDF $f_U(u)$ of the random variable U in Problem 4.2.4.

4.3.4■ For a constant parameter $a > 0$, a Rayleigh random variable X has PDF

$$f_X(x) = \begin{cases} a^2 x e^{-a^2 x^2/2} & x > 0, \\ 0 & \text{otherwise.} \end{cases}$$

What is the CDF of X?

4.3.5◆ Random variable X has a PDF of the form $f_X(x) = \frac{1}{2} f_1(x) + \frac{1}{2} f_2(x)$, where

$$f_1(x) = \begin{cases} c_1 & 0 \le x \le 2, \\ 0 & \text{otherwise,} \end{cases}$$

$$f_2(x) = \begin{cases} c_2 e^{-x} & x \ge 0, \\ 0 & \text{otherwise.} \end{cases}$$

What conditions must c_1 and c_2 satisfy so that $f_X(x)$ is a valid PDF?

4.3.6◆◆ For constants a and b, random variable X has PDF

$$f_X(x) = \begin{cases} ax^2 + bx & 0 \le x \le 1, \\ 0 & \text{otherwise.} \end{cases}$$

What conditions on a and b are necessary and sufficient to guarantee that $f_X(x)$ is a valid PDF?

4.4.1● Random variable X has PDF

$$f_X(x) = \begin{cases} 1/4 & -1 \le x \le 3, \\ 0 & \text{otherwise.} \end{cases}$$

Define the random variable Y by $Y = h(X) = X^2$.

(a) Find $E[X]$ and $Var[X]$.

(b) Find $h(E[X])$ and $E[h(X)]$.

(c) Find $E[Y]$ and $Var[Y]$.

4.4.2● Let X be a continuous random variable with PDF

$$f_X(x) = \begin{cases} 1/8 & 1 \le x \le 9, \\ 0 & \text{otherwise.} \end{cases}$$

Let $Y = h(X) = 1/\sqrt{X}$.

(a) Find $E[X]$ and $Var[X]$.

(b) Find $h(E[X])$ and $E[h(X)]$.

(c) Find $E[Y]$ and $Var[Y]$.

4.4.3● Random variable X has CDF

$$F_X(x) = \begin{cases} 0 & x < 0, \\ x/2 & 0 \le x \le 2, \\ 1 & x > 2. \end{cases}$$

(a) What is $E[X]$?

(b) What is $Var[X]$?

4.4.4● The probability density function of random variable Y is

$$f_Y(y) = \begin{cases} y/2 & 0 \le y < 2, \\ 0 & \text{otherwise.} \end{cases}$$

What are $E[Y]$ and $Var[Y]$?

4.4.5● The cumulative distribution function of the random variable Y is

$$F_Y(y) = \begin{cases} 0 & y < -1, \\ (y+1)/2 & -1 \le y \le 1, \\ 1 & y > 1. \end{cases}$$

What are $E[Y]$ and $Var[Y]$?

4.4.6■ The cumulative distribution function of random variable V is

$$F_V(v) = \begin{cases} 0 & v < -5, \\ (v+5)^2/144 & -5 \le v < 7, \\ 1 & v \ge 7. \end{cases}$$

(a) What are $E[V]$ and $Var[V]$?

(b) What is $E[V^3]$?

4.4.7■ The cumulative distribution function of random variable U is

$$F_U(u) = \begin{cases} 0 & u < -5, \\ \frac{u+5}{8} & -5 \le u < -3, \\ \frac{1}{4} & -3 \le u < 3, \\ \frac{3u-7}{8} & 3 \le u < 5, \\ 1 & u \ge 5. \end{cases}$$

(a) What are $E[U]$ and $Var[U]$?

(b) What is $E[2^U]$?

4.4.8■ X is a Pareto (α, μ) random variable, as defined in Appendix A. What is the largest value of n for which the nth moment $E[X^n]$ exists? For all feasible values of n, find $E[X^n]$.

4.5.1● Y is a continuous uniform $(1, 5)$ random variable.

(a) What is $P[Y > E[Y]]$?

(b) What is $P[Y \leq \text{Var}[Y]]$?

4.5.2● The current Y across a 1 kΩ resistor is a continuous uniform $(-10, 10)$ random variable. Find $P[|Y| < 3]$.

4.5.3● Radars detect flying objects by measuring the power reflected from them. The reflected power of an aircraft can be modeled as a random variable Y with PDF

$$f_Y(y) = \begin{cases} \frac{1}{P_0} e^{-y/P_0} & y \geq 0 \\ 0 & \text{otherwise} \end{cases}$$

where $P_0 > 0$ is some constant. The aircraft is correctly identified by the radar if the reflected power of the aircraft is larger than its average value. What is the probability $P[C]$ that an aircraft is correctly identified?

4.5.4● Y is an exponential random variable with variance $\text{Var}[Y] = 25$.

(a) What is the PDF of Y?

(b) What is $E[Y^2]$?

(c) What is $P[Y > 5]$?

4.5.5● The time delay Y (in milliseconds) that your computer needs to connect to an access point is an exponential random variable.

(a) Find $P[Y > E[Y]]$.

(b) Find $P[Y > 2 E[Y]]$.

4.5.6● X is an Erlang (n, λ) random variable with parameter $\lambda = 1/3$ and expected value $E[X] = 15$.

(a) What is the value of the parameter n?

(b) What is the PDF of X?

(c) What is $\text{Var}[X]$?

4.5.7● Y is an Erlang $(n = 2, \lambda = 2)$ random variable.

(a) What is $E[Y]$?

(b) What is $\text{Var}[Y]$?

(c) What is $P[0.5 \leq Y < 1.5]$?

4.5.8■ U is a zero mean continuous uniform random variable. What is $P[U^2 \leq \text{Var}[U]]$?

4.5.9■ U is a continuous uniform random variable such that $E[U] = 10$ and $P[U > 12] = 1/4$. What is $P[U < 9]$?

4.5.10■ X is a continuous uniform $(-5, 5)$ random variable.

(a) What is the PDF $f_X(x)$?

(b) What is the CDF $F_X(x)$?

(c) What is $E[X]$?

(d) What is $E[X^5]$?

(e) What is $E[e^X]$?

4.5.11■ X is a continuous uniform $(-a, a)$ random variable. Find $P[|X| \leq \text{Var}[X]]$.

4.5.12■ X is a uniform random variable with expected value $\mu_X = 7$ and variance $\text{Var}[X] = 3$. What is the PDF of X?

4.5.13■ The probability density function of random variable X is

$$f_X(x) = \begin{cases} (1/2)e^{-x/2} & x \geq 0, \\ 0 & \text{otherwise.} \end{cases}$$

(a) What is $P[1 \leq X \leq 2]$?

(b) What is $F_X(x)$, the cumulative distribution function of X?

(c) What is $E[X]$, the expected value of X?

(d) What is $\text{Var}[X]$, the variance of X?

4.5.14■ Verify parts (b) and (c) of Theorem 4.6 by directly calculating the expected value and variance of a uniform random variable with parameters $a < b$.

4.5.15■ Long-distance calling plan A offers flat-rate service at 10 cents per minute. Calling plan B charges 99 cents for every call under 20 minutes; for calls over 20 minutes, the charge is 99 cents for the first 20 minutes plus 10 cents for every additional minute. (Note that these plans measure your call duration exactly, without rounding to the next minute or even second.) If your long-distance calls have exponential distribution with expected value τ minutes, which plan offers a lower expected cost per call?

4.5.16 In this problem we verify that an Erlang (n, λ) PDF integrates to 1. Let the integral of the nth order Erlang PDF be denoted by

$$I_n = \int_0^\infty \frac{\lambda^n x^{n-1} e^{-\lambda x}}{(n-1)!} \, dx.$$

First, show directly that the Erlang PDF with $n = 1$ integrates to 1 by verifying that $I_1 = 1$. Second, use integration by parts (Appendix B, Math Fact B.10) to show that $I_n = I_{n-1}$.

4.5.17 Calculate the kth moment $E[X^k]$ of an Erlang (n, λ) random variable X. Use your result to verify Theorem 4.10. Hint: Remember that the Erlang $(n + k, \lambda)$ PDF integrates to 1.

4.5.18 In this problem, we outline the proof of Theorem 4.11.

(a) Let X_n denote an Erlang (n, λ) random variable. Use the definition of the Erlang PDF to show that for any $x \geq 0$,

$$F_{X_n}(x) = \int_0^x \frac{\lambda^n t^{n-1} e^{-\lambda t}}{(n-1)!} \, dt.$$

(b) Apply integration by parts (see Appendix B, Math Fact B.10) to this integral to show that for $x \geq 0$,

$$F_{X_n}(x) = F_{X_{n-1}}(x) - \frac{(\lambda x)^{n-1} e^{-\lambda x}}{(n-1)!}.$$

(c) Use the fact that $F_{X_1}(x) = 1 - e^{-\lambda x}$ for $x \geq 0$ to verify the claim of Theorem 4.11.

4.5.19 Prove by induction that an exponential random variable X with expected value $1/\lambda$ has nth moment

$$E[X^n] = \frac{n!}{\lambda^n}.$$

Hint: Use integration by parts (Appendix B, Math Fact B.10).

4.5.20 This problem outlines the steps needed to show that a nonnegative continuous random variable X has expected value

$$E[X] = \int_0^\infty [1 - F_X(x)] \, dx.$$

(a) For any $r \geq 0$, show that

$$r P[X > r] \leq \int_r^\infty x f_X(x) \, dx.$$

(b) Use part (a) to argue that if $E[X] < \infty$, then

$$\lim_{r \to \infty} r P[X > r] = 0.$$

(c) Now use integration by parts (Appendix B, Math Fact B.10) to evaluate

$$\int_0^\infty [1 - F_X(x)] \, dx.$$

4.6.1 The peak temperature T, as measured in degrees Fahrenheit, on a July day in New Jersey is the Gaussian $(85, 10)$ random variable. What is $P[T > 100]$, $P[T < 60]$, and $P[70 \leq T \leq 100]$?

4.6.2 What is the PDF of Z, the standard normal random variable?

4.6.3 Find each probability.

(a) V is a Gaussian $(\mu = 0, \sigma = 2)$ random variable. Find $P[V > 4]$.

(b) W is a Gaussian $(\mu = 2, \sigma = 5)$ random variable. What is $P[W \leq 2]$?

(c) For a Gaussian $(\mu, \sigma = 2)$ random variable X, find $P[X \leq \mu + 1]$.

(d) Y is a Gaussian $(\mu = 50, \sigma = 10)$ random variable. Calculate $P[Y > 65]$.

4.6.4 In each of the following cases, Y is a Gaussian random variable. Find the expected value $\mu = E[Y]$.

(a) Y has standard deviation $\sigma = 10$ and $P[Y \leq 10] = 0.933$.

(b) Y has standard deviation $\sigma = 10$ and $P[Y \leq 0] = 0.067$.

(c) Y has standard deviation σ and $P[Y \leq 10] = 0.977$. (Find μ as a function of σ.)

(d) $P[Y > 5] = 1/2$.

4.6.5 Your internal body temperature T in degrees Fahrenheit is a Gaussian ($\mu = 98.6, \sigma = 0.4$) random variable. In terms of the $\Phi(\cdot)$ function, find $P[T > 100]$. Does this model seem reasonable?

4.6.6 The temperature T in this thermostatically controlled lecture hall is a Gaussian random variable with expected value $\mu = 68$ degrees Fahrenheit. In addition, $P[T < 66] = 0.1587$. What is the variance of T?

4.6.7 X is a Gaussian random variable with $E[X] = 0$ and $P[|X| \leq 10] = 0.1$. What is the standard deviation σ_X?

4.6.8 A function commonly used in communications textbooks for the tail probabilities of Gaussian random variables is the complementary error function, defined as

$$\text{erfc}(z) = \frac{2}{\sqrt{\pi}} \int_z^\infty e^{-x^2} \, dx.$$

Show that

$$Q(z) = \frac{1}{2}\text{erfc}\left(\frac{z}{\sqrt{2}}\right).$$

4.6.9 The peak temperature T, in degrees Fahrenheit, on a July day in Antarctica is a Gaussian random variable with a variance of 225. With probability 1/2, the temperature T exceeds -75 degrees. What is $P[T > 0]$? What is $P[T < -100]$?

4.6.10 A professor pays 25 cents for each blackboard error made in lecture to the student who points out the error. In a career

of n years filled with blackboard errors, the total amount in dollars paid can be approximated by a Gaussian random variable Y_n with expected value $40n$ and variance $100n$. What is the probability that Y_{20} exceeds 1000? How many years n must the professor teach in order that $P[Y_n > 1000] > 0.99$?

4.6.11 Suppose that out of 100 million men in the United States, 23,000 are at least 7 feet tall. Suppose that the heights of U.S. men are independent Gaussian random variables with a expected value of $5'10''$. Let N equal the number of men who are at least $7'6''$ tall.

(a) Calculate σ_X, the standard deviation of the height of U.S. men.

(b) In terms of the $\Phi(\cdot)$ function, what is the probability that a randomly chosen man is at least 8 feet tall?

(c) What is the probability that no man alive in the United States today is at least $7'6''$ tall?

(d) What is $E[N]$?

4.6.12 In this problem, we verify that for $x \geq 0$,

$$\Phi(x) = \frac{1}{2} + \frac{1}{2}\text{erf}\left(\frac{x}{\sqrt{2}}\right).$$

(a) Let Y have a Gaussian $(0, 1/\sqrt{2})$ distribution and show that

$$F_Y(y) = \int_{-\infty}^y f_Y(u) \, du = \frac{1}{2} + \text{erf}(y).$$

(b) Observe that $Z = \sqrt{2}Y$ is Gaussian $(0, 1)$ and show that

$$\Phi(z) = F_Z(z) = F_Y\left(\frac{z}{\sqrt{2}}\right).$$

4.6.13 This problem outlines the steps needed to show that the Gaussian PDF integrates to unity. For a Gaussian (μ, σ) random variable W, we will show that

$$I = \int_{-\infty}^\infty f_W(w) \, dw = 1.$$

(a) Use the substitution $x = (w - \mu)/\sigma$ to show that

$$I = \frac{1}{\sqrt{2\pi}} \int_{-\infty}^{\infty} e^{-x^2/2} \, dx.$$

(b) Show that

$$I^2 = \frac{1}{2\pi} \int_{-\infty}^{\infty} \int_{-\infty}^{\infty} e^{-(x^2+y^2)/2} \, dx \, dy.$$

(c) Change to polar coordinates to show that $I^2 = 1$.

4.6.14♦ At time $t = 0$, the price of a stock is a constant k dollars. At time $t > 0$ the price of a stock is a Gaussian random variable X with $\mathrm{E}[X] = k$ and $\mathrm{Var}[X] = t$. At time t, a *Call Option at Strike k* has value $V = (X - k)^+$, where the operator $(\cdot)^+$ is defined as $(z)^+ = \max(z, 0)$.

(a) Find the expected value $\mathrm{E}[V]$.

(b) Suppose you can buy the call option for d dollars at time $t = 0$. At time t, you can sell the call for V dollars and earn a profit (or loss perhaps) of $R = V - d$ dollars. Let d_0 denote the value of d such that $\mathrm{P}[R > 0] = 1/2$. Your strategy is that you buy the option if $d \leq d_0$ so that your probability of a profit is $\mathrm{P}[R > 0] \geq 1/2$. Find d_0.

(c) Let d_1 denote the value of d such that $\mathrm{E}[R] = 0.01 \times d$. Now your strategy is to buy the option if $d \leq d_1$ so that your expected return is at least one percent of the option cost. Find d_1.

(d) Are the strategies "Buy the option if $d \leq d_0$" and "Buy the option if $d \leq d_1$" reasonable strategies?

4.6.15♦ In mobile radio communications, the radio channel can vary randomly. In particular, in communicating with a fixed transmitter power over a "Rayleigh fading" channel, the receiver signal-to-noise ratio Y is an exponential random variable with expected value γ. Moreover, when $Y = y$, the probability of an error in decoding a transmitted bit is $P_e(y) = Q(\sqrt{2y})$ where $Q(\cdot)$ is the standard normal complementary CDF.

The average probability of bit error, also known as the bit error rate or BER, is

$$\overline{P}_e = \mathrm{E}[P_e(Y)] = \int_{-\infty}^{\infty} Q(\sqrt{2y}) f_Y(y) \, dy.$$

Find a simple formula for the BER $\overline{P}_e$ as a function of the average SNR γ.

4.6.16♦♦ At time $t = 0$, the price of a stock is a constant k dollars. At some future time $t > 0$, the price X of the stock is a uniform $(k - t, k + t)$ random variable. At this time t, a *Put Option at Strike k* (which is the right to sell the stock at price k) has value $(k - X)^+$ dollars where the operator $(\cdot)^+$ is defined as $(z)^+ = \max(z, 0)$. Similarly a *Call Option at Strike k* (the right to buy the stock at price k) at time t has value $(X - k)^+$.

(a) At time 0, you sell the put and receive d dollars. At time t, you purchase the put for $(k - X)^+$ dollars to cancel your position. Your gain is

$$R = g_p(X) = d - (k - X)^+.$$

Find the central moments $\mathrm{E}[R]$ and $\mathrm{Var}[R]$.

(b) In a short straddle, you sell the put for d dollars and you also sell the call for d dollars. At a future time $t > 0$, you purchase the put for $(k - X)^+$ dollars and the call for $(X - k)^+$ dollars to cancel both positions. Your gain on the put is $g_p(X) = d - (k - X)^+$ dollars and your gain on the call is $g_c(X) = d - (X - k)^+$ dollars. Your net gain is

$$R' = g_p(X) + g_c(X).$$

Find the expected value $\mathrm{E}[R']$ and variance $\mathrm{Var}[R']$.

(c) Explain why selling the straddle might be attractive compared to selling just the put or just the call.

4.6.17♦♦ Continuing Problem 4.6.16, suppose you sell the straddle at time $t = 0$ and liquidate your position at time t, generating a profit (or perhaps a loss) R'. Find the

PDF $f_{R'}(r)$ of R'. Suppose d is sufficiently large that $E[R'] > 0$. Would you be interested in selling the short straddle? Are you getting something, namely $E[R']$ dollars, for nothing?

4.7.1● Let X be a random variable with CDF

$$F_X(x) = \begin{cases} 0 & x < -1, \\ x/3 + 1/3 & -1 \le x < 0, \\ x/3 + 2/3 & 0 \le x < 1, \\ 1 & 1 \le x. \end{cases}$$

Sketch the CDF and find

(a) $P[X < -1]$ and $P[X \le -1]$,

(b) $P[X < 0]$ and $P[X \le 0]$,

(c) $P[0 < X \le 1]$ and $P[0 \le X \le 1]$.

4.7.2● Let X be a random variable with CDF

$$F_X(x) = \begin{cases} 0 & x < -1, \\ x/4 + 1/2 & -1 \le x < 1, \\ 1 & 1 \le x. \end{cases}$$

Sketch the CDF and find

(a) $P[X < -1]$ and $P[X \le -1]$.

(b) $P[X < 0]$ and $P[X \le 0]$.

(c) $P[X > 1]$ and $P[X \ge 1]$.

4.7.3● For random variable X of Problem 4.7.2, find $f_X(x)$, $E[X]$, and $Var[X]$.

4.7.4● X is Bernoulli random variable with expected value p. What is the PDF $f_X(x)$?

4.7.5● X is a geometric random variable with expected value $1/p$. What is the PDF $f_X(x)$?

4.7.6■ When you make a phone call, the line is busy with probability 0.2 and no one answers with probability 0.3. The random variable X describes the conversation time (in minutes) of a phone call that is answered. X is an exponential random variable with $E[X] = 3$ minutes. Let the random variable W denote the conversation time (in seconds) of all calls ($W = 0$ when the line is busy or there is no answer.)

(a) What is $F_W(w)$?

(b) What is $f_W(w)$?

(c) What are $E[W]$ and $Var[W]$?

4.7.7■ For 80% of lectures, Professor X arrives on time and starts lecturing with delay $T = 0$. When Professor X is late, the starting time delay T is uniformly distributed between 0 and 300 seconds. Find the CDF and PDF of T.

4.7.8◆ With probability 0.7, the toss of an Olympic shot-putter travels $D = 60 + X$ feet, where X is an exponential random variable with expected value $\mu = 10$. Otherwise, with probability 0.3, a foul is committed by stepping outside of the shot-put circle and we say $D = 0$. What are the CDF and PDF of random variable D?

4.7.9◆ For 70% of lectures, Professor Y arrives on time. When Professor Y is late, the arrival time delay (in minutes) is a continuous uniform $(0, 10)$ random variable. Yet, as soon as Professor Y is 5 minutes late, all the students get up and leave. If a lecture starts when Professor Y arrives and always ends 80 minutes after the scheduled starting time, what is the PDF of T, the length of time that the students observe a lecture.

4.8.1● Write a function `y=quiz31rv(m)` that produces m samples of random variable Y defined in Quiz 4.2.

4.8.2● The Gaussian $(0, 1)$ complementary CDF $Q(z)$, can be approximated by

$$\hat{Q}(z) = \left(\sum_{n=1}^{5} a_n t^n \right) e^{-z^2/2},$$

for $z \ge 0$ where

$$t = \frac{1}{1 + 0.231641888z} \quad a_1 = 0.127414796$$

$$a_2 = -0.142248368 \qquad a_3 = 0.7107068705$$

$$a_4 = -0.7265760135 \qquad a_5 = 0.5307027145.$$

To compare this approximation to $Q(z)$, use MATLAB to graph

$$e(z) = \frac{Q(z) - \hat{Q}(z)}{Q(z)}.$$

4.8.3■ Use `exponentialrv.m` and Theorem 4.9 and to write a MATLAB function `k=georv(p,m)` that generates m samples of a geometric (p) random variable K. Compare the resulting algorithm to the technique employed in Problem 3.9.8 for `geometricrv(p,m)`.

4.8.4■ Applying Equation (4.14) with x replaced by $i\Delta$ and dx replaced by Δ, we obtain

$$P\left[i\Delta < X \leq i\Delta + \Delta\right] = f_X\left(i\Delta\right)\Delta.$$

If we generate a large number n of samples of random variable X, let n_i denote the number of occurrences of the event

$$\{i\Delta < X \leq (i+1)\Delta\}.$$

We would expect that

$$\lim_{n\to\infty} \frac{n_i}{n} = f_X\left(i\Delta\right)\Delta,$$

or equivalently,

$$\lim_{n\to\infty} \frac{n_i}{n\Delta} = f_X\left(i\Delta\right).$$

Use MATLAB to confirm this with $\Delta = 0.01$ for

(a) an exponential $(\lambda = 1)$ random variable X and for $i = 0,\ldots,500$,

(b) a Gaussian $(3,1)$ random variable X and for $i = 0,\ldots,600$.

5

Multiple Random Variables

Chapter 3 and Chapter 4 analyze experiments in which an outcome is one number. Beginning with this chapter, we analyze experiments in which an outcome is a collection of numbers. Each number is a sample value of a random variable. The probability model for such an experiment contains the properties of the individual random variables and it also contains the relationships among the random variables. Chapter 3 considers only discrete random variables and Chapter 4 considers only continuous random variables. The present chapter considers all random variables because a high proportion of the definitions and theorems apply to both discrete and continuous random variables. However, just as with individual random variables, the details of numerical calculations depend on whether random variables are discrete or continuous. Consequently, we find that many formulas come in pairs. One formula, for discrete random variables, contains sums, and the other formula, for continuous random variables, contains integrals.

In this chapter, we consider experiments that produce a collection of random variables, $X_1, X_2, \ldots, X_n$, where n can be any integer. For most of this chapter, we study $n = 2$ random variables: X and Y. A pair of random variables is enough to show the important concepts and useful problem-solving techniques. Moreover, the definitions and theorems we introduce for X and Y generalize to n random variables. These generalized definitions appear near the end of this chapter in Section 5.10.

We also note that a pair of random variables X and Y is the same as the two-dimensional vector $\begin{bmatrix} X & Y \end{bmatrix}'$. Similarly, the random variables $X_1, \ldots, X_n$ can be written as the n dimensional vector $\mathbf{X} = \begin{bmatrix} X_1 & \cdots & X_n \end{bmatrix}'$. Since the components of $\mathbf{X}$ are random variables, $\mathbf{X}$ is called a *random vector*. Thus this chapter begins our study of random vectors. This subject is continued in Chapter 8, which uses techniques of linear algebra to develop further the properties of random vectors.

We begin here with the definition of $F_{X,Y}(x, y)$, the *joint cumulative distribution function* of two random variables, a generalization of the CDF introduced in

Section 3.4 and again in Section 4.2. The joint CDF is a complete probability model for any experiment that produces two random variables. However, it not very useful for analyzing practical experiments. More useful models are $P_{X,Y}(x,y)$, the *joint probability mass function* for two discrete random variables, presented in Sections 5.2 and 5.3, and $f_{X,Y}(x,y)$, the *joint probability density function* of two continuous random variables, presented in Sections 5.4 and 5.5.

Section 5.7 considers functions of two random variables and expectations, respectively. We extend the definition of independent events to define independent random variables. The subject of Section 5.9 is the special case in which X and Y are Gaussian.

Pairs of random variables appear in a wide variety of practical situations. An example is the strength of the signal at a cellular telephone base station receiver (Y) and the distance (X) of the telephone from the base station. Another example of two random variables that we encounter all the time in our research is the signal (X), emitted by a radio transmitter, and the corresponding signal (Y) that eventually arrives at a receiver. In practice we observe Y, but we really want to know X. Noise and distortion prevent us from observing X directly, and we use a probability model to estimate X.

════Example 5.1════

We would like to measure random variable X, but we instead observe

$$Y = X + Z. \tag{5.1}$$

The noise Z prevents us from perfectly observing X. In some settings, Z is an interfering signal. In the simplest setting, Z is just noise inside the circuitry of your measurement device that is unrelated to X. In this case, it is appropriate to assume that the signal and noise are independent; that is, the events $X = x$ and $Z = z$ are independent. This simple model produces three random variables, X, Y and Z, but any pair completely specifies the remaining random variable. Thus we will see that a probability model for the pair (X, Z) or for the pair (X, Y) will be sufficient to analyze experiments related to this system.

5.1 Joint Cumulative Distribution Function

> The joint CDF $F_{X,Y}(x,y) = \mathrm{P}[X \leq x, Y \leq y]$ is a complete probability model for any pair of random variables X and Y.

In an experiment that produces one random variable, events are points or intervals on a line. In an experiment that leads to two random variables X and Y, each outcome (x,y) is a point in a plane and events are points or areas in the plane. Just as the CDF of one random variable, $F_X(x)$, is the probability of the interval to the left of x, the joint CDF $F_{X,Y}(x,y)$ of two random variables is the probability of the area below and to the left of (x,y). This is the infinite region that includes the shaded area in Figure 5.1 and everything below and to the left of it.

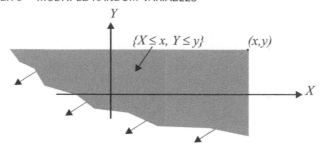

Figure 5.1 The area of the (X, Y) plane corresponding to the joint cumulative distribution function $F_{X,Y}(x, y)$.

━━━━Definition 5.1━━━━Joint Cumulative Distribution Function (CDF)

*The **joint cumulative distribution function** of random variables X and Y is*

$$F_{X,Y}(x, y) = P[X \leq x, Y \leq y].$$

The joint CDF is a complete probability model. The notation is an extension of the notation convention adopted in Chapter 3. The subscripts of F, separated by a comma, are the names of the two random variables. Each name is an uppercase letter. We usually write the arguments of the function as the lowercase letters associated with the random variable names.

The joint CDF has properties that are direct consequences of the definition. For example, we note that the event $\{X \leq x\}$ suggests that Y can have any value so long as the condition on X is met. This corresponds to the joint event $\{X \leq x, Y < \infty\}$. Therefore,

$$F_X(x) = P[X \leq x] = P[X \leq x, Y < \infty] = \lim_{y \to \infty} F_{X,Y}(x, y) = F_{X,Y}(x, \infty). \quad (5.2)$$

We obtain a similar result when we consider the event $\{Y \leq y\}$. The following theorem summarizes some basic properties of the joint CDF.

━━━━Theorem 5.1━━━━

For any pair of random variables, X, Y,

(a) $0 \leq F_{X,Y}(x, y) \leq 1$,

(b) $F_{X,Y}(\infty, \infty) = 1$,

(c) $F_X(x) = F_{X,Y}(x, \infty)$,

(d) $F_Y(y) = F_{X,Y}(\infty, y)$,

(e) $F_{X,Y}(x, -\infty) = 0$,

(f) $F_{X,Y}(-\infty, y) = 0$,

(g) *If $x \leq x_1$ and $y \leq y_1$, then*

$$F_{X,Y}(x, y) \leq F_{X,Y}(x_1, y_1)$$

Although its definition is simple, we rarely use the joint CDF to study probability models. It is easier to work with a probability mass function when the random variables are discrete or with a probability density function if they are continuous. Consider the joint CDF in the following example.

Example 5.2

X years is the age of children entering first grade in a school. Y years is the age of children entering second grade. The joint CDF of X and Y is

$$F_{X,Y}(x, y) = \begin{cases} 0 & x < 5, \\ 0 & y < 6, \\ (x-5)(y-6) & 5 \le x < 6, 6 \le y < 7, \\ y - 6 & x \ge 6, 6 \le y < 7, \\ x - 5 & 5 \le x < 6, y \ge 7, \\ 1 & \text{otherwise.} \end{cases} \tag{5.3}$$

Find $F_X(x)$ and $F_Y(y)$.

Using Theorem 5.1(b) and Theorem 5.1(c), we find

$$F_X(x) = \begin{cases} 0 & x < 5, \\ x - 5 & 5 \le x < 6, \\ 1 & x \ge 6, \end{cases} \qquad F_Y(y) = \begin{cases} 0 & y < 6, \\ y - 6 & 6 \le y < 7, \\ 1 & y \ge 7. \end{cases} \tag{5.4}$$

Referring to Theorem 4.6, we see from Equation (5.4) that X is a continuous uniform $(5, 6)$ random variable and Y is a continuous uniform $(6, 7)$ random variable.

In this example, we need to refer to six different regions in the x, y plane and three different formulas to express a probability model as a joint CDF. Section 5.4 introduces the joint probability density function as another representation of the probability model of a pair of random variables $f_{X,Y}(x, y)$. For childrens' ages X and Y in Example 5.2, we will show in Example 5.6 that the CDF $F_{X,Y}(x, y)$ implies that the joint PDF is the simple expression

$$f_{X,Y}(x, y) = \begin{cases} 1 & 5 \le x < 6, 6 \le y < 7, \\ 0 & \text{otherwise.} \end{cases} \tag{5.5}$$

To get another idea of the complexity of using the joint CDF, try proving the following theorem, which expresses the probability that an outcome is in a rectangle in the X, Y plane in terms of the joint CDF.

Theorem 5.2

$$P[x_1 < X \le x_2, y_1 < Y \le y_2] = F_{X,Y}(x_2, y_2) - F_{X,Y}(x_2, y_1) \\ - F_{X,Y}(x_1, y_2) + F_{X,Y}(x_1, y_1).$$

The steps needed to prove the theorem are outlined in Problem 5.1.5. The theorem says that to find the probability that an outcome is in a rectangle, it is necessary to evaluate the joint CDF at all four corners. When the probability of interest corresponds to a nonrectangular area, using the joint CDF is even more complex.

═══════**Quiz 5.1**═══════

Express the following extreme values of the joint CDF $F_{X,Y}(x, y)$ as numbers or in terms of the CDFs $F_X(x)$ and $F_Y(y)$.

(a) $F_{X,Y}(-\infty, 2)$

(b) $F_{X,Y}(\infty, \infty)$

(c) $F_{X,Y}(\infty, y)$

(d) $F_{X,Y}(\infty, -\infty)$

5.2   Joint Probability Mass Function

> For discrete random variables X and Y, the joint PMF $P_{X,Y}(x, y)$ is the probability that $X = x$ and $Y = y$. It is a complete probability model for X and Y.

Corresponding to the PMF of a single discrete random variable, we have a probability mass function of two variables.

═══════**Definition 5.2**═══════**Joint Probability Mass Function (PMF)**

*The **joint probability mass function** of discrete random variables X and Y is*

$$P_{X,Y}(x, y) = P[X = x, Y = y].$$

For a pair of discrete random variables, the joint PMF $P_{X,Y}(x, y)$ is a complete probability model. For any pair of real numbers, the PMF is the probability of observing these numbers. The notation is consistent with that of the joint CDF. The uppercase subscripts of P, separated by a comma, are the names of the two random variables. We usually write the arguments of the function as the lowercase letters associated with the random variable names. Corresponding to S_X, the range of a single discrete random variable, we use the notation $S_{X,Y}$ to denote the set of possible values of the pair (X, Y). That is,

$$S_{X,Y} = \{(x, y)|P_{X,Y}(x, y) > 0\}. \tag{5.6}$$

Keep in mind that $\{X = x, Y = y\}$ is an event in an experiment. That is, for this experiment, there is a set of observations that leads to both $X = x$ and $Y = y$. For any x and y, we find $P_{X,Y}(x, y)$ by summing the probabilities of all outcomes of the experiment for which $X = x$ and $Y = y$.

There are various ways to represent a joint PMF. We use three of them in the following example: a graph, a list, and a table.

Example 5.3

Test two integrated circuits one after the other. On each test, the possible outcomes are a (accept) and r (reject). Assume that all circuits are acceptable with probability 0.9 and that the outcomes of successive tests are independent. Count the number of acceptable circuits X and count the number of successful tests Y before you observe the first reject. (If both tests are successful, let $Y = 2$.) Draw a tree diagram for the experiment and find the joint PMF $P_{X,Y}(x, y)$.

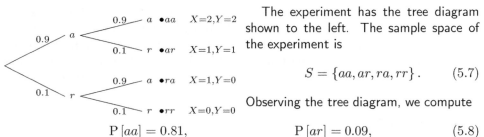

The experiment has the tree diagram shown to the left. The sample space of the experiment is

$$S = \{aa, ar, ra, rr\}. \qquad (5.7)$$

Observing the tree diagram, we compute

$$P\,[aa] = 0.81, \qquad\qquad P\,[ar] = 0.09, \qquad\qquad (5.8)$$
$$P\,[ra] = 0.09, \qquad\qquad P\,[rr] = 0.01. \qquad\qquad (5.9)$$

Each outcome specifies a pair of values X and Y. Let $g(s)$ be the function that transforms each outcome s in the sample space S into the pair of random variables (X, Y). Then

$$g(aa) = (2, 2), \quad g(ar) = (1, 1), \quad g(ra) = (1, 0), \quad g(rr) = (0, 0). \qquad (5.10)$$

For each pair of values x, y, $P_{X,Y}(x, y)$ is the sum of the probabilities of the outcomes for which $X = x$ and $Y = y$. For example, $P_{X,Y}(1, 1) = P[ar]$.

$P_{X,Y}(x,y)$	$y = 0$	$y = 1$	$y = 2$
$x = 0$	0.01	0	0
$x = 1$	0.09	0.09	0
$x = 2$	0	0	0.81

The joint PMF can be represented by the table on left, or, as shown below, as a set of labeled points in the x, y plane where each point is a possible value (probability > 0) of the pair (x, y), or as a simple list:

$$P_{X,Y}(x, y) = \begin{cases} 0.81 & x = 2, y = 2, \\ 0.09 & x = 1, y = 1, \\ 0.09 & x = 1, y = 0, \\ 0.01 & x = 0, y = 0. \\ 0 & \text{otherwise} \end{cases}$$

Note that all of the probabilities add up to 1. This reflects the second axiom of probability (Section 1.2) that states $P[S] = 1$. Using the notation of random variables, we write this as

$$\sum_{x \in S_X} \sum_{y \in S_Y} P_{X,Y}(x, y) = 1. \tag{5.11}$$

As defined in Chapter 3, the range S_X is the set of all values of X with nonzero probability and similarly for S_Y. It is easy to see the role of the first axiom of probability in the PMF: $P_{X,Y}(x, y) \geq 0$ for all pairs x, y. The third axiom, which has to do with the union of mutually exclusive events, takes us to another important property of the joint PMF.

We represent an event B as a region in the X, Y plane. Figure 5.2 shows two examples of events. We would like to find the probability that the pair of random variables (X, Y) is in the set B. When $(X, Y) \in B$, we say the event B occurs. Moreover, we write $P[B]$ as a shorthand for $P[(X, Y) \in B]$. The next theorem says that we can find $P[B]$ by adding the probabilities of all points (x, y) that are in B.

Theorem 5.3

For discrete random variables X and Y and any set B in the X, Y plane, the probability of the event $\{(X, Y) \in B\}$ is

$$P[B] = \sum_{(x,y) \in B} P_{X,Y}(x, y).$$

The following example uses Theorem 5.3.

Example 5.4

Continuing Example 5.3, find the probability of the event B that X, the number of acceptable circuits, equals Y, the number of tests before observing the first failure.
. . . .
Mathematically, B is the event $\{X = Y\}$. The elements of B with nonzero probability are

$$B \cap S_{X,Y} = \{(0,0), (1,1), (2,2)\}. \tag{5.12}$$

Therefore,

$$P[B] = P_{X,Y}(0, 0) + P_{X,Y}(1, 1) + P_{X,Y}(2, 2)$$
$$= 0.01 + 0.09 + 0.81 = 0.91. \tag{5.13}$$

If we view x, y as the outcome of an experiment, then Theorem 5.3 simply says that to find the probability of an event, we sum over all the outcomes in that event. In essence, Theorem 5.3 is a restatement of Theorem 1.4 in terms of random variables X and Y and joint PMF $P_{X,Y}(x, y)$.

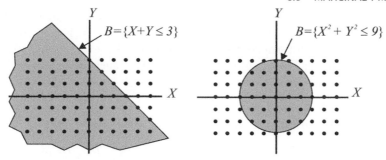

Figure 5.2   Subsets B of the (X,Y) plane. Points $(X,Y) \in S_{X,Y}$ are marked by bullets.

Quiz 5.2

The joint PMF $P_{Q,G}(q,g)$ for random variables Q and G is given in the following table:

$P_{Q,G}(q,g)$	$g=0$	$g=1$	$g=2$	$g=3$
$q=0$	0.06	0.18	0.24	0.12
$q=1$	0.04	0.12	0.16	0.08

Calculate the following probabilities:

(a) $P[Q=0]$

(b) $P[Q=G]$

(c) $P[G>1]$

(d) $P[G>Q]$

5.3   Marginal PMF

> For discrete random variables, the marginal PMFs $P_X(x)$ and $P_Y(y)$ are probability models for the individual random variables X and Y but they do not provide a complete probability model for the pair X,Y.

In an experiment that produces two random variables X and Y, it is always possible to consider one of the random variables, Y, and ignore the other one, X. In this case, we can use the methods of Chapter 3 to analyze the experiment and derive $P_Y(y)$, which contains the probability model for the random variable of interest. On the other hand, if we have already analyzed the experiment to derive the joint PMF $P_{X,Y}(x,y)$, it would be convenient to derive $P_Y(y)$ from $P_{X,Y}(x,y)$ without reexamining the details of the experiment.

To do so, we view x,y as the outcome of an experiment and observe that $P_{X,Y}(x,y)$ is the probability of an outcome. Moreover, $\{Y=y\}$ is an event, so that $P_Y(y) = P[Y=y]$ is the probability of an event. Theorem 5.3 relates the probability of an event to the joint PMF. It implies that we can find $P_Y(y)$ by summing $P_{X,Y}(x, y)$ over all points in $S_{X,Y}$ with the property $Y = y$. In the sum, y is a constant, and each term corresponds to a value of $x \in S_X$. Similarly, we can find $P_X(x)$ by summing $P_{X,Y}(x, y)$ over all points X, Y such that $X = x$. We state this mathematically in the next theorem.

======Theorem 5.4======

For discrete random variables X and Y with joint PMF $P_{X,Y}(x, y)$,

$$P_X(x) = \sum_{y \in S_Y} P_{X,Y}(x, y), \qquad P_Y(y) = \sum_{x \in S_X} P_{X,Y}(x, y).$$

Theorem 5.4 shows us how to obtain the probability model (PMF) of X, and the probability model of Y given a probability model (joint PMF) of X and Y. When a random variable X is part of an experiment that produces two random variables, we sometimes refer to its PMF as a *marginal probability mass function*. This terminology comes from the matrix representation of the joint PMF. By adding rows and columns and writing the results in the margins, we obtain the marginal PMFs of X and Y. We illustrate this by reference to the experiment in Example 5.3.

======Example 5.5======

$P_{X,Y}(x, y)$	$y = 0$	$y = 1$	$y = 2$
$x = 0$	0.01	0	0
$x = 1$	0.09	0.09	0
$x = 2$	0	0	0.81

In Example 5.3, we found that random variables X and Y have the joint PMF shown in this table. Find the marginal PMFs for the random variables X and Y.

. . . .

We note that both X and Y have range $\{0, 1, 2\}$. Theorem 5.4 gives

$$P_X(0) = \sum_{y=0}^{2} P_{X,Y}(0, y) = 0.01 \qquad P_X(1) = \sum_{y=0}^{2} P_{X,Y}(1, y) = 0.18 \qquad (5.14)$$

$$P_X(2) = \sum_{y=0}^{2} P_{X,Y}(2, y) = 0.81 \qquad P_X(x) = 0 \quad x \neq 0, 1, 2 \qquad (5.15)$$

Referring to the table representation of $P_{X,Y}(x, y)$, we observe that each value of $P_X(x)$ is the result of adding all the entries in one row of the table. Similarly, the formula for the PMF of Y in Theorem 5.4, $P_Y(y) = \sum_{x \in S_X} P_{X,Y}(x, y)$, is the sum of all the entries in one column of the table. We display $P_X(x)$ and $P_Y(y)$ by rewriting the table and placing the row sums and column sums in the margins.

$P_{X,Y}(x, y)$	$y = 0$	$y = 1$	$y = 2$	$P_X(x)$
$x = 0$	0.01	0	0	0.01
$x = 1$	0.09	0.09	0	0.18
$x = 2$	0	0	0.81	0.81
$P_Y(y)$	0.10	0.09	0.81	

Thus the column in the right margin shows $P_X(x)$ and the row in the bottom margin shows $P_Y(y)$. Note that the sum of all the entries in the bottom margin is 1 and so is the sum of all the entries in the right margin. This is simply a verification of Theorem 3.1(b), which states that the PMF of any random variable must sum to 1.

Quiz 5.3

The probability mass function $P_{H,B}(h, b)$ for the two random variables H and B is given in the following table. Find the marginal PMFs $P_H(h)$ and $P_B(b)$.

$P_{H,B}(h, b)$	$b = 0$	$b = 2$	$b = 4$
$h = -1$	0	0.4	0.2
$h = 0$	0.1	0	0.1
$h = 1$	0.1	0.1	0

$$(5.16)$$

5.4   Joint Probability Density Function

The most useful probability model of continuous random variables X and Y is the joint PDF $f_{X,Y}(x, y)$. It is a generalization of the PDF of a single random variable.

Definition 5.3━━━━**Joint Probability Density Function (PDF)**

The joint PDF of the continuous random variables X and Y is a function $f_{X,Y}(x, y)$ with the property

$$F_{X,Y}(x, y) = \int_{-\infty}^{x} \int_{-\infty}^{y} f_{X,Y}(u, v)\, dv\, du.$$

Given $F_{X,Y}(x, y)$, Definition 5.3 implies that $f_{X,Y}(x, y)$ is a derivative of the CDF.

Theorem 5.5

$$f_{X,Y}(x, y) = \frac{\partial^2 F_{X,Y}(x, y)}{\partial x\, \partial y}$$

For a single random variable X, the PDF $f_X(x)$ is a measure of probability per unit length. For two random variables X and Y, the joint PDF $f_{X,Y}(x, y)$ measures probability per unit area. In particular, from the definition of the PDF,

$$P[x < X \le x + dx, y < Y \le y + dy] = f_{X,Y}(x, y)\, dx\, dy. \qquad (5.17)$$

Definition 5.3 and Theorem 5.5 demonstrate that the joint CDF $F_{X,Y}(x, y)$ and the joint PDF $f_{X,Y}(x, y)$ represent the same probability model for random variables X

and Y. In the case of one random variable, we found in Chapter 4 that the PDF is typically more useful for problem solving. The advantage is even stronger for a pair of random variables.

====Example 5.6====

Use the joint CDF for childrens' ages X and Y given in Example 5.2 to derive the joint PDF presented in Equation (5.5).

. . . .

Referring to Equation (5.3) for the joint CDF $F_{X,Y}(x, y)$, we must evaluate the partial derivative $\partial^2 F_{X,Y}(x, y)/\partial x\, \partial y$ for each of the six regions specified in Equation (5.3). However, $\partial^2 F_{X,Y}(x, y)/\partial x\, \partial y$ is nonzero only if $F_{X,Y}(x.y)$ is a function of both x and y. In this example, only the region $\{5 \leq x < 6, 6 \leq y < 7\}$ meets this requirement. Over this region,

$$f_{X,Y}(x, y) = \frac{\partial^2}{\partial x\, \partial y} \left[(x - 5)(y - 6)\right] = \frac{\partial}{\partial x}[x - 5]\frac{\partial}{\partial y}[y - 6] = 1. \tag{5.18}$$

Over all other regions, the joint PDF $f_{X,Y}(x, y)$ is zero.

Of course, not every function $f_{X,Y}(x, y)$ is a joint PDF. Properties (e) and (f) of Theorem 5.1 for the CDF $F_{X,Y}(x, y)$ imply corresponding properties for the PDF.

====Theorem 5.6====

A joint PDF $f_{X,Y}(x, y)$ has the following properties corresponding to first and second axioms of probability (see Section 1.2):

(a) $f_{X,Y}(x, y) \geq 0$ *for all* (x, y),   *(b)* $\displaystyle\int_{-\infty}^{\infty} \int_{-\infty}^{\infty} f_{X,Y}(x, y)\, dx\, dy = 1.$

Given an experiment that produces a pair of continuous random variables X and Y, an event A corresponds to a region of the X, Y plane. The probability of A is the double integral of $f_{X,Y}(x, y)$ over the region A of the X, Y plane.

====Theorem 5.7====

The probability that the continuous random variables (X, Y) are in A is

$$\mathrm{P}\left[A\right] = \iint\limits_{A} f_{X,Y}(x, y)\, dx\, dy.$$

====Example 5.7====

Random variables X and Y have joint PDF

$$f_{X,Y}(x, y) = \begin{cases} c & 0 \leq x \leq 5, 0 \leq y \leq 3, \\ 0 & \text{otherwise.} \end{cases} \tag{5.19}$$

Find the constant c and $\mathrm{P}[A] = \mathrm{P}[2 \leq X < 3, 1 \leq Y < 3]$.

. . . .

The large rectangle in the diagram is the area of nonzero probability. Theorem 5.6 states that the integral of the joint PDF over this rectangle is 1:

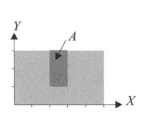

$$1 = \int_0^5 \int_0^3 c\,dy\,dx = 15c. \tag{5.20}$$

Therefore, $c = 1/15$. The small dark rectangle in the diagram is the event $A = \{2 \le X < 3, 1 \le Y < 3\}$. $P[A]$ is the integral of the PDF over this rectangle, which is

$$P[A] = \int_2^3 \int_1^3 \frac{1}{15}\,dv\,du = 2/15. \tag{5.21}$$

This probability model is an example of a pair of random variables uniformly distributed over a rectangle in the X, Y plane.

The following example derives the CDF of a pair of random variables that has a joint PDF that is easy to write mathematically. The purpose of the example is to introduce techniques for analyzing a more complex probability model than the one in Example 5.7. Typically, we extract interesting information from a model by integrating the PDF or a function of the PDF over some region in the X, Y plane. In performing this integration, the most difficult task is to identify the limits. The PDF in the example is very simple, just a constant over a triangle in the X, Y plane. However, to evaluate its integral over the region in Figure 5.1 we need to consider five different situations depending on the values of (x, y). The solution of the example demonstrates the point that the PDF is usually a more concise probability model that offers more insights into the nature of an experiment than the CDF.

Example 5.8

Find the joint CDF $F_{X,Y}(x, y)$ when X and Y have joint PDF

$$f_{X,Y}(x, y) = \begin{cases} 2 & 0 \le y \le x \le 1, \\ 0 & \text{otherwise.} \end{cases} \tag{5.22}$$

....

We can derive the joint CDF using Definition 5.3 in which we integrate the joint PDF $f_{X,Y}(x, y)$ over the area shown in Figure 5.1. To perform the integration it is extremely useful to draw a diagram that clearly shows the area with nonzero probability and then to use the diagram to derive the limits of the integral in Definition 5.3.

The difficulty with this integral is that the nature of the region of integration depends critically on x and y. In this apparently simple example, there are five cases to consider! The five cases are shown in Figure 5.3. First, we note that with $x < 0$ or $y < 0$, the triangle is completely outside the region of integration, as shown in Figure 5.3a. Thus

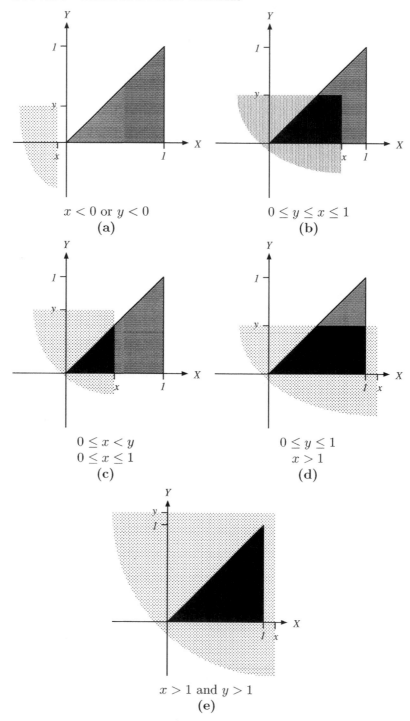

Figure 5.3   Five cases for the CDF $F_{X,Y}(x,y)$ of Example 5.8.

we have $F_{X,Y}(x,y) = 0$ if either $x < 0$ or $y < 0$. Another simple case arises when $x \geq 1$ and $y \geq 1$. In this case, we see in Figure 5.3e that the triangle is completely inside the region of integration, and we infer from Theorem 5.6 that $F_{X,Y}(x,y) = 1$. The other cases we must consider are more complicated. In each case, since $f_{X,Y}(x,y) = 2$ over the triangular region, the value of the integral is two times the indicated area. When (x,y) is inside the area of nonzero probability (Figure 5.3b), the integral is

$$F_{X,Y}(x,y) = \int_0^y \int_v^x 2\,du\,dv = 2xy - y^2 \qquad \text{(Figure 5.3b)}. \qquad (5.23)$$

In Figure 5.3c, (x,y) is above the triangle, and the integral is

$$F_{X,Y}(x,y) = \int_0^x \int_v^x 2\,du\,dv = x^2 \qquad \text{(Figure 5.3c)}. \qquad (5.24)$$

The remaining situation to consider is shown in Figure 5.3d, when (x,y) is to the right of the triangle of nonzero probability, in which case the integral is

$$F_{X,Y}(x,y) = \int_0^y \int_v^1 2\,du\,dv = 2y - y^2 \qquad \text{(Figure 5.3d)} \qquad (5.25)$$

The resulting CDF, corresponding to the five cases of Figure 5.3, is

$$F_{X,Y}(x,y) = \begin{cases} 0 & x < 0 \text{ or } y < 0 & \text{(a)}, \\ 2xy - y^2 & 0 \leq y \leq x \leq 1 & \text{(b)}, \\ x^2 & 0 \leq x < y, 0 \leq x \leq 1 & \text{(c)}, \\ 2y - y^2 & 0 \leq y \leq 1, x > 1 & \text{(d)}, \\ 1 & x > 1, y > 1 & \text{(e)}. \end{cases} \qquad (5.26)$$

In Figure 5.4, the surface plot of $F_{X,Y}(x,y)$ shows that cases (a) through (e) correspond to contours on the "hill" that is $F_{X,Y}(x,y)$. In terms of visualizing the random variables, the surface plot of $F_{X,Y}(x,y)$ is less instructive than the simple triangle characterizing the PDF $f_{X,Y}(x,y)$.

Because the PDF in this example is $f_{X,Y}(x,y) = 2$ over $(x,y) \in S_{X,Y}$, each probability is just two times the area of the region shown in one of the diagrams (either a triangle or a trapezoid). You may want to apply some high school geometry to verify that the results obtained from the integrals are indeed twice the areas of the regions indicated. The approach taken in our solution, integrating over $S_{X,Y}$ to obtain the CDF, works for any PDF.

In Example 5.8, it takes careful study to verify that $F_{X,Y}(x,y)$ is a valid CDF that satisfies the properties of Theorem 5.1, or even that it is defined for all values x and y. Comparing the joint PDF with the joint CDF, we see that the PDF indicates clearly that X, Y occurs with equal probability in all areas of the same size in the triangular region $0 \leq y \leq x \leq 1$. The joint CDF completely hides this simple, important property of the probability model.

In the previous example, the triangular shape of the area of nonzero probability demanded our careful attention. In the next example, the area of nonzero probability is a rectangle. However, the area corresponding to the event of interest is more complicated.

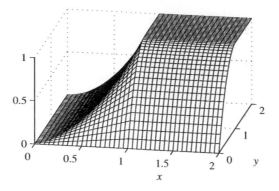

Figure 5.4   A graph of the joint CDF $F_{X,Y}(x,y)$ of Example 5.8.

Example 5.9

As in Example 5.7, random variables X and Y have joint PDF

$$f_{X,Y}(x,y) = \begin{cases} 1/15 & 0 \le x \le 5, 0 \le y \le 3, \\ 0 & \text{otherwise.} \end{cases} \qquad (5.27)$$

What is $P[A] = P[Y > X]$?

....

Applying Theorem 5.7, we integrate $f_{X,Y}(x,y)$ over the part of the X,Y plane satisfying $Y > X$. In this case,

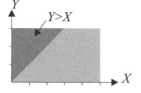

$$P[A] = \int_0^3 \left(\int_x^3 \frac{1}{15} \right) dy\, dx \qquad (5.28)$$

$$= \int_0^3 \frac{3-x}{15}\, dx = \left. -\frac{(3-x)^2}{30} \right|_0^3 = \frac{3}{10}. \qquad (5.29)$$

In this example, it makes little difference whether we integrate first over y and then over x or the other way around. In general, however, an initial effort to decide the simplest way to integrate over a region can avoid a lot of complicated mathematical maneuvering in performing the integration.

Quiz 5.4

The joint probability density function of random variables X and Y is

$$f_{X,Y}(x,y) = \begin{cases} cxy & 0 \le x \le 1, 0 \le y \le 2, \\ 0 & \text{otherwise.} \end{cases} \qquad (5.30)$$

Find the constant c. What is the probability of the event $A = X^2 + Y^2 \le 1$?

5.5   Marginal PDF

> For continuous random variables, the marginal PDFs $f_X(x)$ and $f_Y(y)$ are probability models for the individual random variables X and Y, but they do not provide a complete probability model for the pair X, Y.

Suppose we perform an experiment that produces a pair of random variables X and Y with joint PDF $f_{X,Y}(x, y)$. For certain purposes we may be interested only in the random variable X. We can imagine that we ignore Y and observe only X. Since X is a random variable, it has a PDF $f_X(x)$. It should be apparent that there is a relationship between $f_X(x)$ and $f_{X,Y}(x, y)$. In particular, if $f_{X,Y}(x, y)$ completely summarizes our knowledge of joint events of the form $X = x, Y = y$, then we should be able to derive the PDFs of X and Y from $f_{X,Y}(x, y)$. The situation parallels (with integrals replacing sums) the relationship in Theorem 5.4 between the joint PMF $P_{X,Y}(x, y)$, and the marginal PMFs $P_X(x)$ and $P_Y(y)$. Therefore, we refer to $f_X(x)$ and $f_Y(y)$ as the *marginal probability density functions* of $f_{X,Y}(x, y)$.

Theorem 5.8

If X and Y are random variables with joint PDF $f_{X,Y}(x, y)$,

$$f_X(x) = \int_{-\infty}^{\infty} f_{X,Y}(x, y)\, dy, \qquad f_Y(y) = \int_{-\infty}^{\infty} f_{X,Y}(x, y)\, dx.$$

Proof From the definition of the joint PDF, we can write

$$F_X(x) = P[X \le x] = \int_{-\infty}^{x} \left(\int_{-\infty}^{\infty} f_{X,Y}(u, y)\, dy \right) du. \tag{5.31}$$

Taking the derivative of both sides with respect to x (which involves differentiating an integral with variable limits), we obtain $f_X(x) = \int_{-\infty}^{\infty} f_{X,Y}(x, y)\, dy$. A similar argument holds for $f_Y(y)$.

Example 5.10

The joint PDF of X and Y is

$$f_{X,Y}(x, y) = \begin{cases} 5y/4 & -1 \le x \le 1, x^2 \le y \le 1, \\ 0 & \text{otherwise.} \end{cases} \tag{5.32}$$

Find the marginal PDFs $f_X(x)$ and $f_Y(y)$.

· ·

We use Theorem 5.8 to find the marginal PDF $f_X(x)$. In the figure that accompanies Equation (5.33) below, the gray bowl-shaped region depicts those values of X and Y for which $f_{X,Y}(x, y) > 0$. When $x < -1$ or when $x > 1$, $f_{X,Y}(x, y) = 0$, and therefore $f_X(x) = 0$. For $-1 \le x \le 1$,

$$f_X(x) = \int_{x^2}^{1} \frac{5y}{4}\, dy = \frac{5(1 - x^4)}{8}. \tag{5.33}$$

The complete expression for the marginal PDF of X is

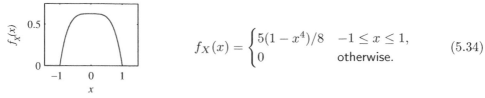

$$f_X(x) = \begin{cases} 5(1 - x^4)/8 & -1 \leq x \leq 1, \\ 0 & \text{otherwise.} \end{cases} \qquad (5.34)$$

For the marginal PDF of Y, we note that for $y < 0$ or $y > 1$, $f_Y(y) = 0$. For $0 \leq y \leq 1$, we integrate over the horizontal bar marked $Y = y$. The boundaries of the bar are $x = -\sqrt{y}$ and $x = \sqrt{y}$. Therefore, for $0 \leq y \leq 1$,

$$f_Y(y) = \int_{-\sqrt{y}}^{\sqrt{y}} \frac{5y}{4}\, dx = \frac{5y}{4} x \Big|_{x=-\sqrt{y}}^{x=\sqrt{y}} = 5y^{3/2}/2. \qquad (5.35)$$

The complete marginal PDF of Y is

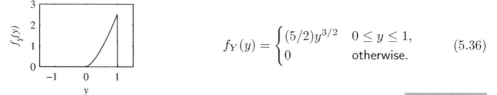

$$f_Y(y) = \begin{cases} (5/2)y^{3/2} & 0 \leq y \leq 1, \\ 0 & \text{otherwise.} \end{cases} \qquad (5.36)$$

Quiz 5.5

The joint probability density function of random variables X and Y is

$$f_{X,Y}(x,y) = \begin{cases} 6(x + y^2)/5 & 0 \leq x \leq 1, 0 \leq y \leq 1, \\ 0 & \text{otherwise.} \end{cases} \qquad (5.37)$$

Find $f_X(x)$ and $f_Y(y)$, the marginal PDFs of X and Y.

5.6   Independent Random Variables

> Random variables X and Y are independent if and only if the events $\{X = x\}$ and $\{Y = y\}$ are independent for all x, y in $S_{X,Y}$. Discrete random variables X and Y are independent if and only if $P_{X,Y}(x,y) = P_X(x)P_Y(y)$. Continuous random variables X and Y are independent if and only if $f_{X,Y}(x,y) = f_X(x)f_Y(y)$.

Chapter 1 presents the concept of independent events. Definition 1.6 states that events A and B are independent if and only if the probability of the intersection is the product of the individual probabilities, $P[AB] = P[A]\,P[B]$.

Applying the idea of independence to random variables, we say that X and Y are independent random variables if and only if the events $\{X = x\}$ and $\{Y = y\}$ are independent for all $x \in S_X$ and all $y \in S_Y$. In terms of probability mass functions and probability density functions, we have the following definition.

=======**Definition 5.4**=======**Independent Random Variables**

Random variables X and Y are **independent** *if and only if*

$$\text{Discrete:} \quad P_{X,Y}(x, y) = P_X(x) P_Y(y);$$

$$\text{Continuous:} \quad f_{X,Y}(x, y) = f_X(x) f_Y(y).$$

=======**Example 5.11**=======

Are the childrens' ages X and Y in Example 5.2 independent?

. . . .

In Example 5.2, we derived the CDFs $F_X(x)$ and $F_Y(y)$, which showed that X is uniform $(5, 6)$ and Y is uniform $(6, 7)$. Thus X and Y have marginal PDFs

$$f_X(x) = \begin{cases} 1 & 5 \leq x < 6, \\ 0 & \text{otherwise,} \end{cases} \qquad f_Y(y) = \begin{cases} 1 & 6 \leq x < 7, \\ 0 & \text{otherwise.} \end{cases} \tag{5.38}$$

Referring to Equation (5.5), we observe that $f_{X,Y}(x, y) = f_X(x) f_Y(y)$. Thus X and Y are independent.

Because Definition 5.4 is an equality of functions, it must be true for all values of x and y.

=======**Example 5.12**=======

$$f_{X,Y}(x, y) = \begin{cases} 4xy & 0 \leq x \leq 1, 0 \leq y \leq 1, \\ 0 & \text{otherwise.} \end{cases}$$

Are X and Y independent?

. . . .

The marginal PDFs of X and Y are

$$f_X(x) = \begin{cases} 2x & 0 \leq x \leq 1, \\ 0 & \text{otherwise,} \end{cases} \qquad f_Y(y) = \begin{cases} 2y & 0 \leq y \leq 1, \\ 0 & \text{otherwise.} \end{cases} \tag{5.39}$$

It is easily verified that $f_{X,Y}(x, y) = f_X(x) f_Y(y)$ for all pairs (x, y), and so we conclude that X and Y are independent.

=======**Example 5.13**=======

$$f_{U,V}(u, v) = \begin{cases} 24uv & u \geq 0, v \geq 0, u + v \leq 1, \\ 0 & \text{otherwise.} \end{cases} \tag{5.40}$$

Are U and V independent?

....

Since $f_{U,V}(u,v)$ looks similar in form to $f_{X,Y}(x,y)$ in the previous example, we might suppose that U and V can also be factored into marginal PDFs $f_U(u)$ and $f_V(v)$. However, this is not the case. Owing to the triangular shape of the region of nonzero probability, the marginal PDFs are

$$f_U(u) = \begin{cases} 12u(1-u)^2 & 0 \le u \le 1, \\ 0 & \text{otherwise,} \end{cases} \qquad f_V(v) = \begin{cases} 12v(1-v)^2 & 0 \le v \le 1, \\ 0 & \text{otherwise.} \end{cases}$$

Clearly, U and V are not independent. Learning U changes our knowledge of V. For example, learning $U = 1/2$ informs us that $P[V \le 1/2] = 1$.

In these two examples, we see that the region of nonzero probability plays a crucial role in determining whether random variables are independent. Once again, we emphasize that to infer that X and Y are independent, it is necessary to verify the functional equalities in Definition 5.4 for all $x \in S_X$ and $y \in S_Y$. There are many cases in which some events of the form $\{X = x\}$ and $\{Y = y\}$ are independent and others are not independent. If this is the case, the random variables X and Y are not independent.

In Examples 5.12 and 5.13, we are given a joint PDF and asked to determine whether the random variables are independent. By contrast, in many applications of probability, the nature of an experiment leads to a model in which X and Y are independent. In these applications we examine an experiment and determine that it is appropriate to model a pair of random variables X and Y as independent. To analyze the experiment, we start with the PDFs $f_X(x)$ and $f_Y(y)$, and then construct the joint PDF $f_{X,Y}(x,y) = f_X(x)f_Y(y)$.

Example 5.14

Consider again the noisy observation model of Example 5.1. Suppose X is a Gaussian $(0, \sigma_X)$ information signal sent by a radio transmitter and $Y = X + Z$ is the output of a low-noise amplifier attached to the antenna of a radio receiver. The noise Z is a Gaussian $(0, \sigma_Z)$ random variable that is generated within the receiver. What is the joint PDF $f_{X,Z}(x,z)$?

....

From the information given, we know that X and Z have PDFs

$$f_X(x) = \frac{1}{\sqrt{2\pi\sigma_X^2}} e^{-x^2/2\sigma_X^2}, \qquad f_Z(z) = \frac{1}{\sqrt{2\pi\sigma_Z^2}} e^{-z^2/2\sigma_Z^2}. \tag{5.41}$$

The signal X depends on the information being transmitted by the sender and the noise Z depends on electrons bouncing around in the receiver circuitry. As there is no reason for these to be related, we model X and Z as independent. Thus, the joint PDF is

$$f_{X,Z}(x,z) = f_X(x)f_Z(z) = \frac{1}{2\pi\sqrt{\sigma_X^2\sigma_Z^2}} e^{-\frac{1}{2}\left(\frac{x^2}{\sigma_X^2} + \frac{z^2}{\sigma_Z^2}\right)}. \tag{5.42}$$

Quiz 5.6

(A) Random variables X and Y in Example 5.3 and random variables Q and G in Quiz 5.2 have joint PMFs:

$P_{X,Y}(x,y)$	$y=0$	$y=1$	$y=2$
$x=0$	0.01	0	0
$x=1$	0.09	0.09	0
$x=2$	0	0	0.81

$P_{Q,G}(q,g)$	$g=0$	$g=1$	$g=2$	$g=3$
$q=0$	0.06	0.18	0.24	0.12
$q=1$	0.04	0.12	0.16	0.08

(a) Are X and Y independent?   (b) Are Q and G independent?

(B) Random variables X_1 and X_2 are independent and identically distributed with probability density function

$$f_X(x) = \begin{cases} x/2 & 0 \le x \le 2, \\ 0 & \text{otherwise.} \end{cases} \tag{5.43}$$

What is the joint PDF $f_{X_1,X_2}(x_1,x_2)$?

5.7   Expected Value of a Function of Two Random Variables

$g(X,Y)$, a function of two random variables, is also a random variable. As with one random variable, it is convenient to calculate the expected value, $\mathrm{E}[g(X,Y)]$, without deriving a probability model of $g(X,Y)$.

There are many situations in which we observe two random variables and use their values to compute a new random variable. For example, we can model the amplitude of the signal transmitted by a radio station as a random variable, X. We can model the attenuation of the signal as it travels to the antenna of a moving car as another random variable, Y. In this case the amplitude of the signal at the radio receiver in the car is the random variable $W = X/Y$.

Formally, we have the following situation. We perform an experiment and observe sample values of two random variables X and Y. Based on our knowledge of the experiment, we have a probability model for X and Y embodied in a joint PMF $P_{X,Y}(x,y)$ or a joint PDF $f_{X,Y}(x,y)$. After performing the experiment, we calculate a sample value of the random variable $W = g(X,Y)$. W is referred to as a derived random variable. This section identifies important properties of the expected value, $\mathrm{E}[W]$. The probability model for W, embodied in $P_W(w)$ or $f_W(w)$, is the subject of Chapter 6.

As with a function of one random variable, we can calculate $\mathrm{E}[W]$ directly from $P_{X,Y}(x,y)$ or $f_{X,Y}(x,y)$ without deriving $P_W(w)$ or $f_W(w)$. Corresponding to Theorems 3.10 and 4.4, we have:

=====Theorem 5.9=====

For random variables X and Y, the expected value of $W = g(X, Y)$ is

$$\text{Discrete:} \quad \mathrm{E}\left[W\right] = \sum_{x \in S_X} \sum_{y \in S_Y} g(x, y) P_{X,Y}(x, y);$$

$$\text{Continuous:} \; \mathrm{E}\left[W\right] = \int_{-\infty}^{\infty} \int_{-\infty}^{\infty} g(x, y) f_{X,Y}(x, y) \, dx \, dy.$$

Theorem 5.9 is surprisingly powerful. For example, it lets us calculate easily the expected value of a linear combination of several functions.

=====Theorem 5.10=====

$$\mathrm{E}\left[a_1 g_1(X, Y) + \cdots + a_n g_n(X, Y)\right] = a_1 \, \mathrm{E}\left[g_1(X, Y)\right] + \cdots + a_n \, \mathrm{E}\left[g_n(X, Y)\right].$$

Proof Let $g(X, Y) = a_1 g_1(X, Y) + \cdots + a_n g_n(X, Y)$. For discrete random variables X, Y, Theorem 5.9 states

$$\mathrm{E}\left[g(X, Y)\right] = \sum_{x \in S_X} \sum_{y \in S_Y} \left(a_1 g_1(x, y) + \cdots + a_n g_n(x, y)\right) P_{X,Y}(x, y). \tag{5.44}$$

We can break the double summation into n weighted double summations:

$$\mathrm{E}\left[g(X, Y)\right] = a_1 \sum_{x \in S_X} \sum_{y \in S_Y} g_1(x, y) P_{X,Y}(x, y) + \cdots + a_n \sum_{x \in S_X} \sum_{y \in S_Y} g_n(x, y) P_{X,Y}(x, y).$$

By Theorem 5.9, the ith double summation on the right side is $\mathrm{E}[g_i(X, Y)]$; thus,

$$\mathrm{E}\left[g(X, Y)\right] = a_1 \, \mathrm{E}\left[g_1(X, Y)\right] + \cdots + a_n \, \mathrm{E}\left[g_n(X, Y)\right]. \tag{5.45}$$

For continuous random variables, Theorem 5.9 says

$$\mathrm{E}\left[g(X, Y)\right] = \int_{-\infty}^{\infty} \int_{-\infty}^{\infty} \left(a_1 g_1(x, y) + \cdots + a_n g_n(x, y)\right) f_{X,Y}(x, y) \, dx \, dy. \tag{5.46}$$

To complete the proof, we express this integral as the sum of n integrals and recognize that each of the new integrals is a weighted expected value, $a_i \, \mathrm{E}[g_i(X, Y)]$.

In words, Theorem 5.10 says that the expected value of a linear combination equals the linear combination of the expected values. We will have many occasions to apply this theorem. The following theorem describes the expected sum of two random variables, a special case of Theorem 5.10.

=====Theorem 5.11=====

For any two random variables X and Y,

$$\mathrm{E}\left[X + Y\right] = \mathrm{E}\left[X\right] + \mathrm{E}\left[Y\right].$$

This theorem implies that we can find the expected sum of two random variables from the separate probability models: $P_X(x)$ and $P_Y(y)$ or $f_X(x)$ and $f_Y(y)$. We do not need a complete probability model embodied in $P_{X,Y}(x, y)$ or $f_{X,Y}(x, y)$.

By contrast, the variance of $X + Y$ depends on the entire joint PMF or joint CDF:

=======Theorem 5.12=======

The variance of the sum of two random variables is

$$\mathrm{Var}\,[X + Y] = \mathrm{Var}\,[X] + \mathrm{Var}\,[Y] + 2\,\mathrm{E}\,[(X - \mu_X)(Y - \mu_Y)]\,.$$

Proof Since $\mathrm{E}[X + Y] = \mu_X + \mu_Y$,

$$\begin{aligned}
\mathrm{Var}[X + Y] &= \mathrm{E}\left[(X + Y - (\mu_X + \mu_Y))^2\right] \\
&= \mathrm{E}\left[((X - \mu_X) + (Y - \mu_Y))^2\right] \\
&= \mathrm{E}\left[(X - \mu_X)^2 + 2(X - \mu_X)(Y - \mu_Y) + (Y - \mu_Y)^2\right]. \quad (5.47)
\end{aligned}$$

We observe that each of the three terms in the preceding expected values is a function of X and Y. Therefore, Theorem 5.10 implies

$$\mathrm{Var}[X + Y] = \mathrm{E}\left[(X - \mu_X)^2\right] + 2\,\mathrm{E}\,[(X - \mu_X)(Y - \mu_Y)] + \mathrm{E}\left[(Y - \mu_Y)^2\right]. \quad (5.48)$$

The first and last terms are, respectively, $\mathrm{Var}[X]$ and $\mathrm{Var}[Y]$.

The expression $\mathrm{E}[(X - \mu_X)(Y - \mu_Y)]$ in the final term of Theorem 5.12 is a parameter of the probability model of X and Y. It reveals important properties of the relationship of X and Y. This quantity appears over and over in practical applications, and it has its own name, *covariance*.

=======Example 5.15=======
A company website has three pages. They require 750 kilobytes, 1500 kilobytes, and 2500 kilobytes for transmission. The transmission speed can be 5 Mb/s for external requests or 10 Mb/s for internal requests. Requests arrive randomly from inside and outside the company independently of page length, which is also random. The probability models for transmision speed, R, and page length, L, are:

$$P_R(r) = \begin{cases} 0.4 & r = 5, \\ 0.6 & r = 10, \\ 0 & \text{otherwise,} \end{cases} \qquad P_L(l) = \begin{cases} 0.3 & l = 750, \\ 0.5 & l = 1500, \\ 0.2 & l = 2500, \\ 0 & \text{otherwise.} \end{cases} \quad (5.49)$$

Write an expression for the transmission time $g(R, L)$ seconds. Derive the expected transmission time $\mathrm{E}[g(R, L)]$. Does $\mathrm{E}[(g(R, L)] = g(\mathrm{E}[R], \mathrm{E}[L])$?

. . . .

The transmission time T seconds is the the page length (in kb) divided by the transmission speed (in kb/s), or $T = 8L/1000R$. Because R and L are independent, $P_{R,L}(r,l) = P_R(r)P_L(l)$ and

$$
\begin{aligned}
\mathrm{E}\left[g(R,L)\right] &= \sum_r \sum_l P_R(r)\, P_L(l)\, \frac{8l}{1000r} \\
&= \frac{8}{1000} \left(\sum_r \frac{P_R(r)}{r} \right) \left(\sum_l P_L(l)\, l \right) \\
&= \frac{8}{1000} \left(\frac{0.4}{5} + \frac{0.6}{10} \right) (0.3(750) + 0.5(1500) + 0.2(2500)) \\
&= 1.652 \text{ s.}
\end{aligned}
\tag{5.50}
$$

By comparison, $\mathrm{E}[R] = \sum_r r P_R(r) = 8$ Mb/s and $\mathrm{E}[L] = \sum_l l P_L(l) = 1475$ kilobytes. This implies

$$
g(\mathrm{E}\left[R\right], \mathrm{E}\left[L\right]) = \frac{8\,\mathrm{E}\left[L\right]}{1000\,\mathrm{E}\left[R\right]} = 1.475 \text{ s} \neq \mathrm{E}\left[g(R,L)\right].
\tag{5.51}
$$

5.8   Covariance, Correlation and Independence

> The covariance $\mathrm{Cov}[X,Y]$, the correlation coefficient $\rho_{X,Y}$, and the correlation $r_{X,Y}$ are parameters of the probability model of X and Y. For independent random variables X and Y, $\mathrm{Cov}[X,Y] = \rho_{X,Y} = 0$.

====Definition 5.5========Covariance

*The **covariance** of two random variables X and Y is*

$$
\mathrm{Cov}\left[X,Y\right] = \mathrm{E}\left[(X - \mu_X)(Y - \mu_Y)\right].
$$

Sometimes, the notation σ_{XY} is used to denote the covariance of X and Y. We have already learned that the expected value parameter, $\mathrm{E}[X]$, is a typical value of X and that the variance parameter, $\mathrm{Var}[X]$, is a single number that describes how samples of X tend to be spread around the expected value $\mathrm{E}[X]$. In an analogous way, the covariance parameter $\mathrm{Cov}[X,Y]$ is a single number that describes how the pair of random variables X and Y vary together.

The key to understanding covariance is the random variable

$$
W = (X - \mu_X)(Y - \mu_Y).
\tag{5.52}
$$

Since $\mathrm{Cov}[X,Y] = \mathrm{E}[W]$, we observe that $\mathrm{Cov}[X,Y] > 0$ tells us that the typical values of $(X - \mu_X)(Y - \mu_Y)$ are positive. However, this is equivalent to saying that $X - \mu_X$ and $Y - \mu_Y$ typically have the same sign. That is, if $X > \mu_X$ then we would typically expect $Y > \mu_Y$; and if $X < \mu_X$ then we would expect to observe $Y < \mu_Y$. In short, if $\mathrm{Cov}[X, Y] > 0$, we would expect X and Y to go up or down together. On the other hand, if $\mathrm{Cov}[X, Y] < 0$, we would expect $X - \mu_X$ and $Y - \mu_Y$ to typically have opposite signs. In this case, when X goes up, Y typically goes down. Finally, if $\mathrm{Cov}[X, Y] \approx 0$, we might expect that the sign of $X - \mu_X$ doesn't provide much of a clue about the sign of $Y - \mu_Y$.

While this casual argument may be reasonably clear, it may also be some-what unsatisfactory. For example, would $\mathrm{Cov}[X, Y] = 0.1$ be fairly described as $\mathrm{Cov}[X, Y] \approx 0$? The answer to this question depends on the measurement units of X and Y.

════Example 5.16════

Suppose we perform an experiment in which we measure X and Y in centimeters (for example the height of two sisters). However, if we change units and measure height in meters, we will perform the same experiment except we observe $\hat{X} = X/100$ and $\hat{Y} = Y/100$. In this case, $\hat{X}$ and $\hat{Y}$ have expected values $\mu_{\hat{X}} = \mu_X/100$ m, $\mu_{\hat{Y}} = \mu_Y/100$ m and

$$\mathrm{Cov}\left[\hat{X}, \hat{Y}\right] = \mathrm{E}\left[(\hat{X} - \mu_{\hat{X}})(\hat{Y} - \mu_{\hat{Y}})\right]$$
$$= \frac{\mathrm{E}\left[(X - \mu_X)(Y - \mu_Y)\right]}{10,000} = \frac{\mathrm{Cov}\left[X, Y\right]}{10,000} \ \mathrm{m}^2. \tag{5.53}$$

Changing the unit of measurement from cm^2 to m^2 reduces the covariance by a factor of $10,000$. However, the tendency of $X - \mu_X$ and $Y - \mu_Y$ to have the same sign is the same as the tendency of $\hat{X} - \mu_{\hat{X}}$ and $\hat{Y} - \mu_{\hat{Y}}$ to have the same sign. (Both are an indication of how likely it is that a girl is taller than average if her sister is taller than average).

A parameter that indicates the relationship of two random variables regardless of measurement units is a normalized version of $\mathrm{Cov}[X, Y]$, called the correlation coefficient.

════Definition 5.6════Correlation Coefficient

*The **correlation coefficient** of two random variables X and Y is*

$$\rho_{X,Y} = \frac{\mathrm{Cov}\left[X, Y\right]}{\sqrt{\mathrm{Var}[X] \, \mathrm{Var}[Y]}} = \frac{\mathrm{Cov}\left[X, Y\right]}{\sigma_X \sigma_Y}.$$

Note that the covariance has units equal to the product of the units of X and Y. Thus, if X has units of kilograms and Y has units of seconds, then $\mathrm{Cov}[X, Y]$ has units of kilogram-seconds. By contrast, $\rho_{X,Y}$ is a dimensionless quantity that is not affected by scale changes.

════Theorem 5.13════

If $\hat{X} = aX + b$ and $\hat{Y} = cY + d$, then
(a) $\rho_{\hat{X}, \hat{Y}} = \rho_{X,Y}$,   *(b)* $\mathrm{Cov}[\hat{X}, \hat{Y}] = ac \, \mathrm{Cov}[X, Y]$.

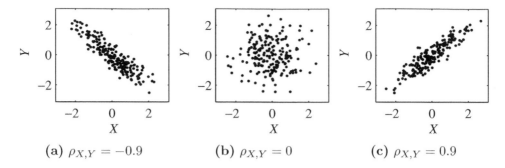

(a) $\rho_{X,Y} = -0.9$ (b) $\rho_{X,Y} = 0$ (c) $\rho_{X,Y} = 0.9$

Figure 5.5   Each graph has 200 samples, each marked by a dot, of the random variable pair (X, Y) such that $E[X] = E[Y] = 0$, $Var[X] = Var[Y] = 1$.

The proof steps are outlined in Problem 5.8.9. Related to this insensitivity of $\rho_{X,Y}$ to scale changes, an important property of the correlation coefficient is that it is bounded by -1 and 1:

━━━━━Theorem 5.14━━━━━

$$-1 \le \rho_{X,Y} \le 1.$$

Proof Let σ_X^2 and σ_Y^2 denote the variances of X and Y, and for a constant a, let $W = X - aY$. Then,

$$\text{Var}[W] = E\left[(X - aY)^2\right] - (E\left[X - aY\right])^2. \tag{5.54}$$

Since $E[X - aY] = \mu_X - a\mu_Y$, expanding the squares yields

$$\text{Var}[W] = E\left[X^2 - 2aXY + a^2 Y^2\right] - \left(\mu_X^2 - 2a\mu_X\mu_Y + a^2\mu_Y^2\right)$$
$$= \text{Var}[X] - 2a\,\text{Cov}\,[X, Y] + a^2\,\text{Var}[Y]. \tag{5.55}$$

Since $\text{Var}[W] \ge 0$ for any a, we have $2a\,\text{Cov}[X, Y] \le \text{Var}[X] + a^2\,\text{Var}[Y]$. Choosing $a = \sigma_X/\sigma_Y$ yields $\text{Cov}[X, Y] \le \sigma_Y\sigma_X$, which implies $\rho_{X,Y} \le 1$. Choosing $a = -\sigma_X/\sigma_Y$ yields $\text{Cov}[X, Y] \ge -\sigma_Y\sigma_X$, which implies $\rho_{X,Y} \ge -1$.

When $\rho_{X,Y} > 0$, we say that X and Y are *positively correlated*, and when $\rho_{X,Y} < 0$ we say X and Y are *negatively correlated*. If $|\rho_{X,Y}|$ is close to 1, say $|\rho_{X,Y}| \ge 0.9$, then X and Y are *highly correlated*. Note that high correlation can be positive or negative. Figure 5.5 shows outcomes of independent trials of an experiment that produces random variables X and Y for random variable pairs with (a) negative correlation, (b) zero correlation, and (c) positive correlation. The following theorem demonstrates that $|\rho_{X,Y}| = 1$ when there is a linear relationship between X and Y.

=====Theorem 5.15=====

If X and Y are random variables such that $Y = aX + b$,

$$\rho_{X,Y} = \begin{cases} -1 & a < 0, \\ 0 & a = 0, \\ 1 & a > 0. \end{cases}$$

The proof is left as an exercise for the reader (Problem 5.5.7). Some examples of positive, negative, and zero correlation coefficients include:

- X is a student's height. Y is the same student's weight. $0 < \rho_{X,Y} < 1$.

- X is the distance of a cellular phone from the nearest base station. Y is the power of the received signal at the cellular phone. $-1 < \rho_{X,Y} < 0$.

- X is the temperature of a resistor measured in degrees Celsius. Y is the temperature of the same resistor measured in Kelvins. $\rho_{X,Y} = 1$.

- X is the gain of an electrical circuit measured in decibels. Y is the attenuation, measured in decibels, of the same circuit. $\rho_{X,Y} = -1$.

- X is the telephone number of a cellular phone. Y is the Social Security number of the phone's owner. $\rho_{X,Y} = 0$.

The *correlation* of two random variables, denoted $r_{X,Y}$, is another parameter of the probability model of X and Y. $r_{X,Y}$ is a close relative of the covariance.

=====Definition 5.7=====Correlation

The **correlation** *of X and Y is $r_{X,Y} = \mathrm{E}[XY]$*

The following theorem contains useful relationships among three expected values: the covariance of X and Y, the correlation of X and Y, and the variance of $X + Y$.

=====Theorem 5.16=====

(a) $\mathrm{Cov}[X,Y] = r_{X,Y} - \mu_X \mu_Y$.

(b) $\mathrm{Var}[X + Y] = \mathrm{Var}[X] + \mathrm{Var}[Y] + 2\,\mathrm{Cov}[X,Y]$.

(c) *If $X = Y$,* $\mathrm{Cov}[X,Y] = \mathrm{Var}[X] = \mathrm{Var}[Y]$ *and* $r_{X,Y} = \mathrm{E}[X^2] = \mathrm{E}[Y^2]$.

Proof Cross-multiplying inside the expected value of Definition 5.5 yields

$$\mathrm{Cov}\,[X,Y] = \mathrm{E}\,[XY - \mu_X Y - \mu_Y X + \mu_X \mu_Y]. \tag{5.56}$$

Since the expected value of the sum equals the sum of the expected values,

$$\mathrm{Cov}\,[X,Y] = \mathrm{E}\,[XY] - \mathrm{E}\,[\mu_X Y] - \mathrm{E}\,[\mu_Y X] + \mathrm{E}\,[\mu_Y \mu_X]. \tag{5.57}$$

Note that in the expression $E[\mu_Y X]$, μ_Y is a constant. Referring to Theorem 3.12, we set $a = \mu_Y$ and $b = 0$ to obtain $E[\mu_Y X] = \mu_Y E[X] = \mu_Y \mu_X$. The same reasoning demonstrates that $E[\mu_X Y] = \mu_X E[Y] = \mu_X \mu_Y$. Therefore,

$$\text{Cov}[X, Y] = E[XY] - \mu_X \mu_Y - \mu_Y \mu_X + \mu_Y \mu_X = r_{X,Y} - \mu_X \mu_Y. \tag{5.58}$$

The other relationships follow directly from the definitions and Theorem 5.12.

Example 5.17

For the integrated circuits tests in Example 5.3, we found in Example 5.5 that the probability model for X and Y is given by the following matrix.

$P_{X,Y}(x, y)$	$y = 0$	$y = 1$	$y = 2$	$P_X(x)$
$x = 0$	0.01	0	0	0.01
$x = 1$	0.09	0.09	0	0.18
$x = 2$	0	0	0.81	0.81
$P_Y(y)$	0.10	0.09	0.81	

Find $r_{X,Y}$ and $\text{Cov}[X, Y]$.

....

By Definition 5.7,

$$r_{X,Y} = E[XY] = \sum_{x=0}^{2} \sum_{y=0}^{2} xy P_{X,Y}(x, y) \tag{5.59}$$

$$= (1)(1)0.09 + (2)(2)0.81 = 3.33. \tag{5.60}$$

To use Theorem 5.16(a) to find the covariance, we find

$$E[X] = (1)(0.18) + (2)(0.81) = 1.80,$$
$$E[Y] = (1)(0.09) + (2)(0.81) = 1.71. \tag{5.61}$$

Therefore, by Theorem 5.16(a), $\text{Cov}[X, Y] = 3.33 - (1.80)(1.71) = 0.252$.

The terms *orthogonal* and *uncorrelated* describe random variables for which $r_{X,Y} = 0$ and random variables for which $\text{Cov}[X, Y] = 0$ respectively.

Definition 5.8   Orthogonal Random Variables
*Random variables X and Y are **orthogonal** if $r_{X,Y} = 0$.*

Definition 5.9   Uncorrelated Random Variables
*Random variables X and Y are **uncorrelated** if $\text{Cov}[X, Y] = 0$.*

This terminology, while widely used, is somewhat confusing, since *orthogonal* means zero correlation and *uncorrelated* means zero covariance.

We have already noted that if X and Y are highly correlated, then observing X tells us a lot about the accompanying observation Y. Graphically, this is visible in Figure 5.5 when we compare the correlated cases (a) and (c) to the uncorrelated case (b). On the other hand, if $\text{Cov}[X, Y] = 0$, it is often the case that learning X tells us little about Y. We have used nearly the same words to describe *independent* random variables X and Y.

The following theorem contains several important properties of expected values of independent random variables. It states that independent random variables are uncorrelated but not necessarily orthogonal.

====**Theorem 5.17**====

For independent random variables X and Y,

(a) $\text{E}[g(X)h(Y)] = \text{E}[g(X)] \, \text{E}[h(Y)]$,

(b) $r_{X,Y} = \text{E}[XY] = \text{E}[X] \, \text{E}[Y]$,

(c) $\text{Cov}[X, Y] = \rho_{X,Y} = 0$,

(d) $\text{Var}[X + Y] = \text{Var}[X] + \text{Var}[Y]$,

Proof We present the proof for discrete random variables. By replacing PMFs and sums with PDFs and integrals we arrive at essentially the same proof for continuous random variables. Since $P_{X,Y}(x, y) = P_X(x)P_Y(y)$,

$$\text{E}[g(X)h(Y)] = \sum_{x \in S_X} \sum_{y \in S_Y} g(x)h(y)P_X(x) \, P_Y(y)$$

$$= \left(\sum_{x \in S_X} g(x)P_X(x) \right) \left(\sum_{y \in S_Y} h(y)P_Y(y) \right) = \text{E}[g(X)] \, \text{E}[h(Y)]. \quad (5.62)$$

If $g(X) = X$, and $h(Y) = Y$, this equation implies $r_{X,Y} = \text{E}[XY] = \text{E}[X] \, \text{E}[Y]$. This equation and Theorem 5.16(a) imply $\text{Cov}[X, Y] = 0$. As a result, Theorem 5.16(b) implies $\text{Var}[X + Y] = \text{Var}[X] + \text{Var}[Y]$. Furthermore, $\rho_{X,Y} = \text{Cov}[X, Y]/(\sigma_X \sigma_Y) = 0$.

These results all follow directly from the joint PMF for independent random variables. We observe that Theorem 5.17(c) states that *independent random variables are uncorrelated*. We will have many occasions to refer to this property. It is important to know that while $\text{Cov}[X, Y] = 0$ is a necessary property for independence, it is not sufficient. There are many pairs of uncorrelated random variables that are *not* independent.

====**Example 5.18**====

For the noisy observation $Y = X + Z$ of Example 5.1, find the covariances $\text{Cov}[X, Z]$ and $\text{Cov}[X, Y]$ and the correlation coefficients $\rho_{X,Z}$ and $\rho_{X,Y}$.

....

We recall from Example 5.1 that the signal X is Gaussian $(0, \sigma_X)$, that the noise Z is Gaussian $(0, \sigma_Z)$, and that X and Z are independent. We know from Theorem 5.17(c) that independence of X and Z implies

$$\text{Cov}\,[X, Z] = \rho_{X,Z} = 0. \tag{5.63}$$

In addition, by Theorem 5.17(d),

$$\text{Var}[Y] = \text{Var}[X] + \text{Var}[Z] = \sigma_X^2 + \sigma_Z^2. \tag{5.64}$$

Since $\text{E}[X] = \text{E}[Z] = 0$, Theorem 5.11 tells us that $\text{E}[Y] = \text{E}[X] + \text{E}[Z] = 0$ and Theorem 5.17)(b) says that $\text{E}[XZ] = \text{E}[X]\,\text{E}[Z] = 0$. This permits us to write

$$\text{Cov}\,[X, Y] = \text{E}\,[XY] = \text{E}\,[X(X + Z)]$$
$$= \text{E}\,[X^2 + XZ] = \text{E}\,[X^2] + \text{E}\,[XZ] = \text{E}\,[X^2] = \sigma_X^2. \tag{5.65}$$

This implies

$$\rho_{X,Y} = \frac{\text{Cov}\,[X, Y]}{\sqrt{\text{Var}[X]\,\text{Var}[Y]}} = \frac{\sigma_X^2}{\sqrt{\sigma_X^2(\sigma_X^2 + \sigma_Z^2)}} = \sqrt{\frac{\sigma_X^2/\sigma_Z^2}{1 + \sigma_X^2/\sigma_Z^2}}. \tag{5.66}$$

We see in Example 5.18 that the covariance between the transmitted signal X and the received signal Y depends on the ratio σ_X^2/σ_Z^2. This ratio, referred to as the *signal-to-noise ratio*, has a strong effect on communication quality. If $\sigma_X^2/\sigma_Z^2 \ll 1$, the correlation of X and Y is weak and the noise dominates the signal at the receiver. Learning y, a sample of the received signal, is not very helpful in determining the corresponding sample of the transmitted signal, x. On the other hand, if $\sigma_X^2/\sigma_Z^2 \gg 1$, the transmitted signal dominates the noise and $\rho_{X,Y} \approx 1$, an indication of a close relationship between X and Y. When there is strong correlation between X and Y, learning y is very helpful in determining x.

Quiz 5.8

(A) Random variables L and T have joint PMF

$P_{L,T}(l, t)$	$t = 40\,\text{sec}$	$t = 60\,\text{sec}$
$l = 1$ page	0.15	0.1
$l = 2$ pages	0.30	0.2
$l = 3$ pages	0.15	0.1.

Find the following quantities.

(a) $\text{E}[L]$ and $\text{Var}[L]$ (b) $\text{E}[T]$ and $\text{Var}[T]$

(c) The covariance $\text{Cov}[L, T]$ (d) The correlation coefficient $\rho_{L,T}$

(B) The joint probability density function of random variables X and Y is

$$f_{X,Y}(x, y) = \begin{cases} xy & 0 \le x \le 1, 0 \le y \le 2, \\ 0 & \text{otherwise.} \end{cases} \tag{5.67}$$

Find the following quantities.

(a) $E[X]$ and $Var[X]$

(b) $E[Y]$ and $Var[Y]$

(c) The covariance $Cov[X, Y]$

(d) The correlation coefficient $\rho_{X,Y}$

5.9 Bivariate Gaussian Random Variables

> The *bivariate Gaussian* PDF of X and Y has five parameters: the expected values and standard deviations of X and Y and the correlation coefficient of X and Y. The marginal PDF of X and the marginal PDF of Y are both Gaussian.

For a PDF representing a family of random variables, one or more parameters define a specific PDF. Properties such as $E[X]$ and $Var[X]$ depend on the parameters. For example, a continuous uniform (a, b) random variable has expected value $(a + b)/2$ and variance $(b - a)^2/12$. For the bivariate Gaussian PDF, the parameters μ_X, μ_Y, σ_X, σ_Y and $\rho_{X,Y}$ are equal to the expected values, standard deviations, and correlation coefficient of X and Y.

Definition 5.10 Bivariate Gaussian Random Variables

*Random variables X and Y have a **bivariate Gaussian PDF** with parameters μ_X, μ_Y, $\sigma_X > 0$, $\sigma_Y > 0$, and $\rho_{X,Y}$ satisfying $-1 < \rho_{X,Y} < 1$ if*

$$f_{X,Y}(x, y) = \frac{\exp\left[-\dfrac{\left(\frac{x-\mu_X}{\sigma_X}\right)^2 - \frac{2\rho_{X,Y}(x-\mu_X)(y-\mu_Y)}{\sigma_X \sigma_Y} + \left(\frac{y-\mu_Y}{\sigma_Y}\right)^2}{2(1 - \rho_{X,Y}^2)}\right]}{2\pi\sigma_X\sigma_Y\sqrt{1 - \rho_{X,Y}^2}},$$

Figure 5.6 illustrates the bivariate Gaussian PDF for $\mu_X = \mu_Y = 0$, $\sigma_X = \sigma_Y = 1$, and three values of $\rho_{X,Y} = \rho$. When $\rho = 0$, the joint PDF has the circular symmetry of a sombrero. When $\rho = 0.9$, the joint PDF forms a ridge over the line $x = y$, and when $\rho = -0.9$ there is a ridge over the line $x = -y$. The ridge becomes increasingly steep as $\rho \to \pm 1$. Adjacent to each PDF, we repeat the graphs in Figure 5.5; each graph shows 200 sample pairs (X, Y) drawn from that bivariate Gaussian PDF. We see that the sample pairs are clustered in the region of the x, y plane where the PDF is large.

To examine mathematically the properties of the bivariate Gaussian PDF, we define

$$\tilde{\mu}_Y(x) = \mu_Y + \rho_{X,Y}\frac{\sigma_Y}{\sigma_X}(x - \mu_X), \qquad \tilde{\sigma}_Y = \sigma_Y\sqrt{1 - \rho_{X,Y}^2}, \qquad (5.68)$$

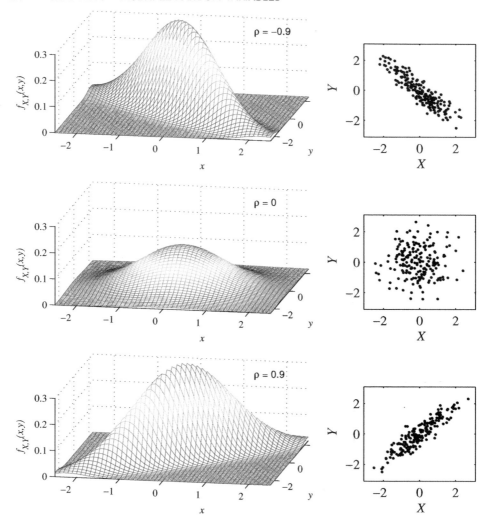

Figure 5.6 The Joint Gaussian PDF $f_{X,Y}(x,y)$ for $\mu_X = \mu_Y = 0$, $\sigma_X = \sigma_Y = 1$, and three values of $\rho_{X,Y} = \rho$. Next to each PDF, we plot 200 sample pairs (X, Y) generated with that PDF.

and manipulate the formula in Definition 5.10 to obtain the following expression for the joint Gaussian PDF:

$$f_{X,Y}(x,y) = \frac{1}{\sigma_X\sqrt{2\pi}}e^{-(x-\mu_X)^2/2\sigma_X^2}\frac{1}{\tilde{\sigma}_Y\sqrt{2\pi}}e^{-(y-\tilde{\mu}_Y(x))^2/2\tilde{\sigma}_Y^2}. \tag{5.69}$$

Equation (5.69) expresses $f_{X,Y}(x,y)$ as the product of two Gaussian PDFs, one with parameters μ_X and σ_X and the other with parameters $\tilde{\mu}_Y$ and $\tilde{\sigma}_Y$. This formula plays a key role in the proof of the following theorem.

====Theorem 5.18====

If X and Y are the bivariate Gaussian random variables in Definition 5.10, X is the Gaussian (μ_X, σ_X) random variable and Y is the Gaussian (μ_Y, σ_Y) random variable:

$$f_X(x) = \frac{1}{\sigma_X \sqrt{2\pi}} e^{-(x-\mu_X)^2/2\sigma_X^2}, \qquad f_Y(y) = \frac{1}{\sigma_Y \sqrt{2\pi}} e^{-(y-\mu_Y)^2/2\sigma_Y^2}.$$

Proof Integrating $f_{X,Y}(x,y)$ in Equation (5.69) over all y, we have

$$
\begin{aligned}
f_X(x) &= \int_{-\infty}^{\infty} f_{X,Y}(x,y)\, dy \\
&= \frac{1}{\sigma_X \sqrt{2\pi}} e^{-(x-\mu_X)^2/2\sigma_X^2} \underbrace{\int_{-\infty}^{\infty} \frac{1}{\tilde{\sigma}_Y \sqrt{2\pi}} e^{-(y-\tilde{\mu}_Y(x))^2/2\tilde{\sigma}_Y^2}\, dy}_{1}
\end{aligned}
\qquad (5.70)
$$

The integral above the bracket equals 1 because it is the integral of a Gaussian PDF. The remainder of the formula is the PDF of the Gaussian (μ_X, σ_X) random variable. The same reasoning with the roles of X and Y reversed leads to the formula for $f_Y(y)$.

The next theorem identifies $\rho_{X,Y}$ in Definition 5.10 as the correlation coefficient of X and Y.

====Theorem 5.19====

Bivariate Gaussian random variables X and Y in Definition 5.10 have correlation coefficient $\rho_{X,Y}$.

The proof of Theorem 5.19 involves algebra that is more easily digested with some insight from Chapter 7; see Section 7.6 for the proof.

From Theorem 5.19, we observe that if X and Y are uncorrelated, then $\rho_{X,Y} = 0$ and, by evaluating the PDF in Definition 5.10 with $\rho_{X,Y} = 0$, we have $f_{X,Y}(x,y) = f_X(x)f_Y(y)$. Thus we have the following theorem.

====Theorem 5.20====

Bivariate Gaussian random variables X and Y are uncorrelated if and only if they are independent.

Another important property of bivariate Gaussian random variables X and Y is that a pair of linear combinations of X and Y forms a pair of bivariate Gaussian random variables.

====Theorem 5.21====

If X and Y are bivariate Gaussian random variables with PDF given by Definition 5.10, and W_1 and W_2 are given by the linearly independent equations

$$W_1 = a_1 X + b_1 Y, \qquad\qquad W_2 = a_2 X + b_2 Y,$$

then W_1 and W_2 are bivariate Gaussian random variables such that

$$\mathrm{E}\,[W_i] = a_i \mu_X + b_i \mu_Y, \qquad\qquad\qquad i = 1, 2,$$
$$\mathrm{Var}[W_i] = a_i^2 \sigma_X^2 + b_i^2 \sigma_Y^2 + 2a_i b_i \rho_{X,Y} \sigma_X \sigma_Y, \qquad\qquad i = 1, 2,$$
$$\mathrm{Cov}\,[W_1, W_2] = a_1 a_2 \sigma_X^2 + b_1 b_2 \sigma_Y^2 + (a_1 b_2 + a_2 b_1) \rho_{X,Y} \sigma_X \sigma_Y.$$

Theorem 5.21 is a special case of Theorem 8.11 when we have $n = 2$ jointly Gaussian random variables. We omit the proof since the proof of Theorem 8.11 for n jointly Gaussian random variables is, with some knowledge of linear algebra, simpler. The requirement that the equations for W_1 and W_2 be "linearly independent" is linear algebra terminology that excludes degenerate cases such as $W_1 = X + 2Y$ and $W_2 = 3X + 6Y$ where $W_2 = 3W_1$ is just a scaled replica of W_1.

Theorem 5.21 is powerful. Even the partial result that W_i by itself is Gaussian is a nontrivial conclusion. When an experiment produces linear combinations of Gaussian random variables, knowing that these combinations are Gaussian simplifies the analysis because all we need to do is calculate the expected values, variances, and covariances of the outputs in order to derive probability models.

Example 5.19

For the noisy observation in Example 5.14, find the PDF of $Y = X + Z$.

. . . .

Since X is Gaussian $(0, \sigma_X)$ and Z is Gaussian $(0, \sigma_Z)$ and X and Z are independent, X and Z are jointly Gaussian. It follows from Theorem 5.21 that Y is Gaussian with $\mathrm{E}[Y] = \mathrm{E}[X] + \mathrm{E}[Z] = 0$ and variance $\sigma_Y^2 = \sigma_X^2 + \sigma_Z^2$. The PDF of Y is

$$f_Y(y) = \frac{1}{\sqrt{2\pi(\sigma_X^2 + \sigma_Z^2)}} e^{-y^2/2(\sigma_X^2 + \sigma_Z^2)}. \tag{5.71}$$

Example 5.20

Continuing Example 5.19, find the joint PDF of X and Y when $\sigma_X = 4$ and $\sigma_Z = 3$.

. . . .

From Theorem 5.21, we know that X and Y are bivariate Gaussian. We also know that $\mu_X = \mu_Y = 0$ and that Y has variance $\sigma_Y^2 = \sigma_X^2 + \sigma_Z^2 = 25$. Substituting $\sigma_X = 4$ and $\sigma_Z = 3$ in the formula for the correlation coefficient derived in Example 5.18, we have

$$\rho_{X,Y} = \sqrt{\frac{\sigma_X^2/\sigma_Z^2}{1 + \sigma_X^2/\sigma_Z^2}} = \frac{4}{5}. \tag{5.72}$$

Applying these parameters to Definition 5.10, we obtain

$$f_{X,Y}(x, y) = \frac{1}{24\pi} e^{-\left(25x^2/16 - 2xy + y^2\right)/18}. \tag{5.73}$$

═══Quiz 5.9═══

Let X and Y be jointly Gaussian $(0,1)$ random variables with correlation coefficient $1/2$. What is the joint PDF of X and Y?

5.10 Multivariate Probability Models

> The probability model of an experiment that produces n random variables can be represented as an n-dimensional CDF. If all of the random variables are discrete, there is a corresponding n-dimensional PMF. If all of the random variables are continuous, there is an n-dimensional PDF. The PDF is the nth partial derivative of the CDF with respect to all n variables. The probability model (CDF, PMF, or PDF) of n independent random variables is the product of the univariate probability models of the n random variables.

This chapter has emphasized probability models of two random variables X and Y. We now generalize the definitions and theorems to experiments that yield an arbitrary number of random variables $X_1, \ldots, X_n$. This section is heavy on n-dimensional definitions and theorems but relatively light on examples. However, the ideas are straightforward extensions of concepts for a pair of random variables. If you have trouble with a theorem or definition, rewrite it for the special case of $n = 2$ random variables. This will yield a familiar result for a pair of random variables.

To express a complete probability model of $X_1, \ldots, X_n$, we define the joint cumulative distribution function.

═══Definition 5.11═══Multivariate Joint CDF

*The **joint CDF** of $X_1, \ldots, X_n$ is*

$$F_{X_1,\ldots,X_n}(x_1,\ldots,x_n) = \mathrm{P}\left[X_1 \leq x_1, \ldots, X_n \leq x_n\right].$$

Definition 5.11 is concise and general. It provides a complete probability model regardless of whether any or all of the X_i are discrete, continuous, or mixed. However, the joint CDF is usually not convenient to use in analyzing practical probability models. Instead, we use the joint PMF or the joint PDF.

═══Definition 5.12═══Multivariate Joint PMF

*The **joint PMF** of the discrete random variables $X_1, \ldots, X_n$ is*

$$P_{X_1,\ldots,X_n}(x_1,\ldots,x_n) = \mathrm{P}\left[X_1 = x_1, \ldots, X_n = x_n\right].$$

=====Definition 5.13=====Multivariate Joint PDF

*The **joint PDF** of the continuous random variables $X_1, \ldots, X_n$ is the function*

$$f_{X_1,\ldots,X_n}(x_1,\ldots,x_n) = \frac{\partial^n F_{X_1,\ldots,X_n}(x_1,\ldots,x_n)}{\partial x_1 \cdots \partial x_n}.$$

Theorems 5.22 and 5.23 indicate that the joint PMF and the joint PDF have properties that are generalizations of the axioms of probability.

=====Theorem 5.22=====

If $X_1, \ldots, X_n$ are discrete random variables with joint PMF $P_{X_1,\ldots,X_n}(x_1,\ldots,x_n)$,

(a) $P_{X_1,\ldots,X_n}(x_1,\ldots,x_n) \geq 0,$

(b) $\displaystyle\sum_{x_1 \in S_{X_1}} \cdots \sum_{x_n \in S_{X_n}} P_{X_1,\ldots,X_n}(x_1,\ldots,x_n) = 1.$

=====Theorem 5.23=====

If $X_1, \ldots, X_n$ have joint PDF $f_{X_1,\ldots,X_n}(x_1,\ldots,x_n)$,

(a) $f_{X_1,\ldots,X_n}(x_1,\ldots,x_n) \geq 0,$

(b) $F_{X_1,\ldots,X_n}(x_1,\ldots,x_n) = \displaystyle\int_{-\infty}^{x_1} \cdots \int_{-\infty}^{x_n} f_{X_1,\ldots,X_n}(u_1,\ldots,u_n)\,du_1 \cdots du_n,$

(c) $\displaystyle\int_{-\infty}^{\infty} \cdots \int_{-\infty}^{\infty} f_{X_1,\ldots,X_n}(x_1,\ldots,x_n)\,dx_1 \cdots dx_n = 1.$

Often we consider an event A described in terms of a property of $X_1, \ldots, X_n$, such as $|X_1 + X_2 + \cdots + X_n| \leq 1$, or $\max_i X_i \leq 100$. To find the probability of the event A, we sum the joint PMF or integrate the joint PDF over all $x_1, \ldots, x_n$ that belong to A.

=====Theorem 5.24=====

The probability of an event A expressed in terms of the random variables $X_1, \ldots, X_n$ is

Discrete: $\mathrm{P}[A] = \displaystyle\sum_{(x_1,\ldots,x_n) \in A} P_{X_1,\ldots,X_n}(x_1,\ldots,x_n);$

Continuous: $\mathrm{P}[A] = \displaystyle\int \cdots \int_{A} f_{X_1,\ldots,X_n}(x_1,\ldots,x_n)\,dx_1\,dx_2 \ldots dx_n.$

Although we have written the discrete version of Theorem 5.24 with a single summation, we must remember that in fact it is a multiple sum over the n variables $x_1, \ldots, x_n$.

======Example 5.21======

Consider a set of n independent trials in which there are r possible outcomes $s_1, \ldots, s_r$ for each trial. In each trial, $P[s_i] = p_i$. Let N_i equal the number of times that outcome s_i occurs over n trials. What is the joint PMF of $N_1, \ldots, N_r$?

· ·

The solution to this problem appears in Theorem 2.9 and is repeated here:

$$P_{N_1,\ldots,N_r}(n_1, \ldots, n_r) = \binom{n}{n_1, \ldots, n_r} p_1^{n_1} p_2^{n_2} \cdots p_r^{n_r}. \tag{5.74}$$

In analyzing an experiment, we might wish to study some of the random variables and ignore other ones. To accomplish this, we can derive marginal PMFs or marginal PDFs that are probability models for a fraction of the random variables in the complete experiment. Consider an experiment with four random variables W, X, Y, Z. The probability model for the experiment is the joint PMF, $P_{W,X,Y,Z}(w, x, y, z)$ or the joint PDF, $f_{W,X,Y,Z}(w, x, y, z)$. The following theorems give examples of marginal PMFs and PDFs.

======Theorem 5.25======

For a joint PMF $P_{W,X,Y,Z}(w, x, y, z)$ of discrete random variables W, X, Y, Z, some marginal PMFs are

$$P_{X,Y,Z}(x, y, z) = \sum_{w \in S_W} P_{W,X,Y,Z}(w, x, y, z),$$

$$P_{W,Z}(w, z) = \sum_{x \in S_X} \sum_{y \in S_Y} P_{W,X,Y,Z}(w, x, y, z),$$

======Theorem 5.26======

For a joint PDF $f_{W,X,Y,Z}(w, x, y, z)$ of continuous random variables W, X, Y, Z, some marginal PDFs are

$$f_{W,X,Y}(w, x, y) = \int_{-\infty}^{\infty} f_{W,X,Y,Z}(w, x, y, z) \, dz,$$

$$f_X(x) = \int_{-\infty}^{\infty} \int_{-\infty}^{\infty} \int_{-\infty}^{\infty} f_{W,X,Y,Z}(w, x, y, z) \, dw \, dy \, dz.$$

Theorems 5.25 and 5.26 can be generalized in a straightforward way to any marginal PMF or marginal PDF of an arbitrary number of random variables. For a probability model described by the set of random variables $\{X_1, \ldots, X_n\}$, each nonempty strict subset of those random variables has a marginal probability model. There are 2^n subsets of $\{X_1, \ldots, X_n\}$. After excluding the entire set and the null set $\varnothing$, we find that there are $2^n - 2$ marginal probability models.

═══**Example 5.22**═══

As in Quiz 5.10, the random variables $Y_1, \ldots, Y_4$ have the joint PDF

$$f_{Y_1,\ldots,Y_4}(y_1, \ldots, y_4) = \begin{cases} 4 & 0 \leq y_1 \leq y_2 \leq 1, 0 \leq y_3 \leq y_4 \leq 1, \\ 0 & \text{otherwise.} \end{cases} \tag{5.75}$$

Find the marginal PDFs $f_{Y_1,Y_4}(y_1, y_4)$, $f_{Y_2,Y_3}(y_2, y_3)$, and $f_{Y_3}(y_3)$.

. . . .

$$f_{Y_1,Y_4}(y_1, y_4) = \int_{-\infty}^{\infty} \int_{-\infty}^{\infty} f_{Y_1,\ldots,Y_4}(y_1, \ldots, y_4)\, dy_2\, dy_3. \tag{5.76}$$

In the foregoing integral, the hard part is identifying the correct limits. These limits will depend on y_1 and y_4. For $0 \leq y_1 \leq 1$ and $0 \leq y_4 \leq 1$,

$$f_{Y_1,Y_4}(y_1, y_4) = \int_{y_1}^{1} \int_{0}^{y_4} 4\, dy_3\, dy_2 = 4(1 - y_1)y_4. \tag{5.77}$$

The complete expression for $f_{Y_1,Y_4}(y_1, y_4)$ is

$$f_{Y_1,Y_4}(y_1, y_4) = \begin{cases} 4(1 - y_1)y_4 & 0 \leq y_1 \leq 1, 0 \leq y_4 \leq 1, \\ 0 & \text{otherwise.} \end{cases} \tag{5.78}$$

Similarly, for $0 \leq y_2 \leq 1$ and $0 \leq y_3 \leq 1$,

$$f_{Y_2,Y_3}(y_2, y_3) = \int_{0}^{y_2} \int_{y_3}^{1} 4\, dy_4\, dy_1 = 4y_2(1 - y_3). \tag{5.79}$$

The complete expression for $f_{Y_2,Y_3}(y_2, y_3)$ is

$$f_{Y_2,Y_3}(y_2, y_3) = \begin{cases} 4y_2(1 - y_3) & 0 \leq y_2 \leq 1, 0 \leq y_3 \leq 1, \\ 0 & \text{otherwise.} \end{cases} \tag{5.80}$$

Lastly, for $0 \leq y_3 \leq 1$,

$$f_{Y_3}(y_3) = \int_{-\infty}^{\infty} f_{Y_2,Y_3}(y_2, y_3)\, dy_2 = \int_{0}^{1} 4y_2(1 - y_3)\, dy_2 = 2(1 - y_3). \tag{5.81}$$

The complete expression is

$$f_{Y_3}(y_3) = \begin{cases} 2(1 - y_3) & 0 \leq y_3 \leq 1, \\ 0 & \text{otherwise.} \end{cases} \tag{5.82}$$

─────

Practical experiments often generates a joint PMF or PDF that is forbiddingly complex. The important exception is an experiment that produces n independent random variables. The following definition extends the definition of independence of two random variables. It states that $X_1, \ldots, X_n$ are independent when the joint PMF or PDF can be factored into a product of n marginal PMFs or PDFs.

Definition 5.14 **N Independent Random Variables**

Random variables $X_1, \ldots, X_n$ *are* **independent** *if for all* $x_1, \ldots, x_n$,

Discrete: $P_{X_1, \ldots, X_n}(x_1, \ldots, x_n) = P_{X_1}(x_1) P_{X_2}(x_2) \cdots P_{X_N}(x_n)$;

Continuous: $f_{X_1, \ldots, X_n}(x_1, \ldots, x_n) = f_{X_1}(x_1) f_{X_2}(x_2) \cdots f_{X_n}(x_n)$.

Independence of n random variables is typically a property of an experiment consisting of n independent subexperiments, in which subexperiment i produces the random variable X_i. If all subexperiments follow the same procedure and have the same observation, all of the X_i have the same PMF or PDF. In this case, we say the random variables X_i are *identically distributed*.

Definition 5.15 **Independent and Identically Distributed (iid)**

$X_1, \ldots, X_n$ *are* **independent and identically distributed (iid)** *if*

Discrete: $P_{X_1, \ldots, X_n}(x_1, \ldots, x_n) = P_X(x_1) P_X(x_2) \cdots P_X(x_n)$;

Continuous: $f_{X_1, \ldots, X_n}(x_1, \ldots, x_n) = f_X(x_1) f_X(x_2) \cdots f_X(x_n)$.

Example 5.23

The random variables $X_1, \ldots, X_n$ have the joint PDF

$$f_{X_1, \ldots, X_n}(x_1, \ldots, x_n) = \begin{cases} 1 & 0 \le x_i \le 1, i = 1, \ldots, n, \\ 0 & \text{otherwise.} \end{cases} \tag{5.83}$$

Let A denote the event that $\max_i X_i \le 1/2$. Find $\mathrm{P}[A]$.

....

We can solve this problem by applying Theorem 5.24:

$$\mathrm{P}[A] = \mathrm{P}\left[\max_i X_i \le 1/2\right] = \mathrm{P}[X_1 \le 1/2, \ldots, X_n \le 1/2]$$

$$= \int_0^{1/2} \cdots \int_0^{1/2} 1 \, dx_1 \cdots dx_n = \frac{1}{2^n}. \tag{5.84}$$

As n grows, the probability that the maximum is less than $1/2$ rapidly goes to 0.

We note that inspection of the joint PDF reveals that $X_1, \ldots, X_4$ are iid continuous uniform $(0, 1)$ random variables. The integration in Equation (5.84) is easy because independence implies

$$
\begin{aligned}
\mathrm{P}[A] &= \mathrm{P}[X_1 \leq 1/2, \ldots, X_n \leq 1/2] \\
&= \mathrm{P}[X_1 \leq 1/2] \times \cdots \times \mathrm{P}[X_n \leq 1/2] = (1/2)^n.
\end{aligned}
\tag{5.85}
$$

Quiz 5.10

The random variables $Y_1, \ldots, Y_4$ have the joint PDF

$$
f_{Y_1, \ldots, Y_4}(y_1, \ldots, y_4) = \begin{cases} 4 & 0 \leq y_1 \leq y_2 \leq 1, 0 \leq y_3 \leq y_4 \leq 1, \\ 0 & \text{otherwise.} \end{cases}
\tag{5.86}
$$

Let C denote the event that $\max_i Y_i \leq 1/2$. Find $\mathrm{P}[C]$.

5.11   MATLAB

> It is convenient to use MATLAB to generate pairs of discrete random variables X and Y with an arbitrary joint PMF. There are no generally applicable techniques for generating sample pairs of a continuous random variable. There are techniques tailored to specific joint PDFs, for example, bivariate Gaussian.

MATLAB is a useful tool for studying experiments that produce a pair of random variables X, Y. Simulation experiments often depend on the generation of sample pairs of random variables with specific probability models. That is, given a joint PMF $P_{X,Y}(x, y)$ or PDF $f_{X,Y}(x, y)$, we need to produce a collection of pairs $\{(x_1, y_1), (x_2, y_2), \ldots, (x_m, y_m)\}$. For finite discrete random variables, we are able to develop some general techniques. For continuous random variables, we give some specific examples.

Discrete Random Variables

We start with the case when X and Y are finite random variables with ranges

$$
S_X = \{x_1, \ldots, x_n\}, \qquad\qquad S_Y = \{y_1, \ldots, y_m\}.
\tag{5.87}
$$

In this case, we can take advantage of MATLAB techniques for surface plots of $g(x, y)$ over the x, y plane. In MATLAB, we represent S_X and S_Y by the n element vector sx and m element vector sy. The function [SX,SY]=ndgrid(sx,sy) produces the pair of $n \times m$ matrices,

$$SX = \begin{bmatrix} x_1 & \cdots & x_1 \\ \vdots & & \vdots \\ x_n & \cdots & x_n \end{bmatrix}, \qquad SY = \begin{bmatrix} y_1 & \cdots & y_m \\ \vdots & & \vdots \\ y_1 & \cdots & y_m \end{bmatrix}. \qquad (5.88)$$

We refer to matrices SX and SY as a *sample space grid* because they are a grid representation of the joint sample space

$$S_{X,Y} = \{(x,y) | x \in S_X, y \in S_Y\}. \qquad (5.89)$$

That is, [SX(i,j) SY(i,j)] is the pair (x_i, y_j).

To complete the probability model, for X and Y, in MATLAB, we employ the $n \times m$ matrix PXY such that PXY(i,j) $= P_{X,Y}(x_i, y_j)$. To make sure that probabilities have been generated properly, we note that [SX(:) SY(:) PXY(:)] is a matrix whose rows list all possible pairs x_i, y_j and corresponding probabilities $P_{X,Y}(x_i, y_j)$.

Given a function $g(x,y)$ that operates on the elements of vectors x and y, the advantage of this grid approach is that the MATLAB function g(SX,SY) will calculate $g(x,y)$ for each $x \in S_X$ and $y \in S_Y$. In particular, g(SX,SY) produces an $n \times m$ matrix with i,jth element $g(x_i, y_j)$.

Example 5.24

An Internet photo developer website prints compressed photo images. Each image file contains a variable-sized image of $X \times Y$ pixels described by the joint PMF

$P_{X,Y}(x,y)$	$y = 400$	$y = 800$	$y = 1200$
$x = 800$	0.2	0.05	0.1
$x = 1200$	0.05	0.2	0.1
$x = 1600$	0	0.1	0.2.

(5.90)

For random variables X, Y, write a script imagepmf.m that defines the sample space grid matrices SX, SY, and PXY.

. . . .

In the script imagepmf.m, the matrix SX has $\begin{bmatrix} 800 & 1200 & 1600 \end{bmatrix}'$ for each column and SY has $\begin{bmatrix} 400 & 800 & 1200 \end{bmatrix}$ for each row. After running imagepmf.m, we can inspect the variables:

```
%imagepmf.m
PXY=[0.2  0.05 0.1; ...
     0.05 0.2  0.1; ...
     0    0.1  0.2];
[SX,SY]=ndgrid([800 1200 1600],...
     [400 800 1200]);
```

```
>> imagepmf; SX
SX =
        800      800      800
       1200     1200     1200
       1600     1600     1600
>> SY
SY =
        400      800     1200
        400      800     1200
        400      800     1200
```

```
>> imagesize
eb =

      319200
sb =

      96000   144000   192000   288000   384000   432000   576000
pb =
      0.2000   0.0500   0.0500   0.3000   0.1000   0.1000   0.2000
```

Figure 5.7   Output resulting from `imagesize.m` in Example 5.25.

Example 5.25

At 24 bits (3 bytes) per pixel, a 10:1 image compression factor yields image files with $B = 0.3XY$ bytes. Find the expected value $\mathrm{E}[B]$ and the PMF $P_B(b)$.

. . . .

```
%imagesize.m
imagepmf;
SB=0.3*(SX.*SY);
eb=sum(sum(SB.*PXY))
sb=unique(SB)'
pb=finitepmf(SB,PXY,sb)'
```

The script `imagesize.m` produces the expected value as `eb`, and produces the PMF, which is represented by the vectors `sb` and `pb`. The 3×3 matrix `SB` has i,jth element $g(x_i, y_j) = 0.3x_i y_j$. The calculation of `eb` is simply a MATLAB implementation of Theorem 5.9. Since some elements of `SB` are identical, `sb=unique(SB)` extracts the unique elements. Although `SB` and `PXY` are both 3×3 matrices, each is stored internally by MATLAB as a 9-element vector. Hence, we can pass `SB` and `PXY` to the `finitepmf()` function, which was designed to handle a finite random variable described by a pair of column vectors. Figure 5.7 shows one result of running the program `imagesize`. The vectors `sb` and `pb` comprise $P_B(b)$. For example, $P_B(288000) = 0.3$.

Random Sample Pairs

For finite random variables X, Y described by S_X, S_Y and joint PMF $P_{X,Y}(x, y)$, or equivalently `SX`, `SY`, and `PXY` in MATLAB, we can generate random sample pairs using the function `finiterv(s,p,m)` defined in Chapter 3. Recall that `x=finiterv(s,p,m)` returned m samples (arranged as a column vector `x`) of a random variable X such that a sample value is `s(i)` with probability `p(i)`. In fact, to support random variable pairs X, Y, the function `w=finiterv(s,p,m)` permits `s` to be a $k \times 2$ matrix where the rows of `s` enumerate all pairs (x, y) with nonzero probability. Given the grid representation `SX`, `SY`, and `PXY`, we generate m sample pairs via

```
xy=finiterv([SX(:) SY(:)],PXY(:),m)
```

In particular, the ith pair, `SX(i),SY(i)`, will occur with probability `PXY(i)`. The output `xy` will be an $m \times 2$ matrix such that each row represents a sample pair x, y.

Example 5.26

Write a function xy=imagerv(m) that generates m sample pairs of the image size random variables X, Y of Example 5.25.

. . . .

The function imagerv uses the imagesize.m script to define the matrices SX, SY, and PXY. It then calls the finiterv.m function. Here is the code imagerv.m and a sample run:

```
function xy = imagerv(m);
imagepmf;
S=[SX(:) SY(:)];
xy=finiterv(S,PXY(:),m);
```

```
>> xy=imagerv(3)
xy =
            800        400
           1200        800
           1600        800
```

Example 5.26 can be generalized to produce sample pairs for any discrete random variable pair X, Y. However, given a collection of, for example, $m = 10,000$ samples of X, Y, it is desirable to be able to check whether the code generates the sample pairs properly. In particular, we wish to check for each $x \in S_X$ and $y \in S_Y$ whether the relative frequency of x, y in m samples is close to $P_{X,Y}(x, y)$. In the following example, we develop a program to calculate a matrix of relative frequencies that corresponds to the matrix PXY.

Example 5.27

Given a list xy of sample pairs of random variables X, Y with MATLAB range grids SX and SY, write a MATLAB function fxy=freqxy(xy,SX,SY) that calculates the relative frequency of every pair x, y. The output fxy should correspond to the matrix [SX(:) SY(:) PXY(:)].

. . . .

```
function fxy = freqxy(xy,SX,SY)
xy=[xy; SX(:) SY(:)];
[U,I,J]=unique(xy,'rows');
N=hist(J,1:max(J))-1;
N=N/sum(N);
fxy=[U N(:)];
fxy=sortrows(fxy,[2 1 3]);
```

The matrix [SX(:) SY(:)] in freqxy has rows that list all possible pairs x, y. We append this matrix to xy to ensure that the new xy has every possible pair x, y. Next, the unique function copies all unique rows of xy to the matrix U and also provides the vector J that indexes the rows of xy in U; that is, xy=U(J). In addition, the number of occurrences of j in J indicates the number of occurrences in xy of row j in U. Thus we use the hist function on J to calculate the relative frequencies. We include the correction factor -1 because we had appended [SX(:) SY(:)] to xy at the start. Lastly, we reorder the rows of fxy because the output of unique produces the rows of U in a different order from [SX(:) SY(:) PXY(:)].

MATLAB provides the function stem3(x,y,z), where x, y, and z are length n vectors, for visualizing a bivariate PMF $P_{X,Y}(x, y)$ or for visualizing relative frequencies of sample values of a pair of random variables. At each position x(i), y(i) on the xy plane, the function draws a stem of height z(i).

Example 5.28

Generate $m = 10,000$ samples of random variables X, Y of Example 5.25. Calculate the relative frequencies and use `stem3` to graph them.
. . . .

The script `imagestem.m` generates the following relative frequency stem plot.

```
%imagestem.m
imagepmf;
xy=imagerv(10000);
fxy=freqxy(xy,SX,SY);
stem3(fxy(:,1),...
    fxy(:,2),fxy(:,3));
xlabel('\it x');
ylabel('\it y');
```

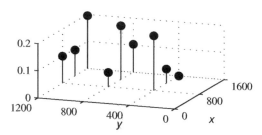

Continuous Random Variables

For continuous random variables, MATLAB can be useful in a variety of ways. Some of these are obvious. For example, a joint PDF $f_{X,Y}(x,y)$ or CDF $F_{X,Y}(x,y)$ can be viewed using the function `plot3`. Figure 5.4 was generated this way. However, for generating sample pairs of continuous random variables, there are no general techniques such as the sample space grids we employed with discrete random variables.

When we introduced continuous random variables in Chapter 4, we also introduced families of widely used random variables. In Section 4.8, we provided a collection of MATLAB functions such as `x=erlangrv(n,lambda,m)` to generate `m` samples from the corresponding PDF. However, for pairs of continuous random variables, we introduced only one family of probability models, namely the bivariate Gaussian random variables X and Y. For the bivariate Gaussian model, we can use Theorem 5.21 and the `randn` function to generate sample values. The command `Z=randn(2,1)` returns the vector $Z = \begin{bmatrix} Z_1 & Z_2 \end{bmatrix}'$ where Z_1 and Z_2 are iid Gaussian $(0,1)$ random variables. Next we form the linear combinations

$$W_1 = \sigma_1 Z_1 \tag{5.91a}$$

$$W_2 = \rho \sigma_2 Z_1 + \sqrt{(1 - \rho^2)\sigma_2^2} Z_2 \tag{5.91b}$$

From Theorem 5.21 we know that W_1 and W_2 are a bivariate Gaussian pair. In addition, from the formulas given in Theorem 5.21, we can show that $\mathrm{E}[W_1] = \mathrm{E}[W_2] = 0$, $\mathrm{Var}[W_1] = \sigma_1^2$, $\mathrm{Var}[W_2] = \sigma_2^2$ and $\rho_{W_1,W_2} = \rho$. This implies that

$$X_1 = W_1 + \mu_1, \qquad\qquad X_2 = W_2 + \mu_2 \tag{5.92}$$

is a pair of bivariate Gaussian random variables with $\mathrm{E}[X_i] = \mu_i$, $\mathrm{Var}[X_i] = \sigma_i^2$, and $\rho_{X_1,X_2} = \rho$. We implement this algorithm that transforms the iid pair Z_1, Z_2

into the bivariate Gaussian pair X_1, X_2 in the MATLAB function

$$xy=gauss2var(mx,my,sdx,sdy,r,m)$$

The output xy is a $2 \times m$ matrix in which each 2-element column is a sample of a bivariate Gaussian pair X, Y with parameters $\mu_X = $ mx, $\mu_Y = $ my, $\sigma_X = $ sdx, $\sigma_Y = $ sdy and covariance $\rho_{X,Y} = $ r.

```
function xy=gauss2rv(mx,sdx,my,sdy,r,m)
mu=[mx my]';
cxy=r*sdx*sdy;
C=[sdx^2 cxy; cxy sdy^2];
xy=gaussvector(mu,C,m);
```

In this code, mu is a $2 \times m$ matrix in which each column holds the pair mx, my. Each column of randn(2,m) is a pair Z_1, Z_2 of independent Gaussian $(0, 1)$ random variables. The calcu-

lation A*randn(2,m) implements Equation (5.91) for m different pairs Z_1, Z_2.

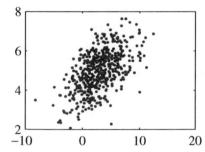

The sample output of gauss2var shown here is produced with the commands

```
>> xy=gauss2rv(3,3,5,1,0.5,500);
>> plot(xy(1,:),xy(2,:),'.');
```

We observe that the center of the cloud is $(\mu_X, \mu_Y) = (3, 5)$. In addition, we note that the X and Y axes are scaled differently because $\sigma_X = 3$ and $\sigma_Y = 1$.

We observe that this example with $\rho_{X,Y} = 0.5$ shows random variables that are less correlated than the examples in Figure 5.5 with $|\rho| = 0.9$.

We note that bivariate Gaussian random variables are a special case of n-dimensional Gaussian random vectors, which are introduced in Chapter 8. Based on linear algebra techniques, Chapter 8 introduces the gaussvector function to generate samples of Gaussian random vectors that generalizes gauss2rv to n dimensions.

Beyond bivariate Gaussian pairs, there exist a variety of techniques for generating sample values of pairs of continuous random variables of specific types. A basic approach is to generate X based on the marginal PDF $f_X(x)$ and then generate Y using a conditional probability model that depends on the value of X. Conditional probability models and MATLAB techniques that employ these models are the subject of Chapter 7.

Problems

Difficulty: ● Easy   ▨ Moderate   ◆ Difficult   ◆◆ Experts Only

5.1.1● Random variables X and Y have the joint CDF

$$F_{X,Y}(x,y) = \begin{cases} (1-e^{-x})(1-e^{-y}) & x \geq 0; \\ & y \geq 0, \\ 0 & \text{ow.} \end{cases}$$

(a) What is $P[X \leq 2, Y \leq 3]$?

(b) What is the marginal CDF, $F_X(x)$?

(c) What is the marginal CDF, $F_Y(y)$?

5.1.2● Express the following extreme values of $F_{X,Y}(x,y)$ in terms of the marginal cumulative distribution functions $F_X(x)$ and $F_Y(y)$.

(a) $F_{X,Y}(x,-\infty)$

(b) $F_{X,Y}(x,\infty)$

(c) $F_{X,Y}(-\infty,\infty)$

(d) $F_{X,Y}(-\infty,y)$

(e) $F_{X,Y}(\infty,y)$

5.1.3■ For continuous random variables X, Y with joint CDF $F_{X,Y}(x,y)$ and marginal CDFs $F_X(x)$ and $F_Y(y)$, find $P[x_1 \leq X < x_2 \cup y_1 \leq Y < y_2]$. This is the probability of the shaded "cross" region in the following diagram.

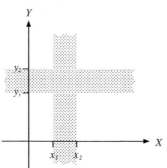

5.1.4■ Random variables X and Y have CDF $F_X(x)$ and $F_Y(y)$. Is $F(x,y) = F_X(x)F_Y(y)$ a valid CDF? Explain your answer.

5.1.5◆ In this problem, we prove Theorem 5.2.

(a) Sketch the following events on the X, Y plane:

$$A = \{X \leq x_1, y_1 < Y \leq y_2\},$$
$$B = \{x_1 < X \leq x_2, Y \leq y_1\},$$
$$C = \{x_1 < X \leq x_2, y_1 < Y \leq y_2\}.$$

(b) Express the probability of the events A, B, and $A \cup B \cup C$ in terms of the joint CDF $F_{X,Y}(x,y)$.

(c) Use the observation that events A, B, and C are mutually exclusive to prove Theorem 5.2.

5.1.6◆ Can the following function be the joint CDF of random variables X and Y?

$$F(x,y) = \begin{cases} 1 - e^{-(x+y)} & x \geq 0, y \geq 0, \\ 0 & \text{otherwise.} \end{cases}$$

5.2.1● Random variables X and Y have the joint PMF

$$P_{X,Y}(x,y) = \begin{cases} cxy & x = 1,2,4; \quad y = 1,3, \\ 0 & \text{otherwise.} \end{cases}$$

(a) What is the value of the constant c?

(b) What is $P[Y < X]$?

(c) What is $P[Y > X]$?

(d) What is $P[Y = X]$?

(e) What is $P[Y = 3]$?

5.2.2● Random variables X and Y have the joint PMF

$$P_{X,Y}(x,y) = \begin{cases} c|x+y| & \begin{matrix} x = -2,0,2; \\ y = -1,0,1, \end{matrix} \\ 0 & \text{otherwise.} \end{cases}$$

(a) What is the value of the constant c?

(b) What is $P[Y < X]$?

(c) What is $P[Y > X]$?

(d) What is $P[Y = X]$?

(e) What is $P[X < 1]$?

5.2.3● Test two integrated circuits. In each test, the probability of rejecting the circuit is p, independent of the other test. Let X be the number of rejects (either 0 or 1) in the first test and let Y be the number of rejects in the second test. Find the joint PMF $P_{X,Y}(x,y)$.

5.2.4■ For two independent flips of a fair coin, let X equal the total number of tails and let Y equal the number of heads on the last flip. Find the joint PMF $P_{X,Y}(x,y)$.

5.2.5■ In Figure 5.2, the axes of the figures are labeled X and Y because the figures depict possible values of the random variables X and Y. However, the figure at the end of Example 5.3 depicts $P_{X,Y}(x,y)$ on axes labeled with lowercase x and y. Should those axes be labeled with the uppercase X and Y? Hint: Reasonable arguments can be made for both views.

5.2.6■ As a generalization of Example 5.3, consider a test of n circuits such that each circuit is acceptable with probability p, independent of the outcome of any other test. Show that the joint PMF of X, the number of acceptable circuits, and Y, the number of acceptable circuits found before observing the first reject, is

$P_{X,Y}(x,y)$

$$= \begin{cases} \binom{n-y-1}{x-y}p^x(1-p)^{n-x} & 0 \le y \le x < n, \\ p^n & x = y = n, \\ 0 & \text{otherwise.} \end{cases}$$

Hint: For $0 \le y \le x < n$, show that

$$\{X = x, Y = y\} = A \cap B \cap C,$$

where

A: The first y tests are acceptable.

B: Test $y + 1$ is a rejection.

C: The remaining $n - y - 1$ tests yield $x - y$ acceptable circuits

5.2.7■ With two minutes left in a five-minute overtime, the score is 0–0 in a Rutgers soccer match versus Villanova. (Note that the overtime is NOT *sudden-death.*) In the next-to-last minute of the game, either (1) Rutgers scores a goal with probability $p = 0.2$, (2) Villanova scores with probability $p = 0.2$, or (3) neither team scores with probability $1 - 2p = 0.6$. If neither team scores in the next-to-last minute, then in the final minute, either (1) Rutgers scores a goal with probability $q = 0.3$, (2) Villanova scores with probability $q = 0.3$, or (3) neither team scores with probability $1 - 2q = 0.4$. However, if a team scores in the next-to-last minute, the trailing team goes for broke so that in the last minute, either (1) the leading team scores with prob-

ability 0.5, or (2) the trailing team scores with probability 0.5. For the final two minutes of overtime:

(a) Sketch a probability tree and construct a table for $P_{R,V}(r,v)$, the joint PMF of R, the number of Rutgers goals scored, and V, the number of Villanova goals scored.

(b) What is the probability $P[T]$ that the overtime ends in a tie?

(c) What is the PMF of R, the number of goals scored by Rutgers?

(d) What is the PMF of G, the total number of goals scored?

5.2.8♦ Each test of an integrated circuit produces an acceptable circuit with probability p, independent of the outcome of the test of any other circuit. In testing n circuits, let K denote the number of circuits rejected and let X denote the number of acceptable circuits (either 0 or 1) in the last test. Find the joint PMF $P_{K,X}(k,x)$.

5.2.9♦ Each test of an integrated circuit produces an acceptable circuit with probability p, independent of the outcome of the test of any other circuit. In testing n circuits, let K denote the number of circuits rejected and let X denote the number of acceptable circuits that appear before the first reject is found. Find the joint PMF $P_{K,X}(k,x)$.

5.3.1● Given the random variables X and Y in Problem 5.2.1, find

(a) The marginal PMFs $P_X(x)$ and $P_Y(y)$,

(b) The expected values $E[X]$ and $E[Y]$,

(c) The standard deviations σ_X and σ_Y.

5.3.2● Given the random variables X and Y in Problem 5.2.2, find

(a) The marginal PMFs $P_X(x)$ and $P_Y(y)$,

(b) The expected values $E[X]$ and $E[Y]$,

(c) The standard deviations σ_X and σ_Y.

5.3.3● For $n = 0, 1, \ldots$ and $0 \le k \le 100$, the joint PMF of random variables N and

K is

$$P_{N,K}(n,k)$$
$$= \frac{100^n e^{-100}}{n!} \binom{100}{k} p^k (1-p)^{100-k}.$$

Otherwise, $P_{N,K}(n,k) = 0$.   Find the marginal PMFs $P_N(n)$ and $P_K(k)$.

5.3.4■ Random variables X and Y have joint PMF

$$P_{X,Y}(x,y) = \begin{cases} 1/21 & x = 0,1,2,3,4,5; \\ & y = 0,1,\ldots,x, \\ 0 & \text{otherwise.} \end{cases}$$

Find the marginal PMFs $P_X(x)$ and $P_Y(y)$ and the expected values $E[X]$ and $E[Y]$.

5.3.5■ Random variables N and K have the joint PMF

$$P_{N,K}(n,k) = \begin{cases} \frac{(1-p)^{n-1}p}{n} & \substack{k=1,\ldots,n; \\ n=1,2,\ldots} \\ 0 & \text{otherwise.} \end{cases}$$

Find the marginal PMFs $P_N(n)$ and $P_K(k)$.

5.3.6◆ Random variables N and K have the joint PMF

$$P_{N,K}(n,k) = \begin{cases} \frac{100^n e^{-100}}{(n+1)!} & \substack{k=0,1,\ldots,n; \\ n=0,1,\ldots} \\ 0 & \text{otherwise.} \end{cases}$$

Find the marginal PMF $P_N(n)$. Show that the marginal PMF $P_K(k)$ satisfies $P_K(k) = P[N > k]/100$.

5.4.1■ Random variables X and Y have joint PDF

$$f_{X,Y}(x,y) = \begin{cases} cxy^2 & 0 \le x \le 1, 0 \le y \le 1, \\ 0 & \text{otherwise.} \end{cases}$$

(a) Find the constant c.

(b) Find $P[X > Y]$ and $P[Y < X^2]$.

(c) Find $P[\min(X,Y) \le 1/2]$.

(d) Find $P[\max(X,Y) \le 3/4]$.

5.4.2■ Random variables X and Y have joint PDF

$$f_{X,Y}(x,y) = \begin{cases} 6e^{-(2x+3y)} & x \ge 0, y \ge 0, \\ 0 & \text{otherwise.} \end{cases}$$

(a) Find $P[X > Y]$ and $P[X + Y \le 1]$.

(b) Find $P[\min(X,Y) \ge 1]$.

(c) Find $P[\max(X,Y) \le 1]$.

5.4.3◆ Random variables X and Y have joint PDF

$$f_{X,Y}(x,y) = \begin{cases} 8xy & 0 \le y \le x \le 1, \\ 0 & \text{otherwise.} \end{cases}$$

Following the method of Example 5.8, find the joint CDF $F_{X,Y}(x,y)$.

5.5.1● Random variables X and Y have the joint PDF

$$f_{X,Y}(x,y) = \begin{cases} 1/2 & -1 \le x \le y \le 1, \\ 0 & \text{otherwise.} \end{cases}$$

Sketch the region of nonzero probability and answer the following questions.

(a) What is $P[X > 0]$?

(b) What is $f_X(x)$?

(c) What is $E[X]$?

5.5.2● Random variables X and Y have joint PDF

$$f_{X,Y}(x,y) = \begin{cases} cx & 0 \le x \le 1, 0 \le y \le 1 \\ 0 & \text{otherwise} \end{cases}$$

(a) Find the constant c.

(b) Find the marginal PDF $f_X(x)$.

(c) Are X and Y independent?   Justify your answer.

5.5.3● X and Y have joint PDF

$$f_{X,Y}(x,y) = \begin{cases} 2 & x+y \le 1, x \ge 0, y \ge 0, \\ 0 & \text{otherwise.} \end{cases}$$

(a) What is the marginal PDF $f_X(x)$?

(b) What is the marginal PDF $f_Y(y)$?

5.5.4■ Over the circle $X^2 + Y^2 \leq r^2$, random variables X and Y have the uniform PDF

$$f_{X,Y}(x,y) = \begin{cases} 1/(\pi r^2) & x^2 + y^2 \leq r^2, \\ 0 & \text{otherwise.} \end{cases}$$

(a) What is the marginal PDF $f_X(x)$?

(b) What is the marginal PDF $f_Y(y)$?

5.5.5■ X and Y are random variables with the joint PDF

$$f_{X,Y}(x,y) = \begin{cases} 5x^2/2 & \begin{matrix} -1 \leq x \leq 1; \\ 0 \leq y \leq x^2 \end{matrix} \\ 0 & \text{otherwise.} \end{cases}$$

(a) What is the marginal PDF $f_X(x)$?

(b) What is the marginal PDF $f_Y(y)$?

5.5.6■ Over the circle $X^2 + Y^2 \leq r^2$, random variables X and Y have the PDF

$$f_{X,Y}(x,y) = \begin{cases} 2\,|xy|/r^4 & x^2 + y^2 \leq r^2, \\ 0 & \text{otherwise.} \end{cases}$$

(a) What is the marginal PDF $f_X(x)$?

(b) What is the marginal PDF $f_Y(y)$?

5.5.7■ For a random variable X, let $Y = aX + b$. Show that if $a > 0$ then $\rho_{X,Y} = 1$. Also show that if $a < 0$, then $\rho_{X,Y} = -1$.

5.5.8■ Random variables X and Y have joint PDF

$$f_{X,Y}(x,y) = \begin{cases} (x + y)/3 & 0 \leq x \leq 1; \\ & 0 \leq y \leq 2, \\ 0 & \text{otherwise.} \end{cases}$$

(a) Find the marginal PDFs $f_X(x)$ and $f_Y(y)$.

(b) What are $E[X]$ and $\text{Var}[X]$?

(c) What are $E[Y]$ and $\text{Var}[Y]$?

5.5.9◆ X and Y have the joint PDF

$$f_{X,Y}(x,y) = \begin{cases} cy & 0 \leq y \leq x \leq 1, \\ 0 & \text{otherwise.} \end{cases}$$

(a) Draw the region of nonzero probability.

(b) What is the value of the constant c?

(c) What is $F_X(x)$?

(d) What is $F_Y(y)$?

(e) What is $P[Y \leq X/2]$?

5.6.1● An ice cream company needs to order ingredients from its suppliers. Depending on the size of the order, the weight of the shipment can be either

1 kg for a small order,

2 kg for a big order.

The company has three different suppliers. The vanilla supplier is 20 miles away. The chocolate supplier is 100 miles away. The strawberry supplier is 300 miles away. An experiment consists of monitoring an order and observing W, the weight of the order, and D, the distance the shipment must be sent. The following probability model describes the experiment:

	van.	choc.	straw.
small	0.2	0.2	0.2
big	0.1	0.2	0.1

(a) What is the joint PMF $P_{W,D}(w,d)$ of the weight and the distance?

(b) Find the expected shipping distance $E[D]$.

(c) Are W and D independent?

5.6.2● A company receives shipments from two factories. Depending on the size of the order, a shipment can be in

1 box for a small order,

2 boxes for a medium order,

3 boxes for a large order.

The company has two different suppliers. Factory Q is 60 miles from the company. Factory R is 180 miles from the company.

An experiment consists of monitoring a shipment and observing B, the number of boxes, and M, the number of miles the shipment travels. The following probability model describes the experiment:

	Factory Q	Factory R
small order	0.3	0.2
medium order	0.1	0.2
large order	0.1	0.1

(a) Find $P_{B,M}(b, m)$, the joint PMF of the number of boxes and the distance.

(b) What is $E[B]$, the expected number of boxes?

(c) Are B and M independent?

5.6.3● Observe 100 independent flips of a fair coin. Let X equal the number of heads in the first 75 flips. Let Y equal the number of heads in the remaining 25 flips. Find $P_X(x)$ and $P_Y(y)$. Are X and Y independent? Find $P_{X,Y}(x, y)$.

5.6.4● Observe independent flips of a fair coin until heads occurs twice. Let X_1 equal the number of flips up to and including the first H. Let X_2 equal the number of additional flips up to and including the second H. What are $P_{X_1}(x_1)$ and $P_{X_2}(x_2)$? Are X_1 and X_2 independent? Find $P_{X_1,X_2}(x_1, x_2)$.

5.6.5● X is the continuous uniform $(0, 2)$ random variable. Y has the continuous uniform $(0, 5)$ PDF, independent of X. What is the joint PDF $f_{X,Y}(x, y)$?

5.6.6● X_1 and X_2 are independent random variables such that X_i has PDF

$$f_{X_i}(x) = \begin{cases} \lambda_i e^{-\lambda_i x} & x \geq 0, \\ 0 & \text{otherwise.} \end{cases}$$

What is $P[X_2 < X_1]$?

5.6.7■ In terms of a positive constant k, random variables X and Y have joint PDF

$$f_{X,Y}(x, y) = \begin{cases} k + 3x^2 & \begin{matrix} -1/2 \leq x \leq 1/2, \\ -1/2 \leq y \leq 1/2, \end{matrix} \\ 0 & \text{otherwise.} \end{cases}$$

(a) What is k?

(b) What is the marginal PDF of X?

(c) What is the marginal PDF of Y?

(d) Are X and Y independent?

5.6.8■ X_1 and X_2 are independent, identically distributed random variables with PDF

$$f_X(x) = \begin{cases} x/2 & 0 \leq x \leq 2, \\ 0 & \text{otherwise.} \end{cases}$$

(a) Find the CDF, $F_X(x)$.

(b) What is $P[X_1 \leq 1, X_2 \leq 1]$, the probability that X_1 and X_2 are both less than or equal to 1?

(c) Let $W = \max(X_1, X_2)$. What is $F_W(1)$, the CDF of W evaluated at $w = 1$?

(d) Find the CDF $F_W(w)$.

5.6.9◆ Prove that random variables X and Y are independent if and only if

$$F_{X,Y}(x, y) = F_X(x) F_Y(y).$$

5.7.1● Continuing Problem 5.6.1, the price per kilogram for shipping the order is one cent per mile. C cents is the shipping cost of one order. What is $E[C]$?

5.7.2● Continuing Problem 5.6.2, the price per mile of shipping each box is one cent per mile the box travels. C cents is the price of one shipment. What is $E[C]$, the expected price of one shipment?

5.7.3● A random ECE sophomore has height X (rounded to the nearest foot) and GPA Y (rounded to the nearest integer). These random variables have joint PMF

$P_{X,Y}(x, y)$	$y = 1$	$y = 2$	$y = 3$	$y = 4$
$x = 5$	0.05	0.1	0.2	0.05
$x = 6$	0.1	0.1	0.3	0.1

Find $E[X + Y]$ and $Var[X + Y]$.

5.7.4● X and Y are independent, identically distributed random variables with PMF

$$P_X(k) = P_Y(k) = \begin{cases} 3/4 & k = 0, \\ 1/4 & k = 20, \\ 0 & \text{otherwise.} \end{cases}$$

Find the following quantities:

$E[X], \quad Var[X],$
$E[X + Y], \quad Var[X + Y], \quad E[XY2^{XY}].$

5.7.5● X and Y are random variables with $E[X] = E[Y] = 0$ and $Var[X] = 1$, $Var[Y] =$

4 and correlation coefficient $\rho = 1/2$. Find $\text{Var}[X + Y]$.

5.7.6● X and Y are random variables such that X has expected value $\mu_X = 0$ and standard deviation $\sigma_X = 3$ while Y has expected value $\mu_Y = 1$ and standard deviation $\sigma_Y = 4$. In addition, X and Y have covariance $\text{Cov}[X, Y] = -3$. Find the expected value and variance of $W = 2X + 2Y$.

5.7.7● Observe independent flips of a fair coin until heads occurs twice. Let X_1 equal the number of flips up to and including the first H. Let X_2 equal the number of additional flips up to and including the second H. Let $Y = X_1 - X_2$. Find $\text{E}[Y]$ and $\text{Var}[Y]$. Hint: Don't try to find $P_Y(y)$.

5.7.8● X_1 and X_2 are independent identically distributed random variables with expected value $\text{E}[X]$ and variance $\text{Var}[X]$.

(a) What is $\text{E}[X_1 - X_2]$?

(b) What is $\text{Var}[X_1 - X_2]$?

5.7.9■ X and Y are identically distributed random variables with $\text{E}[X] = \text{E}[Y] = 0$ and covariance $\text{Cov}[X, Y] = 3$ and correlation coefficient $\rho_{X,Y} = 1/2$. For nonzero constants a and b, $U = aX$ and $V = bY$.

(a) Find $\text{Cov}[U, V]$.

(b) Find the correlation coefficient $\rho_{U,V}$.

(c) Let $W = U + V$. For what values of a and b are X and W uncorrelated?

5.7.10■ True or False: For identically distributed random variables Y_1 and Y_2 with $\text{E}[Y_1] = \text{E}[Y_2] = 0$, $\text{Var}[Y_1 + Y_2] \geq \text{Var}[Y_1]$.

5.7.11■ X and Y are random variables with $\text{E}[X] = \text{E}[Y] = 0$ such that X has standard deviation $\sigma_X = 2$ while Y has standard deviation $\sigma_Y = 4$.

(a) For $V = X - Y$, what are the smallest and largest possible values of $\text{Var}[V]$?

(b) For $W = X - 2Y$, what are the smallest and largest possible values of $\text{Var}[W]$?

5.7.12■ Random variables X and Y have joint PDF

$$f_{X,Y}(x, y) = \begin{cases} 4xy & 0 \leq x \leq 1, 0 \leq y \leq 1, \\ 0 & \text{otherwise.} \end{cases}$$

(a) What are $\text{E}[X]$ and $\text{Var}[X]$?

(b) What are $\text{E}[Y]$ and $\text{Var}[Y]$?

(c) What is $\text{Cov}[X, Y]$?

(d) What is $\text{E}[X + Y]$?

(e) What is $\text{Var}[X + Y]$?

5.7.13■ Random variables X and Y have joint PDF

$$f_{X,Y}(x, y) = \begin{cases} 5x^2/2 & -1 \leq x \leq 1; \\ & 0 \leq y \leq x^2, \\ 0 & \text{otherwise.} \end{cases}$$

Answer the following questions.

(a) What are $\text{E}[X]$ and $\text{Var}[X]$?

(b) What are $\text{E}[Y]$ and $\text{Var}[Y]$?

(c) What is $\text{Cov}[X, Y]$?

(d) What is $\text{E}[X + Y]$?

(e) What is $\text{Var}[X + Y]$?

5.7.14■ Random variables X and Y have joint PDF

$$f_{X,Y}(x, y) = \begin{cases} 2 & 0 \leq y \leq x \leq 1, \\ 0 & \text{otherwise.} \end{cases}$$

(a) What are $\text{E}[X]$ and $\text{Var}[X]$?

(b) What are $\text{E}[Y]$ and $\text{Var}[Y]$?

(c) What is $\text{Cov}[X, Y]$?

(d) What is $\text{E}[X + Y]$?

(e) What is $\text{Var}[X + Y]$?

5.7.15■ A transmitter sends a signal X and a receiver makes the observation $Y = X + Z$, where Z is a receiver noise that is independent of X and $\text{E}[X] = \text{E}[Z] = 0$. Since the average power of the signal is $\text{E}[X^2]$ and the average power of the noise is $E[Z^2]$, a quality measure for the received signal is the signal-to-noise ratio

$$\Gamma = \frac{E\left[X^2\right]}{E\left[Z^2\right]}.$$

How is Γ related to the correlation coefficient $\rho_{X,Y}$?

5.8.1● X and Z are independent random variables with $E[X] = E[Z] = 0$ and variance $\text{Var}[X] = 1$ and $\text{Var}[Z] = 16$. Let $Y = X + Z$. Find the correlation coefficient ρ of X and Y. Are X and Y independent?

5.8.2● For the random variables X and Y in Problem 5.2.1, find

(a) The expected value of $W = Y/X$,

(b) The correlation, $r_{X,Y} = E[XY]$,

(c) The covariance, $\text{Cov}[X,Y]$,

(d) The correlation coefficient, $\rho_{X,Y}$,

(e) The variance of $X + Y$, $\text{Var}[X + Y]$.

(Refer to the results of Problem 5.3.1 to answer some of these questions.)

5.8.3● For the random variables X and Y in Problem 5.2.2 find

(a) The expected value of $W = 2^{XY}$,

(b) The correlation, $r_{X,Y} = E[XY]$,

(c) The covariance, $\text{Cov}[X,Y]$,

(d) The correlation coefficient, $\rho_{X,Y}$,

(e) The variance of $X + Y$, $\text{Var}[X + Y]$.

(Refer to the results of Problem 5.3.2 to answer some of these questions.)

5.8.4● Let H and B be the random variables in Quiz 5.3. Find $r_{H,B}$ and $\text{Cov}[H, B]$.

5.8.5● X and Y are independent random variables with PDFs

$$f_X(x) = \begin{cases} \frac{1}{3}e^{-x/3} & x \geq 0, \\ 0 & \text{otherwise}, \end{cases}$$

$$f_Y(y) = \begin{cases} \frac{1}{2}e^{-y/2} & y \geq 0, \\ 0 & \text{otherwise}. \end{cases}$$

(a) Find the correlation $r_{X,Y}$.

(b) Find the covariance $\text{Cov}[X,Y]$.

5.8.6● The random variables X and Y have joint PMF

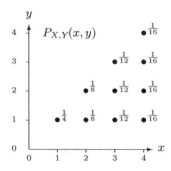

Find

(a) The expected values $E[X]$ and $E[Y]$,

(b) The variances $\text{Var}[X]$ and $\text{Var}[Y]$,

(c) The correlation, $r_{X,Y} = E[XY]$,

(d) The covariance, $\text{Cov}[X,Y]$,

(e) The correlation coefficient, $\rho_{X,Y}$.

5.8.7■ For X and Y with PMF $P_{X,Y}(x,y)$ given in Problem 5.8.6, let $W = \min(X, Y)$ and $V = \max(X, Y)$. Find

(a) The expected values, $E[W]$ and $E[V]$,

(b) The variances, $\text{Var}[W]$ and $\text{Var}[V]$,

(c) The correlation, $r_{W,V}$,

(d) The covariance, $\text{Cov}[W, V]$,

(e) The correlation coefficient, $\rho_{W,V}$.

5.8.8■ Random variables X and Y have joint PDF

$$f_{X,Y}(x, y) = \begin{cases} 1/2 & -1 \leq x \leq y \leq 1, \\ 0 & \text{otherwise}. \end{cases}$$

Find $r_{X,Y}$ and $E[e^{X+Y}]$.

5.8.9■ This problem outlines a proof of Theorem 5.13.

(a) Show that

$$\hat{X} - E[\hat{X}] = a(X - E[X]),$$
$$\hat{Y} - E[\hat{Y}] = c(Y - E[Y]).$$

(b) Use part (a) to show that

$$\text{Cov}\left[\hat{X}, \hat{Y}\right] = ac\, \text{Cov}\left[X, Y\right].$$

(c) Show that $\text{Var}[\hat{X}] = a^2\, \text{Var}[X]$ and $\text{Var}[\hat{Y}] = c^2\, \text{Var}[Y]$.

(d) Combine parts (b) and (c) to relate $\rho_{\hat{X}, \hat{Y}}$ and $\rho_{X, Y}$.

5.8.10◆ Random variables N and K have the joint PMF

$$P_{N,K}(n, k)$$
$$= \begin{cases} (1-p)^{n-1} p/n & k = 1, \ldots, n; \\ & n = 1, 2, \ldots, \\ 0 & \text{otherwise.} \end{cases}$$

Find the marginal PMF $P_N(n)$ and the expected values $E[N]$, $\text{Var}[N]$, $E[N^2]$, $E[K]$, $\text{Var}[K]$, $E[N + K]$, $r_{N,K}$, $\text{Cov}[N, K]$.

5.9.1◉ Random variables X and Y have joint PDF

$$f_{X,Y}(x, y) = ce^{-(x^2/8)-(y^2/18)}.$$

What is the constant c? Are X and Y independent?

5.9.2◉ X is the Gaussian ($\mu = 1, \sigma = 2$) random variable. Y is the Gaussian ($\mu = 2, \sigma = 4$) random variable. X and Y are independent.

(a) What is the PDF of $V = X + Y$?

(b) What is the PDF of $W = 3X + 2Y$?

5.9.3▦ TRUE OR FALSE: X_1 and X_2 are bivariate Gaussian random variables. For any constant y, there exists a constant a such that $P[X_1 + aX_2 \le y] = 1/2$.

5.9.4▦ X_1 and X_2 are identically distributed Gaussian $(0, 1)$ random variables. Moreover, they are jointly Gaussian. Under what conditions are X_1, X_2 and $X_1 + X_2$ identically distributed?

5.9.5▦ Random variables X and Y have joint PDF

$$f_{X,Y}(x, y) = ce^{-(2x^2-4xy+4y^2)}.$$

(a) What are $E[X]$ and $E[Y]$?

(b) Find the correlation coefficient $\rho_{X,Y}$.

(c) What are $\text{Var}[X]$ and $\text{Var}[Y]$?

(d) What is the constant c?

(e) Are X and Y independent?

5.9.6▦ An archer shoots an arrow at a circular target of radius 50 cm. The arrow pierces the target at a random position (X, Y), measured in centimeters from the center of the disk at position $(X, Y) = (0, 0)$. The bullseye is a solid black circle of radius 2 cm, at the center of the target. Calculate the probability $P[B]$ of the event that the archer hits the bullseye under each of the following models:

(a) X and Y are iid continuous uniform $(-50, 50)$ random variables.

(b) The PDF $f_{X,Y}(x, y)$ is uniform over the 50 cm circular target.

(c) X and Y are iid Gaussian ($\mu = 0, \sigma = 10$) random variables.

5.9.7▦ A person's white blood cell (WBC) count W (measured in thousands of cells per microliter of blood) and body temperature T (in degrees Celsius) can be modeled as bivariate Gaussian random variables such that W is Gaussian $(7, 2)$ and T is Gaussian $(37, 1)$. To determine whether a person is sick, first the person's temperature T is measured. If $T > 38$, then the person's WBC count is measured. If $W > 10$, the person is declared ill (event I).

(a) Suppose W and T are uncorrelated. What is $P[I]$? Hint: Draw a tree diagram for the experiment.

(b) Now suppose W and T have correlation coefficient $\rho_{W,T} = 1/\sqrt{2}$. Find the conditional probability $P[I|T = t]$ that a person is declared ill given that the person's temperature is $T = t$.

5.9.8◆ Suppose your grade in a probability course depends on your exam scores X_1 and X_2. The professor, a fan of probability, releases exam scores in a normalized fashion such that X_1 and X_2 are iid Gaussian

($\mu = 0, \sigma = \sqrt{2}$) random variables. Your semester average is $X = 0.5(X_1 + X_2)$.

(a) You earn an A grade if $X > 1$. What is $P[A]$?

(b) To improve his SIRS (Student Instructional Rating Service) score, the professor decides he should award more A's. Now you get an A if $\max(X_1, X_2) > 1$. What is $P[A]$ now?

(c) The professor found out he is unpopular at ratemyprofessor.com and decides to award an A if either $X > 1$ or $\max(X_1, X_2) > 1$. Now what is $P[A]$?

(d) Under criticism of grade inflation from the department chair, the professor adopts a new policy. An A is awarded if $\max(X_1, X_2) > 1$ *and* $\min(X_1, X_2) > 0$. Now what is $P[A]$?

5.9.9◆ Your course grade depends on two test scores: X_1 and X_2. Your score X_i on test i is Gaussian ($\mu = 74, \sigma = 16$) random variable, independent of any other test score.

(a) With equal weighting, grades are determined by $Y = X_1/2 + X_2/2$. You earn an A if $Y \geq 90$. What is $P[A] = P[Y \geq 90]$?

(b) A student asks the professor to choose a weight factor w, $0 \leq w \leq 1$, such that

$$Y = wX_1 + (1 - w)X_2.$$

Find $P[A]$ as a function of the weight w. What value or values of w maximize $P[A] = P[Y \geq 90]$?

(c) A different student proposes that the better exam is the one that should count and that grades should be based on $M = \max(X_1, X_2)$. In a fit of generosity, the professor agrees! Now what is $P[A] = P[M > 90]$?

(d) How generous was the professor? In a class of 100 students, what is the expected increase in the number of A's awarded?

5.9.10◆ Under what conditions on the constants a, b, c, and d is

$$f(x, y) = de^{-(a^2 x^2 + bxy + c^2 y^2)}$$

a joint Gaussian PDF?

5.9.11◆ Show that the joint Gaussian PDF $f_{X,Y}(x, y)$ given by Definition 5.10 satisfies

$$\int_{-\infty}^{\infty} \int_{-\infty}^{\infty} f_{X,Y}(x, y)\, dx\, dy = 1.$$

Hint: Use Equation (5.69) and the result of Problem 4.6.13.

5.9.12◆ Random variables X_1 and X_2 are independent identical Gaussian $(0, 1)$ random variables. Let

$$Y_1 = X_1 \operatorname{sgn}(X_2), \qquad Y_2 = X_2 \operatorname{sgn}(X_1),$$

where

$$\operatorname{sgn}(x) = \begin{cases} 1 & x > 0, \\ -1 & x \leq 0. \end{cases}$$

(a) Find the CDF $F_{Y_1}(y_1)$ in terms of the $\Phi(\cdot)$ function.

(b) Show that Y_1 and Y_2 are both Gaussian random variables.

(c) Are Y_1 and Y_2 bivariate Gaussian random variables?

5.10.1● Every laptop returned to a repair center is classified according its needed repairs: (1) LCD screen, (2) motherboard, (3) keyboard, or (4) other. A random broken laptop needs a type i repair with probability $p_i = 2^{4-i}/15$. Let N_i equal the number of type i broken laptops returned on a day in which four laptops are returned.

(a) Find the joint PMF of N_1, N_2, N_3, N_4.

(b) What is the probability that two laptops require LCD repairs?

(c) What is the probability that more laptops require motherboard repairs than keyboard repairs?

5.10.2● When ordering a personal computer, a customer can add the following features to the basic configuration: (1) additional memory, (2) flat panel display, (3) professional software, and (4) wireless modem. A random computer order has feature i with probability $p_i = 2^{-i}$ independent of other features. In an hour in which three computers are ordered, let N_i equal the number of computers with feature i.

(a) Find the joint PMF

$$P_{N_1,N_2,N_3,N_4}(n_1, n_2, n_3, n_4).$$

(b) What is the probability of selling a computer with no additional features?

(c) What is the probability of selling a computer with at least three additional features?

5.10.3● The random variables $X_1, \ldots, X_n$ have the joint PDF

$$f_{X_1,\ldots,X_n}(x_1, \ldots, x_n) = \begin{cases} 1 & 0 \le x_i \le 1; \\ & i = 1, \ldots, n, \\ 0 & \text{otherwise.} \end{cases}$$

Find

(a) The joint CDF, $F_{X_1,\ldots,X_n}(x_1, \ldots, x_n)$,

(b) $P[\min(X_1, X_2, X_3) \le 3/4]$.

5.10.4● Are N_1, N_2, N_3, N_4 in Problem 5.10.1 independent?

5.10.5■ In a compressed data file of 10,000 bytes, each byte is equally likely to be any one of 256 possible characters $b_0, \ldots, b_{255}$ independent of any other byte. If N_i is the number of times b_i appears in the file, find the joint PMF of $N_0, \ldots, N_{255}$. Also, what is the joint PMF of N_0 and N_1?

5.10.6■ The joint PMF of X, Y, and Z, the number of 1-page, 2-page, and 3-page downloads, respectively, in each of four downloads is given by

$$P_{X,Y,Z}(x, y, z) = \binom{4}{x, y, z} \frac{1}{3^x} \frac{1}{2^y} \frac{1}{6^z}.$$

(a) In a group of four downloads, what is the PMF of the number of 3-page documents?

(b) In a group of four downloads, what is the expected number of 3-page documents?

(c) Given that there are two 3-page documents in a group of four, what is the joint PMF of the number of 1-page documents and the number of 2-page documents?

(d) Given that there are two 3-page documents in a group of four, what is the expected number of 1-page documents?

(e) In a group of four downloads, what is the joint PMF of the number of 1-page documents and the number of 2-page documents?

5.10.7■ X_1, X_2, X_3 are iid exponential (λ) random variables. Find:

(a) the PDF of $V = \min(X_1, X_2, X_3)$,

(b) the PDF of $W = \max(X_1, X_2, X_3)$.

5.10.8■ In a race of 10 sailboats, the finishing times of all boats are iid Gaussian random variables with expected value 35 minutes and standard deviation 5 minutes.

(a) What is the probability that the winning boat will finish the race in less than 25 minutes?

(b) What is the probability that the last boat will cross the finish line in more than 50 minutes?

(c) Given this model, what is the probability that a boat will finish before it starts (negative finishing time)?

5.10.9◆ Random variables $X_1, X_2, \ldots, X_n$ are iid; each X_j has CDF $F_X(x)$ and PDF $f_X(x)$. Consider

$$L_n = \min(X_1, \ldots, X_n)$$
$$U_n = \max(X_1, \ldots, X_n).$$

In terms of $F_X(x)$ and/or $f_X(x)$:

(a) Find the CDF $F_{U_n}(u)$.

(b) Find the CDF $F_{L_n}(l)$.

(c) Find the joint CDF $F_{L_n, U_n}(l, u)$.

5.10.10◆ Suppose you have n suitcases and suitcase i holds X_i dollars where $X_1, X_2, \ldots, X_n$ are iid continuous uniform $(0, m)$ random variables. (Think of a number like one million for the symbol m.) Unfortunately, you don't know X_i until you open suitcase i.

Suppose you can open the suitcases one by one, starting with suitcase n and going down to suitcase 1. After opening suitcase i, you can either accept or reject X_i dollars. If you accept suitcase i, the game ends. If you reject, then you get to choose only from the still unopened suitcases.

What should you do? Perhaps it is not so obvious? In fact, you can decide before the game on a policy, a set of rules to follow. We will specify a policy by a vector $(\tau_1, \ldots, \tau_n)$ of threshold parameters.

- After opening suitcase i, you accept the amount X_i if $X_i \geq \tau_i$.

- Otherwise, you reject suitcase i and open suitcase $i - 1$.

- If you have rejected suitcases n down through 2, then you must accept the amount X_1 in suitcase 1. Thus the threshold $\tau_1 = 0$ since you never reject the amount in the last suitcase.

(a) Suppose you reject suitcases n through $i + 1$, but then you accept suitcase i. Find $E[X_i | X_i \geq \tau_i]$.

(b) Let W_k denote your reward given that there are k unopened suitcases remaining. What is $E[W_1]$?

(c) As a function of τ_k, find a recursive relationship for $E[W_k]$ in terms of τ_k and $E[W_{k-1}]$.

(d) For $n = 4$ suitcases, find the policy $(\tau_1^*, \ldots, \tau_4^*)$, that maximizes $E[W_4]$.

5.10.11◆◆ Given the set $\{U_1, \ldots, U_n\}$ of iid uniform $(0, T)$ random variables, we define

$$X_k = \text{small}_k(U_1, \ldots, U_n)$$

as the kth "smallest" element of the set. That is, X_1 is the minimum element, X_2 is the second smallest, and so on, up to X_n, which is the maximum element of $\{U_1, \ldots, U_n\}$. Note that $X_1, \ldots, X_n$ are known as the *order statistics* of $U_1, \ldots, U_n$. Prove that

$$f_{X_1, \ldots, X_n}(x_1, \ldots, x_n)$$
$$= \begin{cases} n!/T^m & 0 \leq x_1 < \cdots < x_n \leq T, \\ 0 & \text{otherwise.} \end{cases}$$

5.11.1◉ For random variables X and Y in Example 5.25, use MATLAB to generate a list of the form

$$\begin{array}{ccc} x_1 & y_1 & P_{X,Y}(x_1, y_1) \\ x_2 & y_2 & P_{X,Y}(x_2, y_2) \\ \vdots & \vdots & \vdots \end{array}$$

that includes all possible pairs (x, y).

5.11.2◉ For random variables X and Y in Example 5.25, use MATLAB to calculate $E[X]$, $E[Y]$, the correlation $E[XY]$, and the covariance $\text{Cov}[X, Y]$.

5.11.3◉ You generate random variable $W = W$ by typing `W=sum(4*randn(1,2))` in a MATLAB Command window. What is $\text{Var}[W]$?

5.11.4◉ Write `trianglecdfplot.m`, a script that graphs $F_{X,Y}(x, y)$ of Figure 5.4.

5.11.5▦ Problem 5.2.6 extended Example 5.3 to a test of n circuits and identified the joint PDF of X, the number of acceptable circuits, and Y, the number of successful tests before the first reject. Write a MATLAB function

    [SX,SY,PXY]=circuits(n,p)

that generates the sample space grid for the n circuit test. Check your answer against Equation (5.11) for the $p = 0.9$ and $n = 2$ case. For $p = 0.9$ and $n = 50$, calculate the correlation coefficient $\rho_{X,Y}$.

6

Probability Models of Derived Random Variables

There are many situations in which we observe one or more random variables and use their values to compute a new random variable. For example, when voltage across an r_0 ohm resistor is a random variable X, the power dissipated in that resistor is $Y = X^2/r_0$. Circuit designers need a probability model for Y to evaluate the power consumption of the circuit. Similarly, if the amplitude (current or voltage) of a radio signal is X, the received signal power is proportional to $Y = X^2$. A probability model for Y is essential in evaluating the performance of a radio receiver. The output of a limiter or rectifier is another random variable that a circuit designer may need to analyze.

Radio systems also provide practical examples of functions of two random variables. For example, we can describe the amplitude of the signal transmitted by a radio station as a random variable, X. We can describe the attenuation of the signal as it travels to the antenna of a moving car as another random variable, Y. In this case the amplitude of the signal at the radio receiver in the car is the random variable $W = X/Y$. Other practical examples appear in cellular telephone base stations with two antennas. The amplitudes of the signals arriving at the two antennas are modeled as random variables X and Y. The radio receiver connected to the two antennas can use the received signals in a variety of ways.

- It can choose the signal with the larger amplitude and ignore the other one. In this case, the receiver produces the random variable $W = X$ if $|X| > |Y|$ and $W = Y$, otherwise. This is an example of *selection diversity combining*.

- The receiver can add the two signals and use $W = X + Y$. This process is referred to as *equal gain combining* because it treats both signals equally.

- A third alternative is to combine the two signals unequally in order to give less weight to the signal considered to be more distorted. In this case $W = aX + bY$. If a and b are optimized, the receiver performs *maximal ratio combining*.

All three combining processes appear in practical radio receivers.

Formally, we have the following situations.

- We perform an experiment and observe a sample value of random variable X. Based on our knowledge of the experiment, we have a probability model for X embodied in the PMF $P_X(x)$ or PDF $f_X(x)$. After performing the experiment, we calculate a sample value of the random variable $W = g(X)$.

- We perform an experiment and observe a sample value of two random variables X and Y. Based on our knowledge of the experiment, we have a probability model for X and Y embodied in a joint PMF $P_{X,Y}(x, y)$ or a joint PDF $f_{X,Y}(x, y)$. After performing the experiment, we calculate a sample value of the random variable $W = g(X, Y)$.

In both cases, the mathematical problem is to determine the properties of W. Previous chapters address aspects of this problem. Theorem 3.9 provides a formula for $P_W(w)$, the PMF of $W = g(X)$ and Theorem 3.10 provides a formula for $E[W]$ given $P_X(x)$ and $g(X)$. Chapter 4, on continuous random variables, provides, in Theorem 4.4, a formula for $E[W]$ given $f_X(x)$ and $g(X)$ but defers to this chapter examining the probability model of W. Similarly, Chapter 5 examines $E[g(X, Y)]$ but does not explain how to find the PMF or PDF of $W = g(X, Y)$. In this chapter, we develop methods to derive the distribution (PMF, CDF or PDF) of a function of one or two random variables.

Prior chapters have a lot of new ideas and concepts, each illustrated by a relatively small number of examples. In contrast, this chapter has relatively few new concepts but many examples to illustrate the techniques. In particular, Sections 6.2 and 6.3 advocate a single approach: find the CDF $F_W(w) = P[W \leq w]$ by finding those values of X such that $W = g(X) \leq w$. Similarly, Section 6.4 uses the same basic idea: Find those values of X, Y such that $W = g(X, Y) \leq w$. While this idea is simple, the derivations can be complicated.

6.1   PMF of a Function of Two Discrete Random Variables

> $P_W(w)$, the PMF of a function of discrete random variables X and Y is the sum of the probabilities of all sample values (x, y) for which $g(x, y) = w$.

When X and Y are discrete random variables, S_W, the range of W, is a countable set corresponding to all possible values of $g(X, Y)$. Therefore, W is a discrete random variable and has a PMF $P_W(w)$. We can apply Theorem 5.3 to find $P_W(w) = P[W = w]$. Since $\{W = w\}$ is another name for the event $\{g(X, Y) = w\}$, we obtain $P_W(w)$ by adding the values of $P_{X,Y}(x, y)$ corresponding to the x, y pairs for which $g(x, y) = w$.

══════Theorem 6.1══════

For discrete random variables X and Y, the derived random variable $W = g(X, Y)$

has PMF

$$P_W(w) = \sum_{(x,y):g(x,y)=w} P_{X,Y}(x,y).$$

════Example 6.1════

$P_{L,X}(l,x)$	$x=40$	$x=60$
$l=1$	0.15	0.1
$l=2$	0.3	0.2
$l=3$	0.15	0.1

A firm sends out two kinds of newsletters. One kind contains only text and grayscale images and requires 40 cents to print each page. The other kind contains color pictures that cost 60 cents per page. Newsletters can be 1, 2, or 3 pages long. Let the random variable L represent the length of a newsletter in pages. $S_L = \{1,2,3\}$. Let the random variable X represent the cost in cents to print each page. $S_X = \{40, 60\}$. After observing many newsletters, the firm has derived the probability model shown above. Let $W = g(L, X) = LX$ be the total cost in cents of a newsletter. Find the range S_W and the PMF $P_W(w)$.

· ·

$P_{L,X}(l,x)$	$x=40$	$x=60$
$l=1$	0.15 $(W=40)$	0.1 $(W=60)$
$l=2$	0.3 $(W=80)$	0.2 $(W=120)$
$l=3$	0.15 $(W=120)$	0.1 $(W=180)$

For each of the six possible combinations of L and X, we record $W = LX$ under the corresponding entry in the PMF table on the left. The range of W is $S_W = \{40, 60, 80, 120, 180\}$. With the exception of $W = 120$, there is a unique pair L, X such that $W = LX$. For $W = 120$, $P_W(120) = P_{L,X}(3, 40) + P_{L,X}(2, 60)$. The corresponding probabilities are recorded in the second table on the left.

· ·

w	40	60	80	120	180
$P_W(w)$	0.15	0.1	0.3	0.35	0.1

6.2 Functions Yielding Continuous Random Variables

> To obtain the PDF of $W = g(X)$, a continuous function of a continuous random variable, derive the CDF of W and then differentiate. The procedure is straightforward when $g(x, y)$ is a linear function. It is more complex for other functions.

When X and $W = g(X)$ are continuous random variables, we develop a two-step procedure to derive the PDF $f_W(w)$:

1. Find the CDF $F_W(w) = \mathrm{P}[W \leq w]$.

2. The PDF is the derivative $f_W(w) = dF_W(w)/dw$.

This procedure *always* works and is easy to remember. When $g(X)$ is a linear function of X, the method is straightforward. Otherwise, as we shall see in examples, finding $F_W(w)$ can be tricky.

Before proceeding to the examples and theorems, we add one reminder. It is easier to calculate $E[g(X)]$ directly from the PDF $f_X(x)$ using Theorem 4.4 than it is to derive the PDF of $Y = g(X)$ and then use the definition of expected value, Definition 4.4. This section applies to situations in which it is necessary to find a complete probability model of $W = g(X)$.

Example 6.2

In Example 4.2, W centimeters is the location of the pointer on the 1-meter circumference of the circle. Use the solution of Example 4.2 to derive $f_W(w)$.

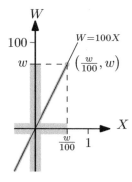

The function $W = 100X$, where X in Example 4.2 is the location of the pointer measured in meters. To find the CDF $F_W(w) = P[W \leq w]$, the first step is to translate the event $\{W \leq w\}$ into an event described by X. Each outcome of the experiment is mapped to an (X, W) pair on the line $W = 100X$. Thus the event $\{W \leq w\}$, shown with gray highlight on the vertical axis, is the same event as $\{X \leq w/100\}$, which is shown with gray highlight on the horizontal axis. Both of these events correspond in the figure to observing an (X, W) pair along the highlighted section of the line $w = g(X) = 100w$. This translation of the event $W = w$ to an event described in terms of X depends only on the function $g(X)$. Specifically, it does not depend on the probability model for X. From the figure, we see that

$$F_W(w) = P[W \leq w] = P[100X \leq w] = P[X \leq w/100] = F_X(w/100). \quad (6.1)$$

The calculation of $F_X(w/100)$ depends on the probability model for X. For this problem, we recall that Example 4.2 derives the CDF of X,

$$F_X(x) = \begin{cases} 0 & x < 0, \\ x & 0 \leq x < 1, \\ 1 & x \geq 1. \end{cases} \quad (6.2)$$

From this result, we can use algebra to find

$$F_W(w) = F_X\left(\frac{w}{100}\right) = \begin{cases} 0 & \frac{w}{100} < 0, \\ \frac{w}{100} & 0 \leq \frac{w}{100} < 1, \\ 1 & \frac{w}{100} \geq 1, \end{cases} = \begin{cases} 0 & w < 0, \\ \frac{w}{100} & 0 \leq w < 100, \\ 1 & w \geq 100. \end{cases} \quad (6.3)$$

We take the derivative of the CDF of W over each of the intervals to find the PDF:

$$f_W(w) = \frac{dF_W(w)}{dw} = \begin{cases} 1/100 & 0 \leq w < 100, \\ 0 & \text{otherwise.} \end{cases} \quad (6.4)$$

We see that W is the uniform $(0, 100)$ random variable.

We use this two-step procedure in the following theorem to generalize Example 6.2 by deriving the CDF and PDF for any scale change and any continuous random variable.

Theorem 6.2

If $W = aX$, where $a > 0$, then W has CDF and PDF

$$F_W(w) = F_X(w/a), \qquad f_W(w) = \frac{1}{a} f_X(w/a).$$

Proof First, we find the CDF of W,

$$F_W(w) = \mathrm{P}\left[aX \le w\right] = \mathrm{P}\left[X \le w/a\right] = F_X(w/a). \tag{6.5}$$

We take the derivative of $F_Y(y)$ to find the PDF:

$$f_W(w) = \frac{dF_W(w)}{dw} = \frac{1}{a} f_X(w/a). \tag{6.6}$$

Theorem 6.2 states that multiplying a random variable by a positive constant stretches $(a > 1)$ or shrinks $(a < 1)$ the original PDF.

Example 6.3

The triangular PDF of X is

$$f_X(x) = \begin{cases} 2x & 0 \le x \le 1, \\ 0 & \text{otherwise.} \end{cases} \tag{6.7}$$

Find the PDF of $W = aX$. Sketch the PDF of W for $a = 1/2, 1, 2$.

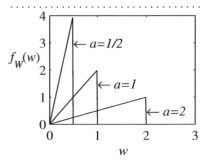

For any $a > 0$, we use Theorem 6.2 to find the PDF:

$$f_W(w) = \frac{1}{a} f_X(w/a)$$
$$= \begin{cases} 2w/a^2 & 0 \le w \le a, \\ 0 & \text{otherwise.} \end{cases} \tag{6.8}$$

As a increases, the PDF stretches horizontally.

For the families of continuous random variables in Sections 4.5 and 4.6, we can use Theorem 6.2 to show that multiplying a random variable by a constant produces a new family member with transformed parameters.

===Theorem 6.3===

$W = aX$, where $a > 0$.

  (a) If X is uniform (b, c), then W is uniform (ab, ac).

  (b) If X is exponential (λ), then W is exponential (λ/a).

  (c) If X is Erlang (n, λ), then W is Erlang $(n, \lambda/a)$.

  (d) If X is Gaussian (μ, σ), then W is Gaussian $(a\mu, a\sigma)$.

The next theorem shows that adding a constant to a random variable simply shifts the CDF and the PDF by that constant.

===Theorem 6.4===

If $W = X + b$,

$$F_W(w) = F_X(w - b), \qquad f_W(w) = f_X(w - b).$$

Proof First, we find the CDF $F_W(w) = P[X + b \le w] = P[X \le w - b] = F_X(w - b)$. We take the derivative of $F_W(w)$ to find the PDF: $f_W(w) = dF_W(w)/dw = f_X(w - b)$.

In contrast to the linear transformations of Theorem 6.2 and Theorem 6.4, the following example is tricky because $g(X)$ transforms more than one value of X to the same W.

===Example 6.4===

Suppose X is the continuous uniform $(-1, 1)$ random variable and $W = X^2$. Find the CDF $F_W(w)$ and PDF $f_W(w)$.

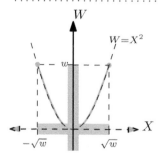

Although X can be negative, W is always nonnegative. Thus $F_W(w) = 0$ for $w < 0$. To find the CDF $F_W(w)$ for $w \ge 0$, the figure on the left shows that the event $\{W \le w\}$, marked with gray highlight on the vertical axis, is the same as the event $\{-\sqrt{w} \le X \le \sqrt{w}\}$ marked on the horizontal axis. Both events correspond to (X, W) pairs on the highlighted segment of the function $W = g(X)$. The corresponding algebra is

$$F_W(w) = P\left[X^2 \le w\right] = P\left[-\sqrt{w} \le X \le \sqrt{w}\right]. \tag{6.9}$$

We can take one more step by writing the probability (6.9) as an integral using the PDF $f_X(x)$:

$$F_W(w) = P\left[-\sqrt{w} \le X \le \sqrt{w}\right] = \int_{-\sqrt{w}}^{\sqrt{w}} f_X(x)\, dx. \tag{6.10}$$

So far, we have used no properties of the PDF $f_X(x)$. However, to evaluate the integral (6.10), we now recall from the problem statement and Definition 4.5 that the PDF of X is

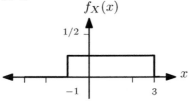

$$f_X(x) = \begin{cases} 1/4 & -1 \le x \le 3, \\ 0 & \text{otherwise.} \end{cases} \tag{6.11}$$

The integral (6.10) is somewhat tricky because the limits depend on the value of w. We first observe that $-1 \le X \le 3$ implies $0 \le W \le 9$. Thus $F_W(w) = 0$ for $w < 0$, and $F_W(w) = 1$ for $w > 9$. For $0 \le w \le 1$,

$$F_W(w) = \int_{-\sqrt{w}}^{\sqrt{w}} \frac{1}{4}\, dx = \frac{\sqrt{w}}{2}. \tag{6.12}$$

For $1 \le w \le 9$,

$$F_W(w) = \int_{-1}^{\sqrt{w}} \frac{1}{4}\, dx = \frac{\sqrt{w}+1}{4}. \tag{6.13}$$

By combining the separate pieces, we can write a complete expression for $F_W(w)$:

$$F_W(w) = \begin{cases} 0 & w < 0, \\ \dfrac{\sqrt{w}}{2} & 0 \le w \le 1, \\ \dfrac{\sqrt{w}+1}{4} & 1 \le w \le 9, \\ 1 & w \ge 9. \end{cases} \tag{6.14}$$

To find $f_W(w)$, we take the derivative of $F_W(w)$ over each interval.

$$f_W(w) = \begin{cases} \dfrac{1}{4\sqrt{w}} & 0 \le w \le 1, \\ \dfrac{1}{8\sqrt{w}} & 1 \le w \le 9, \\ 0 & \text{otherwise.} \end{cases} \tag{6.15}$$

We end this section with a useful application of derived random variables. The following theorem shows how to derive sample values of random variables using the transformation $X = g(U)$ where U is a uniform $(0, 1)$ random variable. In Section 4.8, we used this technique with the MATLAB rand function to generate sample values of a random variable X.

======Theorem 6.5======

Let U be a uniform $(0, 1)$ random variable and let $F(x)$ denote a cumulative distribution function with an inverse $F^{-1}(u)$ defined for $0 < u < 1$. The random variable $X = F^{-1}(U)$ has CDF $F_X(x) = F(x)$.

Proof First, we verify that $F^{-1}(u)$ is a nondecreasing function. To show this, suppose that for $u \geq u'$, $x = F^{-1}(u)$ and $x' = F^{-1}(u')$. In this case, $u = F(x)$ and $u' = F(x')$. Since $F(x)$ is nondecreasing, $F(x) \geq F(x')$ implies that $x \geq x'$. Hence, for the random variable $X = F^{-1}(U)$, we can write

$$F_X(x) = P\left[F^{-1}(U) \leq x\right] = P\left[U \leq F(x)\right] = F(x). \tag{6.16}$$

We observe that the requirement that $F_X(u)$ have an inverse for $0 < u < 1$ limits the applicability of Theorem 6.5. For example, this requirement is not met by the mixed random variables of Section 4.7. A generalizaton of the theorem that does hold for mixed random variables is given in Problem 6.3.13. The following examples demonstrate the utility of Theorem 6.5.

======Example 6.5======

U is the uniform $(0, 1)$ random variable and $X = g(U)$. Derive $g(U)$ such that X is the exponential (1) random variable.
. . . .
The CDF of X is

$$F_X(x) = \begin{cases} 0 & x < 0, \\ 1 - e^{-x} & x \geq 0. \end{cases} \tag{6.17}$$

Note that if $u = F_X(x) = 1 - e^{-x}$, then $x = -\ln(1-u)$. That is, $F_X^{-1}(u) = -\ln(1-u)$ for $0 \leq u < 1$. Thus, by Theorem 6.5,

$$X = g(U) = -\ln(1 - U) \tag{6.18}$$

is the exponential random variable with parameter $\lambda = 1$. Problem 6.2.7 asks the reader to derive the PDF of $X = -\ln(1 - U)$ directly from first principles.

======Example 6.6======

For a uniform $(0, 1)$ random variable U, find a function $g(\cdot)$ such that $X = g(U)$ has a uniform (a, b) distribution.
. . . .
The CDF of X is

$$F_X(x) = \begin{cases} 0 & x < a, \\ (x - a)/(b - a) & a \leq x \leq b, \\ 1 & x > b. \end{cases} \tag{6.19}$$

For any u satisfying $0 \le u \le 1$, $u = F_X(x) = (x-a)/(b-a)$ if and only if

$$x = F_X^{-1}(u) = a + (b-a)u. \tag{6.20}$$

Thus by Theorem 6.5, $X = a + (b-a)U$ is a uniform (a,b) random variable. Note that we could have reached the same conclusion by observing that Theorem 6.3 implies $(b-a)U$ has a uniform $(0, b-a)$ distribution and that Theorem 6.4 implies $a+(b-a)U$ has a uniform $(a, (b-a)+a)$ distribution. Another approach, taken in Problem 6.2.11, is to derive the CDF and PDF of $a + (b-a)U$.

The technique of Theorem 6.5 is particularly useful when the CDF is an easily invertible function. Unfortunately, there are many random variables, including Gaussian and Erlang, in which the CDF and its inverse are difficult to compute. In these cases, we need to develop other methods for transforming sample values of a uniform random vaiable to sample values of a random variable of interest.

Quiz 6.2

X is an exponential (λ) PDF. Show that $Y = \sqrt{X}$ is a Rayleigh random variable (see Appendix A.2). Express the Rayleigh parameter a in terms of the exponential parameter λ.

6.3 Functions Yielding Discrete or Mixed Random Variables

> A hard limiter electronic circuit has two possible output voltages. If the input voltage is a sample value of a continuous random variable, the output voltage is a sample value of a discrete random variable. The output of a soft limiter circuit is a sample value of a mixed random variable. The probability models of the limiters depend on the probability model of the input and on the two limiting voltages.

In Section 6.2, our examples and theorems relate to a continuous random variable derived from two continuous random variables. By contrast, in the following example, the function $g(X)$ transforms a continuous random variable to a discrete random variable.

Example 6.7

Let X be a random variable with CDF $F_X(x)$. Let Y be the output of a clipping circuit, also referred to as a hard limiter, with the characteristic $Y = g(X)$ where

$$g(x) = \begin{cases} 1 & x \le 0, \\ 3 & x > 0. \end{cases} \tag{6.21}$$

Express $F_Y(y)$ and $f_Y(y)$ in terms of $F_X(x)$ and $f_X(x)$.

Before going deeply into the math, it is helpful to think about the nature of the derived random variable Y. The definition of $g(x)$ tells us that Y has only two possible values, $Y = 1$ and $Y = 3$. Thus Y is a discrete random variable. Furthermore, the CDF, $F_Y(y)$, has jumps at $y = 1$ and $y = 3$; it is zero for $y < 1$ and it is one for $y \geq 3$. Our job is to find the heights of the jumps at $y = 1$ and $y = 3$. In particular,

$$F_Y(1) = P[Y \leq 1] = P[X \leq 0] = F_X(0). \tag{6.22}$$

This tells us that the CDF jumps by $F_X(0)$ at $y = 1$. We also know that the CDF has to jump to one at $y = 3$. Therefore, the entire story is

$$F_Y(y) = \begin{cases} 0 & y < 1, \\ F_X(0) & 1 \leq y < 3, \\ 1 & y \geq 3. \end{cases} \tag{6.23}$$

The PDF consists of impulses at $y = 1$ and $y = 3$. The weights of the impulses are the sizes of the two jumps in the CDF: $F_X(0)$ and $1 - F_X(0)$, respectively.

$$f_Y(y) = F_X(0)\,\delta(y - 1) + [1 - F_X(0)]\delta(y - 3).$$

The following example contains a function that transforms continuous random variables to a mixed random variable.

Example 6.8

The output voltage of a microphone is a Gaussian random variable V with expected value $\mu_V = 0$ and standard deviation $\sigma_V = 5$ V. The microphone signal is the input to a soft limiter circuit with cutoff value ± 10 V. The random variable W is the output of the limiter:

$$W = g(V) = \begin{cases} -10 & V < -10, \\ V & -10 \leq V \leq 10, \\ 10 & V > 10. \end{cases} \tag{6.24}$$

What are the CDF and PDF of W?

To find the CDF, we need to find $F_W(w) = P[W \leq w]$ for all values of w. The key is that all possible pairs (V, W) satisfy $W = g(V)$. This implies each w belongs to one of three cases:

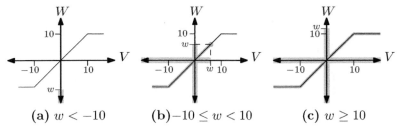

(a) $w < -10$ **(b)** $-10 \leq w < 10$ **(c)** $w \geq 10$

(a) $w < -10$: From the function $W = g(V)$ we see that no possible pairs (V, W) satisfy $W \leq w < -10$. Hence $F_W(w) = P[W \leq w] = 0$ in this case. This is perhaps a roundabout way of observing that $W = -10$ is the minumum possible W.

(b) $-10 \leq w < 10$: In this case we see that the event $\{W \leq w\}$, marked in gray on the vertical axis, corresponds to the event $\{V \leq w\}$, marked in gray on the horizontal axis. The corresponding (V, W) pairs are shown in the highlighted segment of the function $W = g(V)$. In this case, $F_W(w) = P[W \leq w] = P[V \leq w] = F_V(w)$.

(c) $w \geq 10$: Here we see that the event $\{W \leq w\}$ corresponds to all values of V and $P[W \leq w] = P[V < \infty] = 1$. This is another way of saying $W = 10$ is the maximum W.

We combine these separate cases in the CDF

$$F_W(w) = P[W \leq w] = \begin{cases} 0 & w < -10, \\ F_V(w) & -10 \leq w < 10, \\ 1 & w \geq 10. \end{cases} \tag{6.25}$$

These conclusions are based solely on the structure of the limiter function $g(V)$ without regard for the probability model of V. Now we observe that because V is Gaussian $(0, 5)$, Theorem 4.14 states that $F_V(v) = \Phi(v/5)$. Therefore,

$$F_W(w) = \begin{cases} 0 & w < -10, \\ \Phi(w/5) & -10 \leq w \leq 10, \\ 1 & w > 10. \end{cases} \tag{6.26}$$

Note that the CDF jumps from 0 to $\Phi(-10/5) = 0.023$ at $w = -10$ and that it jumps from $\Phi(10/5) = 0.977$ to 1 at $w = 10$. Therefore,

$$f_W(w) = \frac{dF_W(w)}{dw} = \begin{cases} 0.023\delta(w + 10) & w = -10, \\ \dfrac{1}{5\sqrt{2\pi}}e^{-w^2/50} & -10 < w < 10, \\ 0.023\delta(w - 10) & w = 10, \\ 0 & \text{otherwise.} \end{cases} \tag{6.27}$$

======Quiz 6.3======

Random variable X is passed to a hard limiter that outputs Y. The PDF of X and the limiter output Y are

$$f_X(x) = \begin{cases} 1 - x/2 & 0 \le x \le 2, \\ 0 & \text{otherwise,} \end{cases} \qquad Y = \begin{cases} X & X \le 1, \\ 1 & X > 1. \end{cases} \qquad (6.28)$$

(a) What is the CDF $F_X(x)$?

(b) What is P$[Y = 1]$?

(c) What is $F_Y(y)$?

(d) What is $f_Y(y)$?

6.4   Continuous Functions of Two Continuous Random Variables

To obtain the PDF of $W = g(X, Y)$, a continuous function of two continuous random variables, derive the CDF of W and then differentiate. The procedure is straightforward when $g(x, y)$ is a linear function. It is more complex for other functions.

At the start of this chapter, we described three ways radio receivers can use signals from two antennas. These techniques are examples of the following situation. We perform an experiment and observe sample values of two random variables X and Y. After performing the experiment, we calculate a sample value of the random variable $W = g(X, Y)$. Based on our knowledge of the experiment, we have a probability model for X and Y embodied in a joint PMF $P_{X,Y}(x, y)$ or a joint PDF $f_{X,Y}(x, y)$.

In this section, we present methods for deriving a probability model for W. When X and Y are continuous random variables and $g(x, y)$ is a continuous function, $W = g(X, Y)$ is a continuous random variable. To find the PDF, $f_W(w)$, it is usually helpful to first find the CDF $F_W(w)$ and then calculate the derivative. Viewing $\{W \le w\}$ as an event A, we can apply Theorem 5.7.

======Theorem 6.6======

For continuous random variables X and Y, the CDF of $W = g(X, Y)$ is

$$F_W(w) = \mathrm{P}\left[W \le w\right] = \iint\limits_{g(x,y) \le w} f_{X,Y}(x, y) \, dx \, dy.$$

Theorem 6.6 is analogous to our approach in Sections 6.2 and 6.3 for functions $W = g(X)$. There we used the function $g(X)$ to translate the event $\{W \le w\}$ into an event $\{g(X) \le w\}$ that was a subset of the X-axis. We then calculated $F_W(w)$ by integrating $f_X(x)$ over that subset.

In Theorem 6.6, we translate the event $\{g(X, Y) \le w\}$ into a region of the X, Y plane. Integrating the joint PDF $f_{X,Y}(x, y)$ over that region will yield the CDF $F_W(w)$. Once we obtain $F_W(w)$, it is generally straightforward to calculate the derivative $f_W(w) = dF_W(w)/dw$. However, for most functions $g(x, y)$, performing the integration to find $F_W(w)$ can be a tedious process. Fortunately, there are convenient techniques for finding $f_W(w)$ for certain functions that arise in many applications. Section 6.5 and Chapter 9 consider the function, $g(X, Y) = X + Y$. The following theorem addresses $W = \max(X, Y)$, the maximum of two random variables. It follows from the fact that $\{\max(X, Y) \le w\} = \{X \le w\} \cap \{Y \le w\}$.

Theorem 6.7

For continuous random variables X and Y, the CDF of $W = \max(X, Y)$ is

$$F_W(w) = F_{X,Y}(w, w) = \int_{-\infty}^{w} \int_{-\infty}^{w} f_{X,Y}(x, y) \, dx \, dy.$$

Example 6.9

In Examples 5.7 and 5.9, X and Y have joint PDF

$$f_{X,Y}(x, y) = \begin{cases} 1/15 & 0 \le x \le 5, 0 \le y \le 3, \\ 0 & \text{otherwise.} \end{cases} \tag{6.29}$$

Find the PDF of $W = \max(X, Y)$.

....

Because $X \ge 0$ and $Y \ge 0$, $W \ge 0$. Therefore, $F_W(w) = 0$ for $w < 0$. Because $X \le 5$ and $Y \le 3$, $W \le 5$. Thus $F_W(w) = 1$ for $w \ge 5$. For $0 \le w \le 5$, diagrams showing the regions of integration provide a guide to calculating $F_W(w)$. Two cases, $0 \le w \le 3$ and $3 \le w \le 5$, have to be considered separately. When $0 \le w \le 3$, Theorem 6.7 yields

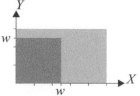

$$F_W(w) = \int_0^w \int_0^w \frac{1}{15} \, dx \, dy = w^2/15. \tag{6.30}$$

Because the joint PDF is uniform, we see this probability is the area w^2 times the value of the joint PDF over that area. When $3 \le w \le 5$, the integral over the region $\{X \le w, Y \le w\}$ is

$$F_W(w) = \int_0^w \left(\int_0^3 \frac{1}{15} \, dy \right) dx = \int_0^w \frac{1}{5} \, dx = w/5, \tag{6.31}$$

which is the area $3w$ times the value of the joint PDF over that area. Combining the parts, we can write the joint CDF:

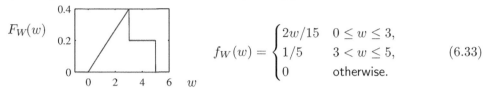

$$F_W(w) = \begin{cases} 0 & w < 0, \\ w^2/15 & 0 \le w \le 3, \\ w/5 & 3 < w \le 5, \\ 1 & w > 5. \end{cases} \qquad (6.32)$$

By taking the derivative, we find the corresponding joint PDF:

$$f_W(w) = \begin{cases} 2w/15 & 0 \le w \le 3, \\ 1/5 & 3 < w \le 5, \\ 0 & \text{otherwise.} \end{cases} \qquad (6.33)$$

In the following example, W is the quotient of two positive numbers.

Example 6.10

X and Y have the joint PDF

$$f_{X,Y}(x,y) = \begin{cases} \lambda\mu e^{-(\lambda x + \mu y)} & x \ge 0, y \ge 0, \\ 0 & \text{otherwise.} \end{cases} \qquad (6.34)$$

Find the PDF of $W = Y/X$.

First we find the CDF:

$$F_W(w) = P\left[Y/X \le w\right] = P\left[Y \le wX\right]. \qquad (6.35)$$

For $w < 0$, $F_W(w) = 0$. For $w \ge 0$, we integrate the joint PDF $f_{X,Y}(x,y)$ over the region of the X,Y plane for which $Y \le wX$, $X \ge 0$, and $Y \ge 0$ as shown:

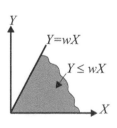

$$\begin{aligned} P\left[Y \le wX\right] &= \int_0^\infty \left(\int_0^{wx} f_{X,Y}(x,y)\, dy \right) dx \\ &= \int_0^\infty \lambda e^{-\lambda x} \left(\int_0^{wx} \mu e^{-\mu y}\, dy \right) dx \\ &= \int_0^\infty \lambda e^{-\lambda x} \left(1 - e^{-\mu wx}\right) dx \\ &= 1 - \frac{\lambda}{\lambda + \mu w}. \end{aligned} \qquad (6.36)$$

Therefore,

$$F_W(w) = \begin{cases} 0 & w < 0, \\ 1 - \dfrac{\lambda}{\lambda + \mu w} & w \ge 0. \end{cases} \qquad (6.37)$$

Differentiating with respect to w, we obtain

$$f_W(w) = \begin{cases} \dfrac{\lambda\mu}{(\lambda + \mu w)^2} & w \geq 0, \\ 0 & \text{otherwise.} \end{cases} \qquad (6.38)$$

Quiz 6.4

(A) A smartphone runs a news application that downloads Internet news every 15 minutes. At the start of a download, the radio modems negotiate a connection speed that depends on the radio channel quality. When the negotiated speed is low, the smartphone reduces the amount of news that it transfers to avoid wasting its battery. The number of kilobytes transmitted, L, and the speed B in kb/s, have the joint PMF

$P_{L,B}(l,b)$	$b = 512$	$b = 1,024$	$b = 2,048$
$l = 256$	0.2	0.1	0.05
$l = 768$	0.05	0.1	0.2
$l = 1536$	0	0.1	0.2

Let T denote the number of seconds needed for the transfer. Express T as a function of L and B. What is the PMF of T?

(B) Find the CDF and the PDF of $W = XY$ when random variables X and Y have joint PDF

$$f_{X,Y}(x,y) = \begin{cases} 1 & 0 \leq x \leq 1, 0 \leq y \leq 1, \\ 0 & \text{otherwise.} \end{cases} \qquad (6.39)$$

6.5   PDF of the Sum of Two Random Variables

> The PDF of the sum of two independent continuous random variables X and Y is the convolution of the PDF of X and the PDF of Y. The PMF of the sum of two independent integer-valued random variables is the discrete convolution of the two PMFs.

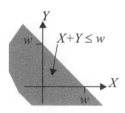

We now examine the sum $W = X + Y$ of two continuous random variables. As we see in Theorem 6.6, the PDF of W depends on the joint PDF $f_{X,Y}(x,y)$. In particular, in the proof of the next theorem, we find the PDF of W using the two-step procedure in which we first find the CDF $F_W(w)$ by integrating the joint PDF $f_{X,Y}(x,y)$ over the region $X + Y \leq w$, as shown.

=======Theorem 6.8=======

The PDF of $W = X + Y$ is

$$f_W(w) = \int_{-\infty}^{\infty} f_{X,Y}(x, w - x)\, dx = \int_{-\infty}^{\infty} f_{X,Y}(w - y, y)\, dy.$$

Proof

$$F_W(w) = P[X + Y \le w] = \int_{-\infty}^{\infty} \left(\int_{-\infty}^{w-x} f_{X,Y}(x,y)\, dy \right) dx. \tag{6.40}$$

Taking the derivative of the CDF to find the PDF, we have

$$f_W(w) = \frac{dF_W(w)}{dw} = \int_{-\infty}^{\infty} \left(\frac{d}{dw} \left(\int_{-\infty}^{w-x} f_{X,Y}(x,y)\, dy \right) \right) dx$$

$$= \int_{-\infty}^{\infty} f_{X,Y}(x, w - x)\, dx. \tag{6.41}$$

By making the substitution $y = w - x$, we obtain

$$f_W(w) = \int_{-\infty}^{\infty} f_{X,Y}(w - y, y)\, dy. \tag{6.42}$$

=======Example 6.11=======

Find the PDF of $W = X + Y$ when X and Y have the joint PDF

$$f_{X,Y}(x, y) = \begin{cases} 2 & 0 \le y \le 1, 0 \le x \le 1, x + y \le 1, \\ 0 & \text{otherwise.} \end{cases} \tag{6.43}$$

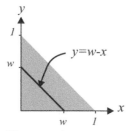

The PDF of $W = X + Y$ can be found using Theorem 6.8. The possible values of X, Y are in the shaded triangular region where $0 \le X + Y = W \le 1$. Thus $f_W(w) = 0$ for $w < 0$ or $w > 1$. For $0 \le w \le 1$, applying Theorem 6.8 yields

$$f_W(w) = \int_0^w 2\, dx = 2w, \qquad 0 \le w \le 1. \tag{6.44}$$

The complete expression for the PDF of W is

$$f_W(w) = \begin{cases} 2w & 0 \le w \le 1, \\ 0 & \text{otherwise.} \end{cases} \tag{6.45}$$

When X and Y are independent, the joint PDF of X and Y is the product of the marginal PDFs $f_{X,Y}(x, y) = f_X(x) f_Y(y)$. Applying Theorem 6.8 to this special case, we obtain the following theorem.

====Theorem 6.9====

When X and Y are independent random variables, the PDF of $W = X + Y$ is

$$f_W(w) = \int_{-\infty}^{\infty} f_X(w-y) f_Y(y) \, dy = \int_{-\infty}^{\infty} f_X(x) f_Y(w-x) \, dx.$$

In Theorem 6.9, we combine two univariate functions, $f_X(\cdot)$ and $f_Y(\cdot)$, in order to produce a third function, $f_W(\cdot)$. The combination in Theorem 6.9, referred to as a *convolution*, arises in many branches of applied mathematics.

When X and Y are independent integer-valued discrete random variables, the PMF of $W = X + Y$ is a convolution (see Problem 6.5.1).

$$P_W(w) = \sum_{k=-\infty}^{\infty} P_X(k) P_Y(w-k). \tag{6.46}$$

You may have encountered convolutions already in studying linear systems. Sometimes, we use the notation $f_W(w) = f_X(x) * f_Y(y)$ to denote convolution.

====Quiz 6.5====

Let X and Y be independent exponential random variables with expected values $E[X] = 1/3$ and $E[Y] = 1/2$. Find the PDF of $W = X + Y$.

6.6 MATLAB

> Theorem 6.5 and the **rand** function can be employed to generate sample values of continuous random variables.

====Example 6.12====

Use Example 6.6 to write a MATLAB function that generates m samples of a uniform (a, b) random variable.

```
function x=uniformrv(a,b,m)
x=a+(b-a)*rand(m,1);
```

Example 6.6 says that $Y = a + (b-a)U$ is a uniform (a, b) random variable. We use this in **uniformrv**.

```
function x=erlangrv(n,lambda,m)
y=exponentialrv(lambda,m*n);
x=sum(reshape(y,m,n),2);
```

Theorem 9.9 will demonstrate that the sum of n independent exponential (λ) random variables is an Erlang random variable. The function **erlangrv** generates m sample values of the Erlang (n, λ) random variable. Note that we first generate nm exponential random variables. The **reshape** function arranges these samples in an $m \times n$ array. Summing across the rows yields m Erlang samples.

```
function x=icdfrv(icdfhandle,m)
%Usage: x=icdfrv(@icdf,m)
%returns m samples of rv X
%with inverse CDF icdf.m
u=rand(m,1);
x=feval(icdfhandle,u);
```

Finally, for a random variable X with an arbitrary CDF $F_X(x)$, we implement the function icdfrv.m, which uses Theorem 6.5 for generating random samples. The key is to define a MATLAB function x=icdfx(u) that calculates $x = F_X^{-1}(u)$. The function icdfx(u) is then passed as an argument to icdfrv.m which generates samples of X. Note that MATLAB passes a function as an argument to another function using a function *handle*, which is a kind of pointer. The following example shows how to use icdfrv.m.

═══════Example 6.13═══════

Write a MATLAB function that uses icdfrv.m to generate samples of Y, the maximum of three pointer spins, in Example 4.5.

· ·

```
function y = icdf3spin(u);
y=u.^(1/3);
```

From Equation (4.18), we see that for $0 \leq y \leq 1$, $F_Y(y) = y^3$. If $u = F_Y(y) = y^3$, then $y = F_Y^{-1}(u) = u^{1/3}$. So we define (and save to disk) icdf3spin.m.

Now, the function call y=icdfrv(@icdf3spin,1000) generates a vector holding 1000 samples of random variable Y. The notation @icdf3spin is the function handle for the function icdf3spin.m.

─────────────────

Keep in mind that for the MATLAB code to run quickly, it is best for the inverse CDF function (icdf3spin.m in the case of the last example) to process the vector u without using a for loop to find the inverse CDF for each element u(i). We also note that this same technique can be extended to cases where the inverse CDF $F_X^{-1}(u)$ does not exist for all $0 \leq u \leq 1$. For example, the inverse CDF does not exist if X is a mixed random variable or if $f_X(x)$ is constant over an interval (a, b). How to use icdfrv.m in these cases is addressed in Problems 6.3.13 and 6.6.4.

═══════Quiz 6.6═══════

Write a MATLAB function V=Vsample(m) that returns m sample of random variable V with PDF

$$f_V(v) = \begin{cases} (v+5)/72 & -5 \leq v \leq 72, \\ 0 & \text{otherwise.} \end{cases} \tag{6.47}$$

Problems

Difficulty: ● Easy ■ Moderate ◆ Difficult ◆◆ Experts Only

6.1.1● Random variables X and Y have joint PMF

$$P_{X,Y}(x,y) = \begin{cases} |x+y|/14 & x = -2, 0, 2; \\ & y = -1, 0, 1, \\ 0 & \text{otherwise.} \end{cases}$$

Find the PMF of $W = X - Y$.

6.1.2● For random variables X and Y in Problem 6.1.1, find the PMF of $W = X + 2Y$.

6.1.3● N is a binomial $(n = 100, p = 0.4)$ random variable. M is a binomial $(n = 50, p = 0.4)$ random variable. Given that M and N are independent, what is the PMF of $L = M + N$?

6.1.4● Let X and Y be discrete random variables with joint PMF $P_{X,Y}(x,y)$ that is zero except when x and y are integers. Let $W = X + Y$ and show that the PMF of W satisfies

$$P_W(w) = \sum_{x=-\infty}^{\infty} P_{X,Y}(x, w - x).$$

6.1.5■ Let X and Y be discrete random variables with joint PMF

$$P_{X,Y}(x,y) = \begin{cases} 0.01 & x = 1, 2 \ldots, 10, \\ & y = 1, 2 \ldots, 10, \\ 0 & \text{otherwise.} \end{cases}$$

What is the PMF of $W = \min(X, Y)$?

6.1.6■ For random variables X and Y in Problem 6.1.5, what is the PMF of $V = \max(X, Y)$?

6.2.1● The voltage X across a 1 Ω resistor is a uniform random variable with parameters 0 and 1. The instantaneous power is $Y = X^2$. Find the CDF $F_Y(y)$ and the PDF $f_Y(y)$ of Y.

6.2.2● X is the Gaussian $(0, 1)$ random variable. Find the CDF of $Y = |X|$ and its expected value E[Y].

6.2.3■ In a 50 km Tour de France time trial, a rider's time T, measured in minutes, is the continuous uniform $(60, 75)$ random variable. Let $V = 3000/T$ denote the rider's speed over the course in km/hr. Find the PDF of V.

6.2.4● In the presence of a headwind of normalized intensity W, your speed on your bike is $V = g(W) = 20 - 10W^{1/3}$ mi/hr. The wind intensity W is the continuous uniform $(-1, 1)$ random variable. (Note: If W is negative, then the headwind is actually a tailwind.) Find the PDF $f_V(v)$.

6.2.5● If X has an exponential (λ) PDF, what is the PDF of $W = X^2$?

6.2.6■ Let X denote the position of the pointer after a spin on a wheel of circumference 1. For that same spin, let Y denote the area within the arc defined by the stopping position of the pointer:

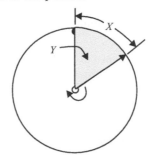

(a) What is the relationship between X and Y?

(b) What is $F_Y(y)$?

(c) What is $f_Y(y)$?

(d) What is E[Y]?

6.2.7■ U is the uniform $(0, 1)$ random variable and $X = -\ln(1 - U)$.

(a) What is $F_X(x)$?

(b) What is $f_X(x)$?

(c) What is E[X]?

6.2.8■ X is the uniform $(0, 1)$ random variable. Find a function $g(x)$ such that the PDF of $Y = g(X)$ is

$$f_Y(y) = \begin{cases} 3y^2 & 0 \le y \le 1, \\ 0 & \text{otherwise.} \end{cases}$$

6.2.9■ An amplifier circuit has power consumption Y that grows nonlinearly with the input signal voltage X. When the input signal is X volts, the instantaneous power consumed by the amplifier is $Y = 20 + 15X^2$ Watts. The input signal X is the continuous uniform $(-1, 1)$ random variable. Find the PDF $f_Y(y)$.

6.2.10■ Use Theorem 6.2 to prove Theorem 6.3.

6.2.11■ For the uniform $(0, 1)$ random variable U, find the CDF and PDF of $Y = a + (b - a)U$ with $a < b$. Show that Y is the uniform (a, b) random variable.

6.2.12■ Theorem 6.5 required the inverse CDF $F^{-1}(u)$ to exist for $0 < u < 1$. Why was it *not* necessary that $F^{-1}(u)$ exist at either $u = 0$ or $u = 1$?

6.2.13◆ X is a continuous random variable. $Y = aX + b$, where $a, b \ne 0$. Prove that

$$f_Y(y) = \frac{f_X((y - b)/a)}{|a|}.$$

Hint: Consider the cases $a < 0$ and $a > 0$ separately.

6.2.14◆ Let continuous random variable X have a CDF $F(x)$ such that $F^{-1}(u)$ exists for all u in $[0, 1]$. Show that $U = F(X)$ is the uniform $(0, 1)$ random variable. Hint: U is a random variable such that when $X = x'$, $U = F(x')$. That is, we evaluate the CDF of X at the observed value of X.

6.3.1■ X has CDF

$$F_X(x) = \begin{cases} 0 & x < -1, \\ x/3 + 1/3 & -1 \le x < 0, \\ x/3 + 2/3 & 0 \le x < 1, \\ 1 & 1 \le x. \end{cases}$$

$Y = g(X)$ where

$$g(X) = \begin{cases} 0 & X < 0, \\ 100 & X \ge 0. \end{cases}$$

(a) What is $F_Y(y)$?

(b) What is $f_Y(y)$?

(c) What is $E[Y]$?

6.3.2■ In a 50 km cycling time trial, a rider's exact time T, measured in minutes, is the continuous uniform $(50, 60)$ random variable. However, a rider's recorded time R in seconds is obtained by rounding *up* T to next whole second. That is, if T is 50 minutes, 27.001 seconds, then $R = 3028$ seconds. On the other hand, if T is exactly 50 minutes 27 seconds, then $R = 3027$. What is the PMF of R?

6.3.3■ The voltage V at the output of a microphone is the continuous uniform $(-1, 1)$ random variable. The microphone voltage is processed by a clipping rectifier with output

$$L = \begin{cases} |V| & |V| \le 0.5, \\ 0.5 & \text{otherwise.} \end{cases}$$

(a) What is $P[L = 0.5]$?

(b) What is $F_L(l)$?

(c) What is $E[L]$?

6.3.4■ U is the uniform random variable with parameters 0 and 2. The random variable W is the output of the clipper:

$$W = g(U) = \begin{cases} U & U \le 1, \\ 1 & U > 1. \end{cases}$$

Find the CDF $F_W(w)$, the PDF $f_W(w)$, and the expected value $E[W]$.

6.3.5 X is a random variable with CDF $F_X(x)$. Let $Y = g(X)$ where

$$g(x) = \begin{cases} 10 & x < 0, \\ -10 & x \geq 0. \end{cases}$$

Express $F_Y(y)$ in terms of $F_X(x)$.

6.3.6 Suppose that a cellular phone costs $30 per month with 300 minutes of use included and that each additional minute of use costs $0.50. The number of minutes you use the phone in a month is an exponential random variable T with with expected value $E[T] = 200$ minutes. The telephone company charges you for exactly how many minutes you use without any rounding of fractional minutes. Let C denote the cost in dollars of one month of service.

(a) What is $P[C = 30]$?

(b) What is the PDF of C?

(c) What is $E[C]$?

6.3.7 The input voltage to a rectifier is the continuous uniform $(0, 1)$ random variable U. The rectifier output is a random variable W defined by

$$W = g(U) = \begin{cases} 0 & U < 0, \\ U & U \geq 0. \end{cases}$$

Find the CDF $F_W(w)$ and the expected value $E[W]$.

6.3.8 Random variable X has PDF

$$f_X(x) = \begin{cases} x/2 & 0 \leq x \leq 2, \\ 0 & \text{otherwise.} \end{cases}$$

X is processed by a clipping circuit with output

$$Y = \begin{cases} 0.5 & X \leq 1, \\ X & X > 1. \end{cases}$$

(a) What is $P[Y = 0.5]$?

(b) Find the CDF $F_Y(y)$.

6.3.9 Given an input voltage V, the output voltage of a half-wave rectifier is given by

$$W = \begin{cases} 0 & V \leq 0, \\ V & 0 < V < 10, \\ 10 & V \geq 10. \end{cases}$$

Suppose the input V is the continuous uniform $(-15, 15)$ random variable. Find the PDF of W.

6.3.10 The current X across a resistor is the continuous uniform $(-2, 2)$ random variable. The power dissipated in the resistor is $Y = 9X^2$ Watts.

(a) Find the CDF and PDF of Y.

(b) A power measurement circuit is range-limited so that its output is

$$W = \begin{cases} Y & Y < 16, \\ 16 & \text{otherwise.} \end{cases}$$

Find the PDF of W.

6.3.11 A defective voltmeter measures small voltages as zero. In particular, when the input voltage is V, the measured voltage is

$$W = \begin{cases} 0 & |V| < 0.6, \\ V & \text{otherwise.} \end{cases}$$

If V is the continuous uniform $(-5, 5)$ random variable, what is the PDF of W?

6.3.12 X is the continuous uniform $(-3, 3)$ random variable. When X is passed through a limiter, the output is the discrete random variable

$$\hat{X} = g(X) = \begin{cases} -c & X < 0 \\ c & X \geq 0 \end{cases}$$

where c is an unspecified positive constant.

(a) What is the PMF $P_{\hat{X}}(x)$ of $\hat{X}$?

(b) When the limiter input is X, the distortion D between the input X and the limiter output $\hat{X}$ is

$$D = d(X) = (X - g(X))^2.$$

In terms of c, find the expected distortion $E[D] = E[d(X)]$. What value of c minimizes $E[D]$?

(c) Y is a Gaussian random variable with the same expected value and variance as X. What is the PDF of Y?

(d) Suppose Y is passed through the limiter yielding the output $\hat{Y} = g(Y)$. The distortion D between the input Y and the limiter output $\hat{Y}$ is

$$D = d(Y) = (Y - g(Y))^2.$$

In terms of c, find the expected distortion $E[D] = E[d(Y)]$. What value of c minimizes $E[D]$?

6.3.13◆◆ In this problem we prove a generalization of Theorem 6.5. Given a random variable X with CDF $F_X(x)$, define

$$\tilde{F}(u) = \min\left\{x | F_X(x) \geq u\right\}.$$

This problem proves that for a continuous uniform $(0, 1)$ random variable U, $\hat{X} = \tilde{F}(U)$ has CDF $F_{\hat{X}}(x) = F_X(x)$.

(a) Show that when $F_X(x)$ is a continuous, strictly increasing function (i.e., X is not mixed, $F_X(x)$ has no jump discontinuities, and $F_X(x)$ has no "flat" intervals (a, b) where $F_X(x) = c$ for $a \leq x \leq b$), then $\tilde{F}(u) = F_X^{-1}(u)$ for $0 < u < 1$.

(b) Show that if $F_X(x)$ has a jump at $x = x_0$, then $\tilde{F}(u) = x_0$ for all u in the interval

$$F_X\left(x_0^-\right) \leq u \leq F_X\left(x_0^+\right).$$

(c) Prove that $\hat{X} = \tilde{F}(U)$ has CDF $F_{\hat{X}}(x) = F_X(x)$.

6.4.1● Random variables X and Y have joint PDF

$$f_{X,Y}(x, y) = \begin{cases} 6xy^2 & 0 \leq x, y \leq 1, \\ 0 & \text{otherwise.} \end{cases}$$

Let $V = \max(X, Y)$. Find the CDF and PDF of V.

6.4.2● For random variables X and Y in Problem 6.4.1, find the CDF and PDF of $W = \min(X, Y)$.

6.4.3■ X and Y have joint PDF

$$f_{X,Y}(x, y) = \begin{cases} 2 & x \geq 0, y \geq 0, x + y \leq 1, \\ 0 & \text{otherwise.} \end{cases}$$

(a) Are X and Y independent?

(b) Let $U = \min(X, Y)$. Find the CDF and PDF of U.

(c) Let $V = \max(X, Y)$. Find the CDF and PDF of V.

6.4.4■ Random variables X and Y have joint PDF

$$f_{X,Y}(x, y) = \begin{cases} x + y & 0 \leq x, y \leq 1, \\ 0 & \text{otherwise.} \end{cases}$$

Let $W = \max(X, Y)$.

(a) What is S_W, the range of W?

(b) Find $F_W(w)$ and $f_W(w)$.

6.4.5■ Random variables X and Y have joint PDF

$$f_{X,Y}(x, y) = \begin{cases} 6y & 0 \leq y \leq x \leq 1, \\ 0 & \text{otherwise.} \end{cases}$$

Let $W = Y - X$.

(a) What is S_W, the range of W?

(b) Find $F_W(w)$ and $f_W(w)$.

6.4.6■ Random variables X and Y have joint PDF

$$f_{X,Y}(x, y) = \begin{cases} 2 & 0 \leq y \leq x \leq 1, \\ 0 & \text{otherwise.} \end{cases}$$

Let $W = Y/X$.

(a) What is S_W, the range of W?

(b) Find $F_W(w)$, $f_W(w)$, and $E[W]$.

6.4.7■ Random variables X and Y have joint PDF

$$f_{X,Y}(x, y) = \begin{cases} 2 & 0 \leq y \leq x \leq 1, \\ 0 & \text{otherwise.} \end{cases}$$

Let $W = X/Y$.

(a) What is S_W, the range of W?

(b) Find $F_W(w)$, $f_W(w)$, and $E[W]$.

6.4.8■ In a simple model of a cellular telephone system, a portable telephone is equally likely to be found anywhere in a circular cell of radius 4 km. (See Problem 5.5.4.) Find the CDF $F_R(r)$ and PDF $f_R(r)$ of R, the distance (in km) between the telephone and the base station at the center of the cell.

6.4.9■ X and Y are independent identically distributed Gaussian $(0, 1)$ random variables. Find the CDF of $W = X^2 + Y^2$.

6.4.10■ X is the exponential (2) random variable and Z is the Bernoulli $(1/2)$ random variable that is independent of X. Find the PDF of $Y = ZX$.

6.4.11◆ X is the Gaussian $(0, 1)$ random variable and Z, independent of X, has PMF

$$P_Z(z) = \begin{cases} 1 - p & z = -1, \\ p & z = 1. \end{cases}$$

Find the PDF of $Y = ZX$.

6.4.12◆ You are waiting on the platform of the first stop of a Manhattan subway line. You could ride either a local or express train to your destination, which is the last stop on the line. The waiting time X for the next express train is the exponential random variable with $E[X] = 10$ minutes. The waiting time Y for the next local train is the exponential random variable with $E[Y] = 5$ minutes. Although the arrival times X and Y of the trains are random and independent, the trains' travel times are deterministic; the local train travels from first stop to last stop in exactly 15 minutes while the express travels from first to last stop in exactly 5 minutes.

(a) What is the joint PDF $f_{X,Y}(x, y)$?

(b) Find $P[L]$ that the local train arrives first at the platform?

(c) Suppose you board the first train that arrives. Find the PDF of your waiting time $W = \min(X, Y)$.

(d) The time until the first train (express or local) reaches final stop is $T = \min(X + 5, Y + 15)$. Find $f_T(t)$.

(e) Suppose the local train does arrive first at your platform. Should you board the local train? Justify your answer. (There may be more than one correct answer.)

6.4.13◆ For a constant $a > 0$, random variables X and Y have joint PDF

$$f_{X,Y}(x, y) = \begin{cases} 1/a^2 & 0 \leq x, y \leq a, \\ 0 & \text{otherwise.} \end{cases}$$

Find the CDF and PDF of random variable

$$W = \max\left(\frac{X}{Y}, \frac{Y}{X}\right).$$

Hint: Is it possible to observe $W < 1$?

6.4.14◆ The joint PDF of X and Y is

$$f_{X,Y}(x, y) = \begin{cases} \lambda^2 e^{-\lambda y} & 0 \leq x < y, \\ 0 & \text{otherwise.} \end{cases}$$

What is the PDF of $W = Y - X$?

6.4.15◆◆ Consider random variables X, Y, and W from Problem 6.4.14.

(a) Are W and X independent?

(b) Are W and Y independent?

6.4.16◆◆ X and Y are independent random variables with CDFs $F_X(x)$ and $F_Y(y)$. Let $U = \min(X, Y)$ and $V = \max(X, Y)$.

(a) What is $F_{U,V}(u, v)$?

(b) What is $f_{U,V}(u, v)$?

Hint: To find the joint CDF, let $A = \{U \leq u\}$ and $B = \{V \leq v\}$ and note that $P[AB] = P[B] - P[A^c B]$.

6.5.1● Let X and Y be independent discrete random variables such that $P_X(k) = P_Y(k) = 0$ for all non-integer k. Show that the PMF of $W = X + Y$ satisfies

$$P_W(w) = \sum_{k=-\infty}^{\infty} P_X(k) P_Y(w - k).$$

6.5.2● X and Y have joint PDF

$$f_{X,Y}(x,y) = \begin{cases} 2 & x \geq 0, y \geq 0, x + y \leq 1, \\ 0 & \text{otherwise.} \end{cases}$$

Find the PDF of $W = X + Y$.

6.5.3■ Find the PDF of $W = X + Y$ when X and Y have the joint PDF

$$f_{X,Y}(x,y) = \begin{cases} 2 & 0 \leq x \leq y \leq 1, \\ 0 & \text{otherwise.} \end{cases}$$

6.5.4■ Find the PDF of $W = X + Y$ when X and Y have the joint PDF

$$f_{X,Y}(x,y) = \begin{cases} 1 & 0 \leq x \leq 1, 0 \leq y \leq 1, \\ 0 & \text{otherwise.} \end{cases}$$

6.5.5■ Random variables X and Y are independent exponential random variables with expected values $E[X] = 1/\lambda$ and $E[Y] = 1/\mu$. If $\mu \neq \lambda$, what is the PDF of $W = X + Y$? If $\mu = \lambda$, what is $f_W(w)$?

6.5.6■ Random variables X and Y have joint PDF

$$f_{X,Y}(x,y) = \begin{cases} 8xy & 0 \leq y \leq x \leq 1, \\ 0 & \text{otherwise.} \end{cases}$$

What is the PDF of $W = X + Y$?

6.5.7■ Continuous random variables X and Y have joint PDF $f_{X,Y}(x,y)$. Show that $W = X - Y$ has PDF

$$f_W(w) = \int_{-\infty}^{\infty} f_{X,Y}(y + w, y)\, dy.$$

Use a variable substitution to show

$$f_W(w) = \int_{-\infty}^{\infty} f_{X,Y}(x, x - w)\, dx.$$

6.5.8◆ In this problem we show directly that the sum of independent Poisson random variables is Poisson. Let J and K be independent Poisson random variables with

expected values α and β, respectively, and show that $N = J + K$ is a Poisson random variable with expected value $\alpha + \beta$. Hint: Show that

$$P_N(n) = \sum_{m=0}^{n} P_K(m) P_J(n - m),$$

and then simplify the summation by extracting the sum of a binomial PMF over all possible values.

6.6.1■ Use `icdfrv.m` to write a function `w=wrv1(m)` that generates m samples of random variable W from Problem 4.2.4. Note that $F_W^{-1}(u)$ does not exist for $u = 1/4$; however, you must define a function `icdfw(u)` that returns a value for `icdfw(0.25)`. Does it matter what value you return for `u=0.25`?

6.6.2■ Write a MATLAB function `u=urv(m)` that generates m samples of random variable U defined in Problem 4.4.7.

6.6.3■ For random variable W of Example 6.10, we can generate random samples in two different ways:

1. Generate samples of X and Y and calculate $W = Y/X$.

2. Find the CDF $F_W(w)$ and generate samples using Theorem 6.5.

Write MATLAB functions `w=wrv1(m)` and `w=wrv2(m)` to implement these methods. Does one method run much faster? If so, why? (Use `cputime` to make comparisons.)

6.6.4◆ Write a function `y=deltarv(m)` that returns m samples of the random variable X with PDF

$$F_X(x) = \begin{cases} 0 & x < -1, \\ (x + 1)/4 & -1 \leq x < 1, \\ 1 & x \geq 1. \end{cases}$$

Since $F_X^{-1}(u)$ is not defined for $1/2 \leq u < 1$, use the result of Problem 6.3.13.

7

Conditional Probability Models

In many applications of probability, we have a probability model of an experiment but it is impossible to observe the outcome of the experiment. Instead we observe an event that is related to the outcome. In some applications, the outcome of interest, for example a sample value of random voltage X, can be obscured by random noise N, and we observe only a sample value of $X + N$. In other examples, we obtain information about a random variable before it is possible to observe the random variable. For example, we might learn the nature of an email (whether it contains images or only text) before we observe the number of bytes that need to be transmitted. In another example, we observe that the beginning of a lecture is delayed by two minutes and we want to predict the actual starting time. In these situations, we obtain a conditional probability model by modifying the original probability model (for the voltage, or the email size, or the starting time) to take into account the information gained from the event we have observed.

7.1 Conditioning a Random Variable by an Event

> The conditional PMF $P_{X|B}(x)$ and conditional PDF $P_{X|B}(x)$ are probability models that use the definition of conditional probability, Definition 1.5, to incorporate partial knowledge of the outcome of an experiment. The partial knowledge is that the outcome is $X \in B \subset S_X$.

Recall from Section 1.3 that the conditional probability

$$P[A|B] = P[AB] / P[B] \tag{7.1}$$

is a number that expresses our new knowledge about the occurrence of event A, when we learn that another event B occurs. In this section, we consider an event A

related to the observation of a random variable X. When X is discrete, we usually are interested in $A = \{X = x\}$ for some x. When X is continuous, we may consider $A = \{x_1 < X \leq x_2\}$ or $A = \{x < X \leq x + dx\}$. The conditioning event B contains information about X but not the precise value of X.

====**Example 7.1**====

Let N equal the number of bytes in an email. A conditioning event might be the event I that the email contains an image. A second kind of conditioning would be the event $\{N > 100{,}000\}$, which tells us that the email required more than 100,000 bytes. Both events I and $\{N > 100{,}000\}$ give us information that the email is likely to have many bytes.

Knowledge of the conditioning event B changes the probability of the event A. Given this information and a probability model, we can use Definition 1.5 to find the conditional probability $P[A|B]$. A starting point is the event $A = \{X \leq x\}$; we would find

$$P[A|B] = P[X \leq x|B] \tag{7.2}$$

for all real numbers x. This formula is a function of x. It is the *conditional cumulative distribution function.*

====**Definition 7.1**========**Conditional CDF**

*Given the event B with $P[B] > 0$, the **conditional cumulative distribution function** of X is*

$$F_{X|B}(x) = P[X \leq x|B].$$

The definition of the conditional CDF applies to discrete, continuous, and mixed random variables. However, just as we have found in prior chapters, the conditional CDF is not the most convenient probability model for many calculations. Instead we have definitions for the special cases of discrete X and continuous X that are more useful.

====**Definition 7.2**========**Conditional PMF Given an Event**

*Given the event B with $P[B] > 0$, the **conditional probability mass function** of X is*

$$P_{X|B}(x) = P[X = x|B].$$

In Chapter 4 we defined the PDF of a continuous random variable as the derivative of the CDF. Similarly, with the knowledge that $x \in B$, we define the conditional PDF as the derivative of the conditional CDF.

========Definition 7.3========Conditional PDF Given an Event

*For a random variable X and an event B with $\mathrm{P}[B] > 0$, the **conditional PDF** of X given B is*

$$f_{X|B}(x) = \frac{dF_{X|B}(x)}{dx}.$$

The functions $P_{X|B}(x)$ and $f_{X|B}(x)$ are probability models for a new random variable related to X. Here we have extended our notation convention for probability functions. We continue the old convention that a CDF is denoted by the letter F, a PMF by P, and a PDF by f, with the subscript containing the name of the random variable. However, with a conditioning event, the subscript contains the name of the random variable followed by a vertical bar followed by a statement of the conditioning event. The argument of the function is usually the lowercase letter corresponding to the variable name. The argument is a dummy variable. It could be any letter, so that $P_{X|B}(x)$ and $f_{Y|B}(y)$ are the same functions as $P_{X|B}(u)$ and $f_{Y|B}(v)$. Sometimes we write the function with no specified argument at all: $P_{X|B}(\cdot)$.

When a conditioning event $B \subset S_X$, both $\mathrm{P}[B]$ and $\mathrm{P}[AB]$ in Equation (7.1) are properties of the PMF $P_X(x)$ or PDF $f_X(x)$. Now either the event $A = \{X = x\}$ is contained in the event B or it is not. If X is discrete and $x \in B$, then $\{AB\} = \{X = x\} \cap B = \{X = x\}$ and $\mathrm{P}[X = x, B] = P_X(x)$. Otherwise, if $x \notin B$, then $\{X = x\} \cap B = \varnothing$ and $\mathrm{P}[X = x, B] = 0$. Similar observations apply when X is continuous. The next theorem uses these observations to calculate the conditional probability models.

========Theorem 7.1========

For a random variable X and an event $B \subset S_X$ with $\mathrm{P}[B] > 0$, the conditional PDF of X given B is

$$\text{Discrete:} \quad P_{X|B}(x) = \begin{cases} \dfrac{P_X(x)}{\mathrm{P}[B]} & x \in B, \\ 0 & \text{otherwise} \end{cases} ;$$

$$\text{Continuous:} \quad f_{X|B}(x) = \begin{cases} \dfrac{f_X(x)}{\mathrm{P}[B]} & x \in B, \\ 0 & \text{otherwise.} \end{cases}$$

The theorem states that when we learn that an outcome $x \in B$, the probabilities of all $x \notin B$ are zero in our conditional model, and the probabilities of all $x \in B$ are proportionally higher than they were before we learned $x \in B$.

========Example 7.2========

A website distributes instructional videos on bicycle repair. The length of a video in minutes X has PMF

$$P_X(x) = \begin{cases} 0.15 & x = 1, 2, 3, 4, \\ 0.1 & x = 5, 6, 7, 8, \\ 0 & \text{otherwise.} \end{cases} \qquad (7.3)$$

Suppose the website has two servers, one for videos shorter than five minutes and the other for videos of five or more minutes. What is the PMF of video length in the second server?

....

We seek a conditional PMF for the condition $x \in L = \{5, 6, 7, 8\}$. From Theorem 7.1,

$$P_{X|L}(x) = \begin{cases} \dfrac{P_X(x)}{P[L]} & x = 5, 6, 7, 8, \\ 0 & \text{otherwise.} \end{cases} \qquad (7.4)$$

From the definition of L, we have

$$P[L] = \sum_{x=5}^{8} P_X(x) = 0.4. \qquad (7.5)$$

With $P_X(x) = 0.1$ for $x \in L$,

$$P_{X|L}(x) = \begin{cases} 0.1/0.4 = 0.25 & x = 5, 6, 7, 8, \\ 0 & \text{otherwise.} \end{cases} \qquad (7.6)$$

Thus the lengths of long videos are equally likely. Among the long videos, each length has probability 0.25.

Sometimes instead of a letter such as B or L that denotes the subset of S_X that forms the condition, we write the condition itself in the PMF. In the preceding example we could use the notation $P_{X|X \geq 5}(x)$ for the conditional PMF.

Example 7.3

For the pointer-spinning experiment of Example 4.1, find the conditional PDF of the pointer position for spins in which the pointer stops on the left side of the circle.

....

Let L denote the left side of the circle. In terms of the stopping position, $L = [1/2, 1)$. Recalling from Example 4.4 that the pointer position X has a uniform PDF over $[0, 1)$,

$$P[L] = \int_{1/2}^{1} f_X(x) \, dx = \int_{1/2}^{1} dx = 1/2. \qquad (7.7)$$

Therefore,

$$f_{X|L}(x) = \begin{cases} 2 & 1/2 \leq x < 1, \\ 0 & \text{otherwise.} \end{cases} \qquad (7.8)$$

Example 7.4

Suppose X, the time in integer minutes you wait for a bus, has the discrete uniform PMF

$$P_X(x) = \begin{cases} 1/20 & x = 1, 2, \ldots, 20, \\ 0 & \text{otherwise.} \end{cases} \tag{7.9}$$

Suppose the bus has not arrived by the eighth minute; what is the conditional PMF of your waiting time X?

. . . .

Let A denote the event $X > 8$. Observing that $P[A] = 12/20$, we can write the conditional PMF of X as

$$P_{X|X>8}(x) = \begin{cases} \dfrac{1/20}{12/20} = \dfrac{1}{12} & x = 9, 10, \ldots, 20, \\ 0 & \text{otherwise.} \end{cases} \tag{7.10}$$

Example 7.5

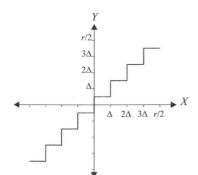

The continuous uniform $(-r/2, r/2)$ random variable X is processed by a b-bit uniform quantizer to produce the quantized output Y. Random variable X is rounded to the nearest quantizer level. With a b-bit quantizer, there are $n = 2^b$ quantization levels. The quantization step size is $\Delta = r/n$, and Y takes on values in the set

$$Q_Y = \left\{ y_{-n/2}, y_{-n/2+1}, \ldots, y_{n/2-1} \right\} \tag{7.11}$$

where $y_i = \Delta/2 + i\Delta$. This relationship is shown for $b = 3$ in the figure on the left. Given the event B_i that $Y = y_i$, find the conditional PDF of X given B_i.

. . . .

In terms of X, we observe that $B_i = \{i\Delta \le X < (i+1)\Delta\}$. Thus,

$$P[B_i] = \int_{i\Delta}^{(i+1)\Delta} f_X(x)\, dx = \frac{\Delta}{r} = \frac{1}{n}. \tag{7.12}$$

By Definition 7.3,

$$f_{X|B_i}(x) = \begin{cases} \dfrac{f_X(x)}{P[B_i]} & x \in B_i, \\ 0 & \text{otherwise,} \end{cases} = \begin{cases} 1/\Delta & i\Delta \le x < (i+1)\Delta, \\ 0 & \text{otherwise.} \end{cases} \tag{7.13}$$

Given B_i, the conditional PDF of X is uniform over the ith quantization interval.

In some applications, we begin with a set of conditional probability models such as the PMFs $P_{X|B_i}(x)$, $i = 1, 2, \ldots, m$, where $B_1, B_2, \ldots, B_m$ is a partition. We then use the law of total probability to find the PMF $P_X(x)$.

====Theorem 7.2====

For random variable X resulting from an experiment with partition $B_1, \ldots, B_m$,

$$\text{Discrete:} \quad P_X(x) = \sum_{i=1}^{m} P_{X|B_i}(x) \, \mathrm{P}\left[B_i\right];$$

$$\text{Continuous:} \quad f_X(x) = \sum_{i=1}^{m} f_{X|B_i}(x) \, \mathrm{P}\left[B_i\right]$$

Proof The theorem follows directly from Theorem 1.9 with $A = \{X = x\}$ for discrete X or $A = \{x < X \leq x + dx\}$ when X is continuous.

====Example 7.6====

Let X denote the number of additional years that a randomly chosen 70-year-old person will live. If the person has high blood pressure, denoted as event H, then X is a geometric $(p = 0.1)$ random variable. Otherwise, if the person's blood pressure is normal, event N, X has a geometric $(p = 0.05)$ PMF. Find the conditional PMFs $P_{X|H}(x)$ and $P_{X|N}(x)$. If 40 percent of all 70-year-olds have high blood pressure, what is the PMF of X?

....

The problem statement specifies the conditional PMFs in words. Mathematically, the two conditional PMFs are

$$P_{X|H}(x) = \begin{cases} 0.1(0.9)^{x-1} & x = 1, 2, \ldots, \\ 0 & \text{otherwise,} \end{cases} \quad P_{X|N}(x) = \begin{cases} 0.05(0.95)^{x-1} & x = 1, 2, \ldots, \\ 0 & \text{otherwise.} \end{cases}$$

Since H, N is a partition, we can use Theorem 7.2 to write

$$\begin{aligned} P_X(x) &= P_{X|H}(x) \, \mathrm{P}\left[H\right] + P_{X|N}(x) \, \mathrm{P}\left[N\right] \\ &= \begin{cases} (0.4)(0.1)(0.9)^{x-1} + (0.6)(0.05)(0.95)^{x-1} & x = 1, 2, \ldots, \\ 0 & \text{otherwise.} \end{cases} \end{aligned} \quad (7.14)$$

====Example 7.7====

Random variable X is a voltage at the receiver of a modem. When symbol "0" is transmitted (event B_0), X is the Gaussian $(-5, 2)$ random variable. When symbol "1" is transmitted (event B_1), X is the Gaussian $(5, 2)$ random variable. Given that symbols "0" and "1" are equally likely to be sent, what is the PDF of X?

....

The problem statement implies that $\mathrm{P}[B_0] = \mathrm{P}[B_1] = 1/2$ and

$$f_{X|B_0}(x) = \frac{1}{2\sqrt{2\pi}} e^{-(x+5)^2/8}, \qquad f_{X|B_1}(x) = \frac{1}{2\sqrt{2\pi}} e^{-(x-5)^2/8}. \qquad (7.15)$$

By Theorem 7.2,

$$\begin{aligned} f_X(x) &= f_{X|B_0}(x)\,\mathrm{P}\,[B_0] + f_{X|B_1}(x)\,\mathrm{P}\,[B_1] \\ &= \frac{1}{4\sqrt{2\pi}}\left(e^{-(x+5)^2/8} + e^{-(x-5)^2/8}\right). \end{aligned} \qquad (7.16)$$

Problem 7.7.1 asks the reader to graph $f_X(x)$ to show its similarity to Figure 4.3.

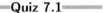

Quiz 7.1

(A) On the Internet, data is transmitted in packets. In a simple model for World Wide Web traffic, the number of packets N needed to transmit a Web page depends on whether the page has graphic images. If the page has images (event I), then N is uniformly distributed between 1 and 50 packets. If the page is just text (event T), then N is uniform between 1 and 5 packets. Assuming a page has images with probability $1/4$, find the

    (a) conditional PMF $P_{N|I}(n)$ (b) conditional PMF $P_{N|T}(n)$

    (c) PMF $P_N(n)$ (d) conditional PMF $P_{N|N\leq 10}(n)$

(B) Y is a continuous uniform $(0, 10)$ random variable. Find the following:

    (a) $\mathrm{P}[Y \leq 6]$ (b) the conditional PDF $f_{Y|Y\leq 6}(y)$

    (c) $\mathrm{P}[Y > 8]$ (d) the conditional PDF $f_{Y|Y>8}(y)$

7.2   Conditional Expected Value Given an Event

> The definitions of conditional expected value $\mathrm{E}[X|B]$ and conditional variance $\mathrm{E}[X|B]$ correspond to the definitions of $\mathrm{E}[X]$ and $\mathrm{Var}[X]$ with $P_{X|B}(x)$ replacing $P_X(x)$ or $f_{X|B}(x)$ replacing $f_X(x)$.

Because the conditioning event B tells us that the outcome of an experiment is an element of B, $P_{X|B}(x)$ or $f_{X|B}(x)$ can be viewed as a PMF or PDF in an experiment with sample space B. This is confirmed by the following theorem, which replaces sample space S with B in Theorem 3.1.

====Theorem 7.3====

Discrete X:

  (a) For any $x \in B$, $P_{X|B}(x) \geq 0$.

  (b) $\sum_{x \in B} P_{X|B}(x) = 1$.

  (c) The conditional probability that X is in the set C is

$$P[C|B] = \sum_{x \in C} P_{X|B}(x).$$

Continuous X:

  (a) For any $x \in B$, $f_{X|B}(x) \geq 0$.

  (b) $\int_B f_{X|B}(x)\,dx = 1$.

  (c) The conditional probability that X is in the set C is

$$P[C|B] = \int_C f_{X|B}(x)\,dx.$$

Conditional probability models have parameters corresponding to the parameters of unconditional probability models.

Therefore, we can compute expected values of the conditional random variable $X|B$ and expected values of functions of $X|B$ in the same way that we compute expected values of X. The only difference is that we use the conditional PMF $P_{X|B}(x)$ or PDF $f_{X|B}(x)$ in place of $P_X(x)$ or $f_X(x)$.

====Definition 7.4====Conditional Expected Value

*The **conditional expected value** of random variable X given condition B is*

$$Discrete: \quad E[X|B] = \sum_{x \in B} x P_{X|B}(x);$$

$$Continuous: E[X|B] = \int_{-\infty}^{\infty} x f_{X|B}(x)\,dx.$$

An alternative notation for $E[X|B]$ is $\mu_{X|B}$.

When we are given the conditional probability models $P_{X|B_i}(x)$ for a partition $B_1, \ldots, B_m$, we can compute the expected value $E[X]$ in terms of the conditional expected values $E[X|B_i]$.

====Theorem 7.4====

For a random variable X resulting from an experiment with partition $B_1, \ldots, B_m$,

$$E[X] = \sum_{i=1}^{m} E[X|B_i] P[B_i].$$

Proof When X is discrete, $\mathrm{E}[X] = \sum_x x P_X(x)$, and we can use Theorem 7.2 to write

$$\mathrm{E}[X] = \sum_x x \sum_{i=1}^m P_{X|B_i}(x)\,\mathrm{P}[B_i]$$

$$= \sum_{i=1}^m \mathrm{P}[B_i] \sum_x x P_{X|B_i}(x) = \sum_{i=1}^m \mathrm{P}[B_i]\,\mathrm{E}[X|B_i]. \tag{7.17}$$

When X is continuous, the proof uses the continuous version of Theorem 7.2 and follows the same logic, with the summation over x replaced by integration.

For a derived random variable $Y = g(X)$, we have the equivalent of Theorem 3.10.

=======**Theorem 7.5**=======

The conditional expected value of $Y = g(X)$ given condition B is

$$Discrete: \quad \mathrm{E}[Y|B] = \mathrm{E}[g(X)|B] = \sum_{x \in B} g(x) P_{X|B}(x);$$

$$Continuous: \mathrm{E}[Y|B] = \mathrm{E}[g(X)|B] = \int_{-\infty}^{\infty} g(x) f_{X|B}(x)\,dx.$$

It follows that the conditional variance and conditional standard deviation conform to Definitions 3.13 and 3.14, with $X|B$ replacing X.

=======**Definition 7.5**=======**Conditional Variance and Standard Deviation**

The conditional variance of X given event B is

$$\mathrm{Var}[X|B] = \mathrm{E}\left[(X - \mu_{X|B})^2 \,|B\right] = \mathrm{E}[X^2|B] - \mu_{X|B}^2.$$

The conditional standard deviation is $\sigma_{X|B} = \sqrt{\mathrm{Var}[X|B]}$.

The conditional variance and conditional standard deviation are useful because they measure the spread of the random variable after we learn the conditioning information B. If the conditional standard deviation $\sigma_{X|B}$ is much smaller than σ_X, then we can say that learning the occurrence of B reduces our uncertainty about X because it shrinks the range of typical values of X.

=======**Example 7.8**=======

Find the conditional expected value, the conditional variance, and the conditional standard deviation for the long videos defined in Example 7.2.

$$E[X|L] = \mu_{X|L} = \sum_{x=5}^{8} x P_{X|L}(x) = 0.25 \sum_{x=5}^{8} x = 6.5 \text{ minutes.} \qquad (7.18)$$

$$E[X^2|L] = 0.25 \sum_{x=5}^{8} x^2 = 43.5 \text{ minutes}^2. \qquad (7.19)$$

$$\text{Var}[X|L] = E[X^2|L] - \mu_{X|L}^2 = 1.25 \text{ minutes}^2. \qquad (7.20)$$

$$\sigma_{X|L} = \sqrt{\text{Var}[X|L]} = 1.12 \text{ minutes.} \qquad (7.21)$$

Quiz 7.2

(A) Continuing Quiz 7.1(A), find

    (a) $E[N|N \le 10]$,   (b) $\text{Var}[N|N \le 10]$.

(B) Continuing Quiz 7.1(B), find

    (a) $E[Y|Y \le 6]$,   (b) $\text{Var}[Y|Y \le 6]$.

7.3   Conditioning Two Random Variables by an Event

> The probability model for random variables X and Y given event B is related to the unconditional probability model for X and Y in the same way that the probability model for X given B is related to the probability model for X. The conditional probability model can be used to find the conditional expected value and conditional variance of $W = g(X, Y)$ in the same way that the unconditional probability model for X and Y is used to compute $E[g(X, Y]$ and $\text{Var}[g(X, Y)]$.

An experiment produces two random variables, X and Y. We learn that the outcome (x, y) is an element of an event, B. We use the information $(x, y) \in B$ to construct a new probability model. If X and Y are discrete, the new model is a conditional joint PMF, the ratio of the joint PMF to $P[B]$. If X and Y are continuous, the new model is a conditional joint PDF, defined as the ratio of the joint PDF to $P[B]$. The definitions of these functions follow from the same intuition as Definition 1.5 for the conditional probability of an event.

═══Definition 7.6═══════Conditional Joint PMF

*For discrete random variables X and Y and an event B with $\mathrm{P}[B] > 0$, the **conditional joint PMF** of X and Y given B is*

$$P_{X,Y|B}(x,y) = \mathrm{P}\left[X = x, Y = y|B\right].$$

The following theorem is an immediate consequence of the definition.

═══Theorem 7.6═══

For any event B, a region of the X, Y plane with $\mathrm{P}[B] > 0$,

$$P_{X,Y|B}(x,y) = \begin{cases} \dfrac{P_{X,Y}(x,y)}{\mathrm{P}[B]} & (x,y) \in B, \\ 0 & otherwise. \end{cases}$$

═══Example 7.9═══

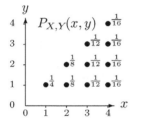

Random variables X and Y have the joint PMF $P_{X,Y}(x,y)$ as shown. Let $B = \{X + Y \leq 4\}$ and find the conditional PMF $P_{X,Y|B}(x,y)$.

· ·

Event $B = \{(1,1), (2,1), (2,2), (3,1)\}$ consists of all points (x, y) such that $x + y \leq 4$. By adding up the probabilities of all outcomes in B, we find

$$\mathrm{P}[B] = P_{X,Y}(1,1) + P_{X,Y}(2,1)$$
$$+ P_{X,Y}(2,2) + P_{X,Y}(3,1) = \frac{7}{12}.$$

The conditional PMF $P_{X,Y|B}(x,y)$ is shown on the left.

In the case of two continuous random variables, we have the following definition of the conditional probability model.

═══Definition 7.7═══════Conditional Joint PDF

Given an event B with $\mathrm{P}[B] > 0$, the conditional joint probability density function of X and Y is

$$f_{X,Y|B}(x,y) = \begin{cases} \dfrac{f_{X,Y}(x,y)}{\mathrm{P}\,[B]} & (x,y) \in B, \\ 0 & \textit{otherwise.} \end{cases}$$

Example 7.10

X and Y are random variables with joint PDF

$$f_{X,Y}(x,y) = \begin{cases} 1/15 & 0 \le x \le 5, 0 \le y \le 3, \\ 0 & \text{otherwise.} \end{cases} \tag{7.22}$$

Find the conditional PDF of X and Y given the event $B = \{X + Y \ge 4\}$.

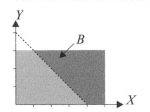

We calculate $\mathrm{P}[B]$ by integrating $f_{X,Y}(x,y)$ over the region B.

$$\mathrm{P}\,[B] = \int_0^3 \int_{4-y}^5 \frac{1}{15}\, dx\, dy$$

$$= \frac{1}{15} \int_0^3 (1+y)\, dy = 1/2. \tag{7.23}$$

Definition 7.7 leads to the conditional joint PDF

$$f_{X,Y|B}(x,y) = \begin{cases} 2/15 & 0 \le x \le 5, 0 \le y \le 3, x+y \ge 4, \\ 0 & \text{otherwise.} \end{cases} \tag{7.24}$$

Corresponding to Theorem 5.9, we have

Theorem 7.7 Conditional Expected Value

For random variables X and Y and an event B of nonzero probability, the conditional expected value of $W = g(X,Y)$ given B is

$$\textit{Discrete:} \quad \mathrm{E}\,[W|B] = \sum_{x \in S_X} \sum_{y \in S_Y} g(x,y) P_{X,Y|B}(x,y)\,;$$

$$\textit{Continuous:}\ \mathrm{E}\,[W|B] = \int_{-\infty}^{\infty} \int_{-\infty}^{\infty} g(x,y) f_{X,Y|B}(x,y)\ dx\, dy.$$

========Example 7.11========

Continuing Example 7.9, find the conditional expected value and the conditional variance of $W = X + Y$ given the event $B = \{X + Y \leq 4\}$.

We recall from Example 7.9 that $P_{X,Y|B}(x, y)$ has four points with nonzero probability: $(1, 1)$, $(1, 2)$, $(1, 3)$, and $(2, 2)$. Their probabilities are $3/7$, $3/14$, $1/7$, and $3/14$, respectively. Therefore,

$$\mathrm{E}\left[W|B\right] = \sum_{x,y}(x + y)P_{X,Y|B}(x, y)$$

$$= 2\left(\frac{3}{7}\right) + 3\left(\frac{3}{14}\right) + 4\left(\frac{1}{7}\right) + 4\left(\frac{3}{14}\right) = \frac{41}{14}. \tag{7.25}$$

Similarly,

$$\mathrm{E}\left[W^2|B\right] = \sum_{x,y}(x + y)^2 P_{X,Y|B}(x, y)$$

$$= 2^2\left(\frac{3}{7}\right) + 3^2\left(\frac{3}{14}\right) + 4^2\left(\frac{1}{7}\right) + 4^2\left(\frac{3}{14}\right) = \frac{131}{14}. \tag{7.26}$$

The conditional variance is

$$\mathrm{Var}\left[W|B\right] = \mathrm{E}\left[W^2|B\right] - (\mathrm{E}\left[W|B\right])^2 = \frac{131}{14} - \left(\frac{41}{14}\right)^2 = \frac{153}{196}. \tag{7.27}$$

========Quiz 7.3========

(A) Random variables L and X have joint PMF

$P_{L,X}(l, x)$	$x = 40$	$x = 60$
$l = 1$	0.15	0.1
$l = 2$	0.3	0.2
$l = 3$	0.15	0.1

$$\tag{7.28}$$

For random variable $V = LX$, we define the event $A = \{V > 80\}$. Find the conditional PMF $P_{L,X|A}(l, x)$. What are $\mathrm{E}[V|A]$ and $\mathrm{Var}[V|A]$?

(B) Random variables X and Y have the joint PDF

$$f_{X,Y}(x, y) = \begin{cases} xy/4000 & 1 \leq x \leq 3, 40 \leq y \leq 60, \\ 0 & \text{otherwise.} \end{cases} \tag{7.29}$$

For random variable $W = XY$, we define the event $B = \{W > 80\}$. Find the conditional joint PDF $f_{X,Y|B}(x, y)$. What are $\mathrm{E}[W|B]$ and $\mathrm{Var}[W|B]$?

7.4   Conditioning by a Random Variable

> When an experiment produces a pair of random variables X and Y, observing a sample value of one of them provides partial information about the other. To incorporate this information in the probability model, we derive new probability models: the conditional PMFs $P_{X|Y}(x|y)$ and $P_{Y|X}(y|x)$ for discrete random variables, as well as the conditional PDFs $f_{X|Y}(x|y)$ and $f_{Y|X}(y|x)$ for continuous random variables.

In Section 7.3, we used the partial knowledge that the outcome of an experiment $(x, y) \in B$ in order to derive a new probability model for the experiment. Now we turn our attention to the special case in which the partial knowledge consists of the value of one of the random variables: either $B = \{X = x\}$ or $B = \{Y = y\}$. Learning $\{Y = y\}$ changes our knowledge of random variables X, Y. We now have complete knowledge of Y and modified knowledge of X. From this information, we derive a modified probability model for X. The new model is either a *conditional PMF of X given Y* or a *conditional PDF of X given Y*. When X and Y are discrete, the conditional PMF and the associated expected value of a function conform to Theorem 7.6 and Theorem 7.7, respectively. However, we adopt the specialized notation $P_{X|Y}(x|y)$ and $E[X|Y]$ corresponding to the more general notation $P_{X,Y|B}(x, y)$ and $E[g(X, Y)|B]$.

=====Definition 7.8=====Conditional PMF

For any event $Y = y$ such that $P_Y(y) > 0$, the **conditional PMF** *of X given $Y = y$ is*

$$P_{X|Y}(x|y) = P[X = x|Y = y].$$

The following theorem contains the relationship between the joint PMF of X and Y and the two conditional PMFs, $P_{X|Y}(x|y)$ and $P_{Y|X}(y|x)$.

=====Theorem 7.8=====

For discrete random variables X and Y with joint PMF $P_{X,Y}(x, y)$, and x and y such that $P_X(x) > 0$ and $P_Y(y) > 0$,

$$P_{X|Y}(x|y) = \frac{P_{X,Y}(x, y)}{P_Y(y)}, \qquad P_{Y|X}(y|x) = \frac{P_{X,Y}(x, y)}{P_X(x)}.$$

Proof Referring to Definition 7.8, Definition 1.5, and Theorem 5.4, we observe that

$$P_{X|Y}(x|y) = P[X = x|Y = y] = \frac{P[X = x, Y = y]}{P[Y = y]} = \frac{P_{X,Y}(x, y)}{P_Y(y)}. \qquad (7.30)$$

The proof of the second part is the same with X and Y reversed.

==========Example 7.12==========

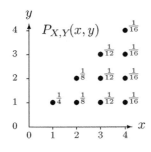

Random variables X and Y have the joint PMF $P_{X,Y}(x,y)$, as given in Example 7.9 and repeated in the accompanying graph. Find the conditional PMF of Y given $X = x$ for each $x \in S_X$.

....

To apply Theorem 7.8, we first find the marginal PMF $P_X(x)$. By Theorem 5.4, $P_X(x) = \sum_{y \in S_Y} P_{X,Y}(x,y)$. For a given $X = x$, we sum the nonzero probablities along the vertical line $X = x$. That is,

$$P_X(x) = \begin{cases} 1/4 & x = 1, \\ 1/8 + 1/8 & x = 2, \\ 1/12 + 1/12 + 1/12 & x = 3, \\ 1/16 + 1/16 + 1/16 + 1/16 & x = 4, \\ 0 & \text{otherwise,} \end{cases} = \begin{cases} 1/4 & x = 1, \\ 1/4 & x = 2, \\ 1/4 & x = 3, \\ 1/4 & x = 4, \\ 0 & \text{otherwise.} \end{cases} \quad (7.31)$$

Theorem 7.8 implies that for $x \in \{1, 2, 3, 4\}$,

$$P_{Y|X}(y|x) = \frac{P_{X,Y}(x,y)}{P_X(x)} = 4P_{X,Y}(x,y). \quad (7.32)$$

For each $x \in \{1, 2, 3, 4\}$, $P_{Y|X}(y|x)$ is a different PMF.

$$P_{Y|X}(y|1) = \begin{cases} 1 & y = 1, \\ 0 & \text{otherwise.} \end{cases} \qquad P_{Y|X}(y|2) = \begin{cases} 1/2 & y \in \{1, 2\}, \\ 0 & \text{otherwise.} \end{cases}$$

$$P_{Y|X}(y|3) = \begin{cases} 1/3 & y \in \{1, 2, 3\}, \\ 0 & \text{otherwise.} \end{cases} \qquad P_{Y|X}(y|4) = \begin{cases} 1/4 & y \in \{1, 2, 3, 4\}, \\ 0 & \text{otherwise.} \end{cases}$$

Given $X = x$, Y is conditionally a discrete uniform $(1, x)$ random variable.

======

By contrast with X and Y discrete, when X and Y are continuous, we cannot apply Section 7.3 directly because $P[B] = P[Y = y] = 0$ as discussed in Chapter 4. Instead, we define a conditional PDF, denoted as $f_{X|Y}(x|y)$, as the ratio of the joint PDF to the marginal PDF.

========Definition 7.9========Conditional PDF

For y such that $f_Y(y) > 0$, the conditional PDF of X given $\{Y = y\}$ is

$$f_{X|Y}(x|y) = \frac{f_{X,Y}(x,y)}{f_Y(y)}.$$

Problem 7.4.12 asks you to verify that $f_{X|Y}(x|y)$ is a conditional density function for X given the conditioning event $y < Y \leq y + \Delta$ in the limit as Δ approaches zero. Definition 7.9 implies

$$f_{Y|X}(y|x) = \frac{f_{X,Y}(x,y)}{f_X(x)}. \tag{7.33}$$

For each y with $f_Y(y) > 0$, the conditional PDF $f_{X|Y}(x|y)$ gives us a new probability model of X. We can use this model in any way that we use $f_X(x)$, the model we have in the absence of knowledge of Y.

Example 7.13

Returning to Example 5.8, random variables X and Y have joint PDF

$$f_{X,Y}(x,y) = \begin{cases} 2 & 0 \leq y \leq x \leq 1, \\ 0 & \text{otherwise.} \end{cases} \tag{7.34}$$

For $0 \leq x \leq 1$, find the conditional PDF $f_{Y|X}(y|x)$. For $0 \leq y \leq 1$, find the conditional PDF $f_{X|Y}(x|y)$.
....

For $0 \leq x \leq 1$, Theorem 5.8 implies

$$f_X(x) = \int_{-\infty}^{\infty} f_{X,Y}(x,y)\, dy = \int_0^x 2\, dy = 2x. \tag{7.35}$$

The conditional PDF of Y given X is

$$f_{Y|X}(y|x) = \frac{f_{X,Y}(x,y)}{f_X(x)} = \begin{cases} 1/x & 0 \leq y \leq x \leq 1, \\ 0 & \text{otherwise.} \end{cases} \tag{7.36}$$

Given $X = x$, we see that Y is the uniform $(0, x)$ random variable. For $0 \leq y \leq 1$, Theorem 5.8 implies

$$f_Y(y) = \int_{-\infty}^{\infty} f_{X,Y}(x,y)\, dx = \int_y^1 2\, dx = 2(1-y). \tag{7.37}$$

Furthermore, Equation (7.33) implies

$$f_{X|Y}(x|y) = \frac{f_{X,Y}(x,y)}{f_Y(y)} = \begin{cases} 1/(1-y) & y \leq x \leq 1, \\ 0 & \text{otherwise.} \end{cases} \tag{7.38}$$

Conditioned on $Y = y$, we see that X is the uniform $(y, 1)$ random variable.

In Example 7.13, we begin with a joint PDF and compute two conditional PDFs. Often in practical situations, we begin with a conditional PDF and a marginal PDF.

Then we use this information to compute the joint PDF and the other conditional PDF. This same approach also works for discrete random variables using PMFs. The necessary formulas are in the following theorems.

Theorem 7.9

For discrete random variables X and Y with joint PMF $P_{X,Y}(x,y)$, and x and y such that $P_X(x) > 0$ and $P_Y(y) > 0$,

$$P_{X,Y}(x,y) = P_{Y|X}(y|x)\, P_X(x) = P_{X|Y}(x|y)\, P_Y(y).$$

Theorem 7.10

For continuous random variables X and Y with joint PDF $f_{X,Y}(x,y)$, and x and y such that $f_X(x) > 0$ and $f_Y(y) > 0$,

$$f_{X,Y}(x,y) = f_{Y|X}(y|x)\, f_X(x) = f_{X|Y}(x|y)\, f_Y(y).$$

Example 7.14

Let R be the uniform $(0,1)$ random variable. Given $R = r$, X is the uniform $(0,r)$ random variable. Find the conditional PDF of R given X.

. . . .

The problem definition states that

$$f_R(r) = \begin{cases} 1 & 0 \le r < 1, \\ 0 & \text{otherwise,} \end{cases} \qquad f_{X|R}(x|r) = \begin{cases} 1/r & 0 \le x < r, \\ 0 & \text{otherwise.} \end{cases} \tag{7.39}$$

It follows from Theorem 7.10 that the joint PDF of R and X is

$$f_{R,X}(r,x) = f_{X|R}(x|r)\, f_R(r) = \begin{cases} 1/r & 0 \le x < r < 1, \\ 0 & \text{otherwise.} \end{cases} \tag{7.40}$$

Now we can find the marginal PDF of X from Theorem 5.8. For $0 < x < 1$,

$$f_X(x) = \int_{-\infty}^{\infty} f_{R,X}(r,x)\, dr = \int_x^1 \frac{dr}{r} = -\ln x. \tag{7.41}$$

By the definition of the conditional PDF,

$$f_{R|X}(r|x) = \frac{f_{R,X}(r,x)}{f_X(x)} = \begin{cases} \frac{1}{-r\ln x} & x \le r \le 1, \\ 0 & \text{otherwise.} \end{cases} \tag{7.42}$$

We recall from Definition 5.4 that discrete random variables X and Y are independent if $P_{X,Y}(x,y) = P_X(x)P_Y(y)$. Similarly, continuous X and Y are independent if $f_{X,Y}(x,y) = f_X(x)f_Y(y)$. We apply these observations to Theorems 7.9 and 7.10.

====Theorem 7.11====

If X and Y are independent,

Discrete:	$P_{X|Y}(x|y) = P_X(x),$	$P_{Y|X}(y|x) = P_Y(y);$
Continuous:	$f_{X|Y}(x|y) = f_X(x),$	$f_{Y|X}(y|x) = f_Y(y).$

The interpretation of Theorem 7.11 is that for independent X and Y, observing $X = x$ does not alter our probability model for Y nor does observing $Y = y$ alter the probability model for X. In particular, when two random variables are independent, each conditional PMF or PDF is identical to a corresponding marginal PMF or PDF.

====Quiz 7.4====

(A) The PMF of random variable X satisfies $P_X(0) = 1 - P_X(2) = 0.4$. The conditional probability model for random variable Y given X is

$$P_{Y|X}(y|0) = \begin{cases} 0.8 & y = 0, \\ 0.2 & y = 1, \\ 0 & \text{otherwise,} \end{cases} \qquad P_{Y|X}(y|2) = \begin{cases} 0.5 & y = 0, \\ 0.5 & y = 1, \\ 0 & \text{otherwise.} \end{cases} \qquad (7.43)$$

(a) What is the probability model for X and Y? Write the joint PMF $P_{X,Y}(x,y)$ as a table.

(b) If $Y = 0$, what is the conditional PMF $P_{X|Y}(x|0)$?

(B) The PDF of X and the conditional PDF of Y given X are

$$f_X(x) = \begin{cases} 3x^2 & 0 < x \le 1, \\ 0 & \text{otherwise,} \end{cases} \qquad f_{Y|X}(y|x) = \begin{cases} 2y/x^2 & 0 \le y \le x, \\ 0 & \text{otherwise.} \end{cases} \qquad (7.44)$$

(a) What is the probability model for X and Y? Find $f_{X,Y}(x,y)$.

(b) If $Y = 1/2$, what is the conditional PDF $f_{X|Y}(x|1/2)$?

7.5   Conditional Expected Value Given a Random Variable

> Random variables X and Y have conditional probability models $P_{X|Y}(x|y)$ or $f_{X|Y}(x|y)$ that have conditional expected value and variance parameters $E[X|Y = y]$ and $\text{Var}[X|Y = y]$.
>
> When we consider $E[X|Y = y]$ as a function of the random observation $Y = y$, we obtain the random variable $E[X|Y]$. The expected value of $E[X|Y]$ is $E[X]$.

For each $y \in S_Y$, the conditional PMF $P_{X|Y}(x|y)$ or conditional PDF $f_{X|Y}(x|y)$ is a modified probability model of X. We can use this model in any way that we use the original $P_X(x)$ or $f_X(x)$, the model we have in the absence of knowledge of Y. Most important, we can find expected values with respect to $P_{X|Y}(x|y)$ or $f_{X|Y}(x|y)$.

=====Definition 7.10=====Conditional Expected Value of a Function

For any $y \in S_Y$, the conditional expected value of $g(X, Y)$ given $Y = y$ is

$$\textit{Discrete:} \quad \mathrm{E}\left[g(X, Y)|Y = y\right] = \sum_{x \in S_X} g(x, y) P_{X|Y}(x|y);$$

$$\textit{Continuous:} \ \mathrm{E}\left[g(X, Y)|Y = y\right] = \int_{-\infty}^{\infty} g(x, y) f_{X|Y}(x|y)\ dx.$$

A special case of Definition 7.10 with $g(x, y) = x$ is the conditional expected value

$$\text{Discrete:} \quad \mathrm{E}\left[X|Y = y\right] = \sum_{x \in S_X} x P_{X|Y}(x|y);$$

$$\text{Continuous:} \ \mathrm{E}\left[X|Y = y\right] = \int_{-\infty}^{\infty} x f_{X|Y}(x|y)\ dx.$$

=====Example 7.15=====

In Example 7.12, we derived conditional PMFs $P_{Y|X}(y|1)$, $P_{Y|X}(y|2)$, $P_{Y|X}(y|3)$, and $P_{Y|X}(y|4)$. Find $E[Y|X = x]$ for $x = 1, 2, 3, 4$.
....
In Example 7.12 we found that given $X = x$, Y was a discrete uniform $(1, x)$ random variable. Since a discrete uniform $(1, x)$ random variable has expected value $(1 + x)/2$,

$$\mathrm{E}\left[Y|X = 1\right] = \frac{1 + 1}{2} = 1, \qquad \mathrm{E}\left[Y|X = 2\right] = \frac{1 + 2}{2} = 1.5, \qquad (7.45)$$

$$\mathrm{E}\left[Y|X = 3\right] = \frac{1 + 3}{2} = 2, \qquad \mathrm{E}\left[Y|X = 4\right] = \frac{1 + 4}{2} = 2.5. \qquad (7.46)$$

Note that in general, the conditional expected value $E[X|Y = y]$ is a function of y

and that $E[Y|X = x]$ is a function of x. However, when X and Y are independent, the observation $Y = y$ provides no information about X; nor does learning $X = x$ inform us about Y. A consequence is that the conditional expected values are the same as the unconditional expected values when X and Y are independent.

═══════ **Theorem 7.12** ═══════

For independent random variables X and Y,

(a) $E[X|Y = y] = E[X]$ *for all $y \in S_Y$,*

(b) $E[Y|X = x] = E[Y]$ *for all $x \in S_X$.*

Proof We present the proof for discrete random variables. By replacing PMFs and sums with PDFs and integrals, we arrive at essentially the same proof for continuous random variables. Since $P_{X|Y}(x|y) = P_X(x)$,

$$E[X|Y = y] = \sum_{x \in S_X} x P_{X|Y}(x|y) = \sum_{x \in S_X} x P_X(x) = E[X]. \tag{7.47}$$

Since $P_{Y|X}(y|x) = P_Y(y)$,

$$E[Y|X = x] = \sum_{y \in S_Y} y P_{Y|X}(y|x) = \sum_{y \in S_Y} y P_Y(y) = E[Y]. \tag{7.48}$$

When we introduced the concept of expected value in Chapters 3 and 4, we observed that $E[X]$ is a property of the probability model of X. This is also true for $E[X|B]$ when $P[B] > 0$. The situation is more complex when we consider $E[X|Y = y]$, the conditional expected value given a random variable. In this case, the conditional expected value is a different number for each possible observation $y \in S_Y$. This implies that $E[X|Y = y]$ is a function of the random variable Y. We use the notation $E[X|Y]$ to denote this function of the random variable Y. Since a function of a random variable is another random variable, we conclude that $E[X|Y]$ *is a random variable!* The following definition may help to clarify this point.

═══════ **Definition 7.11** ═══════ **Conditional Expected Value Function**

The conditional expected value $E[X|Y]$ is a function of random variable Y such that if $Y = y$, then $E[X|Y] = E[X|Y = y]$.

═══════ **Example 7.16** ═══════

For random variables X and Y in Example 5.8, we found in Example 7.13 that the conditional PDF of X given Y is

$$f_{X|Y}(x|y) = \frac{f_{X,Y}(x,y)}{f_Y(y)} = \begin{cases} 1/(1-y) & 0 \le y \le x \le 1, \\ 0 & \text{otherwise.} \end{cases} \tag{7.49}$$

Find the conditional expected values $\mathrm{E}[X|Y=y]$ and $\mathrm{E}[X|Y]$.

Given the conditional PDF $f_{X|Y}(x|y)$, we perform the integration

$$\mathrm{E}\left[X|Y=y\right] = \int_{-\infty}^{\infty} xf_{X|Y}(x|y)\ dx$$

$$= \int_{y}^{1} \frac{1}{1-y} x\, dx = \left.\frac{x^2}{2(1-y)}\right|_{x=y}^{x=1} = \frac{1+y}{2}. \qquad (7.50)$$

Since $\mathrm{E}[X|Y=y] = (1+y)/2$, $\mathrm{E}[X|Y] = (1+Y)/2$.

An interesting property of the random variable $\mathrm{E}[X|Y]$ is its expected value $\mathrm{E}[\mathrm{E}[X|Y]]$. We find $\mathrm{E}[\mathrm{E}[X|Y]]$ in two steps: First we calculate $g(y) = \mathrm{E}[X|Y=y]$, and then we apply Theorem 4.4 to evaluate $\mathrm{E}[g(Y)]$. This two-step process is known as *iterated expectation*.

════════**Theorem 7.13**════════**Iterated Expectation**

$$\mathrm{E}\left[\mathrm{E}\left[X|Y\right]\right] = \mathrm{E}\left[X\right].$$

Proof We consider continuous random variables X and Y and apply Theorem 4.4:

$$\mathrm{E}\left[\mathrm{E}\left[X|Y\right]\right] = \int_{-\infty}^{\infty} \mathrm{E}\left[X|Y=y\right] f_Y(y)\ dy. \qquad (7.51)$$

To obtain this formula from Theorem 4.4, we have used $\mathrm{E}[X|Y=y]$ in place of $g(x)$ and $f_Y(y)$ in place of $f_X(x)$. Next, we substitute the right side of Equation (7.45) for $\mathrm{E}[X|Y=y]$:

$$\mathrm{E}\left[\mathrm{E}\left[X|Y\right]\right] = \int_{-\infty}^{\infty} \left(\int_{-\infty}^{\infty} xf_{X|Y}(x|y)\ dx\right) f_Y(y)\ dy. \qquad (7.52)$$

Rearranging terms in the double integral and reversing the order of integration, we obtain

$$\mathrm{E}\left[\mathrm{E}\left[X|Y\right]\right] = \int_{-\infty}^{\infty} x \int_{-\infty}^{\infty} f_{X|Y}(x|y) f_Y(y)\ dy\, dx. \qquad (7.53)$$

Next, we apply Theorem 7.10 and Theorem 5.8 to infer that the inner integral is $f_X(x)$. Therefore,

$$\mathrm{E}\left[\mathrm{E}\left[X|Y\right]\right] = \int_{-\infty}^{\infty} xf_X(x)\ dx. \qquad (7.54)$$

The proof is complete because the right side of this formula is the definition of $\mathrm{E}[X]$. A similar derivation (using sums instead of integrals) proves the theorem for discrete random variables.

The same derivation can be generalized to any function $g(X)$ of one of the two random variables:

=======Theorem 7.14=======

$$E\left[E\left[g(X)|Y\right]\right] = E\left[g(X)\right].$$

The following formulas apply Theorem 7.14 to discrete and continuous random variables.

Discrete: $E\left[g(X)\right] = E\left[E\left[g(X)|Y\right]\right] = \displaystyle\sum_{y \in S_Y} E\left[g(X)|Y = y\right] P_Y(y)$;

Continuous: $E\left[g(X)\right] = E\left[E\left[g(X)|Y\right]\right] = \displaystyle\int_{-\infty}^{\infty} E\left[g(X)|Y = y\right] f_Y(y)\, dy.$

Theorem 7.14 decomposes the calculation of $E[g(X)]$ into two steps: calculating $E[g(X)|Y = y]$ as a function of Y and then calculating the expected value of the function using the probability model of Y.

=======Quiz 7.5=======

(A) For random variables A and B in Quiz 7.4(A) find:
   (a) $E[Y|X = 2]$,   (b) $\text{Var}[X|Y = 0]$.

(B) For random variables X and Y in Quiz 7.4(B), find:
   (a) $E[Y|X = 1/2]$,   (b) $\text{Var}[X|Y = 1/2]$.

7.6 Bivariate Gaussian Random Variables: Conditional PDFs

For bivariate Gaussian random variables X and Y, the conditional PDFs $f_{X|Y}(x|y)$ and $f_{Y|X}(y|x)$ are Gaussian. $\text{Var}[X|Y] \leq \text{Var}[X]$ and $\text{Var}[Y|X] \leq \text{Var}[Y]$.

Here we return to the bivariate Gaussian random variables X and Y introduced in Section 5.9. Our starting point is the factorized expression for the joint PDF $f_{X,Y}(x, y)$ given in Equation (5.69) and repeated here:

$$f_{X,Y}(x, y) = \frac{1}{\sigma_X \sqrt{2\pi}} e^{-(x - \mu_X)^2 / 2\sigma_X^2} \frac{1}{\tilde{\sigma}_Y \sqrt{2\pi}} e^{-(y - \tilde{\mu}_Y(x))^2 / 2\tilde{\sigma}_Y^2}, \qquad (7.55)$$

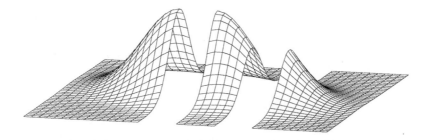

Figure 7.1   Cross-sectional view of the joint Gaussian PDF with $\mu_X = \mu_Y = 0$, $\sigma_X = \sigma_Y = 1$, and $\rho_{X,Y} = 0.9$. Theorem 7.15 confirms that the bell shape of the cross section occurs because the conditional PDF $f_{Y|X}(y|x)$ is Gaussian.

where

$$\tilde{\mu}_Y(x) = \mu_Y + \rho_{X,Y}\frac{\sigma_Y}{\sigma_X}(x - \mu_X), \qquad \tilde{\sigma}_Y = \sigma_Y\sqrt{1 - \rho_{X,Y}^2}. \qquad (7.56)$$

From Theorem 7.10, we know that $f_{X,Y}(x,y) = f_X(x)f_{Y|X}(y|x)$ and we confirmed in Theorem 5.18 that the first factor in (7.55) is the marginal PDF $f_X(x)$. Thus dividing $f_{X,Y}(x,y)$ in Equation (7.55) by $f_X(x)$ we obtain the conditional PDF $f_{Y|X}(y|x)$.

Theorem 7.15

If X and Y are the bivariate Gaussian random variables in Definition 5.10, the conditional PDF of Y given X is

$$f_{Y|X}(y|x) = \frac{1}{\tilde{\sigma}_Y\sqrt{2\pi}}e^{-(y - \tilde{\mu}_Y(x))^2/2\tilde{\sigma}_Y^2},$$

where, given $X = x$, the conditional expected value and variance of Y are

$$\mathrm{E}\,[Y|X = x] = \tilde{\mu}_Y(x) = \mu_Y + \rho_{X,Y}\frac{\sigma_Y}{\sigma_X}(x - \mu_X),$$

$$\mathrm{Var}[Y|X = x] = \tilde{\sigma}_Y^2 = \sigma_Y^2(1 - \rho_{X,Y}^2).$$

Theorem 7.15 demonstrates that given $X = x$, the conditional probability model of Y is Gaussian, with conditional expected value $\mathrm{E}[Y|X = x] = \tilde{\mu}_Y(x)$ and conditional variance $\tilde{\sigma}_Y^2$. The cross sections of Figure 7.1 illustrate the conditional PDF. The figure is a graph of $f_{X,Y}(x,y) = f_{Y|X}(y|x)f_X(x)$. Since X is a constant on each cross section, the cross section is a scaled picture of $f_{Y|X}(y|x)$. As Theorem 7.15 indicates, the cross section has the Gaussian bell shape.

Corresponding to Theorem 7.15, the conditional PDF of X given Y is also Gaussian. This conditional PDF is found by dividing $f_{X,Y}(x,y)$ by $f_Y(y)$ to obtain, after some algebraic manipulations, $f_{X|Y}(x|y)$.

Theorem 7.16

If X and Y are the bivariate Gaussian random variables in Definition 5.10, the conditional PDF of X given Y is

$$f_{X|Y}(x|y) = \frac{1}{\tilde{\sigma}_X \sqrt{2\pi}} e^{-(x - \tilde{\mu}_X(y))^2 / 2\tilde{\sigma}_X^2},$$

where, given $Y = y$, the conditional expected value and variance of X are

$$E[X|Y = y] = \tilde{\mu}_X(y) = \mu_X + \rho_{X,Y} \frac{\sigma_X}{\sigma_Y}(y - \mu_Y),$$

$$\mathrm{Var}[X|Y = y] = \tilde{\sigma}_X^2 = \sigma_X^2(1 - \rho^2).$$

In Theorem 5.19, we asserted that the parameter $\rho_{X,Y}$ in the bivariate Gaussian PDF is the correlation coefficient, but we omitted the proof. Now, with our knowledge of the conditional PDFs, we have the following proof.

Proof (Theorem 5.19) We define $g(X, Y) = (X - \mu_X)(Y - \mu_Y)/(\sigma_X \sigma_Y)$. From Definition 5.5 and Definition 5.6, we have the following formula for the correlation coefficient of any pair of random variables X and Y:

$$E[g(X, Y)] = \frac{E[(X - \mu_X)(Y - \mu_Y)]}{\sigma_X \sigma_Y}. \tag{7.57}$$

We will now show that $E[g(X, Y)] = \rho_{X,Y}$ for bivariate Gaussian random variables X and Y. Using the substitution $f_{X,Y}(x,y) = f_{Y|X}(y|x)f_X(x)$ to evaluate the double integral in the numerator, we obtain

$$E[g(X, Y)] = \frac{1}{\sigma_X \sigma_Y} \int_{-\infty}^{\infty} (x - \mu_X) \left(\int_{-\infty}^{\infty} (y - \mu_Y) f_{Y|X}(y|x)\, dy \right) f_X(x)\, dx$$

$$= \frac{1}{\sigma_X \sigma_Y} \int_{-\infty}^{\infty} (x - \mu_X) E[Y - \mu_Y|X = x] f_X(x)\, dx. \tag{7.58}$$

Because $E[Y|X = x] = \tilde{\mu}_Y(x)$ in Theorem 7.15, it follows that

$$E[Y - \mu_Y|X = x] = \tilde{\mu}_Y(x) - \mu_Y = \rho_{X,Y} \frac{\sigma_Y}{\sigma_X}(x - \mu_X). \tag{7.59}$$

Applying Equation (7.59) to Equation (7.58), we obtain

$$E[g(X, Y)] = \frac{\rho_{X,Y}}{\sigma_X^2} \int_{-\infty}^{\infty} (x - \mu_X)^2 f_X(x)\, dx = \rho_{X,Y}, \tag{7.60}$$

because the integral in the final expression is $\mathrm{Var}[X] = \sigma_X^2$.

Theorem 5.14 states that for any pair of random variables, $|\rho_{X,Y}| < 1$. Introducing this inequality to the formulas for conditional variance in Theorem 7.15 and Theorem 7.16 leads to the following inequalities:

$$\text{Var}\,[Y|X = x] = \sigma_Y^2(1 - \rho_{X,Y}^2) \leq \sigma_Y^2, \tag{7.61}$$

$$\text{Var}\,[X|Y = y] = \sigma_X^2(1 - \rho_{X,Y}^2) \leq \sigma_X^2. \tag{7.62}$$

These formulas state that for $\rho_{X,Y} \neq 0$, learning the value of one of the random variables leads to a model of the other random variable with reduced variance. This suggests that learning the value of Y reduces our uncertainty regarding X.

Quiz 7.6

Let X and Y be jointly Gaussian $(0, 1)$ random variables with correlation coefficient $1/2$. What is the conditional PDF of X given $Y = 2$? What are the conditional expected value and conditional variance $\text{E}[X|Y = 2]$ and $\text{Var}[X|Y = 2]$?

7.7 MATLAB

> To generate sample values of random variables X and Y, use $P_X(x)$ or $f_X(x)$ to generate sample values of X. Then for each sample value x_i, use $P_{Y|X}(y|x_i)$ or $f_{Y|X}(y|x_i)$ to get a sample value of Y.

MATLAB provides the `find` function to identify conditions. We use the `find` function to calculate conditional PMFs for finite random variables.

Example 7.17

Repeating Example 7.2, find the conditional PMF for the length X of a video given event L that the video is long with $X \geq 5$ minutes.

. . . .

```
sx=(1:8)';
px=[0.15*ones(4,1);...
      0.1*ones(4,1)];
sxL=unique(find(sx>=5));
pL=sum(finitepmf(sx,px,sxL));
pxL=finitepmf(sx,px,sxL)/pL;
```

With random variable X defined by `sx` and `px` as in Example 3.36, this code solves this problem. The vector `sxL` identifies the event L, `pL` is the probability $\text{P}[L]$, and `pxL` is the vector of probabilities $P_{X|L}(x_i)$ for each $x_i \in L$.

The conditional PMF and PDF can also be used in MATLAB to simplify the generation of sample pairs (X, Y). For example, when X and Y have the joint PDF $f_{X,Y}(x, y)$, a basic approach is to generate sample values $x_1, \ldots, x_m$ for X using the marginal PDF $f_X(x)$. Then for each sample x_i, we generate y_i using the conditional PDF $f_{Y|X}(y|x_i)$. MATLAB can do this efficiently provided the samples $y_1, \ldots, y_m$ can be generated from $x_1, \ldots, x_m$ using vector-processing techniques, as in the following example.

Example 7.18

Write a function `xy = xytrianglerv(m)` that generates m sample pairs (X, Y) in Example 7.13.

....

In Example 7.13, we found that

$$f_X(x) = \begin{cases} 2x & 0 \le x \le 1, \\ 0 & \text{otherwise,} \end{cases} \qquad f_{Y|X}(y|x) = \begin{cases} 1/x & 0 \le y \le x, \\ 0 & \text{otherwise.} \end{cases} \qquad (7.63)$$

```
function xy = xytrianglerv(m);
x=sqrt(rand(m,1));
y=x.*rand(m,1);
xy=[x y];
```

For $0 \le x \le 1$, we have that $F_X(x) = x^2$. Using Theorem 6.5 to generate sample values of X, we define $u = F_X(x) = x^2$. Then, for $0 < u < 1$, $x = \sqrt{u}$. By Theorem 6.5, if U is uniform $(0, 1)$, then $\sqrt{U}$ has PDF $f_X(x)$. Next, we observe that given $X = x_i$, Y is the uniform $(0, x_i)$ random variable. Given another uniform $(0, 1)$ random variable U_i, Theorem 6.3(a) states that $Y_i = x_i U_i$ is the uniform $(0, x_i)$ random variable. We implement these ideas in the function `xytrianglerv.m`.

Quiz 7.7

For random variables X and Y with joint PMF $P_{X,Y}(x, y)$ given in Example 7.9, write a MATLAB function `xy=dtrianglerv(m)` that generates m sample pairs.

Problems

Difficulty: ● Easy   ■ Moderate   ◆ Difficult   ◆◆ Experts Only

7.1.1● Random variable X has CDF

$$F_X(x) = \begin{cases} 0 & x < -3, \\ 0.4 & -3 \le x < 5, \\ 0.8 & 5 \le x < 7, \\ 1 & x \ge 7. \end{cases}$$

Find the conditional CDF $F_{X|X>0}(x)$ and PMF $P_{X|X>0}(x)$.

7.1.2● X is the discrete uniform $(0, 5)$ random variable. What is $E[X|X \ge E[X]]$?

7.1.3● X has PMF

$$P_X(x) = \binom{4}{x}(1/2)^4.$$

Find $P_{X|B}(x)$ where $B = \{X \ne 0\}$.

7.1.4● In a youth basketball league, a player is fouled in the act of shooting a layup. There is a probability $q = 0.2$ that the layup is good, scoring 2 points. If the layup is good, the player is also awarded 1 free throw, giving the player a chance at a three-point play. If the layup is missed, then (because of the foul) the player is still awarded one point automatically and is also awarded one free throw, enabling a chance to score two points in total. The player makes a free throw with probability $p = 1/2$.

(a) What is the PMF of X, the number of points scored by the player?

(b) Find the conditional PMF $P_{X|T}(x)$ of X given event T that the free throw is good.

7.1.5 Every day you consider going jogging. Before each mile, including the first, you will quit with probability q, independent of the number of miles you have already run. However, you are sufficiently decisive that you never run a fraction of a mile. Also, we say you have run a marathon whenever you run at least 26 miles.

(a) Let M equal the number of miles that you run on an arbitrary day. Find the PMF $P_M(m)$.

(b) Let r be the probability that you run a marathon on an arbitrary day. Find r.

(c) Let J be the number of days in one year (not a leap year) in which you run a marathon. Find the PMF $P_J(j)$. This answer may be expressed in terms of r found in part (b).

(d) Define $K = M - 26$. Let A be the event that you have run a marathon. Find $P_{K|A}(k)$.

7.1.6 A random ECE student has height X in inches given by the PDF

$$f_X(x) = \frac{4e^{-(x-70)^2/8} + e^{-(x-65)^2/8}}{5\sqrt{8\pi}}.$$

(a) Sketch $f_X(x)$ over the interval $60 \le x \le 75$. (For purposes of sketching, note that $\sqrt{8\pi} \approx 5$.)

(b) Find the probability that a random ECE student is less than 5 feet 8 inches tall.

(c) Use conditional PDFs to explain why $f_X(x)$ might be a reasonable model for ECE students.

7.1.7 A test for diabetes is a measurement X of a person's blood sugar level following an overnight fast. For a healthy person, a blood sugar level X in the range of $70 - 110$ mg/dl is considered normal. When a measurement X is used as a test for diabetes, the result is called positive (event T^+) if $X \ge 140$; the test is negative (event T^-) if $X \le 110$, and the test is ambiguous (event T^0) if $110 < X < 140$.

Given that a person is healthy (event H), a blood sugar measurement X is the Gaussian $(90, 20)$ random variable. Given that a person has diabetes, (event D), X is the Gaussian $(60, 40)$ random variable. A randomly chosen person is healthy with probability $P[H] = 0.9$ or has diabetes with probability $P[D] = 0.1$.

(a) What is the conditional PDF $f_{X|H}(x)$?

(b) Calculate the conditional probabilities $P[T^+|H]$, and $P[T^-|H]$.

(c) Find $P[H|T^-]$, the conditional probability that a person is healthy given the event of a negative test.

(d) When a person has an ambiguous test result (T^0), the test is repeated, possibly many times, until either a positive T^+ or negative T^- result is obtained. Let N denote the number of times the test is given. Assuming that for a given person the result of each test is independent of the result of all other tests, find the conditional PMF of N given event H that a person is healthy. Note that $N = 1$ if the person has a positive T^+ or negative T^- result on the first test.

7.1.8 For the quantizer of Example 7.5, the difference $Z = X - Y$ is the quantization error or quantization "noise." As in Example 7.5, assume that X has a uniform $(-r/2, r/2)$ PDF.

(a) Given event B_i that $Y = y_i = \Delta/2 + i\Delta$ and X is in the ith quantization interval, find the conditional PDF of Z.

(b) Show that Z is a uniform random variable. Find the PDF, the expected value, and the variance of Z.

7.1.9 For the quantizer of Example 7.5, we showed in Problem 7.1.8 that the quantization noise Z is a uniform random variable. If X is not uniform, show that Z is nonuniform by calculating the PDF of Z for a simple example.

7.2.1 X is the binomial $(5, 1/2)$ random variable. Find $P_{X|B}(x)$, where the condition $B = \{X \geq \mu_X\}$. What are $E[X|B]$ and Var$[X|B]$?

7.2.2● Random variable X has CDF

$$F_X(x) = \begin{cases} 0 & x < -1, \\ 0.2 & -1 \leq x < 0, \\ 0.7 & 0 \leq x < 1, \\ 1 & x \geq 1. \end{cases}$$

Given $B = \{|X| > 0\}$, find $P_{X|B}(x)$. What are $E[X|B]$ and Var$[X|B]$?

7.2.3● X is the continuous uniform $(-5, 5)$ random variable. Given the event $B = \{|X| \leq 3\}$, find the

(a) conditional PDF, $f_{X|B}(x)$,

(b) conditional expected value, $E[X|B]$,

(c) conditional variance, Var$[X|B]$.

7.2.4● Y is the exponential (0.2) random variable. Given $A = \{Y < 2\}$, find:

(a) $f_{Y|A}(y)$,

(b) $E[Y|A]$.

7.2.5● For the experiment of spinning the pointer three times and observing the maximum pointer position, Example 4.5, find the conditional PDF given the event R that the maximum position is on the right side of the circle. What are the conditional expected value and the conditional variance?

7.2.6▪ The number of pages X in a document has PMF

$$P_X(x) = \begin{cases} 0.15 & x = 1, 2, 3, 4, \\ 0.1 & x = 5, 6, 7, 8, \\ 0 & \text{otherwise.} \end{cases}$$

A firm sends all documents with an even number of pages to printer A and all documents with an odd number of pages to printer B.

(a) Find the conditional PMF of the length X of a document, given the document was sent to A. What are the conditional expected length and standard deviation?

(b) Find the conditional PMF of the length X of a document, given the document was sent to B and had no more than six pages. What are the conditional expected length and standard deviation?

7.2.7▪ Select integrated circuits, test them in sequence until you find the first failure, and then stop. Let N be the number of tests. All tests are independent, with probability of failure $p = 0.1$. Consider the condition $B = \{N \geq 20\}$.

(a) Find the PMF $P_N(n)$.

(b) Find $P_{N|B}(n)$, the conditional PMF of N given that there have been 20 consecutive tests without a failure.

(c) What is $E[N|B]$, the expected number of tests given that there have been 20 consecutive tests without a failure?

7.2.8▪ W is the Gaussian $(0, 4)$ random variable. Given the event $C = \{W > 0\}$, find the conditional PDF, $f_{W|C}(w)$, the conditional expected value, $E[W|C]$, and conditional variance, Var$[W|C]$.

7.2.9▪ The time between telephone calls at a telephone switch is the exponential random variable T with expected value 0.01.

(a) What is $E[T|T > 0.02]$, the conditional expected value of T?

(b) What is Var$[T|T > 0.02]$, the conditional variance of T?

7.2.10▪ As the final rider in the final 60 km time trial of the Tour de France, Roy must finish in time $T \leq 1$ hour to win the Tour. He has the choice of bike made of (1) carbon fiber or (2) titanium. On the carbon fiber bike, his speed V over the course is the continuous uniform random variable with $E[V] = 58$ km/hr and Var$[V] = 12$. On the titanium bike, V is the exponential random variable with $E[V] = 60$ km/hr.

(a) Roy chooses his bike to maximize $P[W]$, the probability he wins the Tour. Which bike does Roy choose and what is $P[W]$?

(b) Suppose instead that Roy flips a fair coin to choose his bike. What is $P[W]$?

7.2.11■ For the distance D of a shot-put toss in Problem 4.7.8, find the conditional PDFs $f_{D|D>0}(d)$ and $f_{D|D\leq70}(d)$.

7.3.1● X and Y are independent identical discrete uniform $(1,10)$ random variables. Let A denote the event that $\min(X,Y)>5$. Find the conditional PMF $P_{X,Y|A}(x,y)$.

7.3.2● Continuing Problem 7.3.1, let B denote the event that $\max(X,Y)\leq 5$. Find the conditional PMF $P_{X,Y|B}(x,y)$.

7.3.3● Random variables X and Y have joint PDF

$$f_{X,Y}(x,y)=\begin{cases}6e^{-(2x+3y)} & x\geq 0, y\geq 0,\\ 0 & \text{otherwise.}\end{cases}$$

Let A be the event that $X+Y\leq 1$. Find the conditional PDF $f_{X,Y|A}(x,y)$.

7.3.4■ N and K have joint PMF

$$P_{N,K}(n,k)=\begin{cases}\frac{(1-p)^{n-1}p}{n} & \begin{array}{l}n=1,2,\dots\\ k=1,\dots,n,\end{array}\\ 0 & \text{otherwise.}\end{cases}$$

Let B denote the event that $N\geq 10$.

(a) Find the conditional PMFs $P_{N|B}(n)$ and $P_{N,K|B}(n,k)$. Which should you find first?

(b) Find the conditional expected values $\mathrm{E}[N|B]$, $\mathrm{E}[K|B]$, $\mathrm{E}[N+K|B]$, $\mathrm{Var}[N|B]$, $\mathrm{Var}[K|B]$, $\mathrm{E}[NK|B]$.

7.3.5■ X and Y have joint PDF

$$f_{X,Y}(x,y)=\begin{cases}(x+y)/3 & \begin{array}{l}0\leq x\leq 1;\\ 0\leq y\leq 2,\end{array}\\ 0 & \text{otherwise.}\end{cases}$$

Let $A=\{Y\leq 1\}$.

(a) What is $\mathrm{P}[A]$?

(b) Find $f_{X,Y|A}(x,y)$.

(c) Find $f_{X|A}(x)$ and $f_{Y|A}(y)$.

7.3.6■ Random variables X and Y have joint PDF

$$f_{X,Y}(x,y)=\begin{cases}(4x+2y)/3 & \begin{array}{l}0\leq x\leq 1;\\ 0\leq y\leq 1,\end{array}\\ 0 & \text{otherwise.}\end{cases}$$

Let $A=\{Y\leq 1/2\}$.

(a) What is $\mathrm{P}[A]$?

(b) Find $f_{X,Y|A}(x,y)$.

(c) Find $f_{X|A}(x)$, and $f_{Y|A}(y)$.

7.3.7■ A study examined whether there was correlation between how much football a person watched and how bald the person was. The time T watching football was measured on a 0, 1, 2 scale such that $T=0$ if a person never watched football, $T=1$ if a person watched football occasionally, and $T=2$ if a person watched a lot of football. Similarly, baldness B was measured on the same scale: $B=0$ for a person with a full head of hair, $B=1$ for a person with thinning hair, and $B=2$ for a person who has not much hair at all. The experiment was to learn B and T for a randomly chosen person, equally likely to be a man (event M) or a woman (event W). The study found that given a person was a man (event M), random variables B and T were conditionally independent. Similarly, given that a person was a woman (event W), B and T were conditionally independent. Moreover, B and T had conditional joint PMFs

b	0	1	2
$P_{B|M}(b)$	0.2	0.3	0.5

t	0	1	2
$P_{T|M}(t)$	0.2	0.2	0.6

b	0	1	2
$P_{B|W}(b)$	0.6	0.3	0.1

t	0	1	2
$P_{T|W}(t)$	0.6	0.2	0.2

(a) Find the conditional PMF $P_{B,T|W}(b,t)$ of B and T given that a person is a woman.

(b) Find the conditional PMF $P_{B,T|M}(b,t)$ of B and T given that a person is a man.

(c) Find the joint PMF $P_{B,T}(b,t)$.

(d) Find the covariance of B and T. Are B and T independent?

7.3.8◆ Random variables X and Y have joint PDF

$$f_{X,Y}(x,y) = \begin{cases} 5x^2/2 & -1 \leq x \leq 1; \\ & 0 \leq y \leq x^2, \\ 0 & \text{otherwise.} \end{cases}$$

Let $A = \{Y \leq 1/4\}$.

(a) Find the conditional PDF $f_{X,Y|A}(x,y)$.

(b) Find $f_{Y|A}(y)$ and $E[Y|A]$.

(c) Find $f_{X|A}(x)$ and $E[X|A]$.

7.3.9◆ X and Y are independent random variables with PDFs

$$f_X(x) = \begin{cases} 2x & 0 \leq x \leq 1, \\ 0 & \text{otherwise,} \end{cases}$$

$$f_Y(y) = \begin{cases} 3y^2 & 0 \leq y \leq 1, \\ 0 & \text{otherwise.} \end{cases}$$

Let $A = \{X > Y\}$.

(a) What are $E[X]$ and $E[Y]$?

(b) What are $E[X|A]$ and $E[Y|A]$?

7.4.1● Given $X = x$,

- Y_1 is Gaussian with conditional expected value x and conditional variance 1.

- Y_2 is Gaussian with conditional expected value x and conditional variance x^2.

Find the conditional PDFs $f_{Y_1|X}(y_1|x)$ and $f_{Y_2|X}(y_2|x)$.

7.4.2● X is the continuous uniform $(0,1)$ random variable. Given $X = x$, Y has a continuous uniform $(0,x)$ PDF. What is the joint PDF $f_{X,Y}(x,y)$? Sketch the region of the X, Y plane for which $f_{X,Y}(x,y) > 0$.

7.4.3● X is the continuous uniform $(0,1)$ random variable. Given $X = x$, Y is conditionally a continuous uniform $(0, 1+x)$ random variable. What is the joint PDF $f_{X,Y}(x,y)$ of X and Y?

7.4.4● Z is a Gaussian $(0,1)$ noise random variable that is independent of X, and

$Y = X + Z$ is a noisy observation of X. What is the conditional PDF $f_{Y|X}(y|x)$?

7.4.5● A business trip is equally likely to take 2, 3, or 4 days. After a d-day trip, the change in the traveler's weight, measured as an integer number of pounds, is a uniform $(-d, d)$ random variable. For one such trip, denote the number of days by D and the change in weight by W. Find the joint PMF $P_{D,W}(d,w)$.

7.4.6● X and Y have joint PDF

$$f_{X,Y}(x,y) = \begin{cases} (4x + 2y)/3 & 0 \leq x \leq 1; \\ & 0 \leq y \leq 1, \\ 0 & \text{otherwise.} \end{cases}$$

(a) For which values of y is $f_{X|Y}(x|y)$ defined? What is $f_{X|Y}(x|y)$?

(b) For which values of x is $f_{Y|X}(y|x)$ defined? What is $f_{Y|X}(y|x)$?

7.4.7● A student's final exam grade depends on how close the student sits to the center of the classroom during lectures. If a student sits r feet from the center of the room, the grade is a Gaussian random variable with expected value $80 - r$ and standard deviation r. If r is a sample value of random variable R, and X is the exam grade, what is $f_{X|R}(x|r)$?

7.4.8■ $Y = ZX$ where X is the Gaussian $(0,1)$ random variable and Z, independent of X, has PMF

$$P_Z(z) = \begin{cases} 1-p & z = -1, \\ p & z = 1. \end{cases}$$

True or False:

(a) Y and Z are independent.

(b) Y and X are independent.

7.4.9■ At the One Top Pizza Shop, mushrooms are the only topping. Curiously, a pizza sold before noon has mushrooms with probability $p = 1/3$ while a pizza sold after noon never has mushrooms. Also, a pizza is equally likely to be sold before noon as after noon. On a day in which 100 pizzas are sold, let N equal the number of pizzas sold before noon and let M equal the number of mushroom pizzas sold during the day. What is the joint PMF $P_{M,N}(m,n)$? Are M and N independent? Hint: Find the conditional PMF of M given N.

7.4.10■ Random variables X and Y have the joint PMF in the following table.

$P_{X,Y}(x,y)$	$y=-1$	$y=0$	$y=1$
$x=-1$	$3/16$	$1/16$	0
$x=0$	$1/6$	$1/6$	$1/6$
$x=1$	0	$1/8$	$1/8$

(a) Are X and Y independent?

(b) The experiment from which X and Y are derived is performed sequentially. First, X is found, then Y is found. In this context, label the conditional branch probabilities of the following tree:

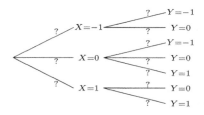

7.4.11■ Flip a coin twice. On each flip, the probability of heads equals p. Let X_i equal the number of heads (either 0 or 1) on flip i. Let $W = X_1 - X_2$ and $Y = X_1 + X_2$. Find $P_{W,Y}(w,y)$, $P_{W|Y}(w|y)$, and $P_{Y|W}(y|w)$.

7.4.12■ Show that

$$\lim_{\Delta \to 0} P\left[x_1 < X \le x_2 | y < Y \le y + \Delta\right]$$

$$= \int_{x_1}^{x_2} f_{X|Y}(x|y)\, dx.$$

Hint: $P[x_1 < X \le x_2, y < Y \le y + \Delta]$ can be written as an integral of $f_{X,Y}(x,y)$.

7.4.13◆ Packets arriving at an Internet router are either voice packets (v) or data packets (d). Each packet is a voice packet

with probability p, independent of any other packet. Observe packets at the Internet router until you see two voice packets. Let M equal the number of packets up to and including the first voice packet. Let N equal the number of packets observed up to and including the second voice packet. Find the conditional PMFs $P_{M|N}(m|n)$ and $P_{N|M}(n|m)$. Interpret your results.

7.4.14◆◆ Suppose you arrive at a bus stop at time 0, and at the end of each minute, with probability p, a bus arrives, or with probability $1 - p$, no bus arrives. Whenever a bus arrives, you board that bus with probability q and depart. Let T equal the number of minutes you stand at a bus stop. Let N be the number of buses that arrive while you wait at the bus stop.

(a) Identify the set of points (n,t) for which $P[N = n, T = t] > 0$.

(b) Find $P_{N,T}(n,t)$.

(c) Find the marginal PMFs $P_N(n)$ and $P_T(t)$.

(d) Find the conditional PMFs $P_{N|T}(n|t)$ and $P_{T|N}(t|n)$.

7.4.15◆◆ Each millisecond at an Internet router, a packet independently arrives with probability p. Each packet is either a data packet (d) with probability q or a video packet (v). Each data packet belongs to an email with probability r. Let N equal the number of milliseconds required to observe the first 100 email packets. Let T equal the number of milliseconds you observe the router waiting for the first email packet. Find the marginal PMF $P_T(t)$ and the conditional PMF $P_{N|T}(n|t)$. Lastly, find the conditional PMF $P_{T|N}(t|n)$.

7.5.1● X and Y have joint PDF

$$f_{X,Y}(x,y) = \begin{cases} 2 & 0 \le y \le x \le 1, \\ 0 & \text{otherwise.} \end{cases}$$

Find the PDF $f_Y(y)$, the conditional PDF $f_{X|Y}(x|y)$, and the conditional expected value $E[X|Y = y]$.

7.5.2 Let random variables X and Y have joint PDF $f_{X,Y}(x,y)$ given in Problem 7.5.1. Find the PDF $f_X(x)$, the conditional PDF $f_{Y|X}(y|x)$, and the conditional expected value $E[Y|X = x]$.

7.5.3 The probability model for random variable A is

$$P_A(a) = \begin{cases} 1/3 & a = -1, \\ 2/3 & a = 1, \\ 0 & \text{otherwise.} \end{cases}$$

The conditional probability model for random variable B given A is:

$$P_{B|A}(b|-1) = \begin{cases} 1/3 & b = 0, \\ 2/3 & b = 1, \\ 0 & \text{otherwise,} \end{cases}$$

$$P_{B|A}(b|1) = \begin{cases} 1/2 & b = 0, \\ 1/2 & b = 1, \\ 0 & \text{otherwise.} \end{cases}$$

(a) What is the probability model for random variables A and B? Write the joint PMF $P_{A,B}(a, b)$ as a table.

(b) If $A = 1$, what is the conditional expected value $E[B|A = 1]$?

(c) If $B = 1$, what is the conditional PMF $P_{A|B}(a|1)$?

(d) If $B = 1$, what is the conditional variance $\text{Var}[A|B = 1]$ of A?

(e) What is the covariance $\text{Cov}[A, B]$?

7.5.4 For random variables A and B given in Problem 7.5.3, let $U = E[B|A]$. Find the PMF $P_U(u)$. What is $E[U] = E[E[B|A]]$?

7.5.5 Random variables N and K have the joint PMF

$$P_{N,K}(n, k) = \begin{cases} \frac{100^n e^{-100}}{(n+1)!} & \begin{matrix} n=0,1,\dots; \\ k=0,1,\dots,n, \end{matrix} \\ 0 & \text{otherwise.} \end{cases}$$

(a) Find the marginal PMF $P_N(n)$ and the conditional PMF $P_{K|N}(k|n)$.

(b) Find the conditional expected value $E[K|N = n]$.

(c) Express the random variable $E[K|N]$ as a function of N and use the iterated expectation to find $E[K]$.

7.5.6 Random variables X and Y have joint PDF

$$f_{X,Y}(x, y) = \begin{cases} 1/2 & -1 \le x \le y \le 1, \\ 0 & \text{otherwise.} \end{cases}$$

(a) What is $f_Y(y)$?

(b) What is $f_{X|Y}(x|y)$?

(c) What is $E[X|Y = y]$?

7.5.7 Over the circle $X^2 + Y^2 \le r^2$, random variables X and Y have the uniform PDF

$$f_{X,Y}(x, y) = \begin{cases} 1/(\pi r^2) & x^2 + y^2 \le r^2, \\ 0 & \text{otherwise.} \end{cases}$$

(a) What is $f_{Y|X}(y|x)$?

(b) What is $E[Y|X = x]$?

7.5.8 (Continuation of Problem 4.6.14) At time $t = 0$, the price of a stock is a constant k dollars. At time $t > 0$ the price of a stock is a Gaussian random variable X with $E[X] = k$ and $\text{Var}[X] = t$. At time t, a *Call Option at Strike* k has value

$$V = (X - k)^+,$$

where the operator $(\cdot)^+$ is defined as $(z)^+ = \max(z, 0)$. Suppose that at the start of each $t = 30$ day month, you can buy the call option at strike k at a price D that is a random variable that fluctuates every month. You decide to buy the call only if the price D is no more than a threshold d^*. What value of the threshold d^* maximizes the expected return $E[R]$?

7.5.9 In a weekly lottery, each $1 ticket sold adds 50 cents to the jackpot that starts at $1 million before any tickets are sold. The jackpot is announced each morning to encourage people to play. On the morning of the ith day before the drawing, the current value of the jackpot J_i is announced.

On that day, the number of tickets sold, N_i, is a Poisson random variable with expected value J_i. Thus, six days before the drawing, the morning jackpot starts at \$1 million and N_6 tickets are sold that day. On the day of the drawing, the announced jackpot is J_0 dollars and N_0 tickets are sold before the evening drawing. What are the expected value and variance of J, the value of the jackpot the instant before the drawing? Hint: Use conditional expectations.

7.6.1 You wish to measure random variable X with expected value $E[X] = 1$ and variance $\text{Var}[X] = 1$, but your measurement procedure yields the noisy observation $Y = X + Z$, where Z is the Gaussian $(0, 2)$ noise that is independent of X.

(a) Find the conditional PDF $f_{Z|X}(z|x)$ of Z given $X = x$.

(b) Find the conditional PDF $f_{Y|X}(y|2)$ of Y given $X = 2$. Hint: Given $X = x$, $Y = x + Z$.

7.6.2 X and Y are jointly Gaussian random variables with $E[X] = E[Y] = 0$ and $\text{Var}[X] = \text{Var}[Y] = 1$. Furthermore, $E[Y|X] = X/2$. Find $f_{X,Y}(x,y)$.

7.6.3 A study of bicycle riders found that a male cyclist's speed X (in miles per hour over a 100-mile "century" ride) and weight Y (kg) could be modeled by a bivariate Gaussian PDF $f_{X,Y}(x,y)$ with parameters $\mu_X = 20$, $\sigma_X = 2$, $\mu_Y = 75$, $\sigma_Y = 5$ and $\rho_{X,Y} = -0.6$. In addition, a female cyclist's speed X' and weight Y' could be modeled by a bivariate Gaussian PDF $f_{X',Y'}(x',y')$ with parameters $\mu_{X'} = 15$, $\sigma_{X'} = 2$, $\mu_{Y'} = 50$, $\sigma_{Y'} = 5$ and $\rho_{X',Y'} = -0.6$. For men and women, the negative correlation of speed and weight reflects the common wisdom that fast cyclists are thin. As it happens, cycling is much more popular among men than women; in a mixed group of cyclists, a cyclist is a male with probability $p = 0.80$.

You suspect it's OK to ignore the differences between men and women since for both groups, weight and speed are negatively correlated, with $\rho = -0.6$. To convince yourself this is OK, you decide to study the speed $\hat{X}$ and weight $\hat{Y}$ of a cyclist randomly chosen from a large mixed group of male and female cyclists. How are $\hat{X}$ and $\hat{Y}$ correlated? Explain your answer.

7.6.4 Let X_1 and X_2 have a bivariate Gaussian PDF with correlation coefficient ρ_{12} such that each X_i is a Gaussian (μ_i, σ_i) random variable. Show that $Y = X_1 X_2$ has variance

$$
\begin{aligned}
\text{Var}[Y] = {}& \sigma_1^2 \sigma_2^2 (1 + \rho_{12}^2) \\
& + \sigma_1^2 \mu_2^2 + \mu_1^2 \sigma_2^2 - \mu_1^2 \mu_2^2.
\end{aligned}
$$

Hints: Look ahead to Problem 9.2.4 and also use the iterated expectation to find

$$ E\left[X_1^2 X_2^2\right] = E\left[E\left[X_1^2 X_2^2 | X_2\right]\right]. $$

7.6.5 Use the iterated expectation for a proof of Theorem 5.19 without integrals.

7.7.1 For the modem receiver voltage X with PDF given in Example 7.7, use MATLAB to plot the PDF and CDF of random variable X. Write a MATLAB function `x=modemrv(m)` that produces m samples of the modem voltage X.

7.7.2 For the quantizer of Example 7.5, we showed in Problem 7.1.9 that the quantization noise Z is nonuniform if X is nonuniform. In this problem, we examine whether it is a reasonable approximation to model the quantization noise as uniform. Consider the special case of a Gaussian $(0, 1)$ random variable X passed through a uniform b-bit quantizer over the interval $(-r/2, r/2)$ with $r = 6$. Does a uniform approximation get better or worse as b increases? Write a MATLAB program to generate histograms for Z to answer this question.

8

Random Vectors

In this chapter, we expand on the concepts presented in Chapter 5. While Chapter 5 introduced the CDF and PDF of n random variables $X_1, \ldots, X_n$, this chapter focuses on the random vector $\mathbf{X} = \begin{bmatrix} X_1 & \cdots & X_n \end{bmatrix}'$. A random vector treats a collection of n random variables as a single entity. Thus, vector notation provides a concise representation of relationships that would otherwise be extremely difficult to represent.

The first section of this chapter presents vector notation for a set of random variables and the associated probability functions. The subsequent sections define marginal probability functions of subsets of n random variables, n independent random variables, independent random vectors, and expected values of functions of n random variables. We then introduce the covariance matrix and correlation matrix, two collections of expected values that play an important role in stochastic processes and in estimation of random variables. The final two sections cover Gaussian random vectors and the application of MATLAB, which is especially useful in working with multiple random variables.

8.1 Vector Notation

> A random vector with n dimensions is a concise representation of a set of n random variables. There is a corresponding notation for the probability model (CDF, PMF, or PDF) of a random vector.

When an experiment produces two or more random variables, vector and matrix notation provide a concise representation of probability models and their properties. This section presents a set of definitions that establish the mathematical notation of random vectors. We use boldface notation $\mathbf{x}$ for a column vector. Row vectors are transposed column vectors; $\mathbf{x}'$ is a row vector. The components of a column vector are, by definition, written in a column. However, to save space, we will often use the transpose of a row vector to display a column vector: $\mathbf{y} = \begin{bmatrix} y_1 & \cdots & y_n \end{bmatrix}'$ is a column vector.

═══**Definition 8.1**═══**Random Vector**

*A **random vector** is a column vector $\mathbf{X} = \begin{bmatrix} X_1 & \cdots & X_n \end{bmatrix}'$. Each X_i is a random variable.*

A random variable is a random vector with $n = 1$. The sample values of the components of a random vector constitute a column vector.

═══**Definition 8.2**═══**Vector Sample Value**

*A **sample value of a random vector** is a column vector $\mathbf{x} = \begin{bmatrix} x_1 & \cdots & x_n \end{bmatrix}'$. The ith component, x_i, of the vector $\mathbf{x}$ is a sample value of a random variable, X_i.*

Following our convention for random variables, the uppercase $\mathbf{X}$ is the random vector and the lowercase $\mathbf{x}$ is a sample value of $\mathbf{X}$. However, we also use boldface capitals such as $\mathbf{A}$ and $\mathbf{B}$ to denote matrices with components that are not random variables. It will be clear from the context whether $\mathbf{A}$ is a matrix of numbers, a matrix of random variables, or a random vector.

The CDF, PMF, or PDF of a random vector is the joint CDF, joint PMF, or joint PDF of the components.

═══**Definition 8.3**═══**Random Vector Probability Functions**

(a) *The **CDF** of a random vector $\mathbf{X}$ is*

$$F_{\mathbf{X}}(\mathbf{x}) = F_{X_1,\ldots,X_n}(x_1,\ldots,x_n).$$

(b) *The **PMF** of a discrete random vector $\mathbf{X}$ is*

$$P_{\mathbf{X}}(\mathbf{x}) = P_{X_1,\ldots,X_n}(x_1,\ldots,x_n).$$

(c) *The **PDF** of a continuous random vector $\mathbf{X}$ is*

$$f_{\mathbf{X}}(\mathbf{x}) = f_{X_1,\ldots,X_n}(x_1,\ldots,x_n).$$

We use similar notation for a function $g(\mathbf{X}) = g(X_1,\ldots,X_n)$ of n random variables and a function $g(\mathbf{x}) = g(x_1,\ldots,x_n)$ of n numbers. Just as we described the relationship of two random variables in Chapter 5, we can explore a pair of random vectors by defining a joint probability model for vectors as a joint CDF, a joint PMF, or a joint PDF.

═══**Definition 8.4**═══**Probability Functions of a Pair of Random Vectors**

For random vectors $\mathbf{X}$ with n components and $\mathbf{Y}$ with m components:

(a) *The **joint CDF** of* $\mathbf{X}$ *and* $\mathbf{Y}$ *is*

$$F_{\mathbf{X},\mathbf{Y}}(\mathbf{x},\mathbf{y}) = F_{X_1,\ldots,X_n,Y_1,\ldots,Y_m}(x_1,\ldots,x_n,y_1,\ldots,y_m);$$

(b) *The **joint PMF** of discrete random vectors* $\mathbf{X}$ *and* $\mathbf{Y}$ *is*

$$P_{\mathbf{X},\mathbf{Y}}(\mathbf{x},\mathbf{y}) = P_{X_1,\ldots,X_n,Y_1,\ldots,Y_m}(x_1,\ldots,x_n,y_1,\ldots,y_m);$$

(c) *The **joint PDF** of continuous random vectors* $\mathbf{X}$ *and* $\mathbf{Y}$ *is*

$$f_{\mathbf{X},\mathbf{Y}}(\mathbf{x},\mathbf{y}) = f_{X_1,\ldots,X_n,Y_1,\ldots,Y_m}(x_1,\ldots,x_n,y_1,\ldots,y_m).$$

The logic of Definition 8.4 is that the pair of random vectors $\mathbf{X}$ and $\mathbf{Y}$ is the same as $\mathbf{W} = \begin{bmatrix} \mathbf{X}' & \mathbf{Y}' \end{bmatrix}' = \begin{bmatrix} X_1 & \cdots & X_n & Y_1 & \cdots & Y_m \end{bmatrix}'$, a concatenation of $\mathbf{X}$ and $\mathbf{Y}$. Thus a probability function of the pair $\mathbf{X}$ and $\mathbf{Y}$ corresponds to the same probability function of $\mathbf{W}$; for example, $F_{\mathbf{X},\mathbf{Y}}(\mathbf{x},\mathbf{y})$ is the same CDF as $F_{\mathbf{W}}(\mathbf{w})$.

If we are interested only in $\mathbf{X} = X_1,\ldots,X_n$, we can use the methods introduced in Section 5.10 to derive a marginal probability model of $X_1,\ldots,X_n$ from the complete probability model for $X_1,\ldots,X_n,Y_1,\ldots,Y_m$. That is, if an experiment produces continuous random vectors $\mathbf{X}$ and $\mathbf{Y}$, then the joint vector PDF $f_{\mathbf{X},\mathbf{Y}}(\mathbf{x},\mathbf{y})$ is a complete probability model, while $f_{\mathbf{X}}(\mathbf{x})$ and $f_{\mathbf{Y}}(\mathbf{y})$ are marginal probability models for $\mathbf{X}$ and $\mathbf{Y}$.

Example 8.1

Random vector $\mathbf{X}$ has PDF

$$f_{\mathbf{X}}(\mathbf{x}) = \begin{cases} 6e^{-\mathbf{a}'\mathbf{x}} & \mathbf{x} \geq 0, \\ 0 & \text{otherwise,} \end{cases} \tag{8.1}$$

where $\mathbf{a} = \begin{bmatrix} 1 & 2 & 3 \end{bmatrix}'$. What is the CDF of $\mathbf{X}$?

Because $\mathbf{a}$ has three components, we infer that $\mathbf{X}$ is a three-dimensional random vector. Expanding $\mathbf{a}'\mathbf{x}$, we write the PDF as a function of the vector components,

$$f_{\mathbf{X}}(\mathbf{x}) = \begin{cases} 6e^{-x_1-2x_2-3x_3} & x_i \geq 0, \\ 0 & \text{otherwise.} \end{cases} \tag{8.2}$$

Applying Definition 8.4, we integrate the PDF with respect to the three variables to obtain

$$F_{\mathbf{X}}(\mathbf{x}) = \begin{cases} (1 - e^{-x_1})(1 - e^{-2x_2})(1 - e^{-3x_3}) & x_i \geq 0, \\ 0 & \text{otherwise.} \end{cases} \tag{8.3}$$

═══Quiz 8.1═══

Discrete random vectors $\mathbf{X} = \begin{bmatrix} x_1 & x_2 & x_3 \end{bmatrix}'$ and $\mathbf{Y} = \begin{bmatrix} y_1 & y_2 & y_3 \end{bmatrix}'$ are related by $\mathbf{Y} = \mathbf{AX}$. Find the joint PMF $P_{\mathbf{Y}}(\mathbf{y})$ if $\mathbf{X}$ has joint PMF

$$P_{\mathbf{X}}(\mathbf{x}) = \begin{cases} (1-p)p^{x_3} & x_1 < x_2 < x_3; \\ & x_1, x_2, x_3 \in \{1, 2, \ldots\}, \\ 0 & \text{otherwise,} \end{cases} \quad \text{and} \quad \mathbf{A} = \begin{bmatrix} 1 & 0 & 0 \\ -1 & 1 & 0 \\ 0 & -1 & 1 \end{bmatrix}.$$

8.2   Independent Random Variables and Random Vectors

> The probability model of the pair of independent random vectors $\mathbf{X}$ and $\mathbf{Y}$ is the product of the probability model of $\mathbf{X}$ and the probability model of $\mathbf{Y}$.

In considering the relationship of a pair of random vectors, we have the following definition of independence:

═══Definition 8.5═══Independent Random Vectors

Random vectors $\mathbf{X}$ and $\mathbf{Y}$ are independent if

$$Discrete: \quad P_{\mathbf{X},\mathbf{Y}}(\mathbf{x}, \mathbf{y}) = P_{\mathbf{X}}(\mathbf{x})\, P_{\mathbf{Y}}(\mathbf{y});$$

$$Continuous: \quad f_{\mathbf{X},\mathbf{Y}}(\mathbf{x}, \mathbf{y}) = f_{\mathbf{X}}(\mathbf{x})\, f_{\mathbf{Y}}(\mathbf{y}).$$

═══Example 8.2═══

As in Example 5.22, random variables $Y_1, \ldots, Y_4$ have the joint PDF

$$f_{Y_1,\ldots,Y_4}(y_1, \ldots, y_4) = \begin{cases} 4 & 0 \le y_1 \le y_2 \le 1, 0 \le y_3 \le y_4 \le 1, \\ 0 & \text{otherwise.} \end{cases} \tag{8.4}$$

Let $\mathbf{V} = \begin{bmatrix} Y_1 & Y_4 \end{bmatrix}'$ and $\mathbf{W} = \begin{bmatrix} Y_2 & Y_3 \end{bmatrix}'$. Are $\mathbf{V}$ and $\mathbf{W}$ independent random vectors?

We first note that the components of $\mathbf{V}$ are $V_1 = Y_1$, and $V_2 = Y_4$. Also, $W_1 = Y_2$, and $W_2 = Y_3$. Therefore,

$$f_{\mathbf{V},\mathbf{W}}(\mathbf{v}, \mathbf{w}) = f_{Y_1,\ldots,Y_4}(v_1, w_1, w_2, v_2) = \begin{cases} 4 & 0 \le v_1 \le w_1 \le 1; \\ & 0 \le w_2 \le v_2 \le 1, \\ 0 & \text{otherwise.} \end{cases} \tag{8.5}$$

Since $\mathbf{V} = \begin{bmatrix} Y_1 & Y_4 \end{bmatrix}'$ and $\mathbf{W} = \begin{bmatrix} Y_2 & Y_3 \end{bmatrix}'$,

$$f_{\mathbf{V}}(\mathbf{v}) = f_{Y_1,Y_4}(v_1,v_2), \qquad\qquad f_{\mathbf{W}}(\mathbf{w}) = f_{Y_2,Y_3}(w_1,w_2). \qquad (8.6)$$

In Example 5.22, we found the marginal PDFs $f_{Y_1,Y_4}(y_1,y_4)$ and $f_{Y_2,Y_3}(y_2,y_3)$ in Equations (5.78) and (5.80). From these marginal PDFs, we have

$$f_{\mathbf{V}}(\mathbf{v}) = \begin{cases} 4(1-v_1)v_2 & 0 \le v_1, v_2 \le 1, \\ 0 & \text{otherwise,} \end{cases} \qquad (8.7)$$

$$f_{\mathbf{W}}(\mathbf{w}) = \begin{cases} 4w_1(1-w_2) & 0 \le w_1, w_2 \le 1, \\ 0 & \text{otherwise.} \end{cases} \qquad (8.8)$$

Therefore,

$$f_{\mathbf{V}}(\mathbf{v})\, f_{\mathbf{W}}(\mathbf{w}) = \begin{cases} 16(1-v_1)v_2 w_1(1-w_2) & 0 \le v_1, v_2, w_1, w_2 \le 1, \\ 0 & \text{otherwise,} \end{cases} \qquad (8.9)$$

which is not equal to $f_{\mathbf{V},\mathbf{W}}(\mathbf{v},\mathbf{w})$. Therefore $\mathbf{V}$ and $\mathbf{W}$ are not independent.

Quiz 8.2

Use the components of $\mathbf{Y} = \begin{bmatrix} Y_1, \ldots, Y_4 \end{bmatrix}'$ in Example 8.2 to construct two independent random vectors $\mathbf{V}$ and $\mathbf{W}$. Prove that $\mathbf{V}$ and $\mathbf{W}$ are independent.

8.3   Functions of Random Vectors

$P_W(w)$, the PMF of $W = g(\mathbf{X})$, a function of discrete random vector $\mathbf{X}$, is the sum of the probabilities of all sample vectors $\mathbf{x}$ for which $g(\mathbf{x}) = w$. To obtain the PDF of W, a function of a continuous random vector, we derive the CDF of W and then differentiate. The expected value of a function of a discrete random vector is the sum over the range of the random vector of the product of the function and the PMF. The expected value of a function of a continuous random vector is the integral over the range of the random vector of the product of the function and the PDF.

Just as we did for one random variable and two random variables, we can derive a random variable $W = g(\mathbf{X})$ that is a function of an arbitrary number of random variables. If W is discrete, the probability model can be calculated as $P_W(w)$, the probability of the event $A = \{W = w\}$ in Theorem 5.24. If W is continuous, the probability model can be expressed as $F_W(w) = P[W \le w]$.

═══Theorem 8.1═══

For random variable $W = g(\mathbf{X})$,

$$\text{Discrete:} \quad P_W(w) = \mathrm{P}[W = w] = \sum_{\mathbf{x}:g(\mathbf{x})=w} P_{\mathbf{X}}(\mathbf{x});$$

$$\text{Continuous:} \quad F_W(w) = \mathrm{P}[W \leq w] = \int \cdots \int_{g(\mathbf{x}) \leq w} f_{\mathbf{X}}(\mathbf{x}) \, dx_1 \cdots dx_n.$$

═══Example 8.3═══

Consider an experiment that consists of spinning the pointer on the wheel of circumference 1 meter in Example 4.1 n times and observing Y_n meters, the maximum position of the pointer in the n spins. Find the CDF and PDF of Y_n.

....

If X_i is the position of the pointer on spin i, then $Y_n = \max\{X_1, X_2, \ldots, X_n\}$. As a result, $Y_n \leq y$ if and only if each $X_i \leq y$. This implies

$$F_{Y_n}(y) = \mathrm{P}[Y_n \leq y] = \mathrm{P}[X_1 \leq y, X_2 \leq y, \ldots X_n \leq y]. \tag{8.10}$$

If we assume the spins to be independent, the events $\{X_1 \leq y\}$, $\{X_2 \leq y\}$, $\ldots$, $\{X_n \leq y\}$ are independent events. Thus

$$F_{Y_n}(y) = \mathrm{P}[X_1 \leq y] \cdots \mathrm{P}[X_n \leq y] = (\mathrm{P}[X \leq y])^n = (F_X(y))^n. \tag{8.11}$$

Example 4.2 derives Equation (4.8):

$$F_X(x) = \begin{cases} 0 & x < 0, \\ x & 0 \leq x < 1, \\ 1 & x \geq 1. \end{cases} \tag{8.12}$$

Equations (8.11) and (8.12) imply that the CDF and corresponding PDF are

$$F_{Y_n}(y) = \begin{cases} 0 & y < 0, \\ y^n & 0 \leq y \leq 1, \\ 1 & y > 1, \end{cases} \qquad f_{Y_n}(y) = \begin{cases} ny^{n-1} & 0 \leq y \leq 1, \\ 0 & \text{otherwise.} \end{cases} \tag{8.13}$$

The following theorem is a generalization of Example 8.3. It expresses the PDF of the maximum and minimum values of a sequence of independent and identically distributed (iid) continuous random variables in terms of the CDF and PDF of the individual random variables.

═════Theorem 8.2═════

Let $\mathbf{X}$ be a vector of n iid continuous random variables, each with CDF $F_X(x)$ and PDF $f_X(x)$.

(a) The CDF and the PDF of $Y = \max\{X_1, \ldots, X_n\}$ are

$$F_Y(y) = (F_X(y))^n, \qquad f_Y(y) = n(F_X(y))^{n-1} f_X(y).$$

(b) The CDF and the PDF of $W = \min\{X_1, \ldots, X_n\}$ are

$$F_W(w) = 1 - (1 - F_X(w))^n, \qquad f_W(w) = n(1 - F_X(w))^{n-1} f_X(w).$$

Proof By definition, $F_Y(y) = P[Y \leq y]$. Because Y is the maximum value of $\{X_1, \ldots, X_n\}$, the event $\{Y \leq y\} = \{X_1 \leq y, X_2 \leq y, \ldots, X_n \leq y\}$. Because all the random variables X_i are iid, $\{Y \leq y\}$ is the intersection of n independent events. Each of the events $\{X_i \leq y\}$ has probability $F_X(y)$. The probability of the intersection is the product of the individual probabilities, which implies the first part of the theorem: $F_Y(y) = (F_X(y))^n$. The second part is the result of differentiating $F_Y(y)$ with respect to y. The derivations of $F_W(w)$ and $f_W(w)$ are similar. They begin with the observations that $F_W(w) = 1 - P[W > w]$ and that the event $\{W > w\} = \{X_1 > w, X_2 > w, \ldots X_n > w\}$, which is the intersection of n independent events, each with probability $1 - F_X(w)$.

In some applications of probability theory, we are interested only in the expected value of a function, not the complete probability model. Although we can always find $E[W]$ by first deriving $P_W(w)$ or $f_W(w)$, it is easier to find $E[W]$ by applying the following theorem.

═════Theorem 8.3═════

For a random vector $\mathbf{X}$, the random variable $g(\mathbf{X})$ has expected value

$$\textit{Discrete:} \quad E[g(\mathbf{X})] = \sum_{x_1 \in S_{X_1}} \cdots \sum_{x_n \in S_{X_n}} g(\mathbf{x}) P_{\mathbf{X}}(\mathbf{x});$$

$$\textit{Continuous:}\ E[g(\mathbf{X})] = \int_{-\infty}^{\infty} \cdots \int_{-\infty}^{\infty} g(\mathbf{x}) f_{\mathbf{X}}(\mathbf{x})\, dx_1 \cdots dx_n.$$

If $W = g(\mathbf{X})$ is the product of n univariate functions and the components of $\mathbf{X}$ are mutually independent, $E[W]$ is a product of n expected values.

═════Theorem 8.4═════

When the components of $\mathbf{X}$ are independent random variables,

$$E[g_1(X_1)g_2(X_2) \cdots g_n(X_n)] = E[g_1(X_1)] E[g_2(X_2)] \cdots E[g_n(X_n)].$$

Proof When $\mathbf{X}$ is discrete, independence implies $P_{\mathbf{X}}(\mathbf{x}) = P_{X_1}(x_1)\cdots P_{X_n}(x_n)$. This implies

$$\mathrm{E}\left[g_1(X_1)\cdots g_n(X_n)\right] = \sum_{x_1 \in S_{X_1}} \cdots \sum_{x_n \in S_{X_n}} g_1(x_1)\cdots g_n(x_n)P_{\mathbf{X}}(\mathbf{x}) \tag{8.14}$$

$$= \left(\sum_{x_1 \in S_{X_1}} g_1(x_1)P_{X_1}(x_1)\right)\cdots\left(\sum_{x_n \in S_{X_n}} g_n(x_n)P_{X_n}(x_n)\right) \tag{8.15}$$

$$= \mathrm{E}\left[g_1(X_1)\right]\mathrm{E}\left[g_2(X_2)\right]\cdots\mathrm{E}\left[g_n(X_n)\right]. \tag{8.16}$$

The derivation is similar for independent continuous random variables.

We have considered the case of a single random variable $W = g(\mathbf{X})$ derived from a random vector $\mathbf{X}$. Some experiments may yield a new random vector $\mathbf{Y}$ with components $Y_1,\ldots,Y_n$ that are functions of the components of $\mathbf{X}$: $Y_k = g_k(\mathbf{X})$. We can derive the PDF of $\mathbf{Y}$ by first finding the CDF $F_{\mathbf{Y}}(\mathbf{y})$ and then applying Definition 5.11. The following theorem demonstrates this technique.

Theorem 8.5

Given the continuous random vector $\mathbf{X}$, define the derived random vector $\mathbf{Y}$ such that $Y_k = aX_k + b$ for constants $a > 0$ and b. The CDF and PDF of $\mathbf{Y}$ are

$$F_{\mathbf{Y}}(\mathbf{y}) = F_{\mathbf{X}}\left(\frac{y_1 - b}{a},\ldots,\frac{y_n - b}{a}\right), \quad f_{\mathbf{Y}}(\mathbf{y}) = \frac{1}{a^n}f_{\mathbf{X}}\left(\frac{y_1 - b}{a},\ldots,\frac{y_n - b}{a}\right).$$

Proof We observe $\mathbf{Y}$ has CDF $F_{\mathbf{Y}}(\mathbf{y}) = \mathrm{P}[aX_1 + b \leq y_1,\ldots,aX_n + b \leq y_n]$. Since $a > 0$,

$$F_{\mathbf{Y}}(\mathbf{y}) = \mathrm{P}\left[X_1 \leq \frac{y_1 - b}{a},\ldots,X_n \leq \frac{y_n - b}{a}\right] = F_{\mathbf{X}}\left(\frac{y_1 - b}{a},\ldots,\frac{y_n - b}{a}\right). \tag{8.17}$$

Definition 5.13 defines the joint PDF of $\mathbf{Y}$,

$$f_{\mathbf{Y}}(\mathbf{y}) = \frac{\partial^n F_{Y_1,\ldots,Y_n}(y_1,\ldots,y_n)}{\partial y_1 \cdots \partial y_n} = \frac{1}{a^n}f_{\mathbf{X}}\left(\frac{y_1 - b}{a},\ldots,\frac{y_n - b}{a}\right). \tag{8.18}$$

Theorem 8.5 is a special case of a transformation of the form $\mathbf{Y} = \mathbf{AX} + \mathbf{b}$. The following theorem is a consequence of the change-of-variable theorem (Appendix B, Math Fact B.13) in multivariable calculus.

Theorem 8.6

If $\mathbf{X}$ is a continuous random vector and $\mathbf{A}$ is an invertible matrix, then $\mathbf{Y} = \mathbf{AX} + \mathbf{b}$ has PDF

$$f_{\mathbf{Y}}(\mathbf{y}) = \frac{1}{|det(\mathbf{A})|}f_{\mathbf{X}}\left(\mathbf{A}^{-1}(\mathbf{y} - \mathbf{b})\right)$$

Proof Let $B = \{\mathbf{y}|\mathbf{y} \leq \tilde{\mathbf{y}}\}$ so that $F_\mathbf{Y}(\tilde{\mathbf{y}}) = \int_B f_\mathbf{Y}(\mathbf{y})\, d\mathbf{y}$. Define the vector transformation $\mathbf{x} = T(\mathbf{y}) = \mathbf{A}^{-1}(\mathbf{y} - \mathbf{b})$. It follows that $\mathbf{Y} \in B$ if and only if $\mathbf{X} \in T(B)$, where $T(B) = \{\mathbf{x}|\mathbf{A}\mathbf{x} + \mathbf{b} \leq \tilde{\mathbf{y}}\}$ is the image of B under transformation T. This implies

$$F_\mathbf{Y}(\tilde{\mathbf{y}}) = P\left[\mathbf{X} \in T(B)\right] = \int_{T(B)} f_\mathbf{X}(\mathbf{x})\, d\mathbf{x} \tag{8.19}$$

By the change-of-variable theorem (Math Fact B.13),

$$F_\mathbf{Y}(\tilde{\mathbf{y}}) = \int_B f_\mathbf{X}\left(\mathbf{A}^{-1}(\mathbf{y} - \mathbf{b})\right) \left|\det\left(\mathbf{A}^{-1}\right)\right| d\mathbf{y} \tag{8.20}$$

where $|\det(\mathbf{A}^{-1})|$ is the absolute value of the determinant of $\mathbf{A}^{-1}$. Definition 8.3 for the CDF and PDF of a random vector combined with Theorem 5.23(b) imply that $f_\mathbf{Y}(\mathbf{y}) = f_\mathbf{X}(\mathbf{A}^{-1}(\mathbf{y} - \mathbf{b}))|\det(\mathbf{A}^{-1})|$. The theorem follows, since $|\det(\mathbf{A}^{-1})| = 1/|\det(\mathbf{A})|$.

=====**Quiz 8.3**=====

(A) A test of light bulbs produced by a machine has three possible outcomes: L, long life; A, average life; and R, reject. The results of different tests are independent. All tests have the following probability model: $P[L] = 0.3$, $P[A] = 0.6$, and $P[R] = 0.1$. Let X_1, X_2, and X_3 be the number of light bulbs that are L, A, and R respectively in five tests. Find the PMF $P_\mathbf{X}(\mathbf{x})$; the marginal PMFs $P_{X_1}(x_1)$, $P_{X_2}(x_2)$, and $P_{X_3}(x_3)$; and the PMF of $W = \max(X_1, X_2, X_3)$.

(B) The random vector $\mathbf{X}$ has PDF

$$f_\mathbf{X}(\mathbf{x}) = \begin{cases} e^{-x_3} & 0 \leq x_1 \leq x_2 \leq x_3, \\ 0 & \text{otherwise.} \end{cases} \tag{8.21}$$

Find the PDF of $\mathbf{Y} = \mathbf{A}\mathbf{X} + \mathbf{b}$. where $\mathbf{A} = \text{diag}[2, 2, 2]$ and $\mathbf{b} = \begin{bmatrix} 4 & 4 & 4 \end{bmatrix}'$.

8.4   Expected Value Vector and Correlation Matrix

The expected value of a random vector is a vector containing the expected values of the components of the vector. The covariance of a random vector is a symmetric matrix containing the variances of the components of the random vector and the covariances of all pairs of random variables in the random vector.

Corresponding to the expected value of a single random variable, the expected value of a random vector is a column vector in which the components are the expected values of the components of the random vector. There is a corresponding definition of the variance and standard deviation of a random vector.

═══════Definition 8.6═══════**Expected Value Vector**

*The **expected value of a random vector** $\mathbf{X}$ is a column vector*

$$\mathrm{E}\left[\mathbf{X}\right] = \boldsymbol{\mu}_{\mathbf{X}} = \left[\mathrm{E}\left[X_1\right] \quad \mathrm{E}\left[X_2\right] \quad \cdots \quad \mathrm{E}\left[X_n\right]\right]'.$$

The correlation and covariance (Definition 5.7 and Definition 5.5) are numbers that contain important information about a pair of random variables. Corresponding information about random vectors is reflected in the set of correlations and the set of covariances of all pairs of components. These sets are referred to as *second-order statistics*. They have a concise matrix notation. To establish the notation, we first observe that for random vectors $\mathbf{X}$ with n components and $\mathbf{Y}$ with m components, the set of all products, $X_i Y_j$, is contained in the $n \times m$ *random matrix* $\mathbf{XY}'$. If $\mathbf{Y} = \mathbf{X}$, the random matrix $\mathbf{XX}'$ contains all products, $X_i X_j$, of components of $\mathbf{X}$.

═══════Example 8.4═══════

If $\mathbf{X} = \left[X_1 \quad X_2 \quad X_3\right]'$, what are the components of $\mathbf{XX}'$?

. . . .

$$\mathbf{XX}' = \begin{bmatrix} X_1 \\ X_2 \\ X_3 \end{bmatrix} \begin{bmatrix} X_1 & X_2 & X_3 \end{bmatrix} = \begin{bmatrix} X_1^2 & X_1 X_2 & X_1 X_3 \\ X_2 X_1 & X_2^2 & X_2 X_3 \\ X_3 X_1 & X_3 X_2 & X_3^2 \end{bmatrix}. \qquad (8.22)$$

In Definition 8.6, we defined the expected value of a random vector as the vector of expected values. This definition can be extended to random matrices.

═══════Definition 8.7═══════**Expected Value of a Random Matrix**

For a random matrix $\mathbf{A}$ with the random variable A_{ij} as its i, jth element, $\mathrm{E}[\mathbf{A}]$ is a matrix with i, jth element $\mathrm{E}[A_{ij}]$.

Applying this definition to the random matrix $\mathbf{XX}'$, we have a concise way to define the correlation matrix of random vector $\mathbf{X}$.

═══════Definition 8.8═══════**Vector Correlation**

*The **correlation of a random vector** $\mathbf{X}$ is an $n \times n$ matrix $\mathbf{R}_{\mathbf{X}}$ with i, jth element $R_X(i, j) = \mathrm{E}[X_i X_j]$. In vector notation,*

$$\mathbf{R}_{\mathbf{X}} = \mathrm{E}\left[\mathbf{XX}'\right].$$

Example 8.5

If $\mathbf{X} = \begin{bmatrix} X_1 & X_2 & X_3 \end{bmatrix}'$, the correlation matrix of $\mathbf{X}$ is

$$\mathbf{R_X} = \begin{bmatrix} \mathrm{E}\left[X_1^2\right] & \mathrm{E}\left[X_1 X_2\right] & \mathrm{E}\left[X_1 X_3\right] \\ \mathrm{E}\left[X_2 X_1\right] & \mathrm{E}\left[X_2^2\right] & \mathrm{E}\left[X_2 X_3\right] \\ \mathrm{E}\left[X_3 X_1\right] & \mathrm{E}\left[X_3 X_2\right] & \mathrm{E}\left[X_3^2\right] \end{bmatrix} = \begin{bmatrix} \mathrm{E}\left[X_1^2\right] & r_{X_1, X_2} & r_{X_1, X_3} \\ r_{X_2, X_1} & \mathrm{E}\left[X_2^2\right] & r_{X_2, X_3} \\ r_{X_3, X_1} & r_{X_3, X_2} & \mathrm{E}\left[X_3^2\right] \end{bmatrix}.$$

The i, jth element of the correlation matrix is the expected value of the random variable $X_i X_j$. The *covariance matrix* of $\mathbf{X}$ is a similar generalization of the covariance of two random variables.

Definition 8.9   **Vector Covariance**

*The **covariance of a random vector** $\mathbf{X}$ is an $n \times n$ matrix $\mathbf{C_X}$ with components $C_X(i,j) = \mathrm{Cov}[X_i, X_j]$. In vector notation,*

$$\mathbf{C_X} = \mathrm{E}\left[(\mathbf{X} - \boldsymbol{\mu_X})(\mathbf{X} - \boldsymbol{\mu_X})'\right]$$

Example 8.6

If $\mathbf{X} = \begin{bmatrix} X_1 & X_2 & X_3 \end{bmatrix}'$, the covariance matrix of $\mathbf{X}$ is

$$\mathbf{C_X} = \begin{bmatrix} \mathrm{Var}[X_1] & \mathrm{Cov}\left[X_1, X_2\right] & \mathrm{Cov}\left[X_1, X_3\right] \\ \mathrm{Cov}\left[X_2, X_1\right] & \mathrm{Var}[X_2] & \mathrm{Cov}\left[X_2, X_3\right] \\ \mathrm{Cov}\left[X_3, X_1\right] & \mathrm{Cov}\left[X_3, X_2\right] & \mathrm{Var}[X_3] \end{bmatrix} \qquad (8.23)$$

Theorem 5.16(a), which connects the correlation and covariance of a pair of random variables, can be extended to random vectors.

Theorem 8.7

For a random vector $\mathbf{X}$ with correlation matrix $\mathbf{R_X}$, covariance matrix $\mathbf{C_X}$, and vector expected value $\boldsymbol{\mu_X}$,

$$\mathbf{C_X} = \mathbf{R_X} - \boldsymbol{\mu_X} \boldsymbol{\mu_X}'.$$

Proof The proof is essentially the same as the proof of Theorem 5.16(a), with vectors replacing scalars. Cross multiplying inside the expectation of Definition 8.9 yields

$$\mathbf{C_X} = \mathrm{E}\left[\mathbf{X}\mathbf{X}' - \mathbf{X}\boldsymbol{\mu_X}' - \boldsymbol{\mu_X}\mathbf{X}' + \boldsymbol{\mu_X}\boldsymbol{\mu_X}'\right]$$
$$= \mathrm{E}\left[\mathbf{X}\mathbf{X}'\right] - \mathrm{E}\left[\mathbf{X}\boldsymbol{\mu_X}'\right] - \mathrm{E}\left[\boldsymbol{\mu_X}\mathbf{X}'\right] + \mathrm{E}\left[\boldsymbol{\mu_X}\boldsymbol{\mu_X}'\right]. \qquad (8.24)$$

Since $E[\mathbf{X}] = \boldsymbol{\mu}_{\mathbf{X}}$ is a constant vector,

$$\mathbf{C}_{\mathbf{X}} = \mathbf{R}_{\mathbf{X}} - E[\mathbf{X}]\,\boldsymbol{\mu}'_{\mathbf{X}} - \boldsymbol{\mu}_{\mathbf{X}}\,E[\mathbf{X}'] + \boldsymbol{\mu}_{\mathbf{X}}\boldsymbol{\mu}'_{\mathbf{X}} = \mathbf{R}_{\mathbf{X}} - \boldsymbol{\mu}_{\mathbf{X}}\boldsymbol{\mu}'_{\mathbf{X}}. \tag{8.25}$$

Example 8.7

Find the expected value $E[\mathbf{X}]$, the correlation matrix $\mathbf{R}_{\mathbf{X}}$, and the covariance matrix $\mathbf{C}_{\mathbf{X}}$ of the two-dimensional random vector $\mathbf{X}$ with PDF

$$f_{\mathbf{X}}(\mathbf{x}) = \begin{cases} 2 & 0 \le x_1 \le x_2 \le 1, \\ 0 & \text{otherwise.} \end{cases} \tag{8.26}$$

. . . .

The elements of the expected value vector are

$$E[X_i] = \int_{-\infty}^{\infty}\int_{-\infty}^{\infty} x_i f_{\mathbf{X}}(\mathbf{x})\, dx_1 dx_2 = \int_0^1 \int_0^{x_2} 2x_i\, dx_1 dx_2, \quad i = 1, 2. \tag{8.27}$$

The integrals are $E[X_1] = 1/3$ and $E[X_2] = 2/3$, so that $\boldsymbol{\mu}_{\mathbf{X}} = E[\mathbf{X}] = \begin{bmatrix} 1/3 & 2/3 \end{bmatrix}'$. The elements of the correlation matrix are

$$E[X_1^2] = \int_{-\infty}^{\infty}\int_{-\infty}^{\infty} x_1^2 f_{\mathbf{X}}(\mathbf{x})\, dx_1 dx_2 = \int_0^1 \int_0^{x_2} 2x_1^2\, dx_1 dx_2, \tag{8.28}$$

$$E[X_2^2] = \int_{-\infty}^{\infty}\int_{-\infty}^{\infty} x_2^2 f_{\mathbf{X}}(\mathbf{x})\, dx_1 dx_2 = \int_0^1 \int_0^{x_2} 2x_2^2\, dx_1 dx_2, \tag{8.29}$$

$$E[X_1 X_2] = \int_{-\infty}^{\infty}\int_{-\infty}^{\infty} x_1 x_2 f_{\mathbf{X}}(\mathbf{x})\, dx_1 dx_2 = \int_0^1 \int_0^{x_2} 2x_1 x_2\, dx_1 dx_2. \tag{8.30}$$

These integrals are $E[X_1^2] = 1/6$, $E[X_2^2] = 1/2$, and $E[X_1 X_2] = 1/4$. Therefore,

$$\mathbf{R}_{\mathbf{X}} = \begin{bmatrix} 1/6 & 1/4 \\ 1/4 & 1/2 \end{bmatrix}. \tag{8.31}$$

We use Theorem 8.7 to find the elements of the covariance matrix.

$$\mathbf{C}_{\mathbf{X}} = \mathbf{R}_{\mathbf{X}} - \boldsymbol{\mu}_{\mathbf{X}}\boldsymbol{\mu}'_{\mathbf{X}} = \begin{bmatrix} 1/6 & 1/4 \\ 1/4 & 1/2 \end{bmatrix} - \begin{bmatrix} 1/9 & 2/9 \\ 2/9 & 4/9 \end{bmatrix} = \begin{bmatrix} 1/18 & 1/36 \\ 1/36 & 1/18 \end{bmatrix}. \tag{8.32}$$

In addition to the correlations and covariances of the elements of one random vector, it is useful to refer to the correlations and covariances of elements of two random vectors.

━━━━━Definition 8.10━━━━━Vector Cross-Correlation

*The **cross-correlation of random vectors**, $\mathbf{X}$ with n components and $\mathbf{Y}$ with m components, is an $n \times m$ matrix $\mathbf{R_{XY}}$ with i, jth element $R_{XY}(i,j) = \mathrm{E}[X_i Y_j]$, or, in vector notation,*

$$\mathbf{R_{XY}} = \mathrm{E}\left[\mathbf{XY'}\right].$$

━━━━━Definition 8.11━━━━━Vector Cross-Covariance

*The **cross-covariance of a pair of random vectors** $\mathbf{X}$ with n components and $\mathbf{Y}$ with m components is an $n \times m$ matrix $\mathbf{C_{XY}}$ with i, jth element $C_{XY}(i,j) = \mathrm{Cov}[X_i, Y_j]$, or, in vector notation,*

$$\mathbf{C_{XY}} = \mathrm{E}\left[(\mathbf{X} - \boldsymbol{\mu_X})(\mathbf{Y} - \boldsymbol{\mu_Y})'\right].$$

To distinguish the correlation or covariance of a random vector from the correlation or covariance of a pair of random vectors, we sometimes use the terminology *autocorrelation* and *autocovariance* when there is one random vector and *cross-correlation* and *cross-covariance* when there is a pair of random vectors. Note that when $\mathbf{X} = \mathbf{Y}$ the autocorrelation and cross-correlation are identical (as are the covariances). Recognizing this identity, some texts use the notation $\mathbf{R_{XX}}$ and $\mathbf{C_{XX}}$ for the correlation and covariance of a random vector.

When $\mathbf{Y}$ is a linear transformation of $\mathbf{X}$, the following theorem states the relationship of the second-order statistics of $\mathbf{Y}$ to the corresponding statistics of $\mathbf{X}$.

━━━━━Theorem 8.8━━━━━

$\mathbf{X}$ *is an n-dimensional random vector with expected value $\boldsymbol{\mu_X}$, correlation $\mathbf{R_X}$, and covariance $\mathbf{C_X}$. The m-dimensional random vector $\mathbf{Y} = \mathbf{AX} + \mathbf{b}$, where $\mathbf{A}$ is an $m \times n$ matrix and $\mathbf{b}$ is an m-dimensional vector, has expected value $\boldsymbol{\mu_Y}$, correlation matrix $\mathbf{R_Y}$, and covariance matrix $\mathbf{C_Y}$ given by*

$$\boldsymbol{\mu_Y} = \mathbf{A}\boldsymbol{\mu_X} + \mathbf{b},$$
$$\mathbf{R_Y} = \mathbf{A R_X A'} + (\mathbf{A}\boldsymbol{\mu_X})\mathbf{b'} + \mathbf{b}(\mathbf{A}\boldsymbol{\mu_X})' + \mathbf{b b'},$$
$$\mathbf{C_Y} = \mathbf{A C_X A'}.$$

Proof We derive the formulas for the expected value and covariance of $\mathbf{Y}$. The derivation for the correlation is similar. First, the expected value of $\mathbf{Y}$ is

$$\boldsymbol{\mu_Y} = \mathrm{E}\left[\mathbf{AX} + \mathbf{b}\right] = \mathbf{A}\,\mathrm{E}\left[\mathbf{X}\right] + \mathrm{E}\left[\mathbf{b}\right] = \mathbf{A}\boldsymbol{\mu_X} + \mathbf{b}. \tag{8.33}$$

It follows that $\mathbf{Y} - \boldsymbol{\mu_Y} = \mathbf{A}(\mathbf{X} - \boldsymbol{\mu_X})$. This implies

$$\begin{aligned}
\mathbf{C_Y} &= \mathrm{E}\left[(\mathbf{A}(\mathbf{X} - \boldsymbol{\mu_X}))(\mathbf{A}(\mathbf{X} - \boldsymbol{\mu_X}))'\right] \\
&= \mathrm{E}\left[\mathbf{A}(\mathbf{X} - \boldsymbol{\mu_X}))(\mathbf{X} - \boldsymbol{\mu_X})'\mathbf{A'}\right] = \mathbf{A}\,\mathrm{E}\left[(\mathbf{X} - \boldsymbol{\mu_X})(\mathbf{X} - \boldsymbol{\mu_X})'\right]\mathbf{A'} = \mathbf{A C_X A'}. \end{aligned} \tag{8.34}$$

========**Example 8.8**========

Given the expected value $\boldsymbol{\mu_X}$, the correlation $\mathbf{R_X}$, and the covariance $\mathbf{C_X}$ of random vector $\mathbf{X}$ in Example 8.7, and $\mathbf{Y} = \mathbf{AX} + \mathbf{b}$, where

$$\mathbf{A} = \begin{bmatrix} 1 & 0 \\ 6 & 3 \\ 3 & 6 \end{bmatrix} \quad \text{and} \quad \mathbf{b} = \begin{bmatrix} 0 \\ -2 \\ -2 \end{bmatrix}, \tag{8.35}$$

find the expected value $\boldsymbol{\mu_Y}$, the correlation $\mathbf{R_Y}$, and the covariance $\mathbf{C_Y}$.

From the matrix operations of Theorem 8.8, we obtain $\boldsymbol{\mu_Y} = \begin{bmatrix} 1/3 & 2 & 3 \end{bmatrix}'$ and

$$\mathbf{R_Y} = \begin{bmatrix} 1/6 & 13/12 & 4/3 \\ 13/12 & 7.5 & 9.25 \\ 4/3 & 9.25 & 12.5 \end{bmatrix}; \quad \mathbf{C_Y} = \begin{bmatrix} 1/18 & 5/12 & 1/3 \\ 5/12 & 3.5 & 3.25 \\ 1/3 & 3.25 & 3.5 \end{bmatrix}. \tag{8.36}$$

The cross-correlation and cross-covariance of two random vectors can be derived using algebra similar to the proof of Theorem 8.8.

========**Theorem 8.9**========

The vectors $\mathbf{X}$ *and* $\mathbf{Y} = \mathbf{AX} + \mathbf{b}$ *have cross-correlation* $\mathbf{R_{XY}}$ *and cross-covariance* $\mathbf{C_{XY}}$ *given by*

$$\mathbf{R_{XY}} = \mathbf{R_X A'} + \boldsymbol{\mu_X} \mathbf{b'}, \qquad \mathbf{C_{XY}} = \mathbf{C_X A'}.$$

In the next example, we see that covariance and cross-covariance matrices allow us to quickly calculate the correlation coefficient between any pair of component random variables.

========**Example 8.9**========

Continuing Example 8.8 for random vectors $\mathbf{X}$ and $\mathbf{Y} = \mathbf{AX} + \mathbf{b}$, calculate

(a) The cross-correlation matrix $\mathbf{R_{XY}}$ and the cross-covariance matrix $\mathbf{C_{XY}}$.

(b) The correlation coefficients ρ_{Y_1,Y_3} and ρ_{X_2,Y_1}.

(a) Direct matrix calculation using Theorem 8.9 yields

$$\mathbf{R_{XY}} = \begin{bmatrix} 1/6 & 13/12 & 4/3 \\ 1/4 & 5/3 & 29/12 \end{bmatrix}; \quad \mathbf{C_{XY}} = \begin{bmatrix} 1/18 & 5/12 & 1/3 \\ 1/36 & 1/3 & 5/12 \end{bmatrix}. \tag{8.37}$$

(b) Referring to Definition 5.6 and recognizing that $\mathrm{Var}[Y_i] = C_\mathbf{Y}(i,i)$, we have

$$\rho_{Y_1,Y_3} = \frac{\mathrm{Cov}\,[Y_1, Y_3]}{\sqrt{\mathrm{Var}[Y_1]\,\mathrm{Var}[Y_3]}} = \frac{C_Y(1,3)}{\sqrt{C_Y(1,1)C_Y(3,3)}} = 0.756 \tag{8.38}$$

Similarly,

$$\rho_{X_2,Y_1} = \frac{\mathrm{Cov}\,[X_2, Y_1]}{\sqrt{\mathrm{Var}[X_2]\,\mathrm{Var}[Y_1]}} = \frac{C_{\mathbf{XY}}(2,1)}{\sqrt{C_\mathbf{X}(2,2)C_\mathbf{Y}(1,1)}} = 1/2. \tag{8.39}$$

Quiz 8.4

The three-dimensional random vector $\mathbf{X} = \begin{bmatrix} X_1 & X_2 & X_3 \end{bmatrix}'$ has PDF

$$f_\mathbf{X}(\mathbf{x}) = \begin{cases} 6 & 0 \le x_1 \le x_2 \ \le x_3 \le 1, \\ 0 & \text{otherwise.} \end{cases} \tag{8.40}$$

Find $E[\mathbf{X}]$ and the correlation and covariance matrices $\mathbf{R_X}$ and $\mathbf{C_X}$.

8.5   Gaussian Random Vectors

> The multivariate Gaussian PDF is a probability model for a vector in which all the components are Gaussian random variables. The parameters of the model are the expected value vector and the covariance matrix of the components. A linear function of a Gaussian random vector is also a Gaussian random vector. The components of the standard normal random vector are mutually independent standard normal random variables.

Multiple Gaussian random variables appear in many practical applications of probability theory. The *multivariate Gaussian distribution* is a probability model for n random variables with the property that the marginal PDFs are all Gaussian. A set of random variables described by the multivariate Gaussian PDF is said to be *jointly Gaussian*. A vector whose components are jointly Gaussian random variables is said to be a *Gaussian random vector*. The PDF of a Gaussian random vector has a particularly concise notation.

Definition 8.12   **Gaussian Random Vector**

$\mathbf{X}$ *is the Gaussian* $(\boldsymbol{\mu}_\mathbf{X}, \mathbf{C_X})$ *random vector with expected value* $\boldsymbol{\mu}_\mathbf{X}$ *and covariance* $\mathbf{C_X}$ *if and only if*

$$f_\mathbf{X}(\mathbf{x}) = \frac{1}{(2\pi)^{n/2}[det(\mathbf{C_X})]^{1/2}} \exp\left(-\frac{1}{2}(\mathbf{x} - \boldsymbol{\mu}_\mathbf{X})'\mathbf{C_X}^{-1}(\mathbf{x} - \boldsymbol{\mu}_\mathbf{X}) \right)$$

where $det(\mathbf{C_X})$, *the determinant of* $\mathbf{C_X}$, *satisfies* $det(\mathbf{C_X}) > 0$.

Definition 8.12 is a generalization of Definition 4.8 and Definition 5.10. When $n = 1$, $\mathbf{C_X}$ and $\mathbf{x} - \boldsymbol{\mu_X}$ are σ_X^2 and $x - \mu_X$, and the PDF in Definition 8.12 reduces to the ordinary Gaussian PDF of Definition 4.8. That is, a 1-dimensional Gaussian (μ, σ^2) random vector is a Gaussian (μ, σ) random variable.[1] In Problem 8.5.8, we ask you to show that for $n = 2$, Definition 8.12 reduces to the bivariate Gaussian PDF in Definition 5.10. The condition that $\det(\mathbf{C_X}) > 0$ is a generalization of the requirement for the bivariate Gaussian PDF that $|\rho_{X,Y}| < 1$. Basically, $\det(\mathbf{C_X}) > 0$ reflects the requirement that no random variable X_i is a linear combination of the other random variables in $\mathbf{X}$.

For a Gaussian random vector $\mathbf{X}$, an important special case is $\text{Cov}[X_i, X_j] = 0$ for all $i \neq j$. In the covariance matrix $\mathbf{C_X}$, the off-diagonal elements are all zero and the ith diagonal element is simply $\text{Var}[X_i] = \sigma_i^2$. In this case, we write $\mathbf{C_X} = \text{diag}[\sigma_1^2, \sigma_2^2, \ldots, \sigma_n^2]$. When the covariance matrix is diagonal, X_i and X_j are uncorrelated for $i \neq j$. In Theorem 5.20, we showed that uncorrelated bivariate Gaussian random variables are independent. The following theorem generalizes this result.

Theorem 8.10

A Gaussian random vector $\mathbf{X}$ has independent components if and only if $\mathbf{C_X}$ is a diagonal matrix.

Proof First, if the components of $\mathbf{X}$ are independent, then for $i \neq j$, X_i and X_j are independent. By Theorem 5.17(c), $\text{Cov}[X_i, X_j] = 0$. Hence the off-diagonal terms of $\mathbf{C_X}$ are all zero. If $\mathbf{C_X}$ is diagonal, then

$$\mathbf{C_X} = \begin{bmatrix} \sigma_1^2 & & \\ & \ddots & \\ & & \sigma_n^2 \end{bmatrix} \quad \text{and} \quad \mathbf{C_X^{-1}} = \begin{bmatrix} 1/\sigma_1^2 & & \\ & \ddots & \\ & & 1/\sigma_n^2 \end{bmatrix}. \tag{8.41}$$

It follows that $\mathbf{C_X}$ has determinant $\det(\mathbf{C_X}) = \prod_{i=1}^n \sigma_i^2$ and that

$$(\mathbf{x} - \boldsymbol{\mu_X})'\mathbf{C_X^{-1}}(\mathbf{x} - \boldsymbol{\mu_X}) = \sum_{i=1}^n \frac{(X_i - \mu_i)^2}{\sigma_i^2}. \tag{8.42}$$

From Definition 8.12, we see that

$$f_{\mathbf{X}}(\mathbf{x}) = \frac{1}{(2\pi)^{n/2} \prod_{i=1}^n \sigma_i^2} \exp\left(-\sum_{i=1}^n (x_i - \mu_i)/2\sigma_i^2\right) \tag{8.43}$$

$$= \prod_{i=1}^n \frac{1}{\sqrt{2\pi\sigma_i^2}} \exp\left(-(x_i - \mu_i)^2/2\sigma_i^2\right). \tag{8.44}$$

Thus $f_{\mathbf{X}}(\mathbf{x}) = \prod_{i=1}^n f_{X_i}(x_i)$, implying $X_1, \ldots, X_n$ are independent.

---

[1]For the Gaussian random variable, we specify parameters μ and σ because they have the same units. However, the PDF of the Gaussian random vector displays $\boldsymbol{\mu}_X$ and $\mathbf{C_X}$ as parameters, and for one dimension $\mathbf{C_X} = \sigma_X^2$

═══Example 8.10═══

Consider the outdoor temperature at a certain weather station. On May 5, the temperature measurements in units of degrees Fahrenheit taken at 6 AM, 12 noon, and 6 PM are all Gaussian random variables, X_1, X_2, X_3, with variance 16 degrees^2. The expected values are 50 degrees, 62 degrees, and 58 degrees respectively. The covariance matrix of the three measurements is

$$\mathbf{C_X} = \begin{bmatrix} 16.0 & 12.8 & 11.2 \\ 12.8 & 16.0 & 12.8 \\ 11.2 & 12.8 & 16.0 \end{bmatrix}. \tag{8.45}$$

(a) Write the joint PDF of X_1, X_2 using the algebraic notation of Definition 5.10.
(b) Write the joint PDF of X_1, X_2 using vector notation.
(c) Write the joint PDF of $\mathbf{X} = \begin{bmatrix} X_1 & X_2 & X_3 \end{bmatrix}'$ using vector notation.

. . . .

(a) First we note that X_1 and X_2 have expected values $\mu_1 = 50$ and $\mu_2 = 62$, variances $\sigma_1^2 = \sigma_2^2 = 16$, and covariance $\text{Cov}[X_1, X_2] = 12.8$. It follows from Definition 5.6 that the correlation coefficient is

$$\rho_{X_1, X_2} = \frac{\text{Cov}[X_1, X_2]}{\sigma_1 \sigma_2} = \frac{12.8}{16} = 0.8. \tag{8.46}$$

From Definition 5.10, the joint PDF is

$$f_{X_1, X_2}(x_1, x_2) = \frac{\exp\left[-\dfrac{(x_1 - 50)^2 - 1.6(x_1 - 50)(x_2 - 62) + (x_2 - 62)^2}{19.2} \right]}{60.3}.$$

(b) Let $\mathbf{W} = \begin{bmatrix} X_1 & X_2 \end{bmatrix}'$ denote a vector representation for random variables X_1 and X_2. From the covariance matrix $\mathbf{C_X}$, we observe that the 2×2 submatrix in the upper left corner is the covariance matrix of the random vector $\mathbf{W}$. Thus

$$\mu_{\mathbf{W}} = \begin{bmatrix} 50 \\ 62 \end{bmatrix}, \qquad \mathbf{C_W} = \begin{bmatrix} 16.0 & 12.8 \\ 12.8 & 16.0 \end{bmatrix}. \tag{8.47}$$

We observe that $\det(\mathbf{C_W}) = 92.16$ and $\det(\mathbf{C_W})^{1/2} = 9.6$. From Definition 8.12, the joint PDF of $\mathbf{W}$ is

$$f_{\mathbf{W}}(\mathbf{w}) = \frac{1}{60.3} \exp\left(-\frac{1}{2}(\mathbf{w} - \mu_{\mathbf{W}})^T \mathbf{C_W}^{-1}(\mathbf{w} - \mu_{\mathbf{W}}) \right). \tag{8.48}$$

(c) Since $\mu_{\mathbf{X}} = \begin{bmatrix} 50 & 62 & 58 \end{bmatrix}'$ and $\det(\mathbf{C_X})^{1/2} = 22.717$, $\mathbf{X}$ has PDF

$$f_{\mathbf{X}}(\mathbf{x}) = \frac{1}{357.8} \exp\left(-\frac{1}{2}(\mathbf{x} - \mu_{\mathbf{X}})^T \mathbf{C_X}^{-1}(\mathbf{x} - \mu_{\mathbf{X}}) \right). \tag{8.49}$$

The following theorem is a generalization of Theorem 4.13. It states that a linear transformation of a Gaussian random vector results in another Gaussian random vector.

═══════**Theorem 8.11**═══════

Given an n-dimensional Gaussian random vector $\mathbf{X}$ with expected value $\boldsymbol{\mu}_{\mathbf{X}}$ and covariance $\mathbf{C}_{\mathbf{X}}$, and an $m \times n$ matrix $\mathbf{A}$ with $\text{rank}(\mathbf{A}) = m$,

$$\mathbf{Y} = \mathbf{A}\mathbf{X} + \mathbf{b}$$

is an m-dimensional Gaussian random vector with expected value $\boldsymbol{\mu}_{\mathbf{Y}} = \mathbf{A}\boldsymbol{\mu}_{\mathbf{X}} + \mathbf{b}$ and covariance $\mathbf{C}_{\mathbf{Y}} = \mathbf{A}\mathbf{C}_{\mathbf{X}}\mathbf{A}'$.

Proof The proof of Theorem 8.8 contains the derivations of $\boldsymbol{\mu}_{\mathbf{Y}}$ and $\mathbf{C}_{\mathbf{Y}}$. Our proof that $\mathbf{Y}$ has a Gaussian PDF is confined to the special case when $m = n$ and $\mathbf{A}$ is an invertible matrix. The case of $m < n$ is addressed in Problem 8.5.14. When $m = n$, we use Theorem 8.6 to write

$$f_{\mathbf{Y}}(\mathbf{y}) = \frac{1}{|\det(\mathbf{A})|} f_{\mathbf{X}}\left(\mathbf{A}^{-1}(\mathbf{y} - \mathbf{b})\right) \tag{8.50}$$

$$= \frac{\exp\left(-\frac{1}{2}[\mathbf{A}^{-1}(\mathbf{y} - \mathbf{b}) - \boldsymbol{\mu}_{\mathbf{X}}]'\mathbf{C}_{\mathbf{X}}^{-1}[\mathbf{A}^{-1}(\mathbf{y} - \mathbf{b}) - \boldsymbol{\mu}_{\mathbf{X}}]\right)}{(2\pi)^{n/2}|\det(\mathbf{A})||\det(\mathbf{C}_{\mathbf{X}})|^{1/2}}. \tag{8.51}$$

In the exponent of $f_{\mathbf{Y}}(\mathbf{y})$, we observe that

$$\mathbf{A}^{-1}(\mathbf{y} - \mathbf{b}) - \boldsymbol{\mu}_{\mathbf{X}} = \mathbf{A}^{-1}[\mathbf{y} - (\mathbf{A}\boldsymbol{\mu}_{\mathbf{X}} + \mathbf{b})] = \mathbf{A}^{-1}(\mathbf{y} - \boldsymbol{\mu}_{\mathbf{Y}}), \tag{8.52}$$

since $\boldsymbol{\mu}_{\mathbf{Y}} = \mathbf{A}\boldsymbol{\mu}_{\mathbf{X}} + \mathbf{b}$. Applying (8.52) to (8.51) yields

$$f_{\mathbf{Y}}(\mathbf{y}) = \frac{\exp\left(-\frac{1}{2}[\mathbf{A}^{-1}(\mathbf{y} - \boldsymbol{\mu}_{\mathbf{Y}})]'\mathbf{C}_{\mathbf{X}}^{-1}[\mathbf{A}^{-1}(\mathbf{y} - \boldsymbol{\mu}_{\mathbf{Y}})]\right)}{(2\pi)^{n/2}|\det(\mathbf{A})||\det(\mathbf{C}_{\mathbf{X}})|^{1/2}}. \tag{8.53}$$

Using the identities $|\det(\mathbf{A})||\det(\mathbf{C}_{\mathbf{X}})|^{1/2} = |\det(\mathbf{A}\mathbf{C}_{\mathbf{X}}\mathbf{A}')|^{1/2}$ and $(\mathbf{A}^{-1})' = (\mathbf{A}')^{-1}$, we can write

$$f_{\mathbf{Y}}(\mathbf{y}) = \frac{\exp\left(-\frac{1}{2}(\mathbf{y} - \boldsymbol{\mu}_{\mathbf{Y}})'(\mathbf{A}')^{-1}\mathbf{C}_{\mathbf{X}}^{-1}\mathbf{A}^{-1}(\mathbf{y} - \boldsymbol{\mu}_{\mathbf{Y}})\right)}{(2\pi)^{n/2}|\det(\mathbf{A}\mathbf{C}_{\mathbf{X}}\mathbf{A}')|^{1/2}}. \tag{8.54}$$

Since $(\mathbf{A}')^{-1}\mathbf{C}_{\mathbf{X}}^{-1}\mathbf{A}^{-1} = (\mathbf{A}\mathbf{C}_{\mathbf{X}}\mathbf{A}')^{-1}$, we see from Equation (8.54) that $\mathbf{Y}$ is a Gaussian vector with expected value $\boldsymbol{\mu}_{\mathbf{Y}}$ and covariance matrix $\mathbf{C}_{\mathbf{Y}} = \mathbf{A}\mathbf{C}_{\mathbf{X}}\mathbf{A}'$.

═══════**Example 8.11**═══════

Continuing Example 8.10, use the formula $Y_i = (5/9)(X_i - 32)$ to convert the three temperature measurements to degrees Celsius.

(a) What is $\boldsymbol{\mu}_{\mathbf{Y}}$, the expected value of random vector $\mathbf{Y}$?

(b) What is $\mathbf{C}_{\mathbf{Y}}$, the covariance of random vector $\mathbf{Y}$?

(c) Write the joint PDF of $\mathbf{Y} = \begin{bmatrix} Y_1 & Y_2 & Y_3 \end{bmatrix}'$ using vector notation.

(a) In terms of matrices, we observe that $\mathbf{Y} = \mathbf{AX} + \mathbf{b}$ where

$$\mathbf{A} = \begin{bmatrix} 5/9 & 0 & 0 \\ 0 & 5/9 & 0 \\ 0 & 0 & 5/9 \end{bmatrix}, \qquad \mathbf{b} = -\frac{160}{9} \begin{bmatrix} 1 \\ 1 \\ 1 \end{bmatrix}. \qquad (8.55)$$

(b) Since $\boldsymbol{\mu}_{\mathbf{X}} = \begin{bmatrix} 50 & 62 & 58 \end{bmatrix}'$, from Theorem 8.11,

$$\boldsymbol{\mu}_{\mathbf{Y}} = \mathbf{A}\boldsymbol{\mu}_{\mathbf{X}} + \mathbf{b} = \begin{bmatrix} 10 \\ 50/3 \\ 130/9 \end{bmatrix}. \qquad (8.56)$$

(c) The covariance of $\mathbf{Y}$ is $\mathbf{C}_{\mathbf{Y}} = \mathbf{AC}_{\mathbf{X}}\mathbf{A}'$. We note that $\mathbf{A} = \mathbf{A}' = (5/9)\mathbf{I}$ where $\mathbf{I}$ is the 3×3 identity matrix. Thus $\mathbf{C}_{\mathbf{Y}} = (5/9)^2\mathbf{C}_{\mathbf{X}}$ and $\mathbf{C}_{\mathbf{Y}}^{-1} = (9/5)^2\mathbf{C}_{\mathbf{X}}^{-1}$. The PDF of $\mathbf{Y}$ is

$$f_{\mathbf{Y}}(\mathbf{y}) = \frac{1}{24.47} \exp\left(-\frac{81}{50}(\mathbf{y} - \boldsymbol{\mu}_{\mathbf{Y}})^T \mathbf{C}_{\mathbf{X}}^{-1}(\mathbf{y} - \boldsymbol{\mu}_{\mathbf{Y}}) \right). \qquad (8.57)$$

A standard normal random vector is a generalization of the standard normal random variable in Definition 4.9.

Definition 8.13   Standard Normal Random Vector

*The n-dimensional **standard normal random vector** $\mathbf{Z}$ is the n-dimensional Gaussian random vector with $\mathrm{E}[\mathbf{Z}] = \mathbf{0}$ and $\mathbf{C}_{\mathbf{Z}} = \mathbf{I}$.*

From Definition 8.13, each component Z_i of $\mathbf{Z}$ has expected value $\mathrm{E}[Z_i] = 0$ and variance $\mathrm{Var}[Z_i] = 1$. Thus Z_i is the Gaussian $(0, 1)$ random variable. In addition, $\mathrm{E}[Z_i Z_j] = 0$ for all $i \neq j$. Since $\mathbf{C}_{\mathbf{Z}}$ is a diagonal matrix, $Z_1, \ldots, Z_n$ are independent.

In many situations, it is useful to transform the Gaussian (μ_X, σ_X) random variable X to the standard normal random variable $Z = (X - \mu_X)/\sigma_X$. For Gaussian vectors, we have a vector transformation to transform $\mathbf{X}$ into a standard normal random vector.

Theorem 8.12

For a Gaussian $(\boldsymbol{\mu}_{\mathbf{X}}, \mathbf{C}_{\mathbf{X}})$ random vector, let $\mathbf{A}$ be an $n \times n$ matrix with the property $\mathbf{AA}' = \mathbf{C}_{\mathbf{X}}$. The random vector

$$\mathbf{Z} = \mathbf{A}^{-1}(\mathbf{X} - \boldsymbol{\mu}_{\mathbf{X}})$$

is a standard normal random vector.

Proof Applying Theorem 8.11 with $\mathbf{A}$ replaced by $\mathbf{A}^{-1}$, and $\mathbf{b} = \mathbf{A}^{-1}\boldsymbol{\mu}_{\mathbf{X}}$, we have that

$\mathbf{Z}$ is a Gaussian random vector with expected value

$$\mathrm{E}\left[\mathbf{Z}\right] = \mathrm{E}\left[\mathbf{A}^{-1}(\mathbf{X} - \boldsymbol{\mu_X})\right] = \mathbf{A}^{-1}\,\mathrm{E}\left[\mathbf{X} - \boldsymbol{\mu_X}\right] = \mathbf{0} \tag{8.58}$$

and covariance

$$\mathbf{C_Z} = \mathbf{A}^{-1}\mathbf{C_X}(\mathbf{A}^{-1})' = \mathbf{A}^{-1}\mathbf{A}\mathbf{A}'(\mathbf{A}')^{-1} = \mathbf{I}. \tag{8.59}$$

The transformation in this theorem is considerably less straightforward than the scalar transformation $Z = (X - \mu_X)/\sigma_X$, because it is necessary to find for a given $\mathbf{C_X}$ a matrix $\mathbf{A}$ with the property $\mathbf{A}\mathbf{A}' = \mathbf{C_X}$. The calculation of $\mathbf{A}$ from $\mathbf{C_X}$ can be achieved by applying the linear algebra procedure *singular value decomposition*. Section 8.6 describes this procedure in more detail and applies it to generating sample values of Gaussian random vectors.

The inverse transform of Theorem 8.12 is particularly useful in computer simulations.

=====Theorem 8.13=====

Given the n-dimensional standard normal random vector $\mathbf{Z}$, an invertible $n \times n$ matrix $\mathbf{A}$, and an n-dimensional vector $\mathbf{b}$,

$$\mathbf{X} = \mathbf{A}\mathbf{Z} + \mathbf{b}$$

is an n-dimensional Gaussian random vector with expected value $\boldsymbol{\mu_X} = \mathbf{b}$ and covariance matrix $\mathbf{C_X} = \mathbf{A}\mathbf{A}'$.

Proof By Theorem 8.11, $\mathbf{X}$ is a Gaussian random vector with expected value

$$\boldsymbol{\mu_X} = \mathrm{E}\left[\mathbf{X}\right] = \mathrm{E}\left[\mathbf{A}\mathbf{Z} + \boldsymbol{\mu_X}\right] = \mathbf{A}\,\mathrm{E}\left[\mathbf{Z}\right] + \mathbf{b} = \mathbf{b}. \tag{8.60}$$

The covariance of $\mathbf{X}$ is

$$\mathbf{C_X} = \mathbf{A}\mathbf{C_Z}\mathbf{A}' = \mathbf{A}\mathbf{I}\mathbf{A}' = \mathbf{A}\mathbf{A}'. \tag{8.61}$$

Theorem 8.13 says that we can transform the standard normal vector $\mathbf{Z}$ into a Gaussian random vector $\mathbf{X}$ whose covariance matrix is of the form $\mathbf{C_X} = \mathbf{A}\mathbf{A}'$. The usefulness of Theorems 8.12 and 8.13 depends on whether we can always find a matrix $\mathbf{A}$ such that $\mathbf{C_X} = \mathbf{A}\mathbf{A}'$. In fact, as we verify below, this is possible for every Gaussian vector $\mathbf{X}$.

=====Theorem 8.14=====

For a Gaussian vector $\mathbf{X}$ with covariance $\mathbf{C_X}$, there always exists a matrix $\mathbf{A}$ such that $\mathbf{C_X} = \mathbf{A}\mathbf{A}'$.

Proof To verify this fact, we connect some simple facts:

- In Problem 8.4.12, we ask you to show that every random vector $\mathbf{X}$ has a positive semidefinite covariance matrix $\mathbf{C_X}$. By Math Fact B.17, every eigenvalue of $\mathbf{C_X}$ is nonnegative.

- The definition of the Gaussian vector PDF requires the existence of $\mathbf{C_X}^{-1}$. Hence, for a Gaussian vector $\mathbf{X}$, all eigenvalues of $\mathbf{C_X}$ are nonzero. From the previous step, we observe that all eigenvalues of $\mathbf{C_X}$ must be positive.

- Since $\mathbf{C_X}$ is a real symmetric matrix, Math Fact B.15 says it has a singular value decomposition (SVD) $\mathbf{C_X} = \mathbf{UDU}'$ where $\mathbf{D} = \text{diag}[d_1, \ldots, d_n]$ is the diagonal matrix of eigenvalues of $\mathbf{C_X}$. Since each d_i is positive, we can define $\mathbf{D}^{1/2} = \text{diag}[\sqrt{d_1}, \ldots, \sqrt{d_n}]$, and we can write

$$\mathbf{C_X} = \mathbf{UD}^{1/2}\mathbf{D}^{1/2}\mathbf{U}' = \left(\mathbf{UD}^{1/2}\right)\left(\mathbf{UD}^{1/2}\right)'. \tag{8.62}$$

We see that $\mathbf{A} = \mathbf{UD}^{1/2}$.

From Theorems 8.12, 8.13, and 8.14, it follows that any Gaussian $(\boldsymbol{\mu_X}, \mathbf{C_X})$ random vector $\mathbf{X}$ can be written as a linear transformation of uncorrelated Gaussian $(0, 1)$ random variables. In terms of the SVD $\mathbf{C_X} = \mathbf{UDU}'$ and the standard normal vector $\mathbf{Z}$, the transformation is

$$\mathbf{X} = \mathbf{UD}^{1/2}\mathbf{Z} + \boldsymbol{\mu_X}. \tag{8.63}$$

We recall that $\mathbf{U}$ has orthonormal columns $\mathbf{u}_1, \ldots, \mathbf{u}_n$. When $\boldsymbol{\mu_X} = \mathbf{0}$, Equation (8.63) can be written as

$$\mathbf{X} = \sum_{i=1}^{n} \sqrt{d_i}\mathbf{u}_i Z_i. \tag{8.64}$$

The interpretation of Equation (8.64) is that a Gaussian random vector $\mathbf{X}$ is a combination of orthogonal vectors $\sqrt{d_i}\mathbf{u}_i$, each scaled by an independent Gaussian random variable Z_i. In a wide variety of problems involving Gaussian random vectors, the transformation from the Gaussian vector $\mathbf{X}$ to the standard normal random vector $\mathbf{Z}$ is the key to an efficient solution. Also, we will see in the next section that Theorem 8.13 is essential in using MATLAB to generate arbitrary Gaussian random vectors.

Quiz 8.5

$\mathbf{Z}$ is the two-dimensional standard normal random vector. The Gaussian random vector $\mathbf{X}$ has components

$$X_1 = 2Z_1 + Z_2 + 2 \quad \text{and} \quad X_2 = Z_1 - Z_2. \tag{8.65}$$

Calculate the expected value vector $\boldsymbol{\mu_X}$ and the covariance matrix $\mathbf{C_X}$.

8.6   MATLAB

MATLAB is especially useful for random vectors. We use a sample space grid to calculate properties of a probability model of a discrete random vector. We use the functions `randn` and `svd` to generate samples of Gaussian random vectors.

As in Section 5.11, we demonstrate two ways of using MATLAB to study random vectors. We first present examples of programs that calculate values of probability functions, in this case the PMF of a discrete random vector and the PDF of a Gaussian random vector. Then we present a program that generates sample values of the Gaussian $(\mu_{\mathbf{X}}, \mathbf{C}_{\mathbf{X}})$ random vector given any $\mu_{\mathbf{X}}$ and $\mathbf{C}_{\mathbf{X}}$.

Probability Functions

The MATLAB approach of using a sample space grid, presented in Section 5.11, can also be applied to finite random vectors $\mathbf{X}$ described by a PMF $P_{\mathbf{X}}(\mathbf{x})$.

====Example 8.12====

Finite random vector $\mathbf{X} = \begin{bmatrix} X_1 & X_2, \cdots & X_5 \end{bmatrix}'$ has PMF

$$P_{\mathbf{X}}(\mathbf{x}) = \begin{cases} k\sqrt{\mathbf{x}'\mathbf{x}} & x_i \in \{-10, -9, \ldots, 10\}\,; \\ & i = 1, 2, \ldots, 5, \\ 0 & \text{otherwise.} \end{cases} \tag{8.66}$$

What is the constant k? Find the expected value and standard deviation of X_3.

Summing $P_{\mathbf{X}}(\mathbf{x})$ over all possible values of $\mathbf{x}$ is the sort of tedious task that MATLAB handles easily. Here are the code and corresponding output:

```
%x5.m
sx=-10:10;
[SX1,SX2,SX3,SX4,SX5]...
    =ndgrid(sx,sx,sx,sx,sx);
P=sqrt(SX1.^2 +SX2.^2+SX3.^2+SX4.^2+SX5.^2);
k=1.0/(sum(sum(sum(sum(sum(P))))));
P=k*P;
EX3=sum(sum(sum(sum(sum(P.*SX3)))));
EX32=sum(sum(sum(sum(sum(P.*(SX3.^2))))));
sigma3=sqrt(EX32-(EX3)^2)
```

```
>> x5
k =
   1.8491e-008
EX3 =
  -3.2960e-017
sigma3 =
   6.3047
>>
```

In fact, by symmetry arguments, it should be clear that $\mathrm{E}[X_3] = 0$. In adding 11^5 terms, MATLAB's finite precision led to a small error on the order of 10^{-17}.

Example 8.12 demonstrates the use of MATLAB to calculate properties of a probability model by performing lots of straightforward calculations. For a continuous random vector $\mathbf{X}$, MATLAB could be used to calculate $\mathrm{E}[g(\mathbf{X})]$ using Theorem 8.3 and numeric integration. One step in such a calculation is computing values of the PDF. The next example performs this function for any Gaussian $(\mu_{\mathbf{X}}, \mathbf{C}_{\mathbf{X}})$ random vector.

====Example 8.13====

Write a MATLAB function `f=gaussvectorpdf(mu,C,x)` that calculates $f_{\mathbf{X}}(\mathbf{x})$ for a Gaussian $(\mu, \mathbf{C})$ random vector.

```
function f=gaussvectorpdf(mu,C,x)
n=length(x);
z=x(:)-mu(:);
f=exp(-z'*inv(C)*z)/...
    sqrt((2*pi)^n*det(C));
```

gaussvectorpdf computes the Gaussian vector PDF $f_{\mathbf{X}}(\mathbf{x})$ of Definition 8.12. Of course, MATLAB makes the calculation simple by providing operators for matrix inverses and determinants.

Sample Values of Gaussian Random Vectors

Gaussian random vectors appear in a wide variety of experiments. Here we present a program that uses the built-in MATLAB function randn to generate sample values of Gaussian $(\boldsymbol{\mu}_{\mathbf{X}}, \mathbf{C}_{\mathbf{X}})$ random vectors. The matrix notation lends itself to concise MATLAB coding. Our approach is based on Theorem 8.13. In particular, we generate a standard normal random vector $\mathbf{Z}$ and, given a covariance matrix $\mathbf{C}$, we use built-in MATLAB functions to calculate a matrix $\mathbf{A}$ such that $\mathbf{C} = \mathbf{AA}'$. By Theorem 8.13, $\mathbf{X} = \mathbf{AZ} + \boldsymbol{\mu}_{\mathbf{X}}$ is a Gaussian $(\boldsymbol{\mu}_{\mathbf{X}}, \mathbf{C})$ vector. Although the MATLAB code for this task will be quite short, it needs some explanation:

- x=randn(m,n) produces an $m \times n$ matrix, with each matrix element a Gaussian $(0, 1)$ random variable. Thus each column of x is a sample vector of standard normal vector $\mathbf{Z}$.

- [U,D,V]=svd(C) is the singular value decomposition (SVD) of matrix C. In math notation, given $\mathbf{C}$, svd produces a diagonal matrix $\mathbf{D}$ of the same dimension as $\mathbf{C}$ and with nonnegative diagonal elements in decreasing order, and unitary matrices $\mathbf{U}$ and $\mathbf{V}$ so that $\mathbf{C} = \mathbf{UDV}'$. Singular value decomposition is a powerful technique that can be applied to any matrix. When $\mathbf{C}$ is a covariance matrix, the singular value decomposition yields $\mathbf{U} = \mathbf{V}$ and $\mathbf{C} = \mathbf{UDU}'$. Just as in the proof of Theorem 8.14, $\mathbf{A} = \mathbf{UD}^{1/2}$.

```
function x=gaussvector(mu,C,m)
[U,D,V]=svd(C);
x=V*(D^(0.5))*randn(n,m)...
    +(mu(:)*ones(1,m));
```

Using MATLAB functions randn and svd, generating Gaussian random vectors is easy. The function x=gaussvector(mu,C,1) produces a Gaussian (mu, C) random vector.

The general form gaussvector(mu,C,m) produces an $n \times m$ matrix where each of the m columns is a Gaussian random vector with expected value mu and covariance C. The reason for defining gaussvector to return m vectors at the same time is that calculating the singular value decomposition is a computationally burdensome step. Instead, we perform the SVD just once, rather than m times.

Quiz 8.6

The daily noon temperature, measured in degrees Fahrenheit, in New Jersey in July can be modeled as a Gaussian random vector $\mathbf{T} = \begin{bmatrix} T_1 & \cdots & T_{31} \end{bmatrix}'$ where T_i is the temperature on the ith day of the month. Suppose that $\mathrm{E}[T_i] = 80$ for all i, and that T_i and T_j have covariance

$$\mathrm{Cov}\,[T_i, T_j] = \frac{36}{1 + |i - j|} \tag{8.67}$$

Define the daily average temperature as

$$Y = \frac{T_1 + T_2 + \cdots + T_{31}}{31}. \tag{8.68}$$

Based on this model, write a MATLAB program `p=julytemps(T)` that calculates $P[Y \geq T]$, the probability that the daily average temperature is at least T degrees.

Further Reading: [WS01] and [PP02] make extensive use of vectors and matrices. To go deeply into vector random variables, students can use [Str98] to gain a firm grasp of principles of linear algebra.

Problems

Difficulty: ● Easy ▦ Moderate ◆ Difficult ◆◆ Experts Only

8.1.1● For random variables $X_1, \ldots, X_n$ in Problem 5.10.3, let $\mathbf{X} = \begin{bmatrix} X_1 & \cdots & X_n \end{bmatrix}'$. What is $f_{\mathbf{X}}(\mathbf{x})$?

8.1.2● Random vector $\mathbf{X}$ has PDF

$$f_{\mathbf{X}}(\mathbf{x}) = \begin{cases} ca'\mathbf{x} & 0 \leq \mathbf{x} \leq 1, \\ 0 & \text{otherwise,} \end{cases}$$

where $\mathbf{a} = \begin{bmatrix} a_1 & \cdots & a_n \end{bmatrix}'$ is a vector with each component $a_i > 0$. What is c?

8.1.3▦ Given $f_{\mathbf{X}}(\mathbf{x})$ with $c = 2/3$ and $a_1 = a_2 = a_3 = 1$ in Problem 8.1.2, find the marginal PDF $f_{X_3}(x_3)$.

8.1.4▦ $\mathbf{X} = \begin{bmatrix} X_1 & X_2 & X_3 \end{bmatrix}'$ has PDF

$$f_{\mathbf{X}}(\mathbf{x}) = \begin{cases} 6 & 0 \leq x_1 \leq x_2 \leq x_3 \leq 1, \\ 0 & \text{otherwise.} \end{cases}$$

Let $\mathbf{U} = \begin{bmatrix} X_1 & X_2 \end{bmatrix}'$, $\mathbf{V} = \begin{bmatrix} X_1 & X_3 \end{bmatrix}'$ and $\mathbf{W} = \begin{bmatrix} X_2 & X_3 \end{bmatrix}'$. Find the marginal PDFs $f_{\mathbf{U}}(\mathbf{u})$, $f_{\mathbf{V}}(\mathbf{v})$ and $f_{\mathbf{W}}(\mathbf{w})$.

8.1.5▦ A wireless data terminal has three messages waiting for transmission. After sending a message, it expects an acknowledgement from the receiver. When it receives the acknowledgement, it transmits the next message. If the acknowledgement does not arrive, it sends the message again. The probability of successful transmission of a message is p independent of other transmissions. Let $\mathbf{K} = \begin{bmatrix} K_1 & K_2 & K_3 \end{bmatrix}'$ be the three-dimensional random vector in which K_i is the total number of transmissions when message i is received successfully. (K_3 is the total number of transmissions used to send all three messages.) Show that

$$P_{\mathbf{K}}(\mathbf{k}) = \begin{cases} p^3(1-p)^{k_3 - 3} & k_1 < k_2 < k_3; \\ & k_i \in \{1, 2 \ldots\}, \\ 0 & \text{otherwise.} \end{cases}$$

8.1.6▦ From the joint PMF $P_{\mathbf{K}}(\mathbf{k})$ in Problem 8.1.5, find the marginal PMFs
(a) $P_{K_1, K_2}(k_1, k_2)$,
(b) $P_{K_1, K_3}(k_1, k_3)$,
(c) $P_{K_2, K_3}(k_2, k_2)$,
(d) $P_{K_1}(k_1)$, $P_{K_2}(k_2)$, and $P_{K_3}(k_3)$.

8.1.7▦ Let $\mathbf{N}$ be the r-dimensional random vector with the multinomial PMF given in Example 5.21 with $n > r > 2$:

$$P_{\mathbf{N}}(\mathbf{n}) = \binom{n}{n_1, \ldots, n_r} p_1^{n_1} \cdots p_r^{n_r}.$$

(a) What is the joint PMF of N_1 and N_2? Hint: Consider a new classification scheme with categories: s_1, s_2, and "other."

(b) Let $T_i = N_1 + \cdots + N_i$. What is the PMF of T_i?

(c) What is the joint PMF of T_1 and T_2?

8.1.8■ The random variables $Y_1, \ldots, Y_4$ have the joint PDF

$$f_{\mathbf{Y}}(\mathbf{y}) = \begin{cases} 24 & 0 \le y_1 \le y_2 \le y_3 \le y_4 \le 1, \\ 0 & \text{otherwise.} \end{cases}$$

Find the marginal PDFs $f_{Y_1,Y_4}(y_1, y_4)$, $f_{Y_1,Y_2}(y_1, y_2)$, and $f_{Y_1}(y_1)$.

8.1.9♦ As a generalization of the message transmission system in Problem 8.1.5, consider a terminal that has n messages to transmit. The components k_i of the n-dimensional random vector $\mathbf{K}$ are the total number of messages transmitted when message i is received successfully.

(a) Find the PMF of $\mathbf{K}$.

(b) For each $j \in \{1, 2, \ldots, n-1\}$, find the marginal PMF $P_{K_1,\ldots,K_j}(k_1, \ldots, k_j)$.

(c) For each $i \in \{1, 2, \ldots, n\}$, find the marginal PMF $P_{K_i}(k_i)$.

Hint: These PMFs are members of a family of discrete random variables in Appendix A.

8.2.1● The n components X_i of random vector $\mathbf{X}$ have $\mathrm{E}[X_i] = 0$ $\mathrm{Var}[X_i] = \sigma^2$. What is the covariance matrix $\mathbf{C_X}$?

8.2.2● The 4-dimensional random vector $\mathbf{X}$ has PDF

$$f_{\mathbf{X}}(\mathbf{x}) = \begin{cases} 1 & 0 \le x_i \le 1, i = 1, 2, 3, 4, \\ 0 & \text{otherwise.} \end{cases}$$

Are the four components of $\mathbf{X}$ independent random variables?

8.2.3● As in Example 8.1, the random vector $\mathbf{X}$ has PDF

$$f_{\mathbf{X}}(\mathbf{x}) = \begin{cases} 6e^{-\mathbf{a}'\mathbf{x}} & \mathbf{x} \ge 0, \\ 0 & \text{otherwise.} \end{cases}$$

where $\mathbf{a} = \begin{bmatrix} 1 & 2 & 3 \end{bmatrix}'$. Are the components of $\mathbf{X}$ independent random variables?

8.2.4● The PDF of the 3-dimensional random vector $\mathbf{X}$ is

$$f_{\mathbf{X}}(\mathbf{x}) = \begin{cases} e^{-x_3} & 0 \le x_1 \le x_2 \le x_3, \\ 0 & \text{otherwise.} \end{cases}$$

Are the components of $\mathbf{X}$ independent random variables?

8.2.5■ The random vector $\mathbf{X}$ has PDF

$$f_{\mathbf{X}}(\mathbf{x}) = \begin{cases} e^{-x_3} & 0 \le x_1 \le x_2 \le x_3, \\ 0 & \text{otherwise.} \end{cases}$$

Find the marginal PDFs $f_{X_1}(x_1)$, $f_{X_2}(x_2)$, and $f_{X_3}(x_3)$.

8.3.1● Discrete random vector $\mathbf{X}$ has PMF $P_{\mathbf{X}}(\mathbf{x})$. Prove that for an invertible matrix $\mathbf{A}$, $\mathbf{Y} = \mathbf{AX} + \mathbf{b}$ has PMF

$$P_{\mathbf{Y}}(\mathbf{y}) = P_{\mathbf{X}}\left(\mathbf{A}^{-1}(\mathbf{y} - \mathbf{b})\right).$$

8.3.2■ In the message transmission problem, Problem 8.1.5, the PMF for the number of transmissions when message i is received successfully is

$$P_{\mathbf{K}}(\mathbf{k}) = \begin{cases} p^3(1-p)^{k_3-3} & k_1 < k_2 < k_3; \\ & k_i \in \{1, 2 \ldots\}, \\ 0 & \text{otherwise.} \end{cases}$$

Let $J_3 = K_3 - K_2$, the number of transmissions of message 3; $J_2 = K_2 - K_1$, the number of transmissions of message 2; and $J_1 = K_1$, the number of transmissions of message 1. Derive a formula for $P_{\mathbf{J}}(\mathbf{j})$, the PMF of the number of transmissions of individual messages.

8.3.3■ In an automatic geolocation system, a dispatcher sends a message to six trucks in a fleet asking their locations. The waiting times for responses from the six trucks are iid exponential random variables, each with expected value 2 seconds.

(a) What is the probability that all six responses will arrive within 5 seconds?

(b) If the system has to locate all six vehicles within 3 seconds, it has to reduce the expected response time of each vehicle. What is the maximum expected response time that will produce a location time for all six vehicles of 3 seconds or less with probability of at least 0.9?

8.3.4♦♦ Let $X_1, \ldots, X_n$ denote n iid random variables with PDF $f_X(x)$ and

CDF $F_X(x)$.   What is the probability $P[X_n = \max\{X_1, \ldots, X_n\}]$?

8.4.1● Random variables X_1 and X_2 have zero expected value and variances $\text{Var}[X_1] = 4$ and $\text{Var}[X_2] = 9$. Their co-variance is $\text{Cov}[X_1, X_2] = 3$.

(a) Find the covariance matrix of $\mathbf{X} = \begin{bmatrix} X_1 & X_2 \end{bmatrix}'$.

(b) Find the covariance matrix of $\mathbf{Y} = \begin{bmatrix} Y_1 & Y_2 \end{bmatrix}'$ given by

$$Y_1 = X_1 - 2X_2,$$
$$Y_2 = 3X_1 + 4X_2.$$

8.4.2● Let $X_1, \ldots, X_n$ be iid random variables with expected value 0, variance 1, and covariance $\text{Cov}[X_i, X_j] = \rho$, for $i \neq j$. Use Theorem 8.8 to find the expected value and variance of the sum $Y = X_1 + \cdots + X_n$.

8.4.3● The two-dimensional random vector $\mathbf{X}$ and the three-dimensional random vector $\mathbf{Y}$ are independent and $E[\mathbf{Y}] = 0$. What is the vector cross-correlation $\mathbf{R_{XY}}$?

8.4.4● The four-dimensional random vector $\mathbf{X}$ has PDF

$$f_{\mathbf{X}}(\mathbf{x}) = \begin{cases} 1 & 0 \leq x_i \leq 1, i = 1, 2, 3, 4 \\ 0 & \text{otherwise.} \end{cases}$$

Find the expected value vector $E[\mathbf{X}]$, the correlation matrix $\mathbf{R_X}$, and the covariance matrix $\mathbf{C_X}$.

8.4.5● The random vector $\mathbf{Y} = \begin{bmatrix} Y_1 & Y_2 \end{bmatrix}'$ has covariance matrix $\mathbf{C_Y} = \begin{bmatrix} 25 & \gamma \\ \gamma & 4 \end{bmatrix}$ where γ is a constant. In terms of γ, what is the correlation coefficient ρ_{Y_1, Y_2} of Y_1 and Y_2? For what values of γ is $\mathbf{C_Y}$ a valid covariance matrix?

8.4.6● In the message transmission system in Problem 8.1.5, the solution to Problem 8.3.2 is a formula for the PMF of $\mathbf{J}$, the number of transmissions of individual messages. For $p = 0.8$, find the expected value vector $E[\mathbf{J}]$, the correlation matrix $\mathbf{R_J}$, and the covariance matrix $\mathbf{C_J}$.

8.4.7■ In the message transmission system in Problem 8.1.5,

$$P_{\mathbf{K}}(\mathbf{k}) = \begin{cases} p^3(1-p)^{k_3-3} & k_1 < k_2 < k_3; \\ & k_i \in \{1, 2, \ldots\}, \\ 0 & \text{otherwise.} \end{cases}$$

For $p = 0.8$, find the expected value vector $E[\mathbf{K}]$, the covariance matrix $\mathbf{C_K}$, and the correlation matrix $\mathbf{R_K}$.

8.4.8■ Random vector $\mathbf{X} = \begin{bmatrix} X_1 & X_2 \end{bmatrix}'$ has PDF

$$f_{\mathbf{X}}(\mathbf{x}) = \begin{cases} 10e^{-5x_1 - 2x_2} & x_1 \geq 0, x_2 \geq 0, \\ 0 & \text{otherwise.} \end{cases}$$

(a) Find $f_{X_1}(x_1)$ and $f_{X_2}(x_2)$.

(b) Derive the expected value vector $\boldsymbol{\mu}_X$ and covariance matrix $\mathbf{C_X}$.

(c) Let $\mathbf{Z} = \mathbf{AX}$, where $\mathbf{A} = \begin{pmatrix} -1 & 1 \\ 1 & 1 \end{pmatrix}$. Find the covariance matrix of $\mathbf{Z}$.

8.4.9■ As in Quiz 5.10 and Example 5.22, the 4-dimensional random vector $\mathbf{Y}$ has PDF

$$f_{\mathbf{Y}}(\mathbf{y}) = \begin{cases} 4 & 0 \leq y_1 \leq y_2 \leq 1; \\ & 0 \leq y_3 \leq y_4 \leq 1, \\ 0 & \text{otherwise.} \end{cases}$$

Find the expected value vector $E[\mathbf{Y}]$, the correlation matrix $\mathbf{R_Y}$, and the covariance matrix $\mathbf{C_Y}$.

8.4.10■ $\mathbf{X} = \begin{bmatrix} X_1 & X_2 \end{bmatrix}'$ is a random vector with $E[\mathbf{X}] = \begin{bmatrix} 0 & 0 \end{bmatrix}'$ and covariance matrix

$$\mathbf{C_X} = \begin{bmatrix} 1 & \rho \\ \rho & 1 \end{bmatrix}.$$

For some ω satisfying $0 \leq \omega \leq 1$, let $Y = \sqrt{\omega}X_1 + \sqrt{1 - \omega}X_2$. What value (or values) of ω will maximize $E[Y^2]$?

8.4.11■ The two-dimensional random vector $\mathbf{Y}$ has PDF

$$f_{\mathbf{Y}}(\mathbf{y}) = \begin{cases} 2 & \mathbf{y} \geq \mathbf{0}, \begin{bmatrix} 1 & 1 \end{bmatrix} \mathbf{y} \leq 1, \\ 0 & \text{otherwise.} \end{cases}$$

Find the expected value vector $E[\mathbf{Y}]$, the correlation matrix $\mathbf{R_Y}$, and the covariance matrix $\mathbf{C_Y}$.

8.4.12◆ Let $\mathbf{X}$ be a random vector with correlation matrix $\mathbf{R_X}$ and covariance matrix $\mathbf{C_X}$. Show that $\mathbf{R_X}$ and $\mathbf{C_X}$ are both positive semidefinite by showing that for any nonzero vector $\mathbf{a}$,

$$\mathbf{a}'\mathbf{R_X}\mathbf{a} \geq 0,$$
$$\mathbf{a}'\mathbf{C_X}\mathbf{a} \geq 0.$$

8.5.1● $\mathbf{X}$ is the 3-dimensional Gaussian random vector with expected value $\boldsymbol{\mu_X} = \begin{bmatrix} 4 & 8 & 6 \end{bmatrix}'$ and covariance

$$\mathbf{C_X} = \begin{bmatrix} 4 & -2 & 1 \\ -2 & 4 & -2 \\ 1 & -2 & 4 \end{bmatrix}.$$

Calculate

(a) the correlation matrix, $\mathbf{R_X}$,

(b) the PDF of the first two components of $\mathbf{X}$, $f_{X_1,X_2}(x_1, x_2)$,

(c) the probability that $X_1 > 8$.

.

8.5.2● $\mathbf{X} = \begin{bmatrix} X_1 & X_2 \end{bmatrix}'$ is the Gaussian random vector with $E[\mathbf{X}] = \begin{bmatrix} 0 & 0 \end{bmatrix}'$ and covariance matrix

$$\mathbf{C_X} = \begin{bmatrix} 1 & 1 \\ 1 & 2 \end{bmatrix}.$$

What is the PDF of $Y = \begin{bmatrix} 2 & 1 \end{bmatrix} \mathbf{X}$?

8.5.3● Given the Gaussian random vector $\mathbf{X}$ in Problem 8.5.1, $\mathbf{Y} = \mathbf{AX} + \mathbf{b}$, where

$$\mathbf{A} = \begin{bmatrix} 1 & 1/2 & 2/3 \\ 1 & -1/2 & 2/3 \end{bmatrix}$$

and $\mathbf{b} = \begin{bmatrix} -4 & -4 \end{bmatrix}'$. Calculate

(a) the expected value $\boldsymbol{\mu_Y}$,

(b) the covariance $\mathbf{C_Y}$,

(c) the correlation $\mathbf{R_Y}$,

(d) the probability that $-1 \leq Y_2 \leq 1$.

8.5.4● Let $\mathbf{X}$ be a Gaussian $(\boldsymbol{\mu_X}, \mathbf{C_X})$ random vector. Given a vector $\mathbf{a}$, find the expected value and variance of $Y = \mathbf{a}'\mathbf{X}$. Is Y a Gaussian random variable?

8.5.5■ Random variables X_1 and X_2 have zero expected value. The random vector $\mathbf{X} = \begin{bmatrix} X_1 & X_2 \end{bmatrix}'$ has a covariance matrix of the form

$$\mathbf{C} = \begin{bmatrix} 1 & \alpha \\ \beta & 4 \end{bmatrix}.$$

(a) For what values of α and β is $\mathbf{C}$ a valid covariance matrix?

(b) For what values of α and β can $\mathbf{X}$ be a Gaussian random vector?

(c) Suppose now that α and β satisfy the conditions in part (b) and $\mathbf{X}$ is a Gaussian random vector. What is the PDF of X_2? What is the PDF of $W = 2X_1 - X_2$?

8.5.6■ The Gaussian random vector $\mathbf{X} = \begin{bmatrix} X_1 & X_2 \end{bmatrix}'$ has expected value $E[\mathbf{X}] = \mathbf{0}$ and covariance matrix $\mathbf{C_X} = \begin{bmatrix} \sigma_1^2 & 1 \\ 1 & \sigma_2^2 \end{bmatrix}$.

(a) Under what conditions on σ_1^2 and σ_2^2 is $\mathbf{C_X}$ a valid covariance matrix?

(b) Suppose $\mathbf{Y} = \begin{bmatrix} Y_1 & Y_2 \end{bmatrix}' = \mathbf{AX}$ where $\mathbf{A} = \begin{bmatrix} 1 & 1 \\ 1 & -1 \end{bmatrix}$. For what values (if any) of σ_1^2 and σ_2^2 are the components Y_1 and Y_2 independent?

8.5.7■ The Gaussian random vector $\mathbf{X} = \begin{bmatrix} X_1 & X_2 \end{bmatrix}'$ has expected value $E[\mathbf{X}] = \mathbf{0}$ and covariance matrix $\mathbf{C_X} = \begin{bmatrix} 2 & 1 \\ 1 & 1 \end{bmatrix}$.

(a) Find the PDF of $W = X_1 + 2X_2$.

(b) Find the PDF $f_\mathbf{Y}(\mathbf{y})$ of $\mathbf{Y} = \mathbf{AX}$ where $\mathbf{A} = \begin{bmatrix} 1 & 1 \\ 1 & -1 \end{bmatrix}$.

8.5.8■ Let $\mathbf{X}$ be a Gaussian random vector with expected value $\begin{bmatrix} \mu_1 & \mu_2 \end{bmatrix}'$ and covariance matrix

$$\mathbf{C_X} = \begin{bmatrix} \sigma_1^2 & \rho\sigma_1\sigma_2 \\ \rho\sigma_1\sigma_2 & \sigma_2^2 \end{bmatrix}.$$

Show that $\mathbf{X}$ has bivariate Gaussian PDF $f_\mathbf{X}(\mathbf{x}) = f_{X_1,X_2}(x_1, x_2)$ given by Definition 5.10.

8.5.9■ $X = \begin{bmatrix} X_1 & X_2 \end{bmatrix}'$ is a Gaussian random vector with $E[\mathbf{X}] = \begin{bmatrix} 0 & 0 \end{bmatrix}'$ and covariance matrix $\mathbf{C_X} = \begin{bmatrix} a & b \\ c & d \end{bmatrix}$.

(a) What conditions must a, b, c, and d satisfy?

(b) Under what conditions (in addition to those in part (a)) are X_1 and X_2 independent?

(c) Under what conditions (in addition to those in part (a)) are X_1 and X_2 identical?

8.5.10■ Let $\mathbf{X}$ be a Gaussian $(\boldsymbol{\mu_X}, \mathbf{C_X})$ random vector. Let $\mathbf{Y} = \mathbf{AX}$ where $\mathbf{A}$ is an $m \times n$ matrix of rank m. By Theorem 8.11, $\mathbf{Y}$ is a Gaussian random vector. Is

$$\mathbf{W} = \begin{bmatrix} \mathbf{X} \\ \mathbf{Y} \end{bmatrix}$$

a Gaussian random vector?

8.5.11■ The 2×2 matrix

$$\mathbf{Q} = \begin{bmatrix} \cos\theta & -\sin\theta \\ \sin\theta & \cos\theta \end{bmatrix}$$

is called a rotation matrix because $\mathbf{y} = \mathbf{Qx}$ is the rotation of $\mathbf{x}$ by the angle θ. Suppose $\mathbf{X} = \begin{bmatrix} X_1 & X_2 \end{bmatrix}'$ is a Gaussian $(\mathbf{0}, \mathbf{C_X})$ vector where $\mathbf{C_X} = \text{diag}[\sigma_1^2, \sigma_2^2]$ and $\sigma_2^2 \geq \sigma_1^2$. Let $\mathbf{Y} = \mathbf{QX}$.

(a) Find the covariance of Y_1 and Y_2. Show that Y_1 and Y_2 are independent for all θ if $\sigma_1^2 = \sigma_2^2$.

(b) Suppose $\sigma_2^2 > \sigma_1^2$. For what values θ are Y_1 and Y_2 independent?

8.5.12◆ $\mathbf{X} = \begin{bmatrix} X_1 & X_2 \end{bmatrix}'$ is a Gaussian $(\mathbf{0}, \mathbf{C_X})$ vector where

$$\mathbf{C_X} = \begin{bmatrix} 1 & \rho \\ \rho & 1 \end{bmatrix}.$$

Thus, depending on the value of the correlation coefficient ρ, the joint PDF of X_1 and X_2 may resemble one of the graphs of Figure 5.6 with $X_1 = X$ and $X_2 = Y$. Show that $\mathbf{X} = \mathbf{QY}$, where $\mathbf{Q}$ is the $\theta = 45°$ rotation matrix (see Problem 8.5.11) and $\mathbf{Y}$

is a Gaussian $(\mathbf{0}, \mathbf{C_Y})$ vector such that

$$\mathbf{C_Y} = \begin{bmatrix} 1+\rho & 0 \\ 0 & 1-\rho \end{bmatrix}.$$

This result verifies, for $\rho \neq 0$, that the PDF of X_1 and X_2 shown in Figure 5.6 is the joint PDF of two independent Gaussian random variables (with variances $1 + \rho$ and $1 - \rho$) rotated by $45°$.

8.5.13◆ An n-dimensional Gaussian vector $\mathbf{W}$ has a block diagonal covariance matrix

$$\mathbf{C_W} = \begin{bmatrix} \mathbf{C_X} & \mathbf{0} \\ \mathbf{0} & \mathbf{C_Y} \end{bmatrix},$$

where $\mathbf{C_X}$ is $m \times m$, $\mathbf{C_Y}$ is $(n-m) \times (n-m)$. Show that $\mathbf{W}$ can be written in terms of component vectors $\mathbf{X}$ and $\mathbf{Y}$ in the form

$$\mathbf{W} = \begin{bmatrix} \mathbf{X} \\ \mathbf{Y} \end{bmatrix},$$

such that $\mathbf{X}$ and $\mathbf{Y}$ are independent Gaussian random vectors.

8.5.14◆◆ In this problem, we extend the proof of Theorem 8.11 to the case when $\mathbf{A}$ is $m \times n$ with $m < n$. For this proof, we assume $\mathbf{X}$ is an n-dimensional Gaussian vector and that we have proved Theorem 8.11 for the case $m = n$. Since the case $m = n$ is sufficient to prove that $\mathbf{Y} = \mathbf{X} + \mathbf{b}$ is Gaussian, it is sufficient to show for $m < n$ that $\mathbf{Y} = \mathbf{AX}$ is Gaussian in the case when $\boldsymbol{\mu_X} = 0$.

(a) Prove there exists an $(n - m) \times n$ matrix $\tilde{\mathbf{A}}$ of rank $n - m$ with the property that $\tilde{\mathbf{A}}\mathbf{A}' = \mathbf{0}$. Hint: Review the Gram–Schmidt procedure.

(b) Let $\hat{\mathbf{A}} = \tilde{\mathbf{A}}\mathbf{C_X}^{-1}$ and define the random vector

$$\bar{\mathbf{Y}} = \begin{bmatrix} \mathbf{Y} \\ \hat{\mathbf{Y}} \end{bmatrix} = \begin{bmatrix} \mathbf{A} \\ \hat{\mathbf{A}} \end{bmatrix}\mathbf{X}.$$

Use Theorem 8.11 for the case $m = n$ to argue that $\bar{\mathbf{Y}}$ is a Gaussian random vector.

(c) Find the covariance matrix $\bar{\mathbf{C}}$ of $\bar{\mathbf{Y}}$. Use the result of Problem 8.5.13 to show that $\mathbf{Y}$ and $\hat{\mathbf{Y}}$ are independent Gaussian random vectors.

8.6.1● Consider the vector $\mathbf{X}$ in Problem 8.5.1 and define $Y = (X_1 + X_2 + X_3)/3$. What is the probability that $Y > 4$?

8.6.2▩ A better model for the sailboat race of Problem 5.10.8 accounts for the fact that all boats are subject to the same randomness of wind and tide. Suppose in the race of ten sailboats, the finishing times X_i are identical Gaussian random variables with expected value 35 minutes and standard deviation 5 minutes. However, for every pair of boats i and j, the finish times X_i and X_j have correlation coefficient $\rho = 0.8$.

(a) What is the covariance matrix of $\mathbf{X} = \begin{bmatrix} X_1 & \cdots & X_{10} \end{bmatrix}'$?

(b) Let

$$Y = \frac{X_1 + X_2 + \cdots + X_{10}}{10}.$$

What are the expected value and variance of Y? What is $P[Y \le 25]$?

8.6.3▩ For the vector of daily temperatures $\begin{bmatrix} T_1 & \cdots & T_{31} \end{bmatrix}'$ and average temperature Y modeled in Quiz 8.6, we wish to estimate the probability of the event

$$A = \left\{ Y \le 82, \min_i T_i \ge 72 \right\}.$$

To form an estimate of A, generate 10,000 independent samples of the vector $\mathbf{T}$ and calculate the relative frequency of A in those trials.

8.6.4◆ We continue Problem 8.6.2 where the vector $\mathbf{X}$ of finish times has correlated components. Let W denote the finish time of the winning boat. We wish to estimate $P[W \le 25]$, the probability that the winning boat finishes in under 25 minutes. To do this, simulate $m = 10,000$ races by generating m samples of the vector $\mathbf{X}$ of finish times. Let $Y_j = 1$ if the winning time in race i is under 25 minutes; otherwise, $Y_j = 0$. Calculate the estimate

$$P\left[W \le 25 \right] \approx \frac{1}{m} \sum_{j=1}^{m} Y_j.$$

8.6.5◆◆ Write a MATLAB program that simulates m runs of the weekly lottery of Problem 7.5.9. For $m = 1000$ sample runs, form a histogram for the jackpot J.

9

Sums of Random Variables

Random variables of the form

$$W_n = X_1 + \cdots + X_n \qquad (9.1)$$

appear repeatedly in probability theory and applications. We could in principle derive the probability model of W_n from the PMF or PDF of $X_1, \ldots, X_n$. However, in many practical applications, the nature of the analysis or the properties of the random variables allow us to apply techniques that are simpler than analyzing a general n-dimensional probability model. In Section 9.1 we consider applications in which our interest is confined to expected values related to W_n, rather than a complete model of W_n. Subsequent sections emphasize techniques that apply when $X_1, \ldots, X_n$ are mutually independent. A useful way to analyze the sum of independent random variables is to transform the PDF or PMF of each random variable to a *moment generating function*.

The central limit theorem reveals a fascinating property of the sum of independent random variables. It states that the CDF of the sum converges to a Gaussian CDF as the number of terms grows without limit. This theorem allows us to use the properties of Gaussian random variables to obtain accurate estimates of probabilities associated with sums of other random variables. In many cases exact calculation of these probabilities is extremely difficult.

9.1 Expected Values of Sums

The expected value of a sum of *any* random variables is the sum of the expected values. The variance of the sum of any random variable is the sum of all the covariances. The variance of the sum of *independent* random variables is the sum of the variances.

The theorems of Section 5.7 can be generalized in a straightforward manner to describe expected values and variances of sums of more than two random variables.

Theorem 9.1

For any set of random variables $X_1, \ldots, X_n$, the sum $W_n = X_1 + \cdots + X_n$ has expected value

$$E[W_n] = E[X_1] + E[X_2] + \cdots + E[X_n].$$

Proof We prove this theorem by induction on n. In Theorem 5.11, we proved $E[W_2] = E[X_1] + E[X_2]$. Now we assume $E[W_{n-1}] = E[X_1] + \cdots + E[X_{n-1}]$. Notice that $W_n = W_{n-1} + X_n$. Since W_n is a sum of the two random variables W_{n-1} and X_n, we know that $E[W_n] = E[W_{n-1}] + E[X_n] = E[X_1] + \cdots + E[X_{n-1}] + E[X_n]$.

Keep in mind that the expected value of the sum equals the sum of the expected values whether or not $X_1, \ldots, X_n$ are independent. For the variance of W_n, we have the generalization of Theorem 5.12:

Theorem 9.2

The variance of $W_n = X_1 + \cdots + X_n$ is

$$\mathrm{Var}[W_n] = \sum_{i=1}^{n} \mathrm{Var}[X_i] + 2 \sum_{i=1}^{n-1} \sum_{j=i+1}^{n} \mathrm{Cov}[X_i, X_j].$$

Proof From the definition of the variance, we can write $\mathrm{Var}[W_n] = E[(W_n - E[W_n])^2]$. For convenience, let μ_i denote $E[X_i]$. Since $W_n = \sum_{i=1}^{n} X_n$ and $E[W_n] = \sum_{i=1}^{n} \mu_i$, we can write

$$\mathrm{Var}[W_n] = E\left[\left(\sum_{i=1}^{n}(X_i - \mu_i)\right)^2\right] = E\left[\sum_{i=1}^{n}(X_i - \mu_i)\sum_{j=1}^{n}(X_j - \mu_j)\right]$$

$$= \sum_{i=1}^{n} \sum_{j=1}^{n} \mathrm{Cov}[X_i, X_j]. \tag{9.2}$$

In terms of the random vector $\mathbf{X} = \begin{bmatrix} X_1 & \cdots & X_n \end{bmatrix}'$, we see that $\mathrm{Var}[W_n]$ is the sum of all the elements of the covariance matrix $\mathbf{C_X}$. Recognizing that $\mathrm{Cov}[X_i, X_i] = \mathrm{Var}[X]$ and $\mathrm{Cov}[X_i, X_j] = \mathrm{Cov}[X_j, X_i]$, we place the diagonal terms of $\mathbf{C_X}$ in one sum and the off-diagonal terms (which occur in pairs) in another sum to arrive at the formula in the theorem.

When $X_1, \ldots, X_n$ are uncorrelated, $\mathrm{Cov}[X_i, X_j] = 0$ for $i \neq j$ and the variance of the sum is the sum of the variances.

═══Theorem 9.3═══

When $X_1, \ldots, X_n$ are uncorrelated,

$$\mathrm{Var}[W_n] = \mathrm{Var}[X_1] + \cdots + \mathrm{Var}[X_n].$$

═══**Example 9.1**═══

$X_0, X_1, X_2, \ldots$ is a sequence of random variables with expected values $\mathrm{E}[X_i] = 0$ and covariances, $\mathrm{Cov}[X_i, X_j] = 0.8^{|i-j|}$. Find the expected value and variance of a random variable Y_i defined as the sum of three consecutive values of the random sequence

$$Y_i = X_i + X_{i-1} + X_{i-2}. \tag{9.3}$$

. . . .

Theorem 9.1 implies that

$$\mathrm{E}[Y_i] = \mathrm{E}[X_i] + \mathrm{E}[X_{i-1}] + \mathrm{E}[X_{i-2}] = 0. \tag{9.4}$$

Applying Theorem 9.2, we obtain for each i,

$$\begin{aligned} \mathrm{Var}[Y_i] &= \mathrm{Var}[X_i] + \mathrm{Var}[X_{i-1}] + \mathrm{Var}[X_{i-2}] \\ &\quad + 2\,\mathrm{Cov}[X_i, X_{i-1}] + 2\,\mathrm{Cov}[X_i, X_{i-2}] + 2\,\mathrm{Cov}[X_{i-1}, X_{i-2}]. \end{aligned} \tag{9.5}$$

We next note that $\mathrm{Var}[X_i] = \mathrm{Cov}[X_i, X_i] = 0.8^{i-i} = 1$ and that

$$\mathrm{Cov}[X_i, X_{i-1}] = \mathrm{Cov}[X_{i-1}, X_{i-2}] = 0.8^1, \qquad \mathrm{Cov}[X_i, X_{i-2}] = 0.8^2. \tag{9.6}$$

Therefore,

$$\mathrm{Var}[Y_i] = 3 \times 0.8^0 + 4 \times 0.8^1 + 2 \times 0.8^2 = 7.48. \tag{9.7}$$

The following example shows how a puzzling problem can be formulated as a question about the sum of a set of dependent random variables.

═══**Example 9.2**═══

At a party of $n \geq 2$ people, each person throws a hat in a common box. The box is shaken and each person blindly draws a hat from the box without replacement. We say a match occurs if a person draws his own hat. What are the expected value and variance of V_n, the number of matches?

. . . .

Let X_i denote an indicator random variable such that

$$X_i = \begin{cases} 1 & \text{person } i \text{ draws his hat,} \\ 0 & \text{otherwise.} \end{cases} \tag{9.8}$$

The number of matches is $V_n = X_1 + \cdots + X_n$. Note that the X_i are generally not independent. For example, with $n = 2$ people, if the first person draws his own hat, then the second person must also draw her own hat. Note that the ith person is equally likely to draw any of the n hats, thus $P_{X_i}(1) = 1/n$ and $E[X_i] = P_{X_i}(1) = 1/n$. Since the expected value of the sum always equals the sum of the expected values,

$$E[V_n] = E[X_1] + \cdots + E[X_n] = n(1/n) = 1. \tag{9.9}$$

To find the variance of V_n, we will use Theorem 9.2. The variance of X_i is

$$\text{Var}[X_i] = E[X_i^2] - (E[X_i])^2 = \frac{1}{n} - \frac{1}{n^2}. \tag{9.10}$$

To find $\text{Cov}[X_i, X_j]$, we observe that

$$\text{Cov}[X_i, X_j] = E[X_i X_j] - E[X_i]E[X_j]. \tag{9.11}$$

Note that $X_i X_j = 1$ if and only if $X_i = 1$ and $X_j = 1$, and $X_i X_j = 0$ otherwise. Thus

$$E[X_i X_j] = P_{X_i, X_j}(1, 1) = P_{X_i|X_j}(1|1) P_{X_j}(1). \tag{9.12}$$

Given $X_j = 1$, that is, the jth person drew his own hat, then $X_i = 1$ if and only if the ith person draws his own hat from the $n-1$ other hats. Hence $P_{X_i|X_j}(1|1) = 1/(n-1)$ and

$$E[X_i X_j] = \frac{1}{n(n-1)}, \qquad \text{Cov}[X_i, X_j] = \frac{1}{n(n-1)} - \frac{1}{n^2}. \tag{9.13}$$

Finally, we can use Theorem 9.2 to calculate

$$\text{Var}[V_n] = n\,\text{Var}[X_i] + n(n-1)\,\text{Cov}[X_i, X_j] = 1. \tag{9.14}$$

That is, both the expected value and variance of V_n are 1, no matter how large n is!

Example 9.3

Continuing Example 9.2, suppose each person immediately returns to the box the hat that he or she drew. What is the expected value and variance of V_n, the number of matches?

. . . .

In this case the indicator random variables X_i are independent and identically distributed (iid) because each person draws from the same bin containing all n hats. The number of matches $V_n = X_1 + \cdots + X_n$ is the sum of n iid random variables. As before, the expected value of V_n is

$$E[V_n] = n\,E[X_i] = 1. \tag{9.15}$$

In this case, the variance of V_n equals the sum of the variances,

$$\text{Var}[V_n] = n\,\text{Var}[X_i] = n\left(\frac{1}{n} - \frac{1}{n^2}\right) = 1 - \frac{1}{n}. \tag{9.16}$$

The remainder of this chapter examines tools for analyzing complete probability models of sums of random variables, with the emphasis on sums of independent random variables.

═════Quiz 9.1═════

Let W_n denote the sum of n independent throws of a fair four-sided die. Find the expected value and variance of W_n.

9.2   Moment Generating Functions

$\phi_X(s)$, the moment generating function (MGF) of a random variable X, is a probability model of X. If X is discrete, the MGF is a transform of the PMF. The MGF of a continuous random variable is a transform of the PDF, similar to a Laplace transform. The n-th moment of X is the n-th derivative of $\phi_X(s)$ evaluated at $s = 0$.

In Section 6.5, we learned in Theorem 6.9 that the PDF of the sum $W_2 = X_1 + X_2$ of independent random variables can be written as the convolution $f_{W_2}(w_2) = \int_{-\infty}^{\infty} f_{X_1}(w_2 - x_2) f_{X_2}(x_2)\, dx_2$. To find the PDF a sum of three independent random variables, $W_3 = X_1 + X_2 + X_3$, we could use Theorem 6.9 to find the PDF of $W_2 = X_1 + X_2$. and then, because $W_3 = W_2 + X_3$ and W_2 and X_3 are independent, we could use Theorem 6.9 again to find the PDF of W_3 from the convolution $f_{W_3}(w_3) = \int_{-\infty}^{\infty} f_{W_2}(w_3 - x_3) f_{X_3}(x_3)\, dx_3$. In principle, we could continue this sequence of convolutions to find the PDF of $W_n = X_1 + \cdots + X_n$ for any n. While this procedure is sound in theory, convolutional integrals are generally tricky, and a sequence of n convolutions is often prohibitively difficult to evaluate by hand. Even MATLAB typically fails to simplify the evaluation of a sequence of convolutions.

In linear system theory, however, convolution in the time domain corresponds to multiplication in the frequency domain with time functions and frequency functions related by the Fourier transform. In probability theory, we can, in a similar way, use transform methods to replace the convolution of PDFs by multiplication of transforms. In the language of probability theory, the transform of a PDF or a PMF is a *moment generating function*.

═════Definition 9.1═════Moment Generating Function (MGF)

*For a random variable X, the **moment generating function (MGF)** of X is*

$$\phi_X(s) = \mathrm{E}\left[e^{sX}\right].$$

Definition 9.1 applies to both discrete and continuous random variables X. What changes in going from discrete X to continuous X is the method of calculating the expected value. When X is a continuous random variable,

$$\phi_X(s) = \int_{-\infty}^{\infty} e^{sx} f_X(x)\ dx. \tag{9.17}$$

For a discrete random variable Y, the MGF is

$$\phi_Y(s) = \sum_{y_i \in S_Y} e^{sy_i} P_Y(y_i). \tag{9.18}$$

Equation (9.17) indicates that the MGF of a continuous random variable is similar to the Laplace transform of a time function. The primary difference between an MGF and a Laplace transform is that the MGF is defined only for real values of s. For a given random variable X, there is a range of possible values of s for which $\phi_X(s)$ exists. The set of values of s for which $\phi_X(s)$ exists is called the *region of convergence*. The definition of the MGF implies that $\phi_X(0) = \mathrm{E}[e^0] = 1$. Thus $s = 0$ is always in the region of convergence. If X is a nonnegative random variable, the region of convergence includes all $s \leq 0$. If X is bounded so that $\mathrm{P}[a < X \leq b] = 1$, then $\phi_X(s)$ exists for all real s. Typically, the region of convergence is an interval around the $s = 0$.

Because the MGF and PMF or PDF form a transform pair, the MGF is also a complete probability model of a random variable. Given the MGF, it is possible to compute the PDF or PMF. Moreover, the derivatives of $\phi_X(s)$ evaluated at $s = 0$ are the moments of X.

Theorem 9.4
A random variable X with MGF $\phi_X(s)$ has nth moment

$$\mathrm{E}[X^n] = \left. \frac{d^n \phi_X(s)}{ds^n} \right|_{s=0}.$$

Proof The first derivative of $\phi_X(s)$ is

$$\frac{d\phi_X(s)}{ds} = \frac{d}{ds} \left(\int_{-\infty}^{\infty} e^{sx} f_X(x)\ dx \right) = \int_{-\infty}^{\infty} x e^{sx} f_X(x)\ dx. \tag{9.19}$$

Evaluating this derivative at $s = 0$ proves the theorem for $n = 1$.

$$\left. \frac{d\phi_X(s)}{ds} \right|_{s=0} = \int_{-\infty}^{\infty} x f_X(x)\ dx = \mathrm{E}[X]. \tag{9.20}$$

Similarly, the nth derivative of $\phi_X(s)$ is

$$\frac{d^n \phi_X(s)}{ds^n} = \int_{-\infty}^{\infty} x^n e^{sx} f_X(x)\ dx. \tag{9.21}$$

The integral evaluated at $s = 0$ is the formula in the theorem statement.

Typically it is easier to calculate the moments of X by finding the MGF and

Random Variable	PMF or PDF	MGF $\phi_X(s)$
Bernoulli (p)	$P_X(x) = \begin{cases} 1-p & x=0 \\ p & x=1 \\ 0 & \text{otherwise} \end{cases}$	$1-p+pe^s$
Binomial (n,p)	$P_X(x) = \binom{n}{x}p^x(1-p)^{n-x}$	$(1-p+pe^s)^n$
Geometric (p)	$P_X(x) = \begin{cases} p(1-p)^{x-1} & x=1,2,\ldots \\ 0 & \text{otherwise} \end{cases}$	$\dfrac{pe^s}{1-(1-p)e^s}$
Pascal (k,p)	$P_X(x) = \binom{x-1}{k-1}p^k(1-p)^{x-k}$	$\left(\dfrac{pe^s}{1-(1-p)e^s}\right)^k$
Poisson (α)	$P_X(x) = \begin{cases} \alpha^x e^{-\alpha}/x! & x=0,1,2,\ldots \\ 0 & \text{otherwise} \end{cases}$	$e^{\alpha(e^s-1)}$
Disc. Uniform (k,l)	$P_X(x) = \begin{cases} \frac{1}{l-k+1} & x=k,\ldots,l \\ 0 & \text{otherwise} \end{cases}$	$\dfrac{e^{sk}-e^{s(l+1)}}{1-e^s}$
Constant (a)	$f_X(x) = \delta(x-a)$	e^{sa}
Uniform (a,b)	$f_X(x) = \begin{cases} \frac{1}{b-a} & a<x<b \\ 0 & \text{otherwise} \end{cases}$	$\dfrac{e^{bs}-e^{as}}{s(b-a)}$
Exponential (λ)	$f_X(x) = \begin{cases} \lambda e^{-\lambda x} & x\geq 0 \\ 0 & \text{otherwise} \end{cases}$	$\dfrac{\lambda}{\lambda-s}$
Erlang (n,λ)	$f_X(x) = \begin{cases} \frac{\lambda^n x^{n-1}e^{-\lambda x}}{(n-1)!} & x\geq 0 \\ 0 & \text{otherwise} \end{cases}$	$\left(\dfrac{\lambda}{\lambda-s}\right)^n$
Gaussian (μ,σ)	$f_X(x) = \frac{1}{\sigma\sqrt{2\pi}}e^{-(x-\mu)^2/2\sigma^2}$	$e^{s\mu+s^2\sigma^2/2}$

Table 9.1 Moment generating function for families of random variables.

differentiating than by integrating $x^n f_X(x)$.

═══Example 9.4═══

X is an exponential random variable with MGF $\phi_X(s) = \lambda/(\lambda - s)$. What are the first and second moments of X? Write a general expression for the nth moment.

The first moment is the expected value:

$$\mathrm{E}\left[X\right] = \left.\frac{d\phi_X(s)}{ds}\right|_{s=0} = \left.\frac{\lambda}{(\lambda - s)^2}\right|_{s=0} = \frac{1}{\lambda}. \tag{9.22}$$

The second moment of X is the mean square value:

$$\mathrm{E}\left[X^2\right] = \left.\frac{d^2\phi_X(s)}{ds^2}\right|_{s=0} = \left.\frac{2\lambda}{(\lambda - s)^3}\right|_{s=0} = \frac{2}{\lambda^2}. \tag{9.23}$$

Proceeding in this way, it should become apparent that the nth moment of X is

$$\mathrm{E}\left[X^n\right] = \left.\frac{d^n\phi_X(s)}{ds^n}\right|_{s=0} = \left.\frac{n!\lambda}{(\lambda - s)^{n+1}}\right|_{s=0} = \frac{n!}{\lambda^n}. \tag{9.24}$$

Table 9.1 presents the MGF for the families of random variables defined in Chapters 3 and 4. The following theorem derives the MGF of a linear transformation of a random variable X in terms of $\phi_X(s)$.

═══Theorem 9.5═══

The MGF of $Y = aX + b$ is $\phi_Y(s) = e^{sb}\phi_X(as)$.

Proof From the definition of the MGF,

$$\phi_Y(s) = \mathrm{E}\left[e^{s(aX+b)}\right] = e^{sb}\,\mathrm{E}\left[e^{(as)X}\right] = e^{sb}\phi_X(as). \tag{9.25}$$

═══Quiz 9.2═══

Random variable K has PMF

$$P_K(k) = \begin{cases} 0.2 & k = 0, \ldots, 4, \\ 0 & \text{otherwise.} \end{cases} \tag{9.26}$$

Use $\phi_K(s)$ to find the first, second, third, and fourth moments of K.

9.3   MGF of the Sum of Independent Random Variables

> Moment generating functions provide a convenient way to determine the probability model of a sum of iid random variables. Using MGFs, we determine that when $W = X_1 + \cdots + X_n$ is a sum of n iid random variables:
>
> - If X_i is Bernoulli (p), W is binomial (n, p).
> - If X_i is Poisson (α), W is Poisson $(n\alpha)$.
> - If X_i is geometric (p), W is Pascal (n, p).
> - If X_i is exponential (λ), W is Erlang (n, λ).
> - If X_i is Gaussian (μ, σ), W is Gaussian $(n\mu, \sqrt{n}\sigma)$.

Moment generating functions are particularly useful for analyzing sums of independent random variables, because if X and Y are independent, the MGF of $W = X + Y$ is the product

$$\phi_W(s) = \mathrm{E}\left[e^{sX}e^{sY}\right] = \mathrm{E}\left[e^{sX}\right]\mathrm{E}\left[e^{sY}\right] = \phi_X(s)\phi_Y(s). \tag{9.27}$$

Theorem 9.6 generalizes this result to a sum of n independent random variables.

Theorem 9.6

For a set of independent random variables $X_1, \ldots, X_n$, the moment generating function of $W = X_1 + \cdots + X_n$ is

$$\phi_W(s) = \phi_{X_1}(s)\phi_{X_2}(s)\cdots\phi_{X_n}(s).$$

When $X_1, \ldots, X_n$ are iid, each with MGF $\phi_{X_i}(s) = \phi_X(s)$,

$$\phi_W(s) = [\phi_X(s)]^n.$$

Proof From the definition of the MGF,

$$\phi_W(s) = \mathrm{E}\left[e^{s(X_1 + \cdots + X_n)}\right] = \mathrm{E}\left[e^{sX_1}e^{sX_2}\cdots e^{sX_n}\right]. \tag{9.28}$$

Here, we have the expected value of a product of functions of independent random variables. Theorem 8.4 states that this expected value is the product of the individual expected values:

$$\mathrm{E}\left[g_1(X_1)g_2(X_2)\cdots g_n(X_n)\right] = \mathrm{E}\left[g_1(X_1)\right]\mathrm{E}\left[g_2(X_2)\right]\cdots\mathrm{E}\left[g_n(X_n)\right]. \tag{9.29}$$

By Equation (9.29) with $g_i(X_i) = e^{sX_i}$, the expected value of the product is

$$\phi_W(s) = \mathrm{E}\left[e^{sX_1}\right]\mathrm{E}\left[e^{sX_2}\right]\cdots\mathrm{E}\left[e^{sX_n}\right] = \phi_{X_1}(s)\phi_{X_2}(s)\cdots\phi_{X_n}(s). \tag{9.30}$$

When $X_1, \ldots, X_n$ are iid, $\phi_{X_i}(s) = \phi_X(s)$ and thus $\phi_W(s) = (\phi_W(s))^n$.

Moment generating functions provide a convenient way to study the properties of sums of independent finite discrete random variables.

═══Example 9.5═══

J and K are independent random variables with probability mass functions

$$
\begin{array}{c|ccc}
j & 1 & 2 & 3 \\
\hline
P_J(j) & 0.2 & 0.6 & 0.2
\end{array}
,\qquad
\begin{array}{c|cc}
k & -1 & 1 \\
\hline
P_K(k) & 0.5 & 0.5
\end{array}
. \tag{9.31}
$$

Find the MGF of $M = J + K$. What are $P_M(m)$ and $E[M^3]$?

. . . .

J and K have have moment generating functions

$$
\phi_J(s) = 0.2e^s + 0.6e^{2s} + 0.2e^{3s}, \qquad \phi_K(s) = 0.5e^{-s} + 0.5e^s. \tag{9.32}
$$

Therefore, by Theorem 9.6, $M = J + K$ has MGF

$$
\phi_M(s) = \phi_J(s)\phi_K(s) = 0.1 + 0.3e^s + 0.2e^{2s} + 0.3e^{3s} + 0.1e^{4s}. \tag{9.33}
$$

The value of $P_M(m)$ at any value of m is the coefficient of e^{ms} in $\phi_M(s)$:

$$
\phi_M(s) = \mathrm{E}\left[e^{sM}\right] = \underbrace{0.1}_{P_M(0)} + \underbrace{0.3}_{P_M(1)} e^s + \underbrace{0.2}_{P_M(2)} e^{2s} + \underbrace{0.3}_{P_M(3)} e^{3s} + \underbrace{0.1}_{P_M(4)} e^{4s}.
$$

From the coefficients of $\phi_M(s)$, we construct the table for the PMF of M:

$$
\begin{array}{c|ccccc}
m & 0 & 1 & 2 & 3 & 4 \\
\hline
P_M(m) & 0.1 & 0.3 & 0.2 & 0.3 & 0.1
\end{array}
.
$$

To find the third moment of M, we differentiate $\phi_M(s)$ three times:

$$
\begin{aligned}
\mathrm{E}\left[M^3\right] &= \left.\frac{d^3\phi_M(s)}{ds^3}\right|_{s=0} \\
&= 0.3e^s + 0.2(2^3)e^{2s} + 0.3(3^3)e^{3s} + 0.1(4^3)e^{4s}\big|_{s=0} = 16.4. \tag{9.34}
\end{aligned}
$$

Besides enabling us to calculate probabilities and moments for sums of discrete random variables, we can also use Theorem 9.6 to derive the PMF or PDF of certain sums of iid random variables. In particular, we use Theorem 9.6 to prove that the sum of independent Poisson random variables is a Poisson random variable, and the sum of independent Gaussian random variables is a Gaussian random variable.

═══Theorem 9.7═══

If $K_1, \ldots, K_n$ are independent Poisson random variables, $W = K_1 + \cdots + K_n$ is a Poisson random variable.

Proof We adopt the notation $\mathrm{E}[K_i] = \alpha_i$ and note in Table 9.1 that K_i has MGF

$$\phi_{K_i}(s) = e^{\alpha_i(e^s-1)}. \tag{9.35}$$

By Theorem 9.6,

$$\phi_W(s) = e^{\alpha_1(e^s-1)}e^{\alpha_2(e^s-1)}\cdots e^{\alpha_n(e^s-1)} = e^{(\alpha_1+\cdots+\alpha_n)(e^s-1)} = e^{(\alpha_T)(e^s-1)} \tag{9.36}$$

where $\alpha_T = \alpha_1 + \cdots + \alpha_n$. Examining Table 9.1, we observe that $\phi_W(s)$ is the moment generating function of the Poisson (α_T) random variable. Therefore,

$$P_W(w) = \begin{cases} \alpha_T^w e^{-\alpha}/w! & w = 0, 1, \ldots, \\ 0 & \text{otherwise.} \end{cases} \tag{9.37}$$

Theorem 9.8

The sum of n independent Gaussian random variables $W = X_1 + \cdots + X_n$ is a Gaussian random variable.

Proof For convenience, let $\mu_i = \mathrm{E}[X_i]$ and $\sigma_i^2 = \mathrm{Var}[X_i]$. Since the X_i are independent, we know that

$$\begin{aligned}
\phi_W(s) &= \phi_{X_1}(s)\phi_{X_2}(s)\cdots\phi_{X_n}(s) \\
&= e^{s\mu_1+\sigma_1^2 s^2/2}e^{s\mu_2+\sigma_2^2 s^2/2}\cdots e^{s\mu_n+\sigma_n^2 s^2/2} \\
&= e^{s(\mu_1+\cdots+\mu_n)+(\sigma_1^2+\cdots+\sigma_n^2)s^2/2}.
\end{aligned} \tag{9.38}$$

From Equation (9.38), we observe that $\phi_W(s)$ is the moment generating function of a Gaussian random variable with expected value $\mu_1 + \cdots + \mu_n$ and variance $\sigma_1^2 + \cdots + \sigma_n^2$.

In general, the sum of independent random variables in one family is a different kind of random variable. The following theorem shows that the Erlang (n, λ) random variable is the sum of n independent exponential (λ) random variables.

Theorem 9.9

If $X_1, \ldots, X_n$ are iid exponential (λ) random variables, then $W = X_1 + \cdots + X_n$ has the Erlang PDF

$$f_W(w) = \begin{cases} \dfrac{\lambda^n w^{n-1} e^{-\lambda w}}{(n-1)!} & w \geq 0, \\ 0 & \text{otherwise.} \end{cases}$$

Proof In Table 9.1 we observe that each X_i has MGF $\phi_X(s) = \lambda/(\lambda-s)$. By Theorem 9.6, W has MGF

$$\phi_W(s) = \left(\frac{\lambda}{\lambda-s}\right)^n. \tag{9.39}$$

Returning to Table 9.1, we see that W has the MGF of an Erlang (n, λ) random variable.

Similar reasoning demonstrates that the sum of n Bernoulli (p) random variables is the binomial (n, p) random variable, and that the sum of k geometric (p) random variables is a Pascal (k, p) random variable.

═══Quiz 9.3═══

(A) Let $K_1, K_2, \ldots, K_m$ be iid discrete uniform random variables with PMF

$$P_K(k) = \begin{cases} 1/n & k = 1, 2, \ldots, n, \\ 0 & \text{otherwise.} \end{cases} \tag{9.40}$$

Find the MGF of $J = K_1 + \cdots + K_m$.

(B) Let $X_1, \ldots, X_n$ be independent Gaussian random variables with $\mathrm{E}[X_i = 0]$ and $\mathrm{Var}[X_i] = i$. Find the PDF of

$$W = \alpha X_1 + \alpha^2 X_2 + \cdots + \alpha^n X_n. \tag{9.41}$$

9.4   Central Limit Theorem

> The central limit theorem states that the CDF of the the sum of n independent random variables converges to a Gaussian CDF as n grows without bound. For values of n encountered in many applications, the approximate Gaussian model provides a very close approximation to the actual model. Using the Gaussian approximation is far more efficient computationally than working with the exact probability model of a sum of random variables.

Probability theory provides us with tools for interpreting observed data. In many practical situations, both discrete PMFs and continuous PDFs approximately follow a *bell-shaped curve*. For example, Figure 9.1 shows the binomial $(n, 1/2)$ PMF for $n = 5$, $n = 10$ and $n = 20$. We see that as n gets larger, the PMF more closely resembles a bell-shaped curve. Recall that in Section 4.6, we encountered a bell-shaped curve as the PDF of a Gaussian random variable. The central limit theorem explains why so many practical phenomena produce data that can be modeled as Gaussian random variables.

We will use the central limit theorem to estimate probabilities associated with the iid sum $W_n = X_1 + \cdots + X_n$. However, as n approaches infinity, $\mathrm{E}[W_n] = n\mu_X$ and $\mathrm{Var}[W_n] = n\,\mathrm{Var}[X]$ approach infinity, which makes it difficult to make a mathematical statement about the convergence of the CDF $F_{W_n}(w)$. Hence, our formal statement of the central limit theorem will be in terms of the standardized random variable

$$Z_n = \frac{\sum_{i=1}^{n} X_i - n\mu_X}{\sqrt{n\sigma_X^2}}. \tag{9.42}$$

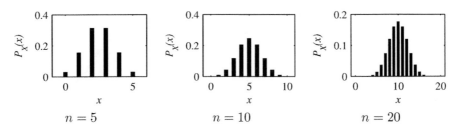

Figure 9.1   The PMF of the X, the number of heads in n coin flips for $n = 5, 10, 20$. As n increases, the PMF more closely resembles a bell-shaped curve.

We say the sum Z_n is standardized since for all n

$$\mathrm{E}\left[Z_n\right] = 0, \qquad \mathrm{Var}[Z_n] = 1. \tag{9.43}$$

Theorem 9.10 **Central Limit Theorem**

Given $X_1, X_2, \ldots$, a sequence of iid random variables with expected value μ_X and variance σ_X^2, the CDF of $Z_n = (\sum_{i=1}^{n} X_i - n\mu_X)/\sqrt{n\sigma_X^2}$ has the property

$$\lim_{n \to \infty} F_{Z_n}(z) = \Phi(z).$$

The proof of this theorem is beyond the scope of this text. In addition to Theorem 9.10, there are other central limit theorems, each with its own statement of the properties of the sums W_n. One remarkable aspect of Theorem 9.10 and its relatives is the fact that there are no restrictions on the nature of the random variables X_i in the sum. They can be continuous, discrete, or mixed. In all cases, the CDF of their sum more and more resembles a Gaussian CDF as the number of terms in the sum increases. Some versions of the central limit theorem apply to sums of sequences X_i that are not even iid.

To use the central limit theorem, we observe that we can express the iid sum $W_n = X_1 + \cdots + X_n$ as

$$W_n = \sqrt{n\sigma_X^2}\, Z_n + n\mu_X. \tag{9.44}$$

The CDF of W_n can be expressed in terms of the CDF of Z_n as

$$F_{W_n}(w) = \mathrm{P}\left[\sqrt{n\sigma_X^2}\, Z_n + n\mu_X \le w\right] = F_{Z_n}\left(\frac{w - n\mu_X}{\sqrt{n\sigma_X^2}}\right). \tag{9.45}$$

For large n, the central limit theorem says that $F_{Z_n}(z) \approx \Phi(z)$. This approximation is the basis for practical applications of the central limit theorem.

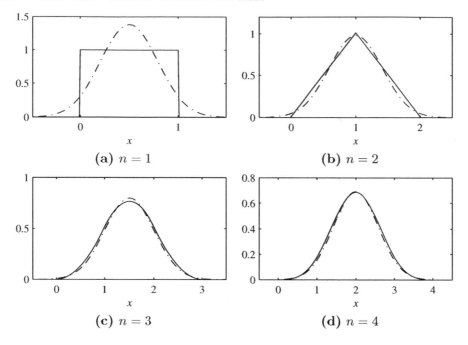

Figure 9.2 The PDF of W_n, the sum of n uniform $(0, 1)$ random variables, and the corresponding central limit theorem approximation for $n = 1, 2, 3, 4$. The solid — line denotes the PDF $f_{W_n}(w)$, and the broken $- \cdot -$ line denotes the Gaussian approximation.

Definition 9.2 Central Limit Theorem Approximation

Let $W_n = X_1 + \cdots + X_n$ be the sum of n iid random variables, each with $\mathrm{E}[X] = \mu_X$ and $\mathrm{Var}[X] = \sigma_X^2$. The central limit theorem approximation to the CDF of W_n is

$$F_{W_n}(w) \approx \Phi\left(\frac{w - n\mu_X}{\sqrt{n\sigma_X^2}}\right).$$

We often call Definition 9.2 a Gaussian approximation for $F_{W_n}(w)$.

Example 9.6

To gain some intuition into the central limit theorem, consider a sequence of iid continuous random variables X_i, where each random variable is uniform $(0,1)$. Let

$$W_n = X_1 + \cdots + X_n. \tag{9.46}$$

Recall that $\mathrm{E}[X] = 0.5$ and $\mathrm{Var}[X] = 1/12$. Therefore, W_n has expected value $\mathrm{E}[W_n] = n/2$ and variance $n/12$. The central limit theorem says that *the CDF of W_n should approach a Gaussian CDF with the same expected value and variance.* Moreover, since W_n is a continuous random variable, we would also expect that the PDF of W_n would converge to a Gaussian PDF. In Figure 9.2, we compare the PDF of W_n to the

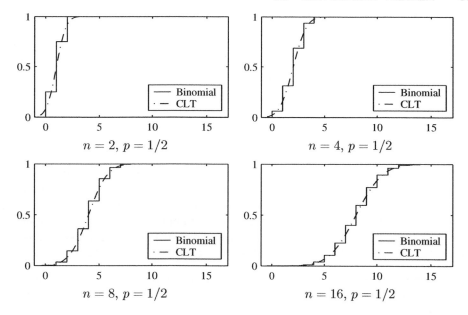

Figure 9.3   The binomial (n, p) CDF and the corresponding central limit theorem approximation for $n = 4, 8, 16, 32$, and $p = 1/2$.

PDF of a Gaussian random variable with the same expected value and variance. First, W_1 is a uniform random variable with the rectangular PDF shown in Figure 9.2(a). This figure also shows the PDF of W_1, a Gaussian random variable with expected value $\mu = 0.5$ and variance $\sigma^2 = 1/12$. Here the PDFs are very dissimilar. When we consider $n = 2$, we have the situation in Figure 9.2(b). The PDF of W_2 is a triangle with expected value 1 and variance $2/12$. The figure shows the corresponding Gaussian PDF. The following figures show the PDFs of $W_3, \ldots, W_6$. The convergence to a bell shape is apparent.

Example 9.7

Now suppose $W_n = X_1 + \cdots + X_n$ is a sum of independent Bernoulli (p) random variables. We know that W_n has the binomial PMF

$$P_{W_n}(w) = \binom{n}{w} p^w (1 - p)^{n-w}. \tag{9.47}$$

No matter how large n becomes, W_n is always a discrete random variable and would have a PDF consisting of impulses. However, the central limit theorem says that the CDF of W_n converges to a Gaussian CDF. Figure 9.3 demonstrates the convergence of the sequence of binomial CDFs to a Gaussian CDF for $p = 1/2$ and four values of n, the number of Bernoulli random variables that are added to produce a binomial random variable. For $n \geq 32$, Figure 9.3 suggests that approximations based on the Gaussian distribution are very accurate.

In addition to helping us understand why we observe bell-shaped curves in so many situations, the central limit theorem makes it possible to perform quick, accurate calculations that would otherwise be extremely complex and time consuming. In these calculations, the random variable of interest is a sum of other random variables, and we calculate the probabilities of events by referring to the corresponding Gaussian random variable. In the following example, the random variable of interest is the average of eight iid uniform random variables. The expected value and variance of the average are easy to obtain. However, a complete probability model is extremely complex (it consists of segments of eighth-order polynomials).

Example 9.8

A compact disc (CD) contains digitized samples of an acoustic waveform. In a CD player with a "one bit digital to analog converter," each digital sample is represented to an accuracy of ± 0.5 mV. The CD player oversamples the waveform by making eight independent measurements corresponding to each sample. The CD player obtains a waveform sample by calculating the average (sample mean) of the eight measurements. What is the probability that the error in the waveform sample is greater than 0.1 mV?

. . . .

The measurements $X_1, X_2, \ldots, X_8$ all have a uniform distribution between $v - 0.5$ mV and $v + 0.5$ mV, where v mV is the exact value of the waveform sample. The compact disk player produces the output $U = W_8/8$, where

$$W_8 = \sum_{i=1}^{8} X_i. \tag{9.48}$$

To find $P[|U - v| > 0.1]$ exactly, we would have to find an exact probability model for W_8, either by computing an eightfold convolution of the uniform PDF of X_i or by using the moment generating function. Either way, the process is extremely complex. Alternatively, we can use the central limit theorem to model W_8 as a Gaussian random variable with $E[W_8] = 8\mu_X = 8v$ mV and variance $\text{Var}[W_8] = 8 \text{Var}[X] = 8/12$. Therefore, U is approximately Gaussian with $E[U] = E[W_8]/8 = v$ and variance $\text{Var}[W_8]/64 = 1/96$. Finally, the error, $U - v$ in the output waveform sample is approximately Gaussian with expected value 0 and variance $1/96$. It follows that

$$P[|U - v| > 0.1] = 2\left[1 - \Phi\left(0.1/\sqrt{1/96}\right)\right] = 0.3272. \tag{9.49}$$

The central limit theorem is particularly useful in calculating events related to binomial random variables. Figure 9.3 from Example 9.7 indicates how the CDF of a sum of n Bernoulli random variables converges to a Gaussian CDF. When n is very high, as in the next two examples, probabilities of events of interest are sums of thousands of terms of a binomial CDF. By contrast, each of the Gaussian approximations requires looking up only one value of the Gaussian CDF $\Phi(x)$.

Example 9.9

Transmit one million bits. Let A denote the event that there are at least 499,000 ones but no more than 501,000 ones. What is $P[A]$?

Let X_i be the value of bit i (either 0 or 1). The number of ones in one million bits is $W = \sum_{i=1}^{10^6} X_i$. Because X_i is a Bernoulli (0.5) random variable, $E[X_i] = 0.5$ and $\text{Var}[X_i] = 0.25$ for all i. Note that $E[W] = 10^6 E[X_i] = 500{,}000$, and $\text{Var}[W] = 10^6 \text{Var}[X_i] = 250{,}000$. Therefore, $\sigma_W = 500$. By the central limit theorem approximation,

$$
\begin{aligned}
P\,[A] &= P\,[W \le 501{,}000] - P\,[W < 499{,}000] \\
&\approx \Phi\left(\frac{501{,}000 - 500{,}000}{500}\right) - \Phi\left(\frac{499{,}000 - 500{,}000}{500}\right) \\
&= \Phi(2) - \Phi(-2) = 0.9544.
\end{aligned}
\tag{9.50}
$$

These examples of using a Gaussian approximation to a binomial probability model contain events that consist of thousands of outcomes. When the events of interest contain a small number of outcomes, the accuracy of the approximation can be improved by accounting for the fact that the Gaussian random variable is continuous whereas the corresponding binomial random variable is discrete.

In fact, using a Gaussian approximation to a discrete random variable is fairly common. We recall that the sum of n Bernoulli random variables is binomial, the sum of n geometric random variables is Pascal, and the sum of n Bernoulli random variables (each with success probability λ/n) approaches a Poisson random variable in the limit as $n \to \infty$. Thus a Gaussian approximation can be accurate for a random variable K that is binomial, Pascal, or Poisson.

Although the central limit theorem approximation provides a useful means of calculating events related to complicated probability models, it has to be used with caution. When the events of interest are confined to outcomes at the edge of the range of a random variable, the central limit theorem approximation can be quite inaccurate. In all of the examples in this section, the random variable of interest has finite range. By contrast, the corresponding Gaussian models have finite probabilities for any range of numbers between $-\infty$ and ∞. Thus in Example 9.8, $P[U - v > 0.5] = 0$, while the Gaussian approximation suggests that $P[U - v > 0.5] = Q(0.5/\sqrt{1/96}) \approx 5 \times 10^{-7}$. Although this is a low probability, there are many applications in which the events of interest have very low probabilities or probabilities very close to 1. In these applications, it is necessary to resort to more complicated methods than a central limit theorem approximation to obtain useful results. In particular, it is often desirable to provide guarantees in the form of an upper bound rather than the approximation offered by the central limit theorem. In the next section, we describe one such method based on the moment generating function.

Quiz 9.4

X milliseconds, the total access time (waiting time + read time) to get one block of information from a computer disk, is the continuous $(0,12)$ random variable. Before performing a certain task, the computer must access 12 different blocks of information from the disk. (Access times for different blocks are independent of one another.) The total access time for all the information is a random variable A milliseconds.

(a) Find the expected value and variance of the access time X.

(b) Find the expected value and standard deviation of the total access time A.

(c) Use the central limit theorem to estimate $P[A > 75 \text{ ms}]$.

(d) Use the central limit theorem to estimate $P[A < 48 \text{ ms}]$.

9.5 MATLAB

MATLAB is convenient for calculating the PMF of the sum of two discrete random variables. To calculate the PMF of the sum of n random variables, run the program for two random variables $n - 1$ times. The central limit theorem suggests a simple way to use a random number generator for the uniform $(0, 1)$ random variable to generate sample values of a Gaussian $(0, 1)$ random variable: Add twelve samples of the uniform $(0, 1)$ random variable and then subtract 6.

As in Sections 5.11 and 8.6, we illustrate two ways of using MATLAB to study random vectors. We first present examples of programs that calculate values of probability functions, in this case the PMF of the sums of independent discrete random variables. Then we present a program that generates sample values of the Gaussian $(0, 1)$ random variable without using the built-in function `randn`.

Probability Functions

The following example produces a MATLAB program for calculating the convolution of two PMFs.

===Example 9.10===

X_1 and X_2 are independent discrete random variables with PMFs

$$P_{X_1}(x) = \begin{cases} 0.04 & x = 1, \ldots, 25, \\ 0 & \text{otherwise,} \end{cases} \qquad P_{X_2}(x) = \begin{cases} \frac{x}{550} & x = 10, 20, \ldots, 100, \\ 0 & \text{otherwise.} \end{cases}$$

What is the PMF of $W = X_1 + X_2$?

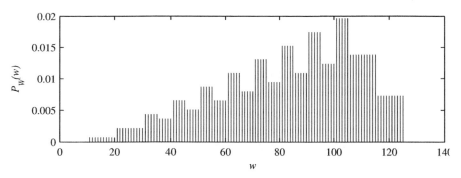

Figure 9.4 The PMF $P_W(w)$ for Example 9.10.

```
%sumx1x2.m
sx1=(1:25);px1=0.04*ones(1,25);
sx2=10*(1:10);px2=sx2/550;
[SX1,SX2]=ndgrid(sx1,sx2);
[PX1,PX2]=ndgrid(px1,px2);
SW=SX1+SX2;PW=PX1.*PX2;
sw=unique(SW);
pw=finitepmf(SW,PW,sw);
pmfplot(sw,pw);
```

As in Example 5.25, sumx1x2.m uses ndgrid to generate a grid for all possible pairs of X_1 and X_2. The matrix SW holds the sum $x_1 + x_2$ for each possible pair x_1, x_2. The probability $P_{X_1,X_2}(x_1,x_2)$ of each such pair is in the matrix PW. For each unique w generated by pairs $x_1 + x_2$, finitepmf finds the probability $P_W(w)$. The graph of $P_W(w)$ appears in Figure 9.4.

```
[SX1,SX2,SX3]=ndgrid(sx1,sx2,sx3);
[PX1,PX2,PX3]=ndgrid(px1,px2,px2);
SW=SX1+SX2+SX3;
PW=PX1.*PX2.*PX3.*PX3;
sw=unique(SW);
pw=finitepmf(SW,PW,sw);
```

The preceding technique extends directly to n independent finite random variables $X_1,\ldots,X_n$ because ndgrid can generate n-dimensional grids. For example, the sum of three random variables can be calculated via the script on the left. However, this technique suffers from the disadvantage that it can generate large matrices. For n random variables such that X_i takes on n_i possible distinct values, SW and PW are square matrices of size $n_1 \times n_2 \times \cdots n_m$. A more efficient technique is to iteratively calculate the PMF of $W_2 = X_1 + X_2$ followed by $W_3 = W_2 + X_3$, $W_4 = W_3 + X_3$. At each step, extracting only the unique values in the range S_{W_n} can economize significantly on memory and computation time.

Sample Values of Gaussian Random Variables

The central limit theorem suggests a simple way to generate samples of the Gaussian (0,1) random variable in computers or calculators without built-in functions like randn. The technique relies on the observation that the sum of 12 independent uniform (0,1) random variables U_i has expected value $12\,\mathrm{E}[U_i] = 6$ and variance

```
>> uniform12(10000);
ans =
  -3.0000 -2.0000 -1.0000        0  1.0000  2.0000  3.0000
   0.0013  0.0228  0.1587  0.5000  0.8413  0.9772  0.9987
   0.0005  0.0203  0.1605  0.5027  0.8393  0.9781  0.9986
>> uniform12(10000);
ans =
  -3.0000 -2.0000 -1.0000        0  1.0000  2.0000  3.0000
   0.0013  0.0228  0.1587  0.5000  0.8413  0.9772  0.9987
   0.0015  0.0237  0.1697  0.5064  0.8400  0.9778  0.9993
```

Figure 9.5   Two sample runs of `uniform12.m`.

$12 \operatorname{Var}[U_i] = 1$. According to the central limit theorem, $X = \sum_{i=1}^{12} U_i - 6$ is approximately Gaussian (0,1).

Example 9.11

Write a MATLAB program to generate $m = 10{,}000$ samples of the random variable $X = \sum_{i=1}^{12} U_i - 6$. Use the data to find the relative frequencies of the following events $\{X \leq T\}$ for $T = -3, -2 \ldots, 3$. Calculate the probabilities of these events when X is a Gaussian $(0, 1)$ random variable.

. . . .

```
function FX=uniform12(m);
x=sum(rand(12,m))-6;
T=(-3:3);FX=(count(x,T)/m)';
[T;phi(T);FX]
```

In `uniform12(m)`, `x` holds the m samples of X. The function `n=count(x,T)` returns `n(i)` as the number of elements of `x` less than or equal to `T(i)`. The output is a three-row table: T on the first row, the true probabilities $P[X \leq T] = \Phi(T)$ second, and the relative frequencies third. Two sample runs of `uniform12` are shown in Figure 9.5. We see that the relative frequencies and the probabilities diverge as T moves farther from zero. In fact this program will never produce a value of $|X| > 6$, no matter how many times it runs. By contrast, $Q(6) = 9.9 \times 10^{-10}$. This suggests that in a set of one billion independent samples of the Gaussian $(0, 1)$ random variable, we can expect two samples with $|X| > 6$, one sample with $X < -6$, and one sample with $X > 6$.

Quiz 9.5

X is the binomial $(100, 0.5)$ random variable and Y is the discrete uniform $(0, 100)$ random variable. Calculate and graph the PMF of $W = X + Y$.

Further Reading: [Dur94] contains a concise, rigorous presentation and proof of the central limit theorem.

Problems

Difficulty: ● Easy   ■ Moderate   ◆ Difficult   ◆◆ Experts Only

9.1.1● X_1 and X_2 are iid random variables with variance $\text{Var}[X]$.

(a) What is $E[X_1 - X_2]$?

(b) What is $\text{Var}[X_1 - X_2]$?

9.1.2● Flip a biased coin 100 times. On each flip, $P[H] = p$. Let X_i denote the number of heads that occur on flip i. What is $P_{X_{33}}(x)$? Are X_1 and X_2 independent? Define $Y = X_1 + X_2 + \cdots + X_{100}$. Describe Y in words. What is $P_Y(y)$? Find $E[Y]$ and $\text{Var}[Y]$.

9.1.3■ A radio program gives concert tickets to the fourth caller with the right answer to a question. Of the people who call, 25% know the answer. Phone calls are independent of one another. The random variable N_r indicates the number of phone calls taken when the rth correct answer arrives. (If the fourth correct answer arrives on the eighth call, then $N_4 = 8$.)

(a) What is the PMF of N_1, the number of phone calls needed to obtain the first correct answer?

(b) What is $E[N_1]$, the expected number of phone calls needed to obtain the first correct answer?

(c) What is the PMF of N_4, the number of phone calls needed to obtain the fourth correct answer? Hint: See Example 3.10.

(d) What is $E[N_4]$? Hint: N_4 can be written as the independent sum $N_4 = K_1 + K_2 + K_3 + K_4$, where each K_i is distributed identically to N_1.

9.1.4■ X_1, X_2 and X_3 are iid continuous uniform random variables. Random variable $Y = X_1 + X_2 + X_3$ has expected value $E[Y] = 0$ and variance $\sigma_Y^2 = 4$. What is the PDF $f_{X_1}(x)$ of X_1?

9.1.5■ Random variables X and Y have joint PDF

$$f_{X,Y}(x,y) = \begin{cases} 2 & x \geq 0, y \geq 0, x + y \leq 1, \\ 0 & \text{otherwise.} \end{cases}$$

What is the variance of $W = X + Y$?

9.2.1● For a constant $a > 0$, a Laplace random variable X has PDF

$$f_X(x) = \frac{a}{2}e^{-a|x|}, \quad -\infty < x < \infty.$$

Calculate the MGF $\phi_X(s)$.

9.2.2■ Random variables J and K have the joint probability mass function

$P_{J,K}(j,k)$	$k = -1$	$k = 0$	$k = 1$
$j = -2$	0.42	0.12	0.06
$j = -1$	0.28	0.08	0.04

(a) What is the MGF of J?

(b) What is the MGF of K?

(c) Find the PMF of $M = J + K$?

(d) What is $E[M^4]$?

9.2.3■ X is the continuous uniform (a,b) random variable. Find the MGF $\phi_X(s)$. Use the MGF to calculate the first and second moments of X.

9.2.4◆ Let X be a Gaussian $(0,\sigma)$ random variable. Use the moment generating function to show that

$$E[X] = 0, \qquad E[X^2] = \sigma^2,$$
$$E[X^3] = 0, \qquad E[X^4] = 3\sigma^4.$$

Let Y be a Gaussian (μ,σ) random variable. Use the moments of X to show that

$$E\left[Y^2\right] = \sigma^2 + \mu^2,$$
$$E\left[Y^3\right] = 3\mu\sigma^2 + \mu^3,$$
$$E\left[Y^4\right] = 3\sigma^4 + 6\mu\sigma^2 + \mu^4.$$

9.2.5◆◆ Random variable K has a discrete uniform $(1,n)$ PMF. Use the MGF $\phi_K(s)$ to find $E[K]$ and $E[K^2]$. Use the first and second moments of K to derive well-known expressions for $\sum_{k=1}^{n} k$ and $\sum_{k=1}^{n} k^2$.

9.3.1● N is the binomial $(100,0.4)$ random variable. M is the binomial $(50,0.4)$ random variable. M and N are independent. What is the PMF of $L = M + N$?

9.3.2 Random variable Y has the moment generating function $\phi_Y(s) = 1/(1-s)$. Random variable V has the moment generating function $\phi_V(s) = 1/(1-s)^4$. Y and V are independent. $W = Y + V$.

(a) What are $E[Y]$, $E[Y^2]$, and $E[Y^3]$?

(b) What is $E[W^2]$?

9.3.3 Let $K_1, K_2, \ldots$ denote a sequence of iid Bernoulli (p) random variables. Let $M = K_1 + \cdots + K_n$.

(a) Find the MGF $\phi_K(s)$.

(b) Find the MGF $\phi_M(s)$.

(c) Use the MGF $\phi_M(s)$ to find $E[M]$ and $\mathrm{Var}[M]$.

9.3.4 Suppose you participate in a chess tournament in which you play n games. Since you are an average player, each game is equally likely to be a win, a loss, or a tie. You collect 2 points for each win, 1 point for each tie, and 0 points for each loss. The outcome of each game is independent of the outcome of every other game. Let X_i be the number of points you earn for game i and let Y equal the total number of points earned over the n games.

(a) Find the moment generating functions $\phi_{X_i}(s)$ and $\phi_Y(s)$.

(b) Find $E[Y]$ and $\mathrm{Var}[Y]$.

9.3.5 At time $t = 0$, you begin counting the arrivals of buses at a depot. The number of buses K_i that arrive between time $i - 1$ minutes and time i minutes has the Poisson PMF

$$P_{K_i}(k) = \begin{cases} 2^k e^{-2}/k! & k = 0, 1, 2, \ldots, \\ 0 & \text{otherwise.} \end{cases}$$

$K_1, K_2, \ldots$ are an iid random sequence. Let $R_i = K_1 + K_2 + \cdots + K_i$ denote the number of buses arriving in the first i minutes.

(a) What is the moment generating function $\phi_{K_i}(s)$?

(b) Find the MGF $\phi_{R_i}(s)$.

(c) Find the PMF $P_{R_i}(r)$. Hint: Compare $\phi_{R_i}(s)$ and $\phi_{K_i}(s)$.

(d) Find $E[R_i]$ and $\mathrm{Var}[R_i]$.

9.3.6 Suppose that during the ith day of December, the energy X_i stored by a solar collector is a Gaussian random variable with expected value $32 - i/4$ kW-hr and standard deviation of 10 kW-hr. Assuming the energy stored each day is independent of any other day, what is the PDF of Y, the total energy stored in the 31 days of December?

9.3.7 $K, K_1, K_2, \ldots$ are iid random variables. Use the MGF of $M = K_1 + \cdots + K_n$ to prove that

(a) $E[M] = n E[K]$.

(b) $E[M^2] = n(n-1)(E[K])^2 + n E[K^2]$.

9.4.1 The waiting time in milliseconds, W, for accessing one record from a computer database is the continuous uniform $(0,10)$ random variable. The read time R (for moving the information from the disk to main memory) is 3 milliseconds. The random variable X milliseconds is the total access time (waiting time + read time) to get one block of information from the disk. Before performing a certain task, the computer must access 12 different blocks of information from the disk. (Access times for different blocks are independent of one another.) The total access time for all the information is a random variable A milliseconds.

(a) What is $E[X]$?

(b) What is $\mathrm{Var}[X]$?

(c) What is $E[A]$?

(d) What is σ_A? time?

(e) Use the central limit theorem to estimate $P[A > 116\mathrm{ms}]$.

(f) Use the central limit theorem to estimate $P[A < 86\mathrm{ms}]$.

9.4.2 Internet packets can be classified as video (V) or as generic data (D). Based on a lot of observations taken by the Internet service provider, we have the following probability model: $P[V] = 3/4$, $P[D] = 1/4$. Data packets and video packets occur independently of one another. The random variable K_n is the number of video packets in a collection of n packets.

(a) What is $E[K_{100}]$, the expected number of video packets in a set of 100 packets?

(b) What is $\sigma_{K_{100}}$?

(c) Use the central limit theorem to estimate $P[K_{100} \geq 18]$.

(d) Use the central limit theorem to estimate $P[16 \leq K_{100} \leq 24]$.

9.4.3● The duration of a cellular telephone call is an exponential random variable with expected value 150 seconds. A subscriber has a calling plan that includes 300 minutes per month at a cost of $30.00 plus $0.40 for each minute that the total calling time exceeds 300 minutes. In a certain month, the subscriber has 120 cellular calls.

(a) Use the central limit theorem to estimate the probability that the subscriber's bill is greater than $36. (Assume that the durations of all phone calls are mutually independent and that the telephone company measures call duration exactly and charges accordingly, without rounding up fractional minutes.)

(b) Suppose the telephone company does charge a full minute for each fractional minute used. Re-calculate your estimate of the probability that the bill is greater than $36.

9.4.4● Let $K_1, K_2, \ldots$ be an iid sequence of Poisson (1) random variables. Let $W_n = K_1 + \cdots + K_n$. Use the improved central limit theorem approximation to estimate $P[W_n = n]$. For $n = 4, 25, 64$, compare the approximation to the exact value of $P[W_n = n]$.

9.4.5● In any one-minute interval, the number of requests for a popular Web page

is a Poisson random variable with expected value 300 requests.

(a) A Web server has a capacity of C requests per minute. If the number of requests in a one-minute interval is greater than C, the server is overloaded. Use the central limit theorem to estimate the smallest value of C for which the probability of overload is less than 0.05.

(b) Use MATLAB to calculate the actual probability of overload for the value of C derived from the central limit theorem.

(c) For the value of C derived from the central limit theorem, what is the probability of overload in a *one-second* interval?

(d) What is the smallest value of C for which the probability of overload in a one-second interval is less than 0.05?

(e) Comment on the application of the central limit theorem to estimate the overload probability in a one-second interval and in a one-minute interval.

9.4.6● Integrated circuits from a certain factory pass a certain quality test with probability 0.8. The outcomes of all tests are mutually independent.

(a) What is the expected number of tests necessary to find 500 acceptable circuits?

(b) Use the central limit theorem to estimate the probability of finding 500 acceptable circuits in a batch of 600 circuits.

(c) Use MATLAB to calculate the actual probability of finding 500 acceptable circuits in a batch of 600 circuits.

(d) Use the central limit theorem to calculate the minimum batch size for finding 500 acceptable circuits with probability 0.9 or greater.

9.4.7■ Internet packets can be classified as video (V) or as generic data (D). Based on a lot of observations taken by the Internet service provider, we have the following probability model: $P[V] = 0.8$, $P[D] = 0.2$. Data packets and video packets occur independently of one another. The random variable K_n is the number of video packets in a collection of n packets.

(a) What is $E[K_{48}]$, the expected number of video packets in a set of 48 packets?

(b) What is $\sigma_{K_{48}}$, the standard deviation of the number of video packets in a set of 48 packets?

(c) Use the central limit theorem to estimate $P[30 \le K_{48} \le 42]$, the probability of between 30 and 42 voice calls in a set of 48 calls.

(d) Use the De Moivre–Laplace formula to estimate $P[30 \le K_{48} \le 42]$.

9.4.8■ In the presence of a headwind of normalized intensity W, your speed on your bike is $V = 20 - 10W^3$ mi/hr. The wind intensity W is a continuous uniform $(-1, 1)$ random variable. Moreover, the wind changes every ten minutes. Let W_i denote the headwind intensity in the ith ten-minute interval. In a five-hour bike ride, with 30 ten-minute intervals, the wind intensities $W_1, \ldots, W_{30}$ are independent and identical to W. The distance you travel is

$$ X = \frac{V_1 + V_2 + \cdots + V_{30}}{6}. $$

Use the CLT to estimate $P[X \ge 95]$.

9.4.9■ An amplifier circuit has power consumption Y that grows nonlinearly with the input signal voltage X. When the input signal is X volts, the instantaneous power consumed by the amplifier is $Y = 20 + 15X^2$ Watts. The input signal X is the continuous uniform $(-1, 1)$ random variable. Sampling the input signal every millisecond over a 100-millisecond interval yields the iid signal samples $X_1, X_2, \ldots, X_{100}$. Over the 100 ms interval, you estimate the average power

of the amplifier as

$$ W = \frac{1}{100} \sum_{i=1}^{100} Y_i $$

where $Y_i = 20 + 15X_i^2$. Use the central limit theorem to estimate $P[W \le 25.4]$.

9.4.10■ In the face of perpetually varying headwinds, cyclists Lance and Ashwin are in a 3000 mile race across America. To maintain a speed of v miles/hour in the presence of a w mi/hr headwind, a cyclist must generate a power output $y = 50 + (v + w - 15)^3$ Watts. During each mile of road, the wind speed W is the continuous uniform $(0, 10)$ random variable independent of the wind speed in any other mile.

(a) Lance rides at constant velocity $v = 15$ mi/hr mile after mile. Let Y denote Lance's power output over a randomly chosen mile. What is $E[Y]$?

(b) Ashwin is less powerful but he is able to ride at constant power $\hat{y}$ Watts in the presence of the same variable headwinds. Use the central limit theorem to find $\hat{y}$ such that Ashwin wins the race with probability $1/2$.

9.4.11◆ Suppose your grade in a probability course depends on 10 weekly quizzes. Each quiz has ten yes/no questions, each worth 1 point. The scoring has no partial credit. Your performance is a model of consistency: On each one-point question, you get the right answer with probability p, independent of the outcome on any other question. Thus your score X_i on quiz i is between 0 and 10. Your average score, $X = \sum_{i=1}^{10} X_i/100$ is used to determine your grade. The course grading has simple letter grades without any curve: A: $X \ge 0.9$, B: $0.8 \le X < 0.9$, C: $0.7 \le X < 0.8$, D: $0.6 \le X < 0.7$ and F: $X < 0.6$. As it happens, you are a borderline B/C student with $p = 0.8$.

(a) What is the PMF of X_i?

(b) Use the central limit theorem to estimate the probability $P[A]$ that your grade is an A.

(c) Suppose now that the course has "attendance quizzes." If you attend a lecture with an attendance quiz, you get credit for a bonus quiz with a score of 10. If you are present for n bonus quizzes, your modified average

$$X' = \frac{10n + \sum_{i=1}^{10} X_i}{10n + 100}$$

is used to calculate your grade: A: $X' \geq 0.9$, B: $0.8 \leq X' < 0.9$, and so on. Given you attend n attendance quizzes, use the central limit theorem to estimate P$[A]$.

(d) Now suppose there are no attendance quizzes and your week 1 quiz is scored an 8. A few hours after the week 1 quiz, you notice that a question was marked incorrectly; your quiz score should have been 9. You appeal to the annoying prof who says "Sorry, all regrade requests must be submitted immediately after receiving your score. But don't worry, the probability it makes a difference is virtually nil." Let U denote the event that your letter grade is unchanged because of the scoring error. Find an exact expression for P$[U]$.

9.5.1● W_n is the number of ones in 10^n independent transmitted bits, each equiprobably 0 or 1. For $n = 3, 4, \ldots$, use the `binomialpmf` function to calculate

$$P\left[0.499 \leq W_n/10^n \leq 0.501\right].$$

What is the largest n for which your MATLAB installation can perform the calcula-

tion? Can you perform the exact calculation of Example 9.9?

9.5.2● Use the MATLAB `plot` function to compare the Erlang (n, λ) PDF to a Gaussian PDF with the same expected value and variance for $\lambda = 1$ and $n = 4, 20, 100$. Why are your results not surprising?

9.5.3● Recreate the plots of Figure 9.3. On the same plots, superimpose the PDF of Y_n, a Gaussian random variable with the same expected value and variance. If X_n denotes the binomial (n, p) random variable, explain why for most integers k, $P_{X_n}(k) \approx f_Y(k)$.

9.5.4● Find the PMF of $W = X_1 + X_2$ in Example 9.10 using the `conv` function.

9.5.5■ Use `uniform12.m` to estimate the probability of a storm surge greater than 7 feet in Example 10.4 based on:

(a) 1000 samples,

(b) 10000 samples.

9.5.6■ X_1, X_2, and X_3 are independent random variables such that X_k has PMF

$$P_{X_k}(x) = \begin{cases} 1/(10k) & x = 1, 2, \ldots, 10k, \\ 0 & \text{otherwise.} \end{cases}$$

Find the PMF of $W = X_1 + X_2 + X_3$.

9.5.7■ Let X and Y denote independent finite random variables described by the probability and range vectors px, sx and py, sy. Write a MATLAB function

`[pw,sw]=sumfinitepmf(px,sx,py,sy)`

such that finite random variable $W = X + Y$ is described by pw and sw.

10

The Sample Mean

Earlier chapters of this book present the properties of probability models. In referring to applications of probability theory, we have assumed prior knowledge of the probability model that governs the outcomes of an experiment. In practice, however, we encounter many situations in which the probability model is not known in advance and experimenters collect data in order to learn about the model. In doing so, they apply principles of *statistical inference*, a body of knowledge that governs the use of measurements to discover the properties of a probability model.

This chapter focuses on the properties of the *sample mean* of a set of data. We refer to independent trials of one experiment, with each trial producing one sample value of a random variable. The sample mean is simply the sum of the sample values divided by the number of trials. We begin by describing the relationship of the sample mean of the data to the expected value of the random variable. We then describe methods of using the sample mean to estimate the expected value.

10.1 Sample Mean: Expected Value and Variance

> The sample mean $M_n(X) = (X_1 + \cdots + X_n)/n$ of n independent observations of random variable X is a random variable with expected value $E[X]$ and variance $\text{Var}[X]/n$.

In this section, we define the *sample mean* of a random variable and identify its expected value and variance. Later sections of this chapter show mathematically how the sample mean converges to a constant as the number of repetitions of an experiment increases. This chapter, therefore, provides the mathematical basis for the statement that although the result of a single experiment is unpredictable, predictable patterns emerge as we collect more and more data.

To define the sample mean, consider repeated independent trials of an experiment. Each trial results in one observation of a random variable, X. After n trials, we have sample values of the n random variables $X_1, \ldots, X_n$, all with the same PDF as X. The sample mean is the numerical average of the observations.

====Definition 10.1====Sample Mean

*For iid random variables $X_1, \ldots, X_n$ with PDF $f_X(x)$, the **sample mean** of X is the random variable*

$$M_n(X) = \frac{X_1 + \cdots + X_n}{n}.$$

The first thing to notice is that $M_n(X)$ is a function of the random variables $X_1, \ldots, X_n$ and is therefore a random variable itself. It is important to distinguish the sample mean, $M_n(X)$, from $E[X]$, which we sometimes refer to as the *mean value* of random variable X. While $M_n(X)$ is a random variable, $E[X]$ is a number. To avoid confusion when studying the sample mean, it is advisable to refer to $E[X]$ as the *expected value* of X, rather than the *mean* of X. The sample mean of X and the expected value of X are closely related. A major purpose of this chapter is to explore the fact that as n increases without bound, $M_n(X)$ predictably approaches $E[X]$. In everyday conversation, this phenomenon is often called the *law of averages*.

The expected value and variance of $M_n(X)$ reveal the most important properties of the sample mean. From our earlier work with sums of random variables in Chapter 9, we have the following result.

====Theorem 10.1====

The sample mean $M_n(X)$ has expected value and variance

$$E[M_n(X)] = E[X], \qquad \text{Var}[M_n(X)] = \frac{\text{Var}[X]}{n}.$$

Proof From Definition 10.1, Theorem 9.1, and the fact that $E[X_i] = E[X]$ for all i,

$$E[M_n(X)] = \frac{1}{n}(E[X_1] + \cdots + E[X_n]) = \frac{1}{n}(E[X] + \cdots + E[X]) = E[X]. \qquad (10.1)$$

Because $\text{Var}[aY] = a^2 \text{Var}[Y]$ for any random variable Y (Theorem 3.15), $\text{Var}[M_n(X)] = \text{Var}[X_1 + \cdots + X_n]/n^2$. Since the X_i are iid, we can use Theorem 9.3 to show

$$\text{Var}[X_1 + \cdots + X_n] = \text{Var}[X_1] + \cdots + \text{Var}[X_n] = n\,\text{Var}[X]. \qquad (10.2)$$

Thus $\text{Var}[M_n(X)] = n\,\text{Var}[X]/n^2 = \text{Var}[X]/n$.

Recall that in Section 3.5, we refer to the expected value of a random variable as a *typical value*. Theorem 10.1 demonstrates that $E[X]$ is a typical value of $M_n(X)$, regardless of n. Furthermore, Theorem 10.1 demonstrates that as n increases without bound, the variance of $M_n(X)$ goes to zero. When we first met the variance, and its square root the standard deviation, we said that they indicate how far a random variable is likely to be from its expected value. Theorem 10.1 suggests that as n approaches infinity, it becomes highly likely that $M_n(X)$ is arbitrarily close to its expected value, $E[X]$. In other words, the sample mean $M_n(X)$ converges to the expected value $E[X]$ as the number of samples n goes to infinity. The rest of this chapter contains the mathematical analysis that describes the nature of this convergence.

Quiz 10.1

X is the exponential (1) random variable; $M_n(X)$ is the sample mean of n independent samples of X. How many samples n are needed to guarantee that the variance of the sample mean $M_n(X)$ is no more than 0.01?

10.2   Deviation of a Random Variable from the Expected Value

> The Chebyshev inequality is an upper bound on the probability $P[|X - \mu_X| > c]$. We use the Chebyshev inequality to derive the Laws of Large Numbers and the parameter-estimation techniques that we study in the next two sections. The Chebyshev inequality is derived from the Markov inequality, a looser upper bound. The Chernoff bound is a more accurate inequality calculated from the complete probability model of X.

The analysis of the convergence of $M_n(X)$ to $E[X]$ begins with a study of the random variable $|X - \mu_X|$, the absolute difference between a random variable X and its expected value. This study leads to the *Chebyshev inequality*, which states that the probability of a large deviation from the expected value is inversely proportional to the square of the deviation. The derivation of the Chebyshev inequality begins with the *Markov inequality*, an upper bound on the probability that a sample value of a nonnegative random variable exceeds the expected value by any arbitrary factor. The Laws of Large Numbers and techniques for parameter estimation, the subject of the next two sections, are a consequence of the Chebyshev inequality.

The *Chernoff bound* is a third inequality used to estimate the probability that a random sample differs substantially from its expected value. The Chernoff bound is more accurate than the Chebyshev and Markov inequalities because it takes into account more information about the probability model of X.

To understand the relationship of the Markov inequality, the Chebyshev inequality, and the Chernoff bound, we consider the example of a storm surge following a hurricane. We assume that the probability model for the random height in feet of storm surges is X, the Gaussian $(5.5, 1)$ random variable, and consider the event $[X \geq 11]$ feet. The probability of this event is very close to zero: $P[X \geq 11] = Q(11 - 5.5) = 1.90 \times 10^{-8}$.

════Theorem 10.2════════Markov Inequality

For a random variable X, such that $P[X < 0] = 0$, and a constant c,

$$P\left[X \geq c^2\right] \leq \frac{E[X]}{c^2}.$$

Proof Since X is nonnegative, $f_X(x) = 0$ for $x < 0$ and

$$E[X] = \int_0^{c^2} x f_X(x) \, dx + \int_{c^2}^\infty x f_X(x) \, dx \geq \int_{c^2}^\infty x f_X(x) \, dx. \qquad (10.3)$$

Since $x \geq c^2$ in the remaining integral,

$$E[X] \geq c^2 \int_{c^2}^\infty f_X(x) \, dx = c^2 \, P\left[X \geq c^2\right]. \qquad (10.4)$$

Keep in mind that the Markov inequality is valid only for nonnegative random variables. As we see in the next example, the bound provided by the Markov inequality can be very loose.

════Example 10.1════

Let X represent the height (in feet) of a storm surge following a hurricane. If the expected height is $E[X] = 5.5$, then the Markov inequality states that an upper bound on the probability of a storm surge at least 11 feet high is

$$P[X \geq 11] \leq 5.5/11 = 1/2. \qquad (10.5)$$

We say the Markov inequality is a loose bound because the probability that a storm surge is higher than 11 feet is essentially zero, while the inequality merely states that it is less than or equal to $1/2$. Although the bound is extremely loose for many random variables, it is tight (in fact, an equation) with respect to some random variables.

════Example 10.2════

Suppose random variable Y takes on the value c^2 with probability p and the value 0 otherwise. In this case, $E[Y] = pc^2$, and the Markov inequality states

$$P\left[Y \geq c^2\right] \leq E[Y]/c^2 = p. \qquad (10.6)$$

Since $P[Y \geq c^2] = p$, we observe that the Markov inequality is in fact an equality in this instance.

The Chebyshev inequality applies the Markov inequality to the nonnegative random variable $(Y - \mu_Y)^2$, derived from any random variable Y.

Theorem 10.3   Chebyshev Inequality

For an arbitrary random variable Y and constant $c > 0$,

$$P\left[|Y - \mu_Y| \geq c\right] \leq \frac{\text{Var}[Y]}{c^2}.$$

Proof In the Markov inequality, Theorem 10.2, let $X = (Y - \mu_Y)^2$. The inequality states

$$P\left[X \geq c^2\right] = P\left[(Y - \mu_Y)^2 \geq c^2\right] \leq \frac{\text{E}\left[(Y - \mu_Y)^2\right]}{c^2} = \frac{\text{Var}[Y]}{c^2}. \tag{10.7}$$

The theorem follows from the fact that $\{(Y - \mu_Y)^2 \geq c^2\} = \{|Y - \mu_Y| \geq c\}$.

Unlike the Markov inequality, the Chebyshev inequality is valid for all random variables. While the Markov inequality refers only to the expected value of a random variable, the Chebyshev inequality also refers to the variance. Because it uses more information about the random variable, the Chebyshev inequality generally provides a tighter bound than the Markov inequality. In particular, when the variance of Y is very small, the Chebyshev inequality says it is unlikely that Y is far away from $\text{E}[Y]$.

Example 10.3

If the height X of a storm surge following a hurricane has expected value $\text{E}[X] = 5.5$ feet and standard deviation $\sigma_X = 1$ foot, use the Chebyshev inequality to to find an upper bound on $P[X \geq 11]$.
....
Since a height X is nonnegative, the probability that $X \geq 11$ can be written as

$$P\left[X \geq 11\right] = P\left[X - \mu_X \geq 11 - \mu_X\right] = P\left[|X - \mu_X| \geq 5.5\right]. \tag{10.8}$$

Now we use the Chebyshev inequality to obtain

$$P\left[X \geq 11\right] = P\left[|X - \mu_X| \geq 5.5\right] \leq \text{Var}[X]/(5.5)^2 = 0.033 \approx 1/30. \tag{10.9}$$

Although this bound is better than the Markov bound, it is also loose. $P[X \geq 11]$ is seven orders of magnitude lower than $1/30$.

The Chernoff bound is an inequality derived from the moment generating function in Definition 9.1. Like the Markov and Chebyshev inequalities, the Chernoff bound is an upper bound on the probability that a sample value of a random variable is greater than some amount. To derive the Chernoff bound we consider the event $P[X \geq c]$ This Chernoff bound is useful when c is large relative to $\text{E}[X]$ and $P[X > c]$ is small.

Theorem 10.4   Chernoff Bound

For an arbitrary random variable X and a constant c,

$$P\left[X \geq c\right] \leq \min_{s \geq 0} e^{-sc}\phi_X(s).$$

Proof In terms of the unit step function, $u(x)$, we observe that

$$P[X \geq c] = \int_c^\infty f_X(x)\,dx = \int_{-\infty}^\infty u(x-c)f_X(x)\,dx. \tag{10.10}$$

For all $s \geq 0$, $u(x-c) \leq e^{s(x-c)}$. This implies

$$P[X \geq c] \leq \int_{-\infty}^\infty e^{s(x-c)}f_X(x)\,dx = e^{-sc}\int_{-\infty}^\infty e^{sx}f_X(x)\,dx = e^{-sc}\phi_X(s). \tag{10.11}$$

This inequality is true for any $s \geq 0$. Hence the upper bound must hold when we choose s to minimize $e^{-sc}\phi_X(s)$.

The Chernoff bound can be applied to any random variable. However, for small values of c, $e^{-sc}\phi_X(s)$ will be minimized by a negative value of s. In this case, the minimizing nonnegative s is $s = 0$, and the Chernoff bound gives the trivial answer $P[X \geq c] \leq 1$.

Example 10.4

If the probability model of the height X, measured in feet, of a storm surge following a hurricane at a certain location is the Gaussian (5.5, 1) random variable, use the Chernoff bound to find an upper bound on $P[X \geq 11]$.
. . . .
In Table 9.1 the MGF of X is

$$\phi_X(s) = e^{(11s+s^2)/2}. \tag{10.12}$$

Thus the Chernoff bound is

$$P[X \geq 11] \leq \min_{s\geq 0} e^{-11s}e^{(11s+s^2)/2} = \min_{s\geq 0} e^{(s^2-11s)/2}. \tag{10.13}$$

To find the minimizing s, it is sufficient to choose s to minimize $h(s) = s^2 - 11s$. Setting the derivative $dh(s)/ds = 2s - 11 = 0$ yields $s = 5.5$. Applying $s = 5.5$ to the bound yields

$$P[X \geq 11] \leq e^{(s^2-11s)/2}\Big|_{s=5.5} = e^{-(5.5)^2/2} = 2.7 \times 10^{-7}. \tag{10.14}$$

Even though the Chernoff bound is 14 times higher than the actual probability, $1 - \Phi(5.5) = 1.90 \times 10^{-8}$, it still conveys the information that a storm surge higher than 11 feet is extremely unlikely. By contrast, the Markov and Chebyshev inequalities provide bounds that suggest that an 11-foot storm surge occurs relatively frequently. The information needed to calculate the three inequalities accounts for the differences in their accuracy. The Markov inequality uses only the expected value, the Chebyshev inequality uses the expected value and the variance, while the much more accurate Chernoff bound is based on knowledge of the complete probability model (expressed as $\phi_X(s)$).

Quiz 10.2

In a subway station, there are exactly enough customers on the platform to fill three trains. The arrival time of the nth train is $X_1 + \cdots + X_n$ where $X_1, X_2, \ldots$

are iid exponential random variables with $E[X_i] = 2$ minutes. Let W equal the time required to serve the waiting customers. For $P[W > 20]$, the probability that W is over twenty minutes,

(a) Use the central limit theorem to find an estimate.

(b) Use the Markov inequality to find an upper bound.

(c) Use the Chebyshev inequality to find an upper bound.

(d) Use the Chernoff bound to find an upper bound.

(e) Use Theorem 4.11 for an exact calculation.

10.3   Laws of Large Numbers

> The sample mean $M_n(X)$ converges to $E[X]$ and the relative frequency of event A converges to $P[A]$ as n, the number of independent trials of an experiment, increases without bound.

When we apply the Chebyshev inequality to $Y = M_n(X)$, we obtain useful insights into the properties of independent samples of a random variable.

=====Theorem 10.5=====Weak Law of Large Numbers (Finite Samples)

For any constant $c > 0$,

(a) $P[|M_n(X) - \mu_X| \geq c] \leq \dfrac{\text{Var}[X]}{nc^2}$,

(b) $P[|M_n(X) - \mu_X| < c] \geq 1 - \dfrac{\text{Var}[X]}{nc^2}$.

Proof Let $Y = M_n(X)$. Theorem 10.1 states that

$$E[Y] = E[M_n(X)] = \mu_X \qquad \text{Var}[Y] = \text{Var}[M_n(X)] = \text{Var}[X]/n. \qquad (10.15)$$

Theorem 10.5(a) follows by applying the Chebyshev inequality (Theorem 10.3) to $Y = M_n(X)$. Theorem 10.5(b) is just a restatement of Theorem 10.5(a), since

$$P[|M_n(X) - \mu_X| \geq c] = 1 - P[|M_n(X) - \mu_X| < c]. \qquad (10.16)$$

In words, Theorem 10.5(a) says that the probability that the sample mean is more than c units from $E[X]$ can be made arbitrarily small by letting the number of samples n become large. By taking the limit as $n \to \infty$, we obtain the infinite limit result in the next theorem.

=====Theorem 10.6=====Weak Law of Large Numbers (Infinite Samples)

If X has finite variance, then for any constant $c > 0$,

(a) $\lim_{n\to\infty} P[|M_n(X) - \mu_X| \geq c] = 0,$

(b) $\lim_{n\to\infty} P[|M_n(X) - \mu_X| < c] = 1.$

In parallel to Theorem 10.5, Theorems 10.6(a) and 10.6(b) are equivalent statements because

$$P[|M_n(X) - \mu_X| \geq c] = 1 - P[|M_n(X) - \mu_X| < c]. \qquad (10.17)$$

In words, Theorem 10.6(b) says that the probability that the sample mean is within $\pm c$ units of $E[X]$ goes to one as the number of samples approaches infinity.

Since c can be arbitrarily small (e.g., 10^{-2000}), both Theorem 10.5(a) and Theorem 10.6(b) can be interpreted as saying that the sample mean converges to $E[X]$ as the number of samples increases without bound. The weak law of large numbers is a very general result because it holds for all random variables X with finite variance. Moreover, we do not need to know any of the parameters, such as the expected value or variance, of random variable X.

The adjective *weak* in the weak law of large numbers suggests that there is also a strong law. They differ in the nature of the convergence of $M_n(X)$ to μ_X. The convergence in Theorem 10.6 is an example of *convergence in probability*.

Definition 10.2   Convergence in Probability

The random sequence Y_n converges in probability to a constant y if for any $\epsilon > 0$,

$$\lim_{n\to\infty} P[|Y_n - y| \geq \epsilon] = 0.$$

The weak law of large numbers (Theorem 10.6) is an example of convergence in probability in which $Y_n = M_n(X)$, $y = E[X]$, and $\epsilon = c$.

The *strong law of large numbers* states that *with probability* 1, the sequence $M_1, M_2, \ldots$ has the limit μ_X. Mathematicians use the terms *convergence almost surely*, *convergence almost always*, and *convergence almost everywhere* as synonyms for convergence with probability 1. The difference between the strong law and the weak law of large numbers is subtle and rarely arises in practical applications of probability theory.

As we will see in the next theorem, the weak law of large numbers validates the relative frequency interpretation of probabilities. Consider an arbitrary event A from an experiment. To examine $P[A]$ we define the indicator random variable

$$X_A = \begin{cases} 1 & \text{if event } A \text{ occurs,} \\ 0 & \text{otherwise.} \end{cases} \qquad (10.18)$$

Since X_A is a Bernoulli random variable with success probability $P[A]$, $E[X_A] = P[A]$. Since general properties of the expected value of a random variable apply to $E[X_A]$, we can apply the law of large numbers to samples of the indicator X_A:

$$\hat{P}_n(A) = M_n(X_A) = \frac{X_{A1} + X_{A2} + \cdots + X_{An}}{n}. \qquad (10.19)$$

Since X_{Ai} just counts whether event A occurred on trial i, $\hat{P}_n(A)$ is the *relative frequency* of event A in n trials. Since $\hat{P}_n(A)$ is the sample mean of X_A, we will see that the properties of the sample mean explain the mathematical connection between relative frequencies and probabilities.

> **Theorem 10.7**
>
> As $n \to \infty$, the relative frequency $\hat{P}_n(A)$ converges to $P[A]$; for any constant $c > 0$,
>
> $$\lim_{n \to \infty} P\left[\left|\hat{P}_n(A) - P[A]\right| \geq c\right] = 0.$$

Proof The proof follows from Theorem 10.6 since $\hat{P}_n(A) = M_n(X_A)$ is the sample mean of the indicator X_A, which has expected value $E[X_A] = P[A]$ and variance $Var[X_A] = P[A](1 - P[A])$.

Theorem 10.7 is a mathematical version of the statement that as the number of observations grows without limit, the relative frequency of any event approaches the probability of the event.

> **Quiz 10.3**
>
> $X_1, \ldots, X_n$ are n iid samples of the Bernoulli $(p = 0.8)$ random variable X.
>
> (a) Find $E[X]$ and $Var[X]$.
>
> (b) What is $Var[M_{100}(X)]$?
>
> (c) Use Theorem 10.5 to find α such that
>
> $$P[|M_{100}(X) - p| \geq 0.05] \leq \alpha.$$
>
> (d) How many samples n are needed to guarantee
>
> $$P[|M_n(X) - p| \geq 0.1] \leq 0.05.$$

10.4   Point Estimates of Model Parameters

> $\hat{R}$, an estimate of a parameter, r, of a probability model is unbiased if $E[\hat{R}] = r$. A sequence of estimates $\hat{R}_1, \hat{R}_2, \ldots$ is consistent if $\lim_{n \to \infty} \hat{R}_n = r$. The sample mean is an unbiased estimator of μ_X. The sequence of sample means is consistent. The sample variance is a biased estimator of $Var[X]$.

In the remainder of this chapter, we consider experiments performed in order to obtain information about a probability model. To do so, investigators usually derive probability models from practical measurements. Later, they use the models in ways described throughout this book. How to obtain a model in the first place is a major subject in statistical inference. In this section we briefly introduce the subject by studying estimates of the expected value and the variance of a random variable.

The general problem is estimation of a *parameter* of a probability model. A parameter is any number that can be calculated from the probability model. For example, for an arbitrary event A, $P[A]$ is a model parameter.

The techniques we study in this chapter rely on the properties of the sample mean $M_n(X)$. Depending on the definition of the random variable X, we can use the sample mean to describe any parameter of a probability model. We focus on a *point estimate*, a number that is as close as possible to the parameter to be estimated.

Properties of Point Estimates

Before presenting estimation methods based on the sample mean, we introduce three properties of point estimates: *bias, consistency,* and *accuracy*. We will see that the sample mean is an unbiased, consistent estimator of the expected value of a random variable. By contrast, we will find that the sample variance is a biased estimate of the variance of a random variable. One measure of the accuracy of an estimate is the *mean square error*, the expected squared difference between an estimate and the estimated parameter.

Consider an experiment that produces observations of sample values of the random variable X. We perform an indefinite number of independent trials of the experiment. The observations are sample values of the random variables $X_1, X_2, \ldots$, all with the same probability model as X. Assume that r is a parameter of the probability model. We use the observations $X_1, X_2, \ldots$ to produce a sequence of estimates of r. The estimates $\hat{R}_1, \hat{R}_2, \ldots$ are all random variables. $\hat{R}_1$ is a function of X_1. $\hat{R}_2$ is a function of X_1 and X_2, and in general $\hat{R}_n$ is a function of $X_1, X_2, \ldots, X_n$. When the sequence of estimates $\hat{R}_1, \hat{R}_2, \ldots$ converges in probability to r, we say the estimator is *consistent*.

=====Definition 10.3=====Consistent Estimator
*The sequence of estimates $\hat{R}_1, \hat{R}_2, \ldots$ of parameter r is **consistent** if for any $\epsilon > 0$,*

$$\lim_{n \to \infty} P\left[\left|\hat{R}_n - r\right| \geq \epsilon\right] = 0.$$

Another property of an estimate, $\hat{R}$, is *bias*. Remember that $\hat{R}$ is a random variable. Of course, we would like $\hat{R}$ to be close to the true parameter value r with high probability. In repeated experiments, however, sometimes $\hat{R} < r$ and other times $\hat{R} > r$. Although $\hat{R}$ is random, it would be undesirable if $\hat{R}$ was either typically less than r or typically greater than r. To be precise, we would like $\hat{R}$ to be *unbiased*.

=====Definition 10.4=====Unbiased Estimator
*An estimate, $\hat{R}$, of parameter r is **unbiased** if $E[\hat{R}] = r$; otherwise, $\hat{R}$ is **biased**.*

Unlike consistency, which is a property of a sequence of estimators, bias (or lack of bias) is a property of a single estimator $\hat{R}$. The concept of *asymptotic bias* applies to a sequence of estimators $\hat{R}_1, \hat{R}_2, \ldots$ such that each $\hat{R}_n$ is biased with the bias diminishing toward zero for large n. This type of sequence is *asymptotically unbiased*.

=======Definition 10.5=======Asymptotically Unbiased Estimator

*The sequence of estimators $\hat{R}_n$ of parameter r is **asymptotically unbiased** if*

$$\lim_{n \to \infty} \mathrm{E}[\hat{R}_n] = r.$$

The mean square error is an important measure of the accuracy of a point estimate. We first encountered the mean square error in Section 3.8; however, in that chapter, we were estimating the value of a random variable. That is, we were guessing a deterministic number as a prediction of a random variable that we had yet to observe. Here we use the same mean square error metric, but we are using a random variable derived from experimental trials to estimate a deterministic but unknown parameter.

=======Definition 10.6=======Mean Square Error

*The **mean square error** of estimator $\hat{R}$ of parameter r is*

$$e = \mathrm{E}\left[(\hat{R} - r)^2\right].$$

Note that when $\hat{R}$ is an unbiased estimate of r and $\mathrm{E}[\hat{R}] = r$, the mean square error is the variance of $\hat{R}$. For a sequence of unbiased estimates, it is enough to show that the mean square error goes to zero to prove that the estimator is consistent.

=======Theorem 10.8=======

If a sequence of unbiased estimates $\hat{R}_1, \hat{R}_2, \ldots$ of parameter r has mean square error $e_n = \mathrm{Var}[\hat{R}_n]$ satisfying $\lim_{n \to \infty} e_n = 0$, then the sequence $\hat{R}_n$ is consistent.

Proof Since $\mathrm{E}[\hat{R}_n] = r$, we apply the Chebyshev inequality to $\hat{R}_n$. For any constant $\epsilon > 0$,

$$P\left[\left|\hat{R}_n - r\right| \geq \epsilon\right] \leq \frac{\mathrm{Var}[\hat{R}_n]}{\epsilon^2}. \tag{10.20}$$

In the limit of large n, we have

$$\lim_{n \to \infty} P\left[\left|\hat{R}_n - r\right| \geq \epsilon\right] \leq \lim_{n \to \infty} \frac{\mathrm{Var}[\hat{R}_n]}{\epsilon^2} = 0. \tag{10.21}$$

=======Example 10.5=======

In any interval of k seconds, the number N_k of packets passing through an Internet router is a Poisson random variable with expected value $\mathrm{E}[N_k] = kr$ packets. Let

$\hat{R}_k = N_k/k$ denote an estimate of the parameter r packets/second. Is each estimate $\hat{R}_k$ an unbiased estimate of r? What is the mean square error e_k of the estimate $\hat{R}_k$? Is the sequence of estimates $\hat{R}_1, \hat{R}_2, \ldots$ consistent?

....

First, we observe that $\hat{R}_k$ is an unbiased estimator since

$$E[\hat{R}_k] = E[N_k/k] = E[N_k]/k = r. \tag{10.22}$$

Next, we recall that since N_k is Poisson, $\mathrm{Var}[N_k] = kr$. This implies

$$\mathrm{Var}[\hat{R}_k] = \mathrm{Var}\left[\frac{N_k}{k}\right] = \frac{\mathrm{Var}[N_k]}{k^2} = \frac{r}{k}. \tag{10.23}$$

Because $\hat{R}_k$ is unbiased, the mean square error of the estimate is the same as its variance: $e_k = r/k$. In addition, since $\lim_{k \to \infty} \mathrm{Var}[\hat{R}_k] = 0$, the sequence of estimators $\hat{R}_k$ is consistent by Theorem 10.8.

Point Estimates of the Expected Value

To estimate $r = E[X]$, we use $\hat{R}_n = M_n(X)$, the sample mean. Since Theorem 10.1 tells us that $E[M_n(X)] = E[X]$, the sample mean is unbiased.

Theorem 10.9

The sample mean $M_n(X)$ is an unbiased estimate of $E[X]$.

Because the sample mean is unbiased, the mean square difference between $M_n(x)$ and $E[X]$ is $\mathrm{Var}[M_n(X)]$, given in Theorem 10.1:

Theorem 10.10

The sample mean estimator $M_n(X)$ has mean square error

$$e_n = E\left[(M_n(X) - E[X])^2\right] = \mathrm{Var}[M_n(X)] = \frac{\mathrm{Var}[X]}{n}.$$

In the terminology of statistical inference, $\sqrt{e_n}$, the standard deviation of the sample mean, is referred to as the *standard error* of the estimate. The standard error gives an indication of how far we should expect the sample mean to deviate from the expected value. In particular, when X is a Gaussian random variable (and $M_n(X)$ is also Gaussian), Problem 10.4.1 asks you to show that

$$P[E[X] - \sqrt{e_n} \le M_n(X) \le E[X] + \sqrt{e_n}] = 2\Phi(1) - 1 \approx 0.68. \tag{10.24}$$

In words, Equation (10.24) says there is roughly a two-thirds probability that the sample mean is within one standard error of the expected value. This same conclusion is approximately true when n is large and the central limit theorem says that $M_n(X)$ is approximately Gaussian.

══════Example 10.6══════

How many independent trials n are needed to guarantee that $\hat{P}_n(A)$, the relative frequency estimate of $P[A]$, has standard error ≤ 0.1?

Since the indicator X_A has variance $\text{Var}[X_A] = P[A](1 - P[A])$, Theorem 10.10 implies that the mean square error of $M_n(X_A)$ is

$$e_n = \frac{\text{Var}[X]}{n} = \frac{P[A](1 - P[A])}{n}. \tag{10.25}$$

We need to choose n large enough to guarantee $\sqrt{e_n} \leq 0.1$ ($e_n \leq 0.01$) even though we don't know $P[A]$. We use the fact that $p(1 - p) \leq 0.25$ for all $0 \leq p \leq 1$. Thus, $e_n \leq 0.25/n$. To guarantee $e_n \leq 0.01$, we choose $n = 0.25/0.01 = 25$ trials.

Theorem 10.10 demonstrates that the standard error of the estimate of $E[X]$ converges to zero as n grows without bound. The following theorem states that this implies that the sequence of sample means is a consistent estimator of $E[X]$.

══════Theorem 10.11══════

If X has finite variance, then the sample mean $M_n(X)$ is a sequence of consistent estimates of $E[X]$.

Proof By Theorem 10.10, the mean square error of $M_n(X)$ satisfies

$$\lim_{n \to \infty} \text{Var}[M_n(X)] = \lim_{n \to \infty} \frac{\text{Var}[X]}{n} = 0. \tag{10.26}$$

By Theorem 10.8, the sequence $M_n(X)$ is consistent.

Theorem 10.11 is a restatement of the weak law of large numbers (Theorem 10.6) in the language of parameter estimation.

Point Estimates of the Variance

When the unknown parameter is $r = \text{Var}[X]$, we have two cases to consider. Because $\text{Var}[X] = E[(X - \mu_X)^2]$ depends on the expected value, we consider separately the situation when $E[X]$ is known and when $E[X]$ is an unknown parameter estimated by $M_n(X)$.

Suppose we know that $E[X] = 0$. In this case, $\text{Var}[X] = E[X^2]$ and estimation of the variance is straightforward. If we define $Y = X^2$, we can view the estimation of $E[X^2]$ from the samples X_i as the estimation of $E[Y]$ from the samples $Y_i = X_i^2$. That is, the sample mean of Y can be written as

$$M_n(Y) = \frac{1}{n}\left(X_1^2 + \cdots + X_n^2\right). \tag{10.27}$$

Assuming that $\text{Var}[Y]$ exists, the weak law of large numbers implies that $M_n(Y)$ is a consistent, unbiased estimator of $E[X^2] = \text{Var}[X]$.

When $E[X]$ is a known quantity μ_X, we know $\text{Var}[X] = E[(X - \mu_X)^2]$. In this case, we can use the sample mean of $W = (X - \mu_X)^2$ to estimate $\text{Var}[X]$.,

$$M_n(W) = \frac{1}{n} \sum_{i=1}^{n} (X_i - \mu_X)^2 . \tag{10.28}$$

If $\text{Var}[W]$ exists, $M_n(W)$ is a consistent, unbiased estimate of $\text{Var}[X]$.

When the expected value μ_X is unknown, the situation is more complicated because the variance of X depends on μ_X. We cannot use Equation (10.28) if μ_X is unknown. In this case, we replace the expected value μ_X by the sample mean $M_n(X)$.

Definition 10.7 Sample Variance
The sample variance of n independent observations of random variable X is

$$V_n(X) = \frac{1}{n} \sum_{i=1}^{n} (X_i - M_n(X))^2 .$$

In contrast to the sample mean, the sample variance is a *biased* estimate of $\text{Var}[X]$.

Theorem 10.12

$$E[V_n(X)] = \frac{n-1}{n} \text{Var}[X].$$

Proof Substituting Definition 10.1 of the sample mean $M_n(X)$ into Definition 10.7 of sample variance and expanding the sums, we derive

$$V_n = \frac{1}{n} \sum_{i=1}^{n} X_i^2 - \frac{1}{n^2} \sum_{i=1}^{n} \sum_{j=1}^{n} X_i X_j . \tag{10.29}$$

Because the X_i are iid, $E[X_i^2] = E[X^2]$ for all i, and $E[X_i] E[X_j] = \mu_X^2$. By Theorem 5.16(a),

$$E[X_i X_j] = \text{Cov}[X_i, X_j] + E[X_i] E[X_j] = \text{Cov}[X_i, X_j] + \mu_X^2 . \tag{10.30}$$

Combining these facts, the expected value of V_n in Equation (10.29) is

$$E[V_n] = E[X^2] - \frac{1}{n^2} \sum_{i=1}^{n} \sum_{j=1}^{n} \left(\text{Cov}[X_i, X_j] + \mu_X^2 \right)$$

$$= \text{Var}[X] - \frac{1}{n^2} \sum_{i=1}^{n} \sum_{j=1}^{n} \text{Cov}[X_i, X_j] . \tag{10.31}$$

Since the double sum has n^2 terms, $\sum_{i=1}^{n} \sum_{j=1}^{n} \mu_X^2 = n^2 \mu_X^2$. Of the n^2 covariance terms, there are n terms of the form $\text{Cov}[X_i, X_i] = \text{Var}[X]$, while the remaining covariance terms are all 0 because X_i and X_j are independent for $i \neq j$. This implies

$$E[V_n] = \text{Var}[X] - \frac{1}{n^2}(n\,\text{Var}[X]) = \frac{n-1}{n}\text{Var}[X]. \qquad (10.32)$$

However, by Definition 10.5, $V_n(X)$ is asymptotically unbiased because

$$\lim_{n \to \infty} E[V_n(X)] = \lim_{n \to \infty} \frac{n-1}{n}\text{Var}[X] = \text{Var}[X]. \qquad (10.33)$$

Although $V_n(X)$ is a biased estimate, Theorem 10.12 suggests the derivation of an unbiased estimate.

Theorem 10.13

The estimate

$$V_n'(X) = \frac{1}{n-1}\sum_{i=1}^{n}(X_i - M_n(X))^2$$

is an unbiased estimate of $\text{Var}[X]$.

Proof Using Definition 10.7, we have

$$V_n'(X) = \frac{n}{n-1}V_n(X), \qquad (10.34)$$

and

$$E[V_n'(X)] = \frac{n}{n-1}E[V_n(X)] = \text{Var}[X]. \qquad (10.35)$$

Comparing the two estimates of $\text{Var}[X]$, we observe that as n grows without limit, the two estimates converge to the same value. However, for $n = 1$, $M_1(X) = X_1$ and $V_1(X) = 0$. By contrast, $V_1'(X)$ is undefined. Because the variance is a measure of the spread of a probability model, it is impossible to obtain an estimate of the spread from only one observation. Thus the estimate $V_1(X) = 0$ is completely illogical. On the other hand, the unbiased estimate of variance based on two observations can be written as $V_2' = (X_1 - X_2)^2/2$, which clearly reflects the spread (mean square difference) of the observations.

To go further and evaluate the consistency of the sequence $V_1'(X), V_2'(X), \ldots$ is a surprisingly difficult problem. It is explored in Problem 10.4.5.

Quiz 10.4

X is the continuous uniform $(-1, 1)$ random variable. Find the mean square error, $E[(\text{Var}[X] - V_{100}(X))^2]$, of the sample variance estimate of $\text{Var}[X]$, based on 100 independent observations of X.

10.5 MATLAB

> MATLAB can help us visualize the mathematical techniques and estimation procedures presented in this chapter. One MATLAB program generates samples of $M_n(X)$ as a function of n and compares $M_n(X)$ with the parameter value of the probability model used in the simulation.

The new ideas in this chapter — namely, the convergence of the sample mean, the Chebyshev inequality, and the weak law of large numbers — are largely theoretical. The application of these ideas relies on mathematical techniques for discrete and continuous random variables and sums of random variables that were introduced in prior chapters. As a result, in terms of MATLAB, this chapter breaks little new ground. Nevertheless, it is instructive to use MATLAB to simulate the convergence of the sample mean $M_n(X)$. In particular, for a random variable X, we can view a set of iid samples $X_1, \ldots, X_n$ as a random vector $\mathbf{X} = \begin{bmatrix} X_1 & \cdots & X_n \end{bmatrix}'$. This vector of iid samples yields a vector of sample mean values $\mathbf{M}(\mathbf{X}) = \begin{bmatrix} M_1(X) & M_2(X) & \cdots & M_n(X) \end{bmatrix}'$ where

$$M_k(X) = \frac{X_1 + \cdots + X_k}{k} \tag{10.36}$$

We call a graph of the sequence $M_k(X)$ versus k a *sample mean trace*. By graphing the sample mean trace as a function of n we can observe the convergence of the point estimate $M_k(X)$ to $\mathrm{E}[X]$. By graphing multiple sample mean traces, we can observe the convergence properties of the sample mean.

════ **Example 10.7** ════

Write a MATLAB function bernoullitraces(n,m,p) to generate m sample mean traces, each of length n, for the sample mean of a Bernoulli (p) random variable.

. . . .

```
function MN=bernoullitraces(n,m,p);
x=reshape(bernoullirv(p,m*n),n,m);
nn=(1:n)'*ones(1,m);
MN=cumsum(x)./nn;
stderr=sqrt(p*(1-p))./sqrt((1:n)');
plot(1:n,0.5+stderr,...
    1:n,0.5-stderr,1:n,MN);
```

In bernoullitraces, each column of x is an instance of a random vector $\mathbf{X}$ with iid Bernoulli (p) components. Similarly, each column of MN is an instance of the vector $\mathbf{M}(\mathbf{X})$.

The output graphs each column of MN as a function of the number of trials n. In addition, we calculate the standard error $\sqrt{e_k}$ and overlay graphs of $p - \sqrt{e_k}$ and $p + \sqrt{e_k}$. Equation (10.24) says that at each step k, we should expect to see roughly two-thirds of the sample mean traces in the range

$$p - \sqrt{e_k} \le M_k(X) \le p + \sqrt{e_k}. \tag{10.37}$$

A sample graph of bernoullitraces(50,40,0.5) is shown in Figure 10.1. The figure shows how at any given step, approximately two thirds of the sample mean traces are within one standard error of the expected value.

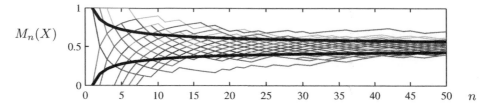

Figure 10.1 Sample output of `bernoullitraces.m`, including the deterministic standard error graphs. The graph shows how at any given step, about two thirds of the sample means are within one standard error of the true mean.

Quiz 10.5

Generate $m = 1000$ traces (each of length $n = 100$) of the sample mean of a Bernoulli (p) random variable. At each step k, calculate M_k and the number of traces, such that M_k is within one standard error of the expected value p. Graph $T_k = M_k/m$ as a function of k. Explain your results.

Further Reading: [Dur94] contains concise, rigorous presentations and proofs of the laws of large numbers. [WS01] covers parameter estimation for both scalar and vector random variables and stochastic processes.

Problems

Difficulty: ● Easy   ▦ Moderate   ◆ Difficult   ◆◆ Experts Only

10.1.1● $X_1, \ldots, X_n$ is an iid sequence of exponential random variables, each with expected value 5.

(a) What is $\mathrm{Var}[M_9(X)]$, the variance of the sample mean based on nine trials?

(b) What is $\mathrm{P}[X_1 > 7]$, the probability that one outcome exceeds 7?

(c) Use the central limit theorem to estimate $\mathrm{P}[M_9(X) > 7]$, the probability that the sample mean of nine trials exceeds 7.

10.1.2● $X_1, \ldots, X_n$ are independent uniform random variables with expected value $\mu_X = 7$ and variance $\mathrm{Var}[X] = 3$.

(a) What is the PDF of X_1?

(b) What is $\mathrm{Var}[M_{16}(X)]$, the variance of the sample mean based on 16 trials?

(c) What is $\mathrm{P}[X_1 > 9]$, the probability that one outcome exceeds 9?

(d) Would you expect $\mathrm{P}[M_{16}(X) > 9]$ to be bigger or smaller than $\mathrm{P}[X_1 > 9]$? To check your intuition, use the central limit theorem to estimate $\mathrm{P}[M_{16}(X) > 9]$.

10.1.3▦ X is a uniform $(0, 1)$ random variable. $Y = X^2$. What is the standard error of the estimate of μ_Y based on 50 independent samples of X?

10.1.4▦ Let $X_1, X_2, \ldots$ denote a sequence of independent samples of a random variable X with variance $\mathrm{Var}[X]$. We define a new random sequence $Y_1, Y_2, \ldots$ as $Y_1 = X_1 - X_2$ and $Y_n = X_{2n-1} - X_{2n}$.

(a) Find $E[Y_n]$ and $Var[Y_n]$.

(b) Find the expected value and variance of $M_n(Y)$.

10.2.1● The weight of a randomly chosen Maine black bear has expected value $E[W] = 500$ pounds and standard deviation $\sigma_W = 100$ pounds. Use the Chebyshev inequality to upper bound the probability that the weight of a randomly chosen bear is more than 200 pounds from the expected value of the weight.

10.2.2● For an arbitrary random variable X, use the Chebyshev inequality to show that the probability that X is more than k standard deviations from its expected value $E[X]$ satisfies

$$P[|X - E[X]| \geq k\sigma] \leq \frac{1}{k^2}.$$

For a Gaussian random variable Y, use the $\Phi(\cdot)$ function to calculate the probability that Y is more than k standard deviations from its expected value $E[Y]$. Compare the result to the upper bound based on the Chebyshev inequality.

10.2.3■ Elevators arrive randomly at the ground floor of an office building. Because of a large crowd, a person will wait for time W in order to board the third arriving elevator. Let X_1 denote the time (in seconds) until the first elevator arrives and let X_i denote the time between the arrival of elevator $i - 1$ and i. Suppose X_1, X_2, X_3 are independent uniform $(0, 30)$ random variables. Find upper bounds to the probability W exceeds 75 seconds using

(a) the Markov inequality,

(b) the Chebyshev inequality,

(c) the Chernoff bound.

10.2.4■ Let X equal the arrival time of the third elevator in Problem 10.2.3. Find the exact value of $P[W \geq 75]$. Compare your answer to the upper bounds derived in Problem 10.2.3.

10.2.5■ In a game with two dice, the event *snake eyes* refers to both six-sided dice showing one spot. Let R denote the number of dice rolls needed to observe the third occurrence of *snake eyes*. Find

(a) the upper bound to $P[R \geq 250]$ based on the Markov inequality,

(b) the upper bound to $P[R \geq 250]$ based on the Chebyshev inequality,

(c) the exact value of $P[R \geq 250]$.

10.2.6■ Use the Chernoff bound to show that the Gaussian $(0, 1)$ random variable Z satisfies

$$P[Z \geq c] \leq e^{-c^2/2}.$$

For $c = 1, 2, 3, 4, 5$, use Table 4.1 and Table 4.2 to compare the Chernoff bound to the true value: $P[Z \geq c] = Q(c)$.

10.2.7■ Use the Chernoff bound to show for a Gaussian (μ, σ) random variable X that

$$P[X \geq c] \leq e^{-(c-\mu)^2/2\sigma^2}.$$

Hint: Apply the result of Problem 10.2.6.

10.2.8■ Let K be a Poisson random variable with expected value α. Use the Chernoff bound to find an upper bound to $P[K \geq c]$. For what values of c do we obtain the trivial upper bound $P[K \geq c] \leq 1$?

10.2.9■ In a subway station, there are exactly enough customers on the platform to fill three trains. The arrival time of the nth train is $X_1 + \cdots + X_n$ where $X_1, X_2, \ldots$ are iid exponential random variables with $E[X_i] = 2$ minutes. Let W equal the time required to serve the waiting customers. Find $P[W > 20]$.

10.2.10■ Let $X_1, \ldots, X_n$ be independent samples of a random variable X. Use the Chernoff bound to show that $M_n(X) = (X_1 + \cdots + X_n)/n$ satisfies

$$P[M_n(X) \geq c] \leq \left(\min_{s \geq 0} e^{-sc} \phi_X(s) \right)^n.$$

10.3.1● Let $X_1, X_2, \ldots$ denote an iid sequence of random variables, each with expected value 75 and standard deviation 15.

(a) How many samples n do we need to guarantee that the sample mean $M_n(X)$ is between 74 and 76 with probability 0.99?

(b) If each X_i has a Gaussian distribution, how many samples n' would we need to guarantee $M_{n'}(X)$ is between 74 and 76 with probability 0.99?

10.3.2● Let X_A be the indicator random variable for event A with probability $P[A] = 0.8$. Let $\hat{P}_n(A)$ denote the relative frequency of event A in n independent trials.

(a) Find $E[X_A]$ and $Var[X_A]$.

(b) What is $Var[\hat{P}_n(A)]$?

(c) Use the Chebyshev inequality to find the confidence coefficient $1 - \alpha$ such that $\hat{P}_{100}(A)$ is within 0.1 of $P[A]$. In other words, find α such that

$$P\left[\left|\hat{P}_{100}(A) - P[A]\right| \leq 0.1\right] \geq 1 - \alpha.$$

(d) Use the Chebyshev inequality to find out how many samples n are necessary to have $\hat{P}_n(A)$ within 0.1 of $P[A]$ with confidence coefficient 0.95. In other words, find n such that

$$P\left[\left|\hat{P}_n(A) - P[A]\right| \leq 0.1\right] \geq 0.95.$$

10.3.3● $X_1, X_2, \ldots$ is a sequence of iid Bernoulli $(1/2)$ random variables. Consider the random sequence $Y_n = X_1 + \cdots + X_n$.

(a) What is $\lim_{n \to \infty} P[|Y_{2n} - n| \leq \sqrt{n/2}]$?

(b) What does the weak law of large numbers say about Y_{2n}?

10.3.4■ In communication systems, the error probability $P[E]$ may be difficult to calculate; however it may be easy to derive an upper bound of the form $P[E] \leq \epsilon$. In this case, we may still want to estimate $P[E]$ using the relative frequency $\hat{P}_n(E)$ of E in n trials. In this case, show that

$$P\left[\left|\hat{P}_n(E) - P[E]\right| \geq c\right] \leq \frac{\epsilon}{nc^2}.$$

10.3.5■ A factory manufactures chocolate chip cookies on an assembly line. Each cookie is sprinkled with K chips from a very large vat of chips, where K is Poisson with $E[K] = 10$, independent of the number on any other cookie. Imagine you are a chip in the vat and you are sprinkled onto a cookie. Let J denote the number of chips (including you) in your cookie. What is the PMF of J? Hint: Suppose n cookies have been made such that N_k cookies have k chips. You are just one of the $\sum_{k=0}^{\infty} kN_k$ chips used in the n cookies.

10.3.6◆ In this problem, we develop a weak law of large numbers for a correlated sequence $X_1, X_2, \ldots$ of identical random variables. In particular, each X_i has expected value $E[X_i] = \mu$, and the random sequence has covariance function

$$C_X[m, k] = Cov[X_m, X_{m+k}] = \sigma^2 a^{|k|}$$

where a is a constant such that $|a| < 1$. For this correlated random sequence, we can define the sample mean of n samples as

$$M_n = \frac{X_1 + \cdots + X_n}{n}.$$

(a) Use Theorem 9.2 to show that

$$Var[X_1 + \cdots X_n] \leq n\sigma^2 \left(\frac{1+a}{1-a}\right).$$

(b) Use the Chebyshev inequality to show that for any $c > 0$,

$$P[|M_n - \mu| \geq c] \leq \frac{\sigma^2(1+a)}{n(1-a)c^2}.$$

(c) Use part (b) to show that for any $c > 0$,

$$\lim_{n \to \infty} P[|M_n - \mu| \geq c] = 0.$$

10.3.7◆◆ In the Gaussian Movie DataBase (GMDB), reviewers like you rate movies with Gaussian scores. In particular, the first person to rate a movie assigns a Gaussian $(q, 1)$ review score X_1, where r_0 is the true "quality" of the movie. After n reviews, a movie's rating is $R_n = \sum_{i=1}^{n} X_i/n$. Strangely enough, in the GMDB, reviewers are influenced by prior reviews; if after $n-1$ reviews a movie is rated $R_{n-1} = r$, the nth review n will rate the movie X_n, a Gaussian $(r, 1)$ random variable, *conditionally* independent of $X_1, \ldots, X_{n-1}$ given $R_{n-1} = r$.

(a) Find $E[R_n]$.

(b) Find the PDF $f_{R_n}(r)$. Hint: You may have unresolved parameters in this answer.

(c) Find $\text{Var}[R_n]$. Hint: Find $E[R_n^2 | R_{n-1}]$.

(d) Interpret your results as $n \to \infty$? Does the law of large numbers apply here?

10.4.1● When X is Gaussian, verify Equation (10.24), which states that the sample mean is within one standard error of the expected value with probability 0.68.

10.4.2■ Suppose the sequence of estimates $\hat{R}_n$ is biased but asymptotically unbiased. If $\lim_{n \to \infty} \text{Var}[\hat{R}_n] = 0$, is the sequence $\hat{R}_n$ consistent?

10.4.3■ An experimental trial produces random variables X_1 and X_2 with correlation $r = E[X_1 X_2]$. To estimate r, we perform n independent trials and form the estimate

$$\hat{R}_n = \frac{1}{n} \sum_{i=1}^{n} X_1(i) X_2(i),$$

where $X_1(i)$ and $X_2(i)$ are samples of X_1 and X_2 on trial i. Show that if $\text{Var}[X_1 X_2]$ is finite, then $\hat{R}_1, \hat{R}_2, \ldots$ is an unbiased, consistent sequence of estimates of r.

10.4.4■ An experiment produces random vector $\mathbf{X} = \begin{bmatrix} X_1 & \cdots & X_k \end{bmatrix}'$ with expected value $\boldsymbol{\mu}_{\mathbf{X}} = \begin{bmatrix} \mu_1 & \cdots & \mu_k \end{bmatrix}'$. The ith component of $\mathbf{X}$ has variance $\text{Var}[X_i] = \sigma_i^2$. To estimate $\boldsymbol{\mu}_{\mathbf{X}}$, we perform n independent trials such that $\mathbf{X}(i)$ is the sample of $\mathbf{X}$ on trial

i, and we form the vector mean

$$\mathbf{M}(n) = \frac{1}{n} \sum_{i=1}^{n} \mathbf{X}(i).$$

(a) Show $\mathbf{M}(n)$ is unbiased by showing $E[\mathbf{M}(n)] = \boldsymbol{\mu}_{\mathbf{X}}$.

(b) Show that the sequence of estimates $\mathbf{M}_n$ is consistent by showing that for any constant $c > 0$,

$$\lim_{n \to \infty} P\left[\max_{j=1,\ldots,k} |M_j(n) - \mu_j| \geq c \right] = 0.$$

Hint: Let $A_i = \{|M_i(n) - \mu_i| \geq c\}$ and apply the union bound (see Problem 1.2.11) to upper bound $P[A_1 \cup A_2 \cup \cdots \cup A_k]$. Then apply the Chebyshev inequality.

10.4.5◆ Given the iid samples $X_1, X_2, \ldots$ of X, define the sequence $Y_1, Y_2, \ldots$ by

$$Y_k = \left(X_{2k-1} - \frac{X_{2k-1} + X_{2k}}{2} \right)^2 + \left(X_{2k} - \frac{X_{2k-1} + X_{2k}}{2} \right)^2.$$

Note that each Y_k is an example of V_2', an estimate of the variance of X using two samples, given in Theorem 10.13. Show that if $E[X^k] < \infty$ for $k = 1, 2, 3, 4$, then the sample mean $M_n(Y)$ is a consistent, unbiased estimate of $\text{Var}[X]$.

10.4.6◆◆ An experiment produces a Gaussian random vector $\mathbf{X} = \begin{bmatrix} X_1 & \cdots & X_k \end{bmatrix}'$ with $E[\mathbf{X}] = \mathbf{0}$ and correlation matrix $\mathbf{R} = E[\mathbf{XX}']$. To estimate $\mathbf{R}$, we perform n independent trials, yielding the iid sample vectors $\mathbf{X}(1), \mathbf{X}(2), \ldots, \mathbf{X}(n)$, and form the sample correlation matrix

$$\hat{\mathbf{R}}(n) = \frac{1}{n} \sum_{m=1}^{n} \mathbf{X}(m) \mathbf{X}'(m).$$

(a) Show $\hat{\mathbf{R}}(n)$ is unbiased by showing $E[\hat{\mathbf{R}}(n)] = \mathbf{R}$.

(b) Show that the sequence of estimates $\hat{\mathbf{R}}(n)$ is consistent by showing that every element $\hat{R}_{ij}(n)$ of the matrix $\hat{\mathbf{R}}$ converges to R_{ij}. That is, show that for any $c > 0$,

$$\lim_{n \to \infty} P\left[\max_{i,j} \left|\hat{R}_{ij} - R_{ij}\right| \geq c\right] = 0.$$

Hint: Extend the technique used in Problem 10.4.4. You will need to use the result of Problem 7.6.4 to show that $\text{Var}[X_i X_j]$ is finite.

10.5.1 X is the Bernoulli $(1/2)$ random variable. The sample mean $M_n(X)$ has standard error

$$e_n = \sqrt{\frac{\text{Var}[X]}{n}} = \frac{1}{2\sqrt{n}}.$$

The probability that $M_n(X)$ is within one standard error of p is

$$p_n = P\left[\frac{1}{2} - \frac{1}{2\sqrt{n}} \leq M_n(X) \leq \frac{1}{2} + \frac{1}{2\sqrt{n}}\right].$$

Use the `binomialcdf` function to calculate the exact probability p_n as a function of n. What is the source of the unusual sawtooth pattern? Compare your results to the solution of Quiz 10.5.

10.5.2 Recall that an exponential (λ) random variable X has

$$E[X] = 1/\lambda,$$
$$\text{Var}[X] = 1/\lambda^2.$$

Thus, to estimate λ from n independent samples $X_1, \ldots, X_n$, either of the following techniques should work.

(a) Calculate the sample mean $M_n(X)$ and form the estimate $\hat{\lambda} = 1/M_n(X)$.

(b) Calculate the unbiased variance estimate $V_n'(X)$ of Theorem 10.13 and form the estimate $\tilde{\lambda} = 1/\sqrt{V_n'(X)}$.

Use MATLAB to simulate the calculation $\hat{\lambda}$ and $\tilde{\lambda}$ for $m = 1000$ experimental trials to determine which estimate is better.

10.5.3 $\mathbf{X}$ is 10-dimensional Gaussian $(\mathbf{0}, \mathbf{I})$ random vector. Since $E[\mathbf{X}] = \mathbf{0}$, $\mathbf{R_X} = \mathbf{C_X} = \mathbf{I}$. We will use the method of Problem 10.4.6 and estimate $\mathbf{R_X}$ using the sample correlation matrix

$$\hat{\mathbf{R}}(n) = \frac{1}{n} \sum_{m=1}^{n} \mathbf{X}(m)\mathbf{X}'(m).$$

For $n \in \{10, 100, 1000, 10{,}000\}$, construct a MATLAB simulation to estimate

$$P\left[\max_{i,j} \left|\hat{R}_{ij} - I_{ij}\right| \geq 0.05\right].$$

10.5.4 In terms of parameter a, random variable X has CDF

$$F_X(x) = \begin{cases} 0 & x < a - 1, \\ 1 - \frac{1}{[x-(a-2)]^2} & x \geq a - 1. \end{cases}$$

(a) Show that $E[X] = a$ by showing that $E[X - (a - 2)] = 2$.

(b) Generate $m = 100$ traces of the sample mean $M_n(X)$ of length $n = 1000$. Do you observe convergence of the sample mean to $E[X] = a$?

11

Hypothesis Testing

Some of the most important applications of probability theory involve reasoning in the presence of uncertainty. In these applications, we analyze the observations of an experiment in order to make a decision. When the decision is based on the properties of random variables, the reasoning is referred to as *statistical inference*. In Chapter 10, we introduced point estimation as a type of statistical inference for model parameters. In this chapter, we introduce *hypothesis testing*, another type of inference.

Statistical inference is a broad, deep subject with a very large body of theoretical knowledge and practical techniques. It has its own extensive literature and a vast collection of practical techniques, many of them valuable secrets of companies and governments. Chapter 10, this chapter, and Chapter 12 provide an introductory view of the subject of statistical inference. Our aim is to indicate to readers how the fundamentals of probability theory presented in the earlier chapters can be used to make accurate decisions in the presence of uncertainty.

Like probability theory, the theory of statistical inference refers to an experiment consisting of a procedure and observations. In all statistical inference methods, there is also a set of possible decisions and a means of measuring the accuracy of a decision. A statistical inference method assigns a decision to each possible outcome of the experiment. Therefore, a statistical inference method consists of three steps: Perform an experiment, observe an outcome, state a decision. The aim of the assignment is to achieve the highest possible accuracy.

Decision The observations result from one of M hypothetical probability models:
$H_0, H_1, \ldots, H_{M-1}$.

Accuracy Measure Probability that the decision is H_i when the true model is H_j for $i, j = 0, 1, \ldots, M - 1$.

The assignment of decisions to outcomes is based on probability theory. The accuracy measure associates a cost with each type of wrong decision.

Example 11.1

Suppose $X_1, \ldots, X_n$ are iid samples of an exponential (λ) random variable X with unknown parameter λ. Using the observations $X_1, \ldots, X_n$, each of the statistical inference methods can answer questions regarding the unknown λ. For each of the methods, we state the underlying assumptions of the method and a question that can be addressed by the method.

Decision Assuming λ is a constant, does λ equal $\lambda_0 = 2.5$, $\lambda_1 = 3.5$, or $\lambda_2 = 4.5$?

Accuracy Measure Probability that the decision is $\lambda = \lambda_i$ when the true model is λ_j for $i, j = 0, 1, 2$.

To decide the question in Example 11.1, we have to state in advance which values of $X_1, \ldots, X_n$ produce each possible decision. For the hypothesis test in Example 11.1, the decision must be one of the numbers 2.5, 3.5, or 4.5. This is an example of a ternary hypothesis test. A ternary test is a special case of the M-ary hypothesis test with $M = 3$. We study M-ary hypothesis tests in Section 11.2. We start with binary hypothesis tests in the next section.

11.1   Binary Hypothesis Testing

> A binary hypothesis test creates a partition $\{A_0, A_1\}$ for an experiment. When an outcome is in H_0, the decision is to accept hypothesis H_0. Otherwise the decision is to accept H_1. The quality measure of a test is related to the probability of a false alarm (decide H_1 when H_0 is true) and the probability of a miss (decide H_0 when H_1 is true.)

In a binary hypothesis test, there are two hypothetical probability models, H_0 and H_1, and two possible decisions: *accept* H_0 as the true model, and *accept* H_1. There is also a probability model for H_0 and H_1, conveyed by the numbers $\mathrm{P}[H_0]$ and $\mathrm{P}[H_1] = 1 - \mathrm{P}[H_0]$. These numbers are referred to as the *a priori probabilities* or *prior probabilities* of H_0 and H_1. They reflect the state of knowledge about the probability model before an outcome is observed.

The complete experiment for a binary hypothesis test consists of two subexperiments. The first subexperiment chooses a probability model from sample space $S' = \{H_0, H_1\}$. The probability models H_0 and H_1 have the same sample space, S. The second subexperiment produces an observation corresponding to an outcome, $s \in S$. When the observation leads to a random vector $\mathbf{X}$, we call $\mathbf{X}$ the *decision statistic*. Often, the decision statistic is simply a random variable X. When the decision statistic $\mathbf{X}$ is discrete, the probability models are conditional probability mass functions $P_{\mathbf{X}|H_0}(\mathbf{x})$ and $P_{\mathbf{X}|H_1}(\mathbf{x})$. When $\mathbf{X}$ is a continuous random vector, the probability models are conditional probability density functions $f_{\mathbf{X}|H_0}(\mathbf{x})$ and $f_{\mathbf{X}|H_1}(\mathbf{x})$. In the terminology of statistical inference, these functions are referred to as *likelihood functions*. For example, $f_{\mathbf{X}|H_0}(\mathbf{x})$ is the likelihood of $\mathbf{x}$ given H_0.

The test design divides S into two sets, A_0 and $A_1 = A_0^c$. If the outcome $s \in A_0$, the decision is *accept H_0*. Otherwise, the decision is *accept H_1*. In a binary hypothesis test, two kinds of errors are possible. Statisticians refer to them as *Type I errors* and *Type II errors* with the following definitions:

- **Type I Error** *False Rejection*: Reject H_0 when H_0 is true.

- **Type II Error** *False Acceptance*: Accept H_0 when H_0 is false.

The accuracy measure of the test consists of the corresponding two error probabilities. $P[A_1|H_0]$ is the probability of a Type I error. It is the probability of accepting H_1 when H_0 is the true probability model. Similarly, $P[A_0|H_1]$ is the probability of accepting H_0 when H_1 is the true probability model. It corresponds to the probability of a Type II error.

Example 11.2

One electrical engineering application of binary hypothesis testing relates to a radar system. The transmitter sends out a signal, and it is the job of the receiver to decide whether a target is present. To make this decision, the receiver examines the received signal to determine whether it contains a reflected version of the transmitted signal. The hypothesis H_0 corresponds to the situation in which there is no target. H_1 corresponds to the presence of a target. In the terminology of radar, a Type I error (decide target present when there is no target) is referred to as a *false alarm* and a Type II error (decide no target when there is a target present) is referred to as a *miss*.

The design of a binary hypothesis test represents a trade-off between the two error probabilities, $P_{FA} = P[A_1|H_0]$ and $P_{MISS} = P[A_0|H_1]$. To understand the trade-off, consider an extreme design in which $A_0 = S$ consists of the entire sample space and $A_1 = \varnothing$ is the empty set. In this case, $P_{FA} = 0$ and $P_{MISS} = 1$. Now let A_1 expand to include an increasing proportion of the outcomes in S. As A_1 expands, P_{FA} increases and P_{MISS} decreases. At the other extreme, $A_0 = \varnothing$, which implies $P_{MISS} = 0$. In this case, $A_1 = S$ and $P_{FA} = 1$.

A graph representing the possible values of P_{FA} and P_{MISS} is referred to as a *receiver operating curve (ROC)*. Examples appear in Figure 11.1. A receiver operating curve displays P_{MISS} as a function of P_{FA} for all possible A_0 and A_1. The graph on the left represents probability models with a continuous sample space S. In the graph on the right, S is a discrete set and the receiver operating curve consists of a collection of isolated points in the P_{FA}, P_{MISS} plane. At the top left corner of the graph, the point $(0, 1)$ corresponds to $A_0 = S$ and $A_1 = \varnothing$. When we move one outcome from A_0 to A_1, we move to the next point on the curve. Moving downward along the curve corresponds to taking more outcomes from A_0 and putting them in A_1 until we arrive at the lower right corner $(1, 0)$, where all the outcomes are in A_1.

Example 11.3

The noise voltage in a radar detection system is a Gaussian $(0, 1)$ random variable, N. When a target is present, the received signal is $X = v + N$ volts with $v \geq 0$.

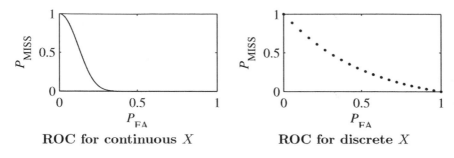

Figure 11.1   Continuous and discrete examples of a receiver operating curve (ROC).

Otherwise the received signal is $X = N$ volts. Periodically, the detector performs a binary hypothesis test, with H_0 as the hypothesis *no target* and H_1 as the hypothesis *target present*. The acceptance sets for the test are $A_0 = \{X \leq x_0\}$ and $A_1 = \{X > x_0\}$. Draw the receiver operating curves of the radar system for the three target voltages $v = 0, 1, 2$ volts.

· ·

To derive a receiver operating curve, it is necessary to find P_{MISS} and P_{FA} as functions of x_0. To perform the calculations, we observe that under hypothesis H_0, $X = N$ is a Gaussian $(0, \sigma)$ random variable. Under hypothesis H_1, $X = v + N$ is a Gaussian (v, σ) random variable. Therefore,

$$P_{\text{MISS}} = \text{P}\left[A_0 | H_1\right] = \text{P}\left[X \leq x_0 | H_1\right] = \Phi(x_0 - v) \tag{11.1}$$

$$P_{\text{FA}} = \text{P}\left[A_1 | H_0\right] = \text{P}\left[X > x_0 | H_0\right] = 1 - \Phi(x_0). \tag{11.2}$$

Figure 11.2(a) shows P_{MISS} and P_{FA} as functions of x_0 for $v = 0$, $v = 1$, and $v = 2$ volts. Note that there is a single curve for P_{FA} since the probability of a false alarm does not depend on v. The same data also appears in the corresponding receiver operating curves of Figure 11.2(b). When $v = 0$, the received signal is the same regardless of whether or not a target is present. In this case, $P_{\text{MISS}} = 1 - P_{\text{FA}}$. As v increases, it is easier for the detector to distinguish between the two targets. We see that the ROC improves as v increases. That is, we can choose a value of x_0 such that both P_{MISS} and P_{FA} are lower for $v = 2$ than for $v = 1$.

In a practical binary hypothesis test, it is necessary to adopt one test (a specific A_0) and a corresponding trade-off between P_{FA} and P_{MISS}. There are many approaches to selecting A_0. In the radar application, the cost of a miss (ignoring a threatening target) could be far higher than the cost of a false alarm (causing the operator to take an unnecessary precaution). This suggests that the radar system should operate with a low value of x_0 to produce a low P_{MISS} even though this will produce a relatively high P_{FA}. The remainder of this section describes four methods of choosing A_0.

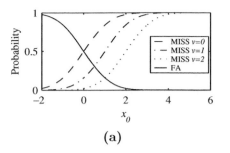
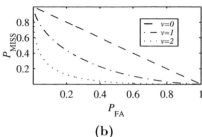

(a)   (b)

Figure 11.2   **(a)** The probability of a miss and the probability of a false alarm as a function of the threshold x_0 for Example 11.3. **(b)** The corresponding receiver operating curve for the system. We see that the ROC improves as v increases.

Maximum A posteriori Probability (MAP) Test

Example 11.4

A modem transmits a binary signal to another modem. Based on a noisy measurement, the receiving modem must choose between hypothesis H_0 (the transmitter sent a 0) and hypothesis H_1 (the transmiiter sent a 1). A false alarm occurs when a 0 is sent but a 1 is detected at the receiver. A miss occurs when a 1 is sent but a 0 is detected. For both types of error, the cost is the same; one bit is detected incorrectly.

The maximum a posteriori probability test minimizes P_{ERR}, the total probability of error of a binary hypothesis test. The law of total probability, Theorem 1.8, relates P_{ERR} to the a priori probabilities of H_0 and H_1 and to the two conditional error probabilities, $P_{\mathrm{FA}} = P[A_1|H_0]$ and $P_{\mathrm{MISS}} = P[A_0|H_1]$:

$$P_{\mathrm{ERR}} = P[A_1|H_0]P[H_0] + P[A_0|H_1]P[H_1]. \tag{11.3}$$

When the two types of errors have the same cost, as in Example 11.4, minimizing P_{ERR} is a sensible strategy. The following theorem specifies the binary hypothesis test that produces the minimum possible P_{ERR}.

Theorem 11.1   Maximum A posteriori Probability (MAP) Test

Given a binary hypothesis-testing experiment with outcome s, the following rule leads to the lowest possible value of P_{ERR}:

$$s \in A_0 \ \text{if } P[H_0|s] \geq P[H_1|s]; \qquad s \in A_1 \ \text{otherwise.}$$

Proof To create the partition $\{A_0, A_1\}$, it is necessary to place every element $s \in S$ in either A_0 or A_1. Consider the effect of a specific value of s on the sum in Equation (11.3). Either s will contribute to the first (A_1) or second (A_0) term in the sum. By placing each

s in the term that has the lower value for the specific outcome s, we create a partition that minimizes the entire sum. Thus we have the rule

$$s \in A_0 \text{ if } \mathrm{P}\left[s|H_1\right]\mathrm{P}\left[H_1\right] \leq \mathrm{P}\left[s|H_0\right]\mathrm{P}\left[H_0\right]; \qquad s \in A_1 \text{ otherwise.} \qquad (11.4)$$

From Bayes' theorem (Theorem 1.10), the left side of the inequality is $\mathrm{P}[H_1|s]\,\mathrm{P}[s]$ and the right side of the inequality is $\mathrm{P}[H_0|s]\,\mathrm{P}[s]$. Therefore the inequality is identical to $\mathrm{P}[H_0|s]\,\mathrm{P}[s] \geq \mathrm{P}[H_1|s]\,\mathrm{P}[s]$, which simplifies to the inequality in the theorem.

Note that $\mathrm{P}[H_0|s]$ and $\mathrm{P}[H_1|s]$ are referred to as the *a posteriori* probabilities of H_0 and H_1. Just as the a priori probabilities $\mathrm{P}[H_0]$ and $\mathrm{P}[H_1]$ reflect our knowledge of H_0 and H_1 prior to performing an experiment, $\mathrm{P}[H_0|s]$ and $\mathrm{P}[H_1|s]$ reflect our knowledge after observing s. Theorem 11.1 states that in order to minimize P_{ERR} it is necessary to accept the hypothesis with the higher a posteriori probability. A test that follows this rule is a *maximum a posteriori probability (MAP)* hypothesis test. In such a test, A_0 contains all outcomes s for which $\mathrm{P}[H_0|s] > \mathrm{P}[H_1|s]$, and A_1 contains all outcomes s for which $\mathrm{P}[H_1|s] > \mathrm{P}[H_0|s]$. If $\mathrm{P}[H_0|s] = \mathrm{P}[H_1|s]$, the assignment of s to either A_0 or A_1 does not affect P_{ERR}. In Theorem 11.1, we arbitrarily assign s to A_0 when the a posteriori probabilities are equal. We would have the same probability of error if we assign s to A_1 for all outcomes that produce equal a posteriori probabilities or if we assign some outcomes with equal a posteriori probabilities to A_0 and others to A_1.

Equation (11.4) is another statement of the MAP decision rule. It contains the three probability models that are assumed to be known:

- The a priori probabilities of the hypotheses: $\mathrm{P}[H_0]$ and $\mathrm{P}[H_1]$,
- The likelihood function of H_0: $\mathrm{P}[s|H_0]$,
- The likelihood function of H_1: $\mathrm{P}[s|H_1]$.

When the outcomes of an experiment yield a random vector $\mathbf{X}$ as the decision statistic, we can express the MAP rule in terms of conditional PMFs or PDFs. If $\mathbf{X}$ is discrete, we take $\mathbf{X} = \mathbf{x}_i$ to be the outcome of the experiment. If the sample space S of the experiment is continuous, we interpret the conditional probabilities by assuming that each outcome corresponds to the random vector $\mathbf{X}$ in the small volume $\mathbf{x} \leq \mathbf{X} < \mathbf{x} + d\mathbf{x}$ with probability $f_{\mathbf{X}}(\mathbf{x})d\mathbf{x}$. Thus in terms of the random variable X, we have the following version of the MAP hypothesis test.

Theorem 11.2

For an experiment that produces a random vector $\mathbf{X}$, the MAP hypothesis test is

$$\textit{Discrete:} \quad \mathbf{x} \in A_0 \textit{ if } \frac{P_{\mathbf{X}|H_0}(\mathbf{x})}{P_{\mathbf{X}|H_1}(\mathbf{x})} \geq \frac{\mathrm{P}\left[H_1\right]}{\mathrm{P}\left[H_0\right]}; \qquad \mathbf{x} \in A_1 \textit{ otherwise;}$$

$$\textit{Continuous: } \mathbf{x} \in A_0 \textit{ if } \frac{f_{\mathbf{X}|H_0}(\mathbf{x})}{f_{\mathbf{X}|H_1}(\mathbf{x})} \geq \frac{\mathrm{P}\left[H_1\right]}{\mathrm{P}\left[H_0\right]}; \qquad \mathbf{x} \in A_1 \textit{ otherwise.}$$

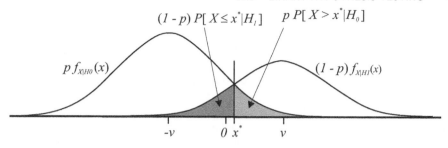

$(1-p)P[X \le x^*|H_1]$ $pP[X > x^*|H_0]$

$pf_{X|H0}(x)$ $(1-p)f_{X|H1}(x)$

$-v$ $0\ x^*$ v

Figure 11.3   Decision regions for Example 11.5.

In these formulas, the ratio of conditional probabilities is referred to as a *likelihood ratio*. The formulas state that in order to perform a binary hypothesis test, we observe the outcome of an experiment, calculate the likelihood ratio on the left side of the formula, and compare it with a constant on the right side of the formula. We can view the likelihood ratio as the evidence, based on an observation, in favor of H_0. If the likelihood ratio is greater than 1, H_0 is more likely than H_1. The ratio of prior probabilities, on the right side, is the evidence, prior to performing the experiment, in favor of H_1. Therefore, Theorem 11.2 states that accepting H_0 is the better decision if the evidence in favor of H_0, based on the experiment, outweighs the prior evidence in favor of accepting H_1.

In many practical hypothesis tests, including the following example, it is convenient to compare the logarithms of the two ratios.

Example 11.5

With probability p, a digital communications system transmits a 0. It transmits a 1 with probability $1 - p$. The received signal is either $X = -v + N$ volts, if the transmitted bit is 0, or $v + N$ volts, if the transmitted bit is 1. The voltage $\pm v$ is the information component of the received signal, and N, a Gaussian $(0, \sigma)$ random variable, is the noise component. Given the received signal X, what is the minimum probability of error rule for deciding whether 0 or 1 was sent?

. . . .

With 0 transmitted, X is the Gaussian $(-v, \sigma)$ random variable. With 1 transmitted, X is the Gaussian (v, σ) random variable. With H_i denoting the hypothesis that bit i was sent, the likelihood functions are

$$f_{X|H_0}(x) = \frac{1}{\sqrt{2\pi\sigma^2}}e^{-(x+v)^2/2\sigma^2}, \qquad f_{X|H_1}(x) = \frac{1}{\sqrt{2\pi\sigma^2}}e^{-(x-v)^2/2\sigma^2}. \qquad (11.5)$$

Since $P[H_0] = p$, the likelihood ratio test of Theorem 11.2 becomes

$$x \in A_0 \text{ if } \frac{e^{-(x+v)^2/2\sigma^2}}{e^{-(x-v)^2/2\sigma^2}} \ge \frac{1-p}{p}; \qquad x \in A_1 \text{ otherwise.} \qquad (11.6)$$

Taking the logarithm of both sides and simplifying yields

$$x \in A_0 \text{ if } x \le x^* = \frac{\sigma^2}{2v}\ln\left(\frac{p}{1-p}\right); \qquad x \in A_1 \text{ otherwise.} \qquad (11.7)$$

When $p = 1/2$, the threshold $x^* = 0$ and the decision depends only on whether the evidence in the received signal favors 0 or 1, as indicated by the sign of x. When $p \neq 1/2$, the prior information shifts the decision threshold x^*. The shift favors 1 ($x^* < 0$) if $p < 1/2$. The shift favors 0 ($x^* > 0$) if $p > 1/2$. The influence of the prior information also depends on the signal-to-noise voltage ratio, $2v/\sigma$. When the ratio is relatively high, the information in the received signal is reliable and the received signal has relatively more influence than the prior information (x^* closer to 0). When $2v/\sigma$ is relatively low, the prior information has relatively more influence.

In Figure 11.3, the threshold x^* is the value of x for which the two likelihood functions, each multiplied by a prior probability, are equal. The probability of error is the sum of the shaded areas. Compared to all other decision rules, the threshold x^* produces the minimum possible P_{ERR}.

Example 11.6

Find the error probability of the communications system of Example 11.5.

Applying Equation (11.3), we can write the probability of an error as

$$P_{\text{ERR}} = p\, P\left[X > x^*|H_0\right] + (1-p)\, P\left[X < x^*|H_1\right]. \tag{11.8}$$

Given H_0, X is Gaussian $(-v, \sigma)$. Given H_1, X is Gaussian (v, σ). Consequently,

$$
\begin{aligned}
P_{\text{ERR}} &= pQ\left(\frac{x^* + v}{\sigma}\right) + (1-p)\Phi\left(\frac{x^* - v}{\sigma}\right) \\
&= pQ\left(\frac{\sigma}{2v}\ln\frac{p}{1-p} + \frac{v}{\sigma}\right) + (1-p)\Phi\left(\frac{\sigma}{2v}\ln\frac{p}{1-p} - \frac{v}{\sigma}\right). \tag{11.9}
\end{aligned}
$$

This equation shows how the prior information, represented by $\ln[(1-p)/p]$, and the power of the noise in the received signal, represented by σ, influence P_{ERR}.

Example 11.7

At a computer disk drive factory, the manufacturing failure rate is the probability that a randomly chosen new drive fails the first time it is powered up. Normally, the production of drives is very reliable, with a failure rate $q_0 = 10^{-4}$. However, from time to time there is a production problem that causes the failure rate to jump to $q_1 = 10^{-1}$. Let H_i denote the hypothesis that the failure rate is q_i.

Every morning, an inspector chooses drives at random from the previous day's production and tests them. If a failure occurs too soon, the company stops production and checks the critical part of the process. Production problems occur at random once every ten days, so that $P[H_1] = 0.1 = 1 - P[H_0]$. Based on N, the number of drives tested up to and including the first failure, design a MAP hypothesis test. Calculate the conditional error probabilities P_{FA} and P_{MISS} and the total error probability P_{ERR}.

Given a failure rate of q_i, N is a geometric random variable (see Example 3.8) with expected value $1/q_i$. That is, $P_{N|H_i}(n) = q_i(1-q_i)^{n-1}$ for $n = 1, 2, \ldots$ and $P_{N|H_i}(n) = 0$ otherwise. Therefore, by Theorem 11.2, the MAP design states

$$n \in A_0 \text{ if } \frac{P_{N|H_0}(n)}{P_{N|H_1}(n)} \geq \frac{P[H_1]}{P[H_0]}; \qquad n \in A_1 \text{ otherwise} \tag{11.10}$$

With some algebra, we find that the MAP design is

$$n \in A_0 \text{ if } n \geq n^* = 1 + \frac{\ln\left(\frac{q_1 \, \mathrm{P}[H_1]}{q_0 \, \mathrm{P}[H_0]}\right)}{\ln\left(\frac{1-q_0}{1-q_1}\right)}; \qquad n \in A_1 \text{ otherwise.} \qquad (11.11)$$

Substituting $q_0 = 10^{-4}$, $q_1 = 10^{-1}$, $\mathrm{P}[H_0] = 0.9$, and $\mathrm{P}[H_1] = 0.1$, we obtain $n^* = 45.8$. Therefore, in the MAP hypothesis test, $A_0 = \{n \geq 46\}$. This implies that the inspector tests at most 45 drives in order to reach a decision about the failure rate. If the first failure occurs before test 46, the company assumes that the failure rate is 10^{-1}. If the first 45 drives pass the test, then $N \geq 46$ and the company assumes that the failure rate is 10^{-4}. The error probabilities are:

$$P_{\mathsf{FA}} = \mathrm{P}\left[N \leq 45 | H_0\right] = F_{N|H_0}(45) = 1 - (1 - 10^{-4})^{45} = 0.0045, \qquad (11.12)$$
$$P_{\mathsf{MISS}} = \mathrm{P}\left[N > 45 | H_1\right] = 1 - F_{N|H_1}(45) = (1 - 10^{-1})^{45} = 0.0087. \qquad (11.13)$$

The total probability of error is

$$P_{\mathsf{ERR}} = \mathrm{P}\left[H_0\right] P_{\mathsf{FA}} + \mathrm{P}\left[H_1\right] P_{\mathsf{MISS}} = 0.0049.$$

We will return to Example 11.7 when we examine other types of tests.

Minimum Cost Test

The MAP test implicitly assumes that both types of errors (miss and false alarm) are equally serious. As discussed in connection with the radar application earlier in this section, this is not the case in many important situations. Consider an application in which $C = C_{10}$ units is the cost of a false alarm (decide H_1 when H_0 is correct) and $C = C_{01}$ units is the cost of a miss (decide H_0 when H_1 is correct). In this situation the expected cost of test errors is

$$\mathrm{E}\left[C\right] = \mathrm{P}\left[A_1 | H_0\right] \mathrm{P}\left[H_0\right] C_{10} + \mathrm{P}\left[A_0 | H_1\right] \mathrm{P}\left[H_1\right] C_{01}. \qquad (11.14)$$

Minimizing $\mathrm{E}[C]$ is the goal of the minimum cost hypothesis test. When the decision statistic is a random vector $\mathbf{X}$, we have the following theorem.

Theorem 11.3 **Minimum Cost Binary Hypothesis Test**

For an experiment that produces a random vector $\mathbf{X}$, the minimum cost hypothesis test is

$$\text{Discrete:} \quad \mathbf{x} \in A_0 \text{ if } \frac{P_{\mathbf{X}|H_0}(\mathbf{x})}{P_{\mathbf{X}|H_1}(\mathbf{x})} \geq \frac{\mathrm{P}\left[H_1\right] C_{01}}{\mathrm{P}\left[H_0\right] C_{10}}; \qquad \mathbf{x} \in A_1 \text{ otherwise;}$$

$$\text{Continuous: } \mathbf{x} \in A_0 \text{ if } \frac{f_{\mathbf{X}|H_0}(\mathbf{x})}{f_{\mathbf{X}|H_1}(\mathbf{x})} \geq \frac{\mathrm{P}\left[H_1\right] C_{01}}{\mathrm{P}\left[H_0\right] C_{10}}; \qquad \mathbf{x} \in A_1 \text{ otherwise.}$$

Proof The function to be minimized, Equation (11.14), is identical to the function to be minimized in the MAP hypothesis test, Equation (11.3), except that $P[H_1]C_{01}$ appears in place of $P[H_1]$ and $P[H_0]C_{10}$ appears in place of $P[H_0]$. Thus the optimum hypothesis test is the test in Theorem 11.2, with $P[H_1]C_{01}$ replacing $P[H_1]$ and $P[H_0]C_{10}$ replacing $P[H_0]$.

In this test we note that only the relative cost C_{01}/C_{10} influences the test, not the individual costs or the units in which cost is measured. A ratio >1 implies that misses are more costly than false alarms. Therefore, a ratio >1 expands A_1, the acceptance set for H_1, making it harder to miss H_1 when it is correct. On the other hand, the same ratio contracts H_0 and increases the false alarm probability, because a false alarm is less costly than a miss.

Example 11.8

Continuing the disk drive test of Example 11.7, the factory produces 1000 disk drives per hour and 10,000 disk drives per day. The manufacturer sells each drive for $100. However, each defective drive is returned to the factory and replaced by a new drive. The cost of replacing a drive is $200, consisting of $100 for the replacement drive and an additional $100 for shipping, customer support, and claims processing. Further note that remedying a production problem results in 30 minutes of lost production. Based on the decision statistic N, the number of drives tested up to and including the first failure, what is the minimum cost test?

. . . .

Based on the given facts, the cost C_{10} of a false alarm is 30 minutes (5000 drives) of lost production, or roughly $50,000. On the other hand, the cost C_{01} of a miss is that 10% of the daily production will be returned for replacement. For 1000 drives returned at $200 per drive, the expected cost is $200,000. The minimum cost test is

$$n \in A_0 \text{ if } \frac{P_{N|H_0}(n)}{P_{N|H_1}(n)} \geq \frac{P[H_1]C_{01}}{P[H_0]C_{10}}; \qquad n \in A_1 \text{ otherwise.} \qquad (11.15)$$

Performing the same substitutions and simplifications as in Example 11.7 yields

$$n \in A_0 \text{ if } n \geq n^* = 1 + \frac{\ln\left(\frac{q_1 P[H_1]C_{01}}{q_0 P[H_0]C_{10}}\right)}{\ln\left(\frac{1-q_0}{1-q_1}\right)} = 58.92; \qquad n \in A_1 \text{ otherwise.} \qquad (11.16)$$

Therefore, in the minimum cost hypothesis test, $A_0 = \{n \geq 59\}$. An inspector tests at most 58 disk drives to reach a decision regarding the state of the factory. If 58 drives pass the test, then $A_0 = \{N \geq 59\}$, and the failure rate is assumed to be 10^{-4}. The error probabilities are:

$$P_{\mathsf{FA}} = P[N \leq 58|H_0] = F_{N|H_0}(58) = 1 - (1 - 10^{-4})^{58} = 0.0058, \qquad (11.17)$$

$$P_{\mathsf{MISS}} = P[N \geq 59|H_1] = 1 - F_{N|H_1}(58) = (1 - 10^{-1})^{58} = 0.0022. \qquad (11.18)$$

The average cost (in dollars) of this rule is

$$\begin{aligned} E[C_{\mathsf{MC}}] &= P[H_0] P_{\mathsf{FA}} C_{10} + P[H_1] P_{\mathsf{MISS}} C_{01} \\ &= (0.9)(0.0058)(50,000) + (0.1)(0.0022)(200,000) = 305. \end{aligned} \qquad (11.19)$$

By comparison, the MAP test, which minimizes the probability of an error rather than the expected cost, has error probabilities $P_{FA} = 0.0045$ and $P_{MISS} = 0.0087$ and the expected cost

$$\mathrm{E}\left[C_{\mathrm{MAP}}\right] = (0.9)(0.0045)(50{,}000) + (0.1)(0.0087)(200{,}000) = 376.50. \quad (11.20)$$

The effect of the high cost of a miss has been to reduce the miss probability from 0.0087 to 0.0022. However, the false alarm probability rises from 0.0047 in the MAP test to 0.0058 in the minimum cost test. A savings of $\$376.50 - \$305 = \$71.50$ may not seem very large. The reason is that both the MAP test and the minimum cost test work very well. By comparison, for a "no test" policy that skips testing altogether, each day that the failure rate is $q_1 = 0.1$ will result, on average, in 1000 returned drives at an expected cost of $\$200,000$. Since such days occur with probability $\mathrm{P}[H_1] = 0.1$, the expected cost of a "no test" policy is $\$20,000$ per day.

Neyman–Pearson Test

Given an observation, the MAP test minimizes the probability of accepting the wrong hypothesis and the minimum cost test minimizes the cost of errors. However, the MAP test requires that we know the a priori probabilities $\mathrm{P}[H_i]$ of the competing hypotheses, and the minimum cost test requires that we know in addition the relative costs of the two types of errors. In many situations, these costs and a priori probabilities are difficult or even impossible to specify. In this case, an alternative approach would be to specify a tolerable level for either the false alarm or miss probability. This idea is the basis for the Neyman–Pearson test. The Neyman–Pearson test minimizes P_{MISS} subject to the false alarm probability constraint $P_{\mathrm{FA}} = \alpha$, where α is a constant that indicates our tolerance of false alarms. Because $P_{\mathrm{FA}} = \mathrm{P}[A_1|H_0]$ and $P_{\mathrm{MISS}} = \mathrm{P}[A_0|H_1]$ are conditional probabilities, the test does not require knowledge of the a priori probabilities $\mathrm{P}[H_0]$ and $\mathrm{P}[H_1]$. We first describe the Neyman–Pearson test when the decision statistic is a continous random vector $\mathbf{X}$.

═══Theorem 11.4═══Neyman–Pearson Binary Hypothesis Test

Based on the decision statistic $\mathbf{X}$, *a continuous random vector, the decision rule that minimizes* P_{MISS}, *subject to the constraint* $P_{FA} = \alpha$, *is*

$$\mathbf{x} \in A_0 \ \text{if} \ L(\mathbf{x}) = \frac{f_{\mathbf{X}|H_0}(\mathbf{x})}{f_{\mathbf{X}|H_1}(\mathbf{x})} \geq \gamma; \qquad \mathbf{x} \in A_1 \ \text{otherwise},$$

where γ *is chosen so that* $\int_{L(\mathbf{x}) < \gamma} f_{\mathbf{X}|H_0}(\mathbf{x}) \, d\mathbf{x} = \alpha$.

Proof Using the Lagrange multiplier method, we define the Lagrange multiplier λ and the function

$$G = P_{\text{MISS}} + \lambda(P_{\text{FA}} - \alpha)$$

$$= \int_{A_0} f_{\mathbf{X}|H_1}(\mathbf{x}) \, d\mathbf{x} + \lambda \left(1 - \int_{A_0} f_{\mathbf{X}|H_0}(\mathbf{x}) \, d\mathbf{x} - \alpha \right)$$

$$= \int_{A_0} \left(f_{\mathbf{X}|H_1}(\mathbf{x}) - \lambda f_{\mathbf{X}|H_0}(\mathbf{x}) \right) \, d\mathbf{x} + \lambda(1 - \alpha). \tag{11.21}$$

For a given λ and α, we see that G is minimized if A_0 includes all $\mathbf{x}$ satisfying

$$f_{\mathbf{X}|H_1}(\mathbf{x}) - \lambda f_{\mathbf{X}|H_0}(\mathbf{x}) \leq 0. \tag{11.22}$$

Note that λ is found from the constraint $P_{\text{FA}} = \alpha$. Moreover, we observe that Equation (11.21) implies $\lambda > 0$; otherwise, $f_{\mathbf{X}|H_0}(\mathbf{x}) - \lambda f_{\mathbf{X}|H_1}(\mathbf{x}) > 0$ for all $\mathbf{x}$ and $A_0 = \varnothing$, the empty set, would minimize G. In this case, $P_{\text{FA}} = 1$, which would violate the constraint that $P_{\text{FA}} = \alpha$. Since $\lambda > 0$, we can rewrite the inequality (11.22) as $L(\mathbf{x}) \geq 1/\lambda = \gamma$.

In the radar system of Example 11.3, the decision statistic was a random variable X and the receiver operating curves (ROCs) of Figure 11.2 were generated by adjusting a threshold x_0 that specified the sets $A_0 = \{X \leq x_0\}$ and $A_1 = \{X > x_0\}$. Example 11.3 did not question whether this rule finds the best ROC, that is, the best trade-off between P_{MISS} and P_{FA}. The Neyman–Pearson test finds the best ROC. For each specified value of $P_{\text{FA}} = \alpha$; the Neyman–Pearson test identifies the decision rule that minimizes P_{MISS}.

In the Neyman–Pearson test, an increase in γ decreases P_{MISS} but increases P_{FA}. When the decision statistic $\mathbf{X}$ is a continuous random vector, we can choose γ so that false alarm probability is exactly α. This may not be possible when $\mathbf{X}$ is discrete. In the discrete case, we have the following version of the Neyman–Pearson test.

Theorem 11.5   Discrete Neyman–Pearson Test

Based on the decision statistic $\mathbf{X}$, a discrete random vector, the decision rule that minimizes P_{MISS}, subject to the constraint $P_{\text{FA}} \leq \alpha$, is

$$\mathbf{x} \in A_0 \ \text{if} \ L(\mathbf{x}) = \frac{P_{\mathbf{X}|H_0}(\mathbf{x})}{P_{\mathbf{X}|H_1}(\mathbf{x})} \geq \gamma; \qquad \mathbf{x} \in A_1 \ \text{otherwise},$$

where γ is the largest possible value such that $\sum_{L(\mathbf{x}) < \gamma} P_{\mathbf{X}|H_0}(\mathbf{x}) \leq \alpha$.

Example 11.9

Continuing the disk drive factory test of Example 11.7, design a Neyman–Pearson test such that the false alarm probability satisfies $P_{\text{FA}} \leq \alpha = 0.01$. Calculate the resulting miss and false alarm probabilities.
. . . .
The Neyman–Pearson test is

$$n \in A_0 \text{ if } L(n) = \frac{P_{N|H_0}(n)}{P_{N|H_1}(n)} \geq \gamma; \qquad n \in A_1 \text{ otherwise.} \qquad (11.23)$$

We see from Equation (11.10) that this is the same as the MAP test with $P[H_1]/P[H_0]$ replaced by γ. Thus, just like the MAP test, the Neyman–Pearson test must be a threshold test of the form

$$n \in A_0 \text{ if } n \geq n^*; \qquad n \in A_1 \text{ otherwise.} \qquad (11.24)$$

Some algebra would allow us to find the threshold n^* in terms of the parameter γ. However, this is unnecessary. It is simpler to choose n^* directly so that the test meets the false alarm probability constraint

$$P_{\text{FA}} = P\left[N \leq n^* - 1 | H_0\right] = F_{N|H_0}(n^* - 1) = 1 - (1 - q_0)^{n^* - 1} \leq \alpha. \qquad (11.25)$$

This implies

$$n^* \leq 1 + \frac{\ln(1 - \alpha)}{\ln(1 - q_0)} = 1 + \frac{\ln(0.99)}{\ln(0.9)} = 101.49. \qquad (11.26)$$

Thus, we can choose $n^* = 101$ and still meet the false alarm probability constraint. The error probabilities are:

$$P_{\text{FA}} = P\left[N \leq 100 | H_0\right] = 1 - (1 - 10^{-4})^{100} = 0.00995, \qquad (11.27)$$

$$P_{\text{MISS}} = P\left[N \geq 101 | H_1\right] = (1 - 10^{-1})^{100} = 2.66 \cdot 10^{-5}. \qquad (11.28)$$

We see that tolerating a one percent false alarm probability effectively reduces the probability of a miss to 0 (on the order of one miss per 100 years) but raises the expected cost to

$$E\left[C_{\text{NP}}\right] = (0.9)(0.01)(50{,}000) + (0.1)(2.66 \cdot 10^{-5})(200{,}000) = \$450.53.$$

Although the Neyman–Pearson test minimizes neither the overall probability of a test error nor the expected cost $E[C]$, it may seem preferable to both the MAP test and the minimum cost test because customers will judge the quality of the disk drives and the reputation of the factory based on the number of defective drives that are shipped. Compared to the other tests, the Neyman–Pearson test results in a much lower miss probability and far fewer defective drives being shipped. However, it seems far too conservative, performing 101 tests before deciding that the factory is functioning correctly.

Maximum Likelihood Test

Similar to the Neyman–Pearson test, the *maximum likelihood (ML) test* is another method that avoids the need for a priori probabilities. Under the ML approach, for each outcome s we decide the hypothesis H_i for which $P[s|H_i]$ is largest. The idea behind choosing a hypothesis to maximize the probability of the observation is to avoid making assumptions about costs and a priori probabilities $P[H_i]$. The resulting decision rule, called the *maximum likelihood (ML)* rule, can be written mathematically as:

========Definition 11.1========Maximum Likelihood Decision Rule

For a binary hypothesis test based on the experimental outcome $s \in S$, the maximum likelihood (ML) decision rule is

$$s \in A_0 \ \text{if} \ P\left[s|H_0\right] \geq P\left[s|H_1\right]; \qquad s \in A_1 \ \text{otherwise.}$$

Comparing Theorem 11.1 and Definition 11.1, we see that in the absence of information about the a priori probabilities $P[H_i]$, we have adopted a maximum likelihood decision rule that is the same as the MAP rule under the assumption that hypotheses H_0 and H_1 occur with equal probability. In essence, in the absence of a priori information, the ML rule assumes that all hypotheses are equally likely. By comparing the likelihood ratio to a threshold equal to 1, the ML hypothesis test is neutral about whether H_0 has a higher probability than H_1 or vice versa.

When the decision statistic of the experiment is a random vector $\mathbf{X}$, we can express the ML rule in terms of conditional PMFs or PDFs, just as we did for the MAP rule.

========Theorem 11.6========

If an experiment produces a random vector $\mathbf{X}$, the ML decision rule states

$$\text{Discrete:} \quad \mathbf{x} \in A_0 \ \text{if} \ \frac{P_{\mathbf{X}|H_0}(\mathbf{x})}{P_{\mathbf{X}|H_1}(\mathbf{x})} \geq 1; \quad \mathbf{x} \in A_1 \ \text{otherwise;}$$

$$\text{Continuous:} \ \mathbf{x} \in A_0 \ \text{if} \ \frac{f_{\mathbf{X}|H_0}(\mathbf{x})}{f_{\mathbf{X}|H_1}(\mathbf{x})} \geq 1; \quad \mathbf{x} \in A_1 \ \text{otherwise.}$$

Comparing Theorem 11.6 to Theorem 11.4, when $\mathbf{X}$ is continuous, or Theorem 11.5, when $\mathbf{X}$ is discrete, we see that the maximum likelihood test is the same as the Neyman–Pearson test with parameter $\gamma = 1$. This guarantees that the maximum likelihood test is optimal in the limited sense that no other test can reduce P_{MISS} for the same P_{FA}.

In practice, we use the ML hypothesis test in many applications. It is almost as effective as the MAP hypothesis test when the experiment that produces outcome s is reliable in the sense that P_{ERR} for the ML test is low. To see why this is true, examine the decision rule in Example 11.5. When the signal-to-noise ratio $2v/\sigma$ is high, the right side of Equation (11.7) is close to 0 unless one of the a priori probabilities p or $1 - p$ is close to zero (in which case the logarithm on the right side is a low negative number or a high positive number, indicating strong prior knowledge that the transmitted bit is 0 or 1. When the right side is nearly 0, usually the case in binary communication, the evidence produced by the received signal has much more influence on the decision than the a priori information and the result of the MAP hypothesis test is close to the result of the ML hypothesis test.

Example 11.10

Continuing the disk drive test of Example 11.7, design the maximum likelihood test for the factory status based on the decision statistic N, the number of drives tested up to and including the first failure.

....

The ML hypothesis test corresponds to the MAP test with $P[H_0] = P[H_1] = 0.5$. In this case, Equation (11.11) implies $n^* = 66.62$ or $A_0 = \{n \geq 67\}$. The conditional error probabilities and the cost of the ML decision rule are

$$P_{\mathsf{FA}} = P\left[N \leq 66 | H_0\right] = 1 - (1 - 10^{-4})^{66} = 0.0066,$$
$$P_{\mathsf{MISS}} = P\left[N \geq 67 | H_1\right] = (1 - 10^{-1})^{66} = 9.55 \cdot 10^{-4},$$
$$E\left[C_{\mathsf{ML}}\right] = (0.9)(0.0066)(50{,}000) + (0.1)(9.55 \cdot 10^{-4})(200{,}000) = \$316.10.$$

For the ML test, $P_{\mathsf{ERR}} = 0.0060$. Comparing the MAP rule with the ML rule, we see that the prior information used in the MAP rule makes it more difficult to reject the null hypothesis. We need only 46 good drives in the MAP test to accept H_0, while in the ML test, the first 66 drives have to pass. The ML design, which does not take into account the fact that the failure rate is usually low, is more susceptible to false alarms than the MAP test. Even though the error probability is higher for the ML test, the cost is lower because a costly miss occurs very infrequently (only once every four months). The cost of the ML test is only \$11.10 more than the minimum cost. This is because the a priori probabilities suggest avoiding false alarms because the factory functions correctly, while the costs suggest avoiding misses, because each one is very expensive. Because these two prior considerations balance each other, the ML test, which ignores both of them, is very similar to the minimum cost test.

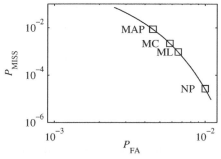

Table 11.1 compares the four binary hypothesis tests (MAP, MC, ML and NP) for the disk drive example. In addition, the receiver operating curve (shown on the left) associated with the decision statistic N, the number of tests up to and including the first failure, shows the performance trade-off between these tests. Even though it uses less prior information than the other tests, the ML test might be a good choice because the cost of testing is nearly minimum and the miss probability is very low. The consequence of a false alarm is likely to be an examination of the manufacturing process to find out if something is wrong. A miss, on the other hand (deciding the factory is functioning properly when 10% of the drives are defective) could be harmful to sales in the long run.

Test	Objective	# tests	P_{FA}	P_{MISS}	Cost
MAP	Minimize probability of incorrect decision	45	4.5×10^{-3}	8.7×10^{-3}	$365
MC	Minimize expected cost	58	5.8×10^{-3}	2.2×10^{-3}	$305
ML	Maximize likelihood; ignore costs and a priori probabilities	67	6.6×10^{-3}	9.6×10^{-4}	$316
NP	Minimize P_{MISS} for given P_{FA}	101	1.0×10^{-2}	2.7×10^{-5}	$451

Table 11.1 Comparison of the maximum a posteriori probability (MAP), minimum cost (MC), maximum likelihood (ML), and Neyman–Pearson (NP) tests at the disk drive factory. Tests are ordered by **# tests**, the maximum number of tests required by each method.

====**Quiz 11.1**====

In an optical communications system, the photodetector output is a Poisson random variable K, either with an expected value of 10,000 photons (hypothesis H_0) or with an expected value of 1,000,000 photons (hypothesis H_1). Given that both hypotheses are equally likely, design a MAP hypothesis test using observed values of random variable K.

11.2   Multiple Hypothesis Test

A multiple hypothesis test is a generalization of a binary hypothesis test from 2 to M hypotheses. As in the binary test, observing an outcome in A_i corresponds to accepting the hypothesis H_i. The accuracy of a multiple hypothesis test is embodied in a matrix of conditional probabilities of deciding H_i when H_j is the correct hypothesis. A maximum a posteriori (MAP) test takes into account a priori probabilities and observations to maximize the probability of a correct decision. A maximum likelihood (ML) test uses only observations. The two tests coincide when all hypotheses are equally likely a priori.

There are many applications in which an experiment can conform to more than two known probability models, all with the same sample space S. A multiple hypothesis test is a generalization of a binary hypothesis test. There are M hypothetical probability models: $H_0, H_1, \cdots, H_{M-1}$. We perform an experiment, and based on the outcome, we come to the decision that a certain H_m is the true probability model. The design of the test consists of dividing S into a partition $A_0, A_1, \cdots, A_{M-1}$, such that the decision is accept H_i if $s \in A_i$. The accuracy measure of the experiment consists of M^2 conditional probabilities, $\mathrm{P}[A_i|H_j]$, $i,j = 0,1,2,\cdots,M-1$. The M probabilities, $\mathrm{P}[A_i|H_i]$, $i = 0,1,\cdots,M-1$ are probabilities of correct decisions.

Example 11.11

A computer modem is capable of transmitting 16 different signals. Each signal represents a sequence of four bits in the digital bit stream at the input to the modem. The modem receiver examines the received signal and produces four bits in the bit stream at the output of the modem. The design of the modem considers the task of the receiver to be a test of 16 hypotheses $H_0, H_1, \ldots, H_{15}$, where H_0 represents 0000, H_1 represents 0001, $\cdots$, and H_{15} represents 1111. The sample space of the experiment is an ensemble of possible received signals. The test design places each outcome s in a set A_i such that the event $s \in A_i$ leads to the output of the four-bit sequence corresponding to H_i.

For a multiple hypothesis test, the MAP hypothesis test and the ML hypothesis test are generalizations of the tests in Theorem 11.1 and Definition 11.1. Minimizing the probability of error corresponds to maximizing the probability of a correct decision,

$$P_{\text{CORRECT}} = \sum_{i=0}^{M-1} \mathrm{P}\left[A_i|H_i\right]\mathrm{P}\left[H_i\right]. \tag{11.29}$$

Theorem 11.7   MAP Multiple Hypothesis Test

Given a multiple hypothesis testing experiment with outcome s, the following rule leads to the highest possible value of $P_{CORRECT}$:

$$s \in A_m \ \ if \ \mathrm{P}\left[H_m|s\right] \geq \mathrm{P}\left[H_j|s\right] \ for \ all \ j = 0,1,2,\ldots,M-1.$$

As in binary hypothesis testing, we can apply Bayes' theorem to derive a decision rule based on the probability models (likelihood functions) corresponding to the hypotheses and the a priori probabilities of the hypotheses. Therefore, corresponding to Theorem 11.2, we have the following generalization of the MAP binary hypothesis test.

Theorem 11.8

For an experiment that produces a random variable X, the MAP multiple hypothesis test is

Discrete: $x_i \in A_m$ *if* $\mathrm{P}\left[H_m\right]P_{X|H_m}(x_i) \geq \mathrm{P}\left[H_j\right]P_{X|H_j}(x_i)$ *for all j;*

Continuous: $x \in A_m$ *if* $\mathrm{P}\left[H_m\right]f_{X|H_m}(x) \geq \mathrm{P}\left[H_j\right]f_{X|H_j}(x)$ *for all j.*

If information about the a priori probabilities of the hypotheses is not available, a maximum likelihood hypothesis test is appropriate.

═══════Definition 11.2═══════ML Multiple Hypothesis Test

A maximum likelihood test of multiple hypotheses has the decision rule

$$s \in A_m \text{ if } \mathrm{P}\left[s|H_m\right] \geq \mathrm{P}\left[s|H_j\right] \text{ for all } j.$$

The ML hypothesis test corresponds to the MAP hypothesis test when all hypotheses H_i have equal probability.

═══════**Example 11.12**═══════

In a quaternary phase shift keying (QPSK) communications system, the transmitter sends one of four equally likely symbols $\{s_0, s_1, s_2, s_3\}$. Let H_i denote the hypothesis that the transmitted signal was s_i. When s_i is transmitted, a QPSK receiver produces the vector $\mathbf{X} = \begin{bmatrix} X_1 & X_2 \end{bmatrix}'$ such that

$$X_1 = \sqrt{E}\cos(i\pi/2 + \pi/4) + N_1, \qquad X_2 = \sqrt{E}\sin(i\pi/2 + \pi/4) + N_2, \qquad (11.30)$$

where N_1 and N_2 are iid Gaussian $(0, \sigma)$ random variables that characterize the receiver noise and E is the average energy per symbol. Based on the receiver output $\mathbf{X}$, the receiver must decide which symbol was transmitted. Design a hypothesis test that maximizes the probability of correctly deciding which symbol was sent.

. . . .

Since the four hypotheses are equally likely, both the MAP and ML tests maximize the probability of a correct decision. To derive the ML hypothesis test, we need to calculate the conditional joint PDFs $f_{\mathbf{X}|H_i}(\mathbf{x})$. Given H_i, N_1 and N_2 are independent and thus X_1 and X_2 are independent. That is, using $\theta_i = i\pi/2 + \pi/4$, we can write

$$\begin{aligned} f_{\mathbf{X}|H_i}(\mathbf{x}) &= f_{X_1|H_i}(x_1)\, f_{X_2|H_i}(x_2) \\ &= \frac{1}{2\pi\sigma^2} e^{-(x_1 - \sqrt{E}\cos\theta_i)^2/2\sigma^2} e^{-(x_2 - \sqrt{E}\sin\theta_i)^2/2\sigma^2} \\ &= \frac{1}{2\pi\sigma^2} e^{-[(x_1 - \sqrt{E}\cos\theta_i)^2 + (x_2 - \sqrt{E}\sin\theta_i)^2]/2\sigma^2}. \end{aligned} \qquad (11.31)$$

We must assign each possible outcome $\mathbf{x}$ to an acceptance set A_i. From Definition 11.2, the acceptance sets A_i for the ML multiple hypothesis test must satisfy

$$\mathbf{x} \in A_i \text{ if } f_{\mathbf{X}|H_i}(\mathbf{x}) \geq f_{\mathbf{X}|H_j}(\mathbf{x}) \text{ for all } j. \qquad (11.32)$$

Equivalently, the ML acceptance sets are given by the rule that $\mathbf{x} \in A_i$ if for all j,

$$(x_1 - \sqrt{E}\cos\theta_i)^2 + (x_2 - \sqrt{E}\sin\theta_i)^2 \leq (x_1 - \sqrt{E}\cos\theta_j)^2 + (x_2 - \sqrt{E}\sin\theta_j)^2.$$

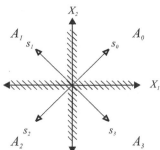

Defining the signal vectors

$$\mathbf{s}_i = \left[\sqrt{E}\cos\theta_i \quad \sqrt{E}\sin\theta_i\right]', \tag{11.33}$$

we can write the ML rule as

$$\mathbf{x} \in A_i \text{ if } \|\mathbf{x} - \mathbf{s}_i\|^2 \leq \|\mathbf{x} - \mathbf{s}_j\|^2, \tag{11.34}$$

where $\|\mathbf{u}\|^2 = u_1^2 + u_2^2$ denotes the square of the Euclidean length of two-dimensional vector $\mathbf{u}$. In Equation (11.34), the acceptance set A_i is the set of all vectors $\mathbf{x}$ that are closest to the vector $\mathbf{s}_i$. These acceptance sets $\{A_0, A_1, A_2, A_3\}$ are the four quadrants (with boundaries marked by shaded bars) shown on the left. In communications textbooks, the space of vectors $\mathbf{x}$ is called the *signal space*, the set of vectors $\{\mathbf{s}_1, \ldots, \mathbf{s}_4\}$ is called the *signal constellation*, and the acceptance sets A_i are called *decision regions*.

Quiz 11.2

For the QPSK communications system of Example 11.12, what is the probability that the receiver makes an error and decodes the wrong symbol?

11.3 MATLAB

> MATLAB programs generate sample values of known probability models in order to compute sample values of derived random variables that appear in hypothesis tests. The programs use the derived sample values in simulations and calculate relative frequencies of events such as misses and false alarms.

In the examples of this chapter, we have chosen experiments with simple probability models in order to highlight the concepts and characteristic properties of hypothesis tests. MATLAB greatly extends our ability to design and evaluate hypothesis tests, especially in practical problems where exact analysis of the probability model becomes too complex. For example, MATLAB can easily perform probability of error calculations and graph receiver operating curves. In addition, there are many cases in which analysis can identify the acceptance sets of a hypothesis test but calculation of the error probabilities is overly complex. In this case, MATLAB can simulate repeated trials of the hypothesis test.

The following example presents a situation frequently encountered by communications engineers. Details of a practical system create probability models that are hard to analyze mathematically. Instead, engineers use MATLAB and other software tools to simulate operation of the systems of interest. Simulation data provides estimates of system performance for each of several design alternatives. This example is similar to Example 11.5, with the added complication that an amplifier in the receiver produces a fraction of the square of the signal plus noise. In this example, there is a well-known probability model for the noise N, but the models for the derived random variables $-v + N + d(-v + N)^2$ and $v + N + d(v + N)^2$ are difficult to derive.

To study this test, we write a MATLAB program that generates m sample values of N. For each sample of N, the program calculates the two functions of N, performs a binary hypothesis test, and determines whether the test results in a hit or false alarm. It reports the relative frequencies of hits and false alarms as estimates of P_{MISS} and P_{FA}.

Example 11.13

A digital communications system transmits either a bit $B = 0$ or $B = 1$ with probability $1/2$. The internal circuitry of the receiver results in a "squared distortion" such that received signal (measured in volts) is either

$$X = \begin{cases} -v + N + d(-v + N)^2 & B = 0, \\ v + N + d(v + N)^2 & B = 1, \end{cases} \tag{11.35}$$

where N, the noise, is Gaussian $(0, 1)$. For each bit transmitted, the receiver produces an output $\hat{B} = 0$ if $X \leq T$ and an output $\hat{B} = 1$, otherwise. Simulate the transmission of 20,000 bits through this system with $v = 1.5$ volts, $d = 0.5$ and the following values of the decision threshold: $T = -0.5, -0.2, 0, 0.2, 0.5$ volts. Which choice of T produces the lowest probability of error? Can you find a value of T that does a better job?

Since each bit is transmitted and received independently of the others, the program sqdistor transmits $m = 10,000$ zeroes to estimate $P[\hat{B} = 1 | B = 0]$, the probability of 1 received given 0 transmitted, for each of the thresholds. It then transmits $m = 10,000$ ones to estimate $P[\hat{B} = 0 | B = 1]$. The average probability of error is

$$P_{\text{ERR}} = 0.5\, P\left[\hat{B} = 1 | B = 0\right] + 0.5\, P\left[\hat{B} = 0 | B = 1\right]. \tag{11.36}$$

```
function y=sqdistor(v,d,m,T)
%P(error) for m bits tested
%transmit +-v, add N & d(v+N)^2
%receive 1 if x>T, otherwise 0
x=(v+randn(m,1));
[XX,TT]=ndgrid(x,T(:));
P01=sum((XX+d*(XX.^2)< TT),1)/m;
x= -v+randn(m,1);
[XX,TT]=ndgrid(x,T(:));
P10=sum((XX+d*(XX.^2)>TT),1)/m;
y=0.5*(P01+P10);
```

By defining the grid matrices XX and TT, we can test each candidate value of T for the same set of noise variables. We observe the output in Figure 11.4. Because of the bias induced by the squared distortion term, $T = 0.5$ is best among the candidate values of T. However, the data suggests that a value of T greater than 0.5 might work better. Problem 11.3.3 examines this possibility.

The problems for this section include a collection of hypothesis testing problems that can be solved using MATLAB but are too difficult to solve by hand. The

```
>> T
T =
    -0.5000   -0.2000          0    0.2000    0.5000
>> Pe=sqdistor(1.5,0.5,10000,T)
Pe =
     0.5000    0.2733    0.2265    0.1978    0.1762
```

Figure 11.4 Average error rate for the squared distortion communications system of Example 11.13.

solutions are built on the MATLAB methods developed in prior chapters; however, the necessary MATLAB calculations and simulations are typically problem specific.

═══Quiz 11.3═══

For the communications system of Example 11.13 with squared distortion, we can define the miss and false alarm probabilities as

$$P_{\text{MISS}} = P_{01} = \text{P}\left[\hat{B} = 0|B = 1\right], \qquad P_{\text{FA}} = P_{10} = \text{P}\left[\hat{B} = 1|B = 0\right]. \quad (11.37)$$

Modify the program `sqdistor` in Example 11.13 to produce receiver operating curves for the parameters $v = 3$ volts and $d = 0.1$, 0.2, and 0.3. Hint: The points on the ROC correspond to different values of the threshold T volts.

Further Reading: [Kay98] provides detailed, readable coverage of hypothesis testing. [Hay01] presents detection of digital communications signals as a hypothesis test. A collection of challenging homework problems for sections 11.2 and 11.3 are based on bit detection for code division multiple access (CDMA) communications systems. The authoritative treatment of this subject can be found in [Ver98].

Problems _____

Difficulty: ● Easy   ▨ Moderate   ♦ Difficult   ♦♦ Experts Only

11.1.1● In a random hour, the number of call attempts N at a telephone switch has a Poisson distribution with an expected value of either α_0 (hypothesis H_0) or α_1 (hypothesis H_1). For a priori probabilities $\text{P}[H_i]$, find the MAP and ML hypothesis testing rules given the observation of N.

11.1.2▨ A cellular telephone company is upgrading its network to a new (N) transmission system one area at a time, but they do not announce where the upgrades take place. You have the task of determining whether certain areas have been upgraded. You have decided to use an application in your smartphone to measure the ping time (how long it takes to receive a response to a certain message) in each area. The new system is faster than the old (O) one. It has on average shorter ping times. The ping time, in milliseconds of a new transmission system is the exponential (60) random variable N. The ping time of an old system is an exponential random variable O with expected value $\mu_O > 60$ ms. The null hypothesis of a binary hypothesis test is H_0: The transmission system is the new system. The alternative hypothesis is H_1: The transmission system is the old system. The probability of a new system is $P[N] = 0.8$. The probability of an old system is $P[O] = 0.2$. A binary hypothesis test measures T milliseconds, the result of one ping test. The decision is H_0 if $T \leq t_0$ ms. Otherwise, the decision is H_1.

(a) Write a formula for the false alarm probability as a function of t_0 and μ_O.

(b) Write a formula for the miss probability as a function of t_0 and μ_O.

(c) Calculate the maximum likelihood decision time $t_0 = t_{ML}$ for $\mu_O = 120$ ms and $\mu_O = 200$ ms.

(d) Do you think that t_{MAP}, the maximum a posteriori decision time, is greater than or less than t_{ML}? Explain your answer.

(e) Calculate the maximum a posteriori probability decision time $t_0 = t_{MAP}$ for $\mu_O = 120$ ms and $\mu_O = 200$ ms.

(f) Draw the receiver operating curves for $\mu_O = 120$ ms and $\mu_O = 200$ ms.

11.1.3■ An automatic doorbell system rings a bell whenever it detects someone at the door. The system uses a photodetector such that if a person is present, hypothesis H_1, the photodetector output N is a Poisson random variable with an expected value of 1300 photons. Otherwise, if no one is there, hypothesis H_0, the photodetector output is a Poisson random variable with an expected value of 1000. Devise a Neyman–Pearson test for the presence of someone outside the door such that the false alarm probability is $\alpha \leq 10^{-6}$. What is minimum value of P_{MISS}?

11.1.4■ In the radar system of Example 11.3, $P[H_1] = 0.01$. In the case of a false alarm, the system issues an unnecessary alert at the cost of $C_{10} = 1$ unit. The cost of a miss is $C_{01} = 10^4$ units because the target could cause a lot of damage. When the target is present, the voltage

is $X = 4+N$, a Gaussian $(4, 1)$ random variable. When there is no target present, the voltage is $X = N$, the Gaussian $(0, 1)$ random variable. In a binary hypothesis test, the acceptance sets are $A_0 = \{X \leq x_0\}$ and $A_1 = \{X > x_0\}$.

(a) For the MAP hypothesis test, find the decision threshold $x_0 = x_{MAP}$, the error probabilities P_{FA} and P_{MISS}, and the average cost $E[C]$.

(b) Compare the MAP test performance against the minimum cost hypothesis test.

11.1.5■ In the radar system of Example 11.3, show that the ROC in Figure 11.2 is the result of a Neyman–Pearson test. That is, show that the Neyman–Pearson test is a threshold test with acceptance set $A_0 = \{X \leq x_0\}$. How is x_0 related to the false alarm probability α?

11.1.6◆ A system administrator (and part-time spy) at a classified research facility wishes to use a gateway router for covert communication of research secrets to an outside accomplice. The sysadmin covertly communicates a bit W for every n transmitted packets. To signal $W = 0$, the router does nothing while n regular packets are sent out through the gateway as a Poisson process of rate λ_0 packets/sec. To signal $W = 1$ the sysadmin injects additional fake outbound packets so that n outbound packets are sent as a Poisson process of rate $2\lambda_0$. The secret communication bits are equiprobable in that $P[W = 1] = P[W = 0] = 1/2$. The sysadmin's accomplice (outside the gateway) monitors the outbound packet transmission process by observing the vector $\mathbf{X} = [X_1, X_2, \ldots, X_n]$ of packet interarrival times and guessing the bit W every n packets.

(a) Find the conditional PDFs $f_{\mathbf{X}|W=0}(\mathbf{x})$ and $f_{\mathbf{X}|W=1}(\mathbf{x})$.

(b) What are the MAP and ML hypothesis tests for the accomplice to guess either hypothesis H_0 that $W = 0$ or hypothesis H_1 that $W = 1$?

(c) Let $\hat{W}$ denote the decision of the ML hypothesis test. Use the Chernoff bound to upper bound the error probability $P[\hat{W} = 0|W = 1]$.

11.1.7◆ The ping time, in milliseconds, of a new transmission system, described in Problem 11.1.2 is the exponential (60) random variable N. The ping time of an old system is the exponential (120) random variable O. The null hypothesis of a binary hypothesis test is H_0: The transmission system is the new system. The alternative hypothesis is H_1: The transmission system is the old system. The probability of a new system is $P[N] = 0.8$. The probability of an old system $P[O] = 0.2$. A binary hypothesis test performs k ping tests and calculates $M_n(T)$, the sample mean of the ping time. The decision is H_0 if $M_n(T) \le t_0$ ms. Otherwise, the decision is H_1.

(a) Use the central limit theorem to write a formula for the false alarm probability as a function of t_0 and k.

(b) Use the central limit theorem to write a formula for the miss probability as a function of t_0 and k.

(c) Calculate the maximum likelihood decision time, $t_0 = t_{\mathrm{ML}}$, for $k = 9$ ping tests.

(d) Calculate the maximum a posteriori probability decision time, $t_0 = t_{\mathrm{MAP}}$ for $k = 9$ ping tests.

(e) Draw the receiver operating curves for $k = 9$ ping tests and $k = 16$ ping tests.

11.1.8◆ In this problem, we perform the old/new detection test of Problem 11.1.7, except now we monitor k ping tests and observe whether each ping lasts longer than t_0 ms. The random variable M is the number of pings that last longer than t_0 ms. The decision is H_0 if $M \le m_0$. Otherwise, the decision is H_1.

(a) Write a formula for the false alarm probability as a function of t_0, m_0, and n.

(b) Find the maximum likelihood decision number $m_0 = m_{ML}$ for $t_0 = 4.5$ ms and $k = 16$ ping tests.

(c) Find the maximum a posteriori probability decision number $m_0 = m_{MAP}$ for $t_0 = 4.5$ ms and $k = 16$ ping tests.

(d) Draw the receiver operating curves for $t_0 = 90$ ms and $t_0 = 60$ ms. In both cases let $k = 16$ ping tests.

11.1.9◆ A binary communication system has transmitted signal X, the Bernoulli $(1/2)$ random variable. At the receiver, we observe $Y = VX + W$, where V is a "fading factor" and W is additive noise. Note that V and W are exponential (1) random variables and that X, V, and W are mutually independent. Given the observation Y, we must guess whether $X = 0$ or $X = 1$ was transmitted. Use a binary hypothesis test to determine the rule that minimizes the probability P_e of a decoding error. For the optimum decision rule, calculate P_e.

11.1.10◆ In a BPSK amplify-and-forward relay system, a source transmits a random bit $V \in \{-1, 1\}$ every T seconds to a destination receiver via a set of n relay transmitters. $V = 1$ and $V = -1$ are equally likely. In this communication system, the source transmits during the time period $(0, T/2)$ such that relay i receives

$$X_i = \alpha_i V + W_i, \qquad i = 1, 2, \dots, n,$$

where the W_i are iid Gaussian $(0, 1)$ random variables representing relay i receiver noise. In the time interval $(T/2, T)$, each relay node amplifies and forwards the received source signal. The destination receiver obtains the vector $\mathbf{Y} = \begin{bmatrix} Y_1 & \cdots & Y_n \end{bmatrix}'$ such that

$$Y_i = \beta_i X_i + Z_i, \qquad i = 1, 2, \dots, n,$$

where the Z_i are also iid Gaussian $(0, 1)$ random variables. In the following, assume that the parameters α_i and β_i are all nonnegative. Also, let H_0 denote the hypothesis that $V = -1$ and H_1 the hypothesis $V = 1$.

(a) Suppose you build a suboptimal detector based on the sum $Y = \sum_{i=1}^{n} Y_i$. If $Y > 0$, the receiver guesses H_1; otherwise the receiver guesses H_0. What is the probability of error P_e for this receiver?

(b) Based on the observation $\mathbf{Y}$, now suppose the destination receiver detector performs a MAP test for hypotheses H_0 or H_1. What is the MAP detector rule? Simplify your answer as much as possible. Hint: First find the likelihood functions $f_{\mathbf{Y}|H_i}(y)$.

(c) What is the probability of bit error P_e^* of the MAP detector?

(d) Compare the two detectors when $n = 4$ and

$$(\alpha_1, \beta_1) = (1,1), \quad (\alpha_2, \beta_2) = (10,1),$$
$$(\alpha_3, \beta_3) = (1,10), \quad (\alpha_4, \beta_4) = (10,10).$$

In general, what's bad about the suboptimal detector?

11.1.11 In a BPSK communication system, a source wishes to communicate a random bit $X \in \{-1, 1\}$ to a receiver. Inputs $X = 1$ and $X = -1$ are equally likely. In this system, the source transmits X multiple times. In the ith transmission, the receiver observes $Y_i = X + W_i$, where the W_i are iid Gaussian $(0,1)$ noises, independent of X.

(a) After n transmissions of X, you observe $\mathbf{Y} = \mathbf{y} = \begin{bmatrix} y_1 & \cdots & y_n \end{bmatrix}'$. Find $P[X = 1|\mathbf{Y} = \mathbf{y}]$. Express your answer in terms of the likelihood ratio

$$L(\mathbf{y}) = \frac{f_{\mathbf{Y}|X}(\mathbf{y}|-1)}{f_{\mathbf{Y}|X}(\mathbf{y}|1)}.$$

(b) Suppose after n transmissions, the receiver observes $\mathbf{Y} = \mathbf{y}$ and decides

$$X^* = \begin{cases} 1 & P[X = 1|\mathbf{Y} = \mathbf{y}] > 1/2, \\ -1 & \text{otherwise.} \end{cases}$$

Find the probability of error $P_e = P[X^* \neq X]$ in terms of the $\Phi(\cdot)$ func-

tion. Hint: For P_e calculations, symmetry implies $P_e = P[X^* \neq X|X = 1]$.

(c) Now suppose the system uses ARQ as follows. If $|\hat{X}_n(\mathbf{y})| < 1 - \epsilon$, the receiver requests that X be retransmitted; otherwise the receiver guesses what is transmitted. In particular, if $\hat{X}_n(\mathbf{y}) > 1 - \epsilon$, the receiver guesses $X^* = 1$. If $\hat{X}_n(\mathbf{y}) < -1 + \epsilon$, the receiver guesses $X^* = -1$. Following the receiver's guess, the transmitter starts sending a new bit. Find upper and lower bounds to $P_e = P[X^* \neq X]$. That is, find ϵ_1 and ϵ_2 such that

$$\epsilon_1 \le P_e \le \epsilon_2.$$

11.1.12 Suppose in the disk drive factory of Example 11.7, we can observe K, the number of failed devices out of n devices tested. As in the example, let H_i denote the hypothesis that the failure rate is q_i.

(a) Assuming $q_0 < q_1$, what is the ML hypothesis test based on an observation of K?

(b) What are the conditional probabilities of error $P_{FA} = P[A_1|H_0]$ and $P_{MISS} = P[A_0|H_1]$? Calculate these probabilities for $n = 500, q_0 = 10^{-4}, q_1 = 10^{-2}$.

(c) Compare this test to that considered in Example 11.7. Which test is more reliable? Which test is easier to implement?

11.1.13 Consider a binary hypothesis test in which there is a cost associated with each type of decision. In addition to the cost C_{10}' for a false alarm and C_{01}' for a miss, we also have the costs C_{00}' for correctly deciding hypothesis H_0 and the C_{11}' for correctly deciding hypothesis H_1. Based on the observation of a continuous random vector $\mathbf{X}$, design the hypothesis test that minimizes the total expected cost

$$\begin{aligned} E\left[C'\right] = {} & P\left[A_1|H_0\right] P\left[H_0\right] C_{10}' \\ & + P\left[A_0|H_0\right] P\left[H_0\right] C_{00}' \\ & + P\left[A_0|H_1\right] P\left[H_1\right] C_{01}' \\ & + P\left[A_1|H_1\right] P\left[H_1\right] C_{11}'. \end{aligned}$$

Show that the decision rule that minimizes total cost is the same as decision rule of the minimum cost test in Theorem 11.3, with the costs C_{01} and C_{10} replaced by the differential costs $C'_{01} - C'_{11}$ and $C'_{10} - C'_{00}$.

11.2.1● In a ternary amplitude shift keying (ASK) communications system, there are three equally likely transmitted signals $\{s_0, s_1, s_2\}$. These signals are distinguished by their amplitudes such that if signal s_i is transmitted, the receiver output will be

$$X = a(i - 1) + N,$$

where a is a positive constant and N is a Gaussian $(0, \sigma_N)$ random variable. Based on the output X, the receiver must decode which symbol s_i was transmitted.

(a) What are the acceptance sets A_i for the hypotheses H_i that s_i was transmitted?

(b) What is $P[D_E]$, the probability that the receiver decodes the wrong symbol?

11.2.2● A multilevel QPSK communications system transmits three bits every unit of time. For each possible sequence ijk of three bits, one of eight symbols, $\{s_{000}, s_{001}, \ldots, s_{111}\}$, is transmitted. When signal s_{ijk} is transmitted, the receiver output is

$$X = s_{ijk} + N,$$

where N is a Gaussian $(0, \sigma^2 I)$ random vector. The two-dimensional signal vectors $s_{000}, \ldots, s_{111}$ are

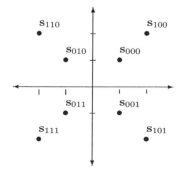

Let H_{ijk} denote the hypothesis that s_{ijk} was transmitted. The receiver output $X =$ $[X_1 \quad X_2]'$ is used to decide the acceptance sets $\{A_{000}, \ldots, A_{111}\}$. If all eight symbols are equally likely, sketch the acceptance sets.

11.2.3■ An M-ary quadrature amplitude modulation (QAM) communications system can be viewed as a generalization of the QPSK system described in Example 11.12. In the QAM system, one of M equally likely symbols $s_0, \ldots, s_{m-1}$ is transmitted every unit of time. When symbol s_i is transmitted, the receiver produces the two-dimensional vector output

$$X = s_i + N,$$

where N has iid Gaussian $(0, \sigma^2)$ components. Based on the output X, the receiver must decide which symbol was transmitted. Design a hypothesis test that maximizes the probability of correctly deciding what symbol was sent. Hint: Following Example 11.12, describe the acceptance set in terms of the vectors

$$\mathbf{x} = \begin{bmatrix} x_1 \\ x_2 \end{bmatrix}, \quad \mathbf{s}_i = \begin{bmatrix} s_{i1} \\ s_{i2} \end{bmatrix}.$$

11.2.4■ Suppose a user of the multilevel QPSK system needs to decode only the third bit k of the message ijk. For $k = 0, 1$, let H_k denote the hypothesis that the third bit was k. What are the acceptance sets A_0 and A_1? What is $P[B_3]$, the probability that the third bit is in error?

11.2.5◆ The QPSK system of Example 11.12 can be generalized to an M-ary phase shift keying (MPSK) system with $M > 4$ equally likely signals. The signal vectors are $\{s_0, \ldots, s_{M-1}\}$, where

$$\mathbf{s}_i = \begin{bmatrix} s_{i1} \\ s_{i2} \end{bmatrix} = \begin{bmatrix} \sqrt{E} \cos \theta_i \\ \sqrt{E} \sin \theta_i \end{bmatrix}$$

and $\theta_i = 2\pi i / M$. When the ith message is sent, the received signal is $X = s_i + N$ where N is a Gaussian $(0, \sigma^2 I)$ noise vector.

(a) Sketch the acceptance set A_i for the hypothesis H_i that s_i was transmitted.

(b) Find the largest value of d such that

$$\{\mathbf{x}|\,\|\mathbf{x} - \mathbf{s}_i\| \le d\} \subset A_i.$$

(c) Use d to find an upper bound for the probability of error.

11.2.6◆ A modem uses QAM (see Problem 11.2.3) to transmit one of 16 symbols, $s_0, \ldots, s_{15}$, every $1/600$ seconds. When signal s_i is transmitted, the receiver output is

$$\mathbf{X} = \mathbf{s}_i + \mathbf{N}.$$

The signal vectors $\mathbf{s}_0, \ldots, \mathbf{s}_{15}$ are

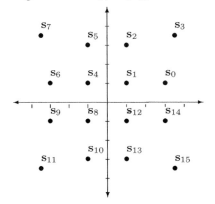

(a) Sketch the acceptance sets based on the receiver outputs X_1, X_2. Hint: Apply the solution to Problem 11.2.3.

(b) Let H_i be the event that symbol s_i was transmitted and let C be the event that the correct symbol is decoded. What is $P[C|H_1]$?

(c) Argue that $P[C] \ge P[C|H_1]$.

11.2.7◆ For the QPSK communications system of Example 11.12, identify the acceptance sets for the MAP hypothesis test when the symbols are not equally likely. Sketch the acceptance sets when $\sigma = 0.8$, $E = 1$, $P[H_0] = 1/2$, and $P[H_1] = P[H_2] = P[H_3] = 1/6$.

11.2.8◆ In a code division multiple access (CDMA) communications system, k users share a radio channel using a set of n-dimensional code vectors $\{\mathbf{S}_1, \ldots, \mathbf{S}_k\}$ to distinguish their signals. The dimensionality factor n is known as the processing gain. Each user i transmits independent data bits X_i such that the vector $\mathbf{X} = \begin{bmatrix} X_1 & \cdots & X_k \end{bmatrix}'$ has iid components with $P_{X_i}(1) = P_{X_i}(-1) = 1/2$. The received signal is

$$\mathbf{Y} = \sum_{i=1}^{k} X_i \sqrt{p_i} \mathbf{S}_i + \mathbf{N},$$

where $\mathbf{N}$ is a Gaussian $(\mathbf{0}, \sigma^2 \mathbf{I})$ noise vector. From the observation $\mathbf{Y}$, the receiver performs a multiple hypothesis test to decode the data bit vector $\mathbf{X}$.

(a) Show that in terms of vectors,

$$\mathbf{Y} = \mathbf{S}\mathbf{P}^{1/2}\mathbf{X} + \mathbf{N},$$

where $\mathbf{S}$ is an $n \times k$ matrix with ith column $\mathbf{S}_i$ and $\mathbf{P}^{1/2} = \text{diag}[\sqrt{p_1}, \ldots, \sqrt{p_k}]$ is a $k \times k$ diagonal matrix.

(b) Given $\mathbf{Y} = \mathbf{y}$, show that the MAP and ML detectors for $\mathbf{X}$ are the same and are given by

$$\mathbf{x}^*(\mathbf{y}) = \arg\min_{\mathbf{x} \in B_k} \left\| \mathbf{y} - \mathbf{S}\mathbf{P}^{1/2}\mathbf{x} \right\|.$$

where B_k is the set of all k dimensional vectors with ± 1 elements.

(c) How many hypotheses does the ML detector need to evaluate?

11.2.9◆ For the CDMA communications system of Problem 11.2.8, a detection strategy known as *decorrelation* applies a transformation to $\mathbf{Y}$ to generate

$$\tilde{\mathbf{Y}} = (\mathbf{S}'\mathbf{S})^{-1}\mathbf{S}'\mathbf{Y} = \mathbf{P}^{1/2}\mathbf{X} + \tilde{\mathbf{N}}$$

where $\tilde{\mathbf{N}} = (\mathbf{S}'\mathbf{S})^{-1}\mathbf{S}'\mathbf{N}$ is still a Gaussian noise vector with expected value $E[\tilde{\mathbf{N}}] = \mathbf{0}$. Decorrelation separates the signals in that the ith component of $\tilde{\mathbf{Y}}$ is

$$\tilde{Y}_i = \sqrt{p_i} X_i + \tilde{N}_i,$$

which is the same as a single-user receiver output of the binary communication system of Example 11.5. For equally likely inputs

$X_i = 1$ and $X_i = -1$, Example 11.5 showed that the optimal (minimum probability of bit error) decision rule based on the receiver output $\tilde{Y}_i$ is

$$\hat{X}_i = \mathrm{sgn}\,(\tilde{Y}_i).$$

Although this technique requires the code vectors $\mathbf{S}_1, \ldots, \mathbf{S}_k$ to be linearly independent, the number of hypotheses that must be tested is greatly reduced in comparison to the optimal ML detector introduced in Problem 11.2.8. In the case of linearly independent code vectors, is the decorrelator optimal? That is, does it achieve the same bit error rate (BER) as the optimal ML detector?

11.3.1 A wireless pressure sensor (buried in the ground) reports a discrete random variable X with range $S_X = \{1, 2, \ldots, 20\}$ to signal the presence of an object. Given an observation X and a threshold x_0, we decide that an object is present (hypothesis H_1) if $X > x_0$; otherwise we decide that no object is present (hypothesis H_0). Under hypothesis H_i, X has conditional PMF

$$P_{X|H_i}(x) = \begin{cases} \frac{(1-p_i)p_i^{x-1}}{1-p_i^{20}} & x = 1, 2, \ldots, 20, \\ 0 & \text{otherwise,} \end{cases}$$

where $p_0 = 0.99$ and $p_1 = 0.9$. Calculate and plot the false alarm and miss probabilities as a function of the detection threshold x_0. Calculate the discrete receiver operating curve (ROC) specified by x_0.

11.3.2 For the binary communications system of Example 11.6, graph the error probability P_{ERR} as a function of p, the probability that the transmitted signal is 0. For the signal-to-noise voltage ratio, consider $v/\sigma \in \{0.1, 1, 10\}$. What values of p minimize P_{ERR}? Why are those values not practical?

11.3.3 For the squared distortion communications system of Example 11.13 with $v = 1.5$ and $d = 0.5$, find the value of T that minimizes P_e.

11.3.4 A poisonous gas sensor reports continuous random variable X. In the presence of toxic gases, hypothesis H_1,

$$f_{X|H_1}(x) = \begin{cases} (x/8)e^{-x^2/16} & x \geq 0, \\ 0 & \text{otherwise.} \end{cases}$$

In the absence of dangerous gases, X has conditional PDF

$$f_{X|H_0}(x) = \begin{cases} (1/2)e^{-x/2} & x \geq 0, \\ 0 & \text{otherwise.} \end{cases}$$

Devise a hypothesis test that determines the presence of poisonous gases. Plot the false alarm and miss probabilities for the test as a function of the decision threshold. Lastly, plot the corresponding receiver operating curve.

11.3.5 Simulate the M-ary PSK system in Problem 11.2.5 for $M = 8$ and $M = 16$. Let $\hat{P}_{\mathrm{ERR}}$ denote the relative frequency of symbol errors in the simulated transmission in 10^5 symbols. For each value of M, graph $\hat{P}_{\mathrm{ERR}}$, as a function of the signal-to-noise power ratio (SNR) $\gamma = E/\sigma^2$. Consider $10 \log_{10} \gamma$, the SNR in dB, ranging from 0 to 30 dB.

11.3.6 In this problem, we evaluate the bit error rate (BER) performance of the CDMA communications system introduced in Problem 11.2.8. In our experiments, we will make the following additional assumptions.

- In practical systems, code vectors are generated pseudorandomly. We will assume the code vectors are random. For each transmitted data vector $\mathbf{X}$, the code vector of user i will be $\mathbf{S}_i = \frac{1}{\sqrt{n}} \begin{bmatrix} S_{i1} & S_{i2} & \cdots & S_{in} \end{bmatrix}'$, where the components S_{ij} are iid random variables such that $P_{S_{ij}}(1) = P_{S_{ij}}(-1) = 1/2$. Note that the factor $1/\sqrt{n}$ is used so that each code vector $\mathbf{S}_i$ has length 1: $\|\mathbf{S}_i\|^2 = \mathbf{S}_i'\mathbf{S}_i = 1$.

- Each user transmits at 6dB SNR. For convenience, assume $P_i = p = 4$ and $\sigma^2 = 1$.

(a) Use MATLAB to simulate a CDMA system with processing gain $n = 16$. For each experimental trial, generate a random set of code vectors $\{S_i\}$, data vector X, and noise vector N. Find the ML estimate x^* and count the number of bit errors; i.e., the number of positions in which $x_i^* \neq X_i$. Use the relative frequency of bit errors as an estimate of the probability of bit error. Consider $k = 2, 4, 8, 16$ users. For each value of k, perform enough trials so that bit errors are generated on 100 independent trials. Explain why your simulations take so long.

(b) For a simpler detector known as the matched filter, when $Y = y$, the detector decision for user i is

$$\hat{x}_i = \text{sgn}\,(S_i'y)$$

where $\text{sgn}\,(x) = 1$ if $x > 0$, $\text{sgn}\,(x) = -1$ if $x < 0$, and otherwise $\text{sgn}\,(x) = 0$. Compare the bit error rate of the matched filter and the maximum likelihood detectors. Note that the matched filter is also called a single user detector since it can detect the bits of user i without the knowledge of the code vectors of the other users.

11.3.7◆◆ For the CDMA system in Problem 11.2.8, we wish to use MATLAB to evaluate the bit error rate (BER) performance of the decorrelater introduced Problem 11.2.9. In particular, we want to estimate P_e, the probability that for a set of randomly chosen code vectors, that a randomly chosen user's bit is decoded incorrectly at the receiver.

(a) For a k user system with a fixed set of code vectors $S_1, \ldots, S_k$, let S denote the matrix with S_i as its ith column. Assuming that the matrix inverse $(S'S)^{-1}$ exists, write an expression for $P_{e,i}(S)$, the probability of error for the transmitted bit of user i, in terms of S and the $Q(\cdot)$ function. For the same fixed set of code vectors S, write an expression for P_e, the proba-

bility of error for the bit of a randomly chosen user.

(b) In the event that $(S'S)^{-1}$ does not exist, we assume the decorrelator flips a coin to guess the transmitted bit of each user. What are $P_{e,i}$ and P_e in this case?

(c) For a CDMA system with processing gain $n = 32$ and k users, each with SNR 6 dB, write a MATLAB program that averages over randomly chosen matrices S to estimate P_e for the decorrelator. Note that unlike the case for Problem 11.3.6, simulating the transmission of bits is not necessary. Graph your estimate $\hat{P}_e$ as a function of k.

11.3.8◆◆ Simulate the multi-level QAM system of Problem 11.2.4. Estimate the probability of symbol error and the probability of bit error as a function of the noise variance σ^2.

11.3.9◆◆ In Problem 11.3.5, we used simulation to estimate the probability of *symbol* error. For transmitting a binary bit stream over an MPSK system, we set each $M = 2^N$ and each transmitted symbol corresponds to N bits. For example, for $M = 16$, we map each four-bit input $b_3b_2b_1b_0$ to one of 16 symbols. A simple way to do this is binary index mapping: transmit s_i when $b_3b_2b_1b_0$ is the binary representation of i. For example, the bit input 1100 is mapped to the transmitted signal s_{12}. Symbol errors in the communication system cause bit errors. For example if s_1 is sent but noise causes s_2 to be decoded, the input bit sequence $b_3b_2b_1b_0 = 0001$ is decoded as $\hat{b}_3\hat{b}_2\hat{b}_1\hat{b}_0 = 0010$, resulting in 2 correct bits and 2 bit errors. In this problem, we use MATLAB to investigate how the mapping of bits to symbols affects the probability of bit error. For our preliminary investigation, it will be sufficient to map the three bits $b_2b_1b_0$ to the $M = 8$ PSK system of Problem 11.2.5.

(a) Find the acceptance sets $\{A_0, \ldots, A_7\}$.

(b) Simulate m trials of the transmission of symbol s_0. Estimate the probabilities

$\{P_{0j}|j = 0, 1, \ldots, 7\}$, that the receiver output is s_j when s_0 was sent. By symmetry, use the set $\{P_{0j}\}$ to determine P_{ij} for all i and j.

(c) Let $\mathbf{b}(i) = \begin{bmatrix} b_2(i) & b_1(i) & b_0(i) \end{bmatrix}$ denote the input bit sequence that is mapped to s_i. Let d_{ij} denote the number of bit positions in which $\mathbf{b}(i)$ and $\mathbf{b}(j)$ differ. For a given mapping, the bit error rate (BER) is

$$\text{BER} = \frac{1}{M} \sum_i \sum_j P_{ij} d_{ij}.$$

(d) Estimate the BER for the binary index mapping.

(e) The Gray code is perhaps the most commonly used mapping:

b	000	001	010	011	100	101	110	111
s_i	s_0	s_1	s_3	s_2	s_7	s_6	s_4	s_5

Does the Gray code reduce the BER compared to the binary index mapping?

11.3.10◆◆ Continuing Problem 11.3.9, in the mapping of the bit sequence $b_2 b_1 b_0$ to the symbols s_i, we wish to determine the probability of error for each input bit b_i. Let q_i denote the probability that bit b_i is decoded in error. Determine q_0, q_1, and q_2 for both the binary index mapping as well as the Gray code mapping.

12

Estimation of a Random Variable

The techniques in Chapters 10 and 11 use the outcomes of experiments to make inferences about probability models. In this chapter we use observations to calculate an approximate value of a sample value of a random variable that has not been observed. The random variable of interest may be unavailable because it is impractical to measure (for example, the temperature of the sun), or because it is obscured by distortion (a signal corrupted by noise), or because it is not available soon enough. We refer to the estimation of future observations as *prediction*. A predictor uses random variables observed in early subexperiments to estimate a random variable produced by a later subexperiment. If X is the random variable to be estimated, we adopt the notation $\hat{X}$ (also a random variable) for the estimate. In most of the chapter, we use the *mean square error*

$$e = \mathrm{E}\left[(X - \hat{X})^2\right] \tag{12.1}$$

as a measure of the quality of the estimate.

Signal estimation is a big subject. To introduce it in one chapter, we confine our attention to the following problems:

- Blind estimation of a random variable

- Estimation of a random variable given an event

- Estimation of a random variable given one other random variable

- Linear estimation of a random variable given a random vector

- Linear estimation of a random vector given another random vector

12.1   Minimum Mean Square Error Estimation

> The estimate of X that minimizes the mean square error is the expected value of X given available information. The optimum blind estimate is $\mathrm{E}[X]$. It uses only the probability model of X. The optimum estimate given $X \in A$ is $\mathrm{E}[X|A]$. The optimum estimate given $Y = y$ is $\mathrm{E}[X|Y = y]$.

An experiment produces a random variable X. However, we are unable to observe X directly. Instead, we observe an event or a random variable that provides partial information about the sample value of X. X can be either discrete or continuous. If X is a discrete random variable, it is possible to use hypothesis testing to estimate X. For each $x_i \in S_X$, we could define hypothesis H_i as the probability model $P_X(x_i) = 1$, $P_X(x) = 0$, $x \neq x_i$. A hypothesis test would then lead us to choose the most probable x_i given our observations. Although this procedure maximizes the probability of determining the correct value of x_i, it does not take into account the consequences of incorrect results. It treats all errors in the same manner, regardless of whether they produce answers that are close to or far from the correct value of X. Section 12.3 describes estimation techniques that adopt this approach. By contrast, the aim of the estimation procedures presented in this section is to find an estimate $\hat{X}$ that, on average, is close to the true value of X, even if the estimate never produces a correct answer. A popular example is an estimate of the number of children in a family. The best estimate, based on available information, might be 2.4 children.

In an estimation procedure, we aim for a low probability that the estimate is far from the true value of X. An accuracy measure that helps us achieve this aim is the mean square error in Equation (12.1). The mean square error is one of many ways of defining the accuracy of an estimate. Two other accuracy measures, which might be appropriate to certain applications, are the expected value of the absolute estimation error $\mathrm{E}[|X - \hat{X}|]$ and the maximum absolute estimation error, $\max |X - \hat{X}|$. In this section, we confine our attention to the mean square error, which is the most widely used accuracy measure because it lends itself to mathematical analysis and often leads to estimates that are convenient to compute. In particular, we use the mean square error accuracy measure to examine three different ways of estimating random variable X. They are distinguished by the information available. We consider three types of information:

- The probability model of X (blind estimation),

- The probability model of X and information that the sample value $x \in A$,

- The probability model of random variables X and Y and information that $Y = y$.

The estimation methods for these three situations are fundamentally the same. Each one implies a probability model for X, which may be a PDF, a conditional PDF, a PMF, or a conditional PMF. In all three cases, the estimate of X that produces the minimum mean square error is the expected value (or conditional expected value) of X calculated with the probability model that incorporates the available information. While the expected value is the best estimate of X, it may be complicated to calculate in a practical application. Many applications derive an easily calculated *linear estimate* of X, the subject of Section 12.2.

Blind Estimation of X

An experiment produces a random variable X. Before the experiment is performed, what is the best estimate of X? This is the *blind estimation* problem because it requires us to make an inference about X in the absence of any observations. Although it is unlikely that we will guess the correct value of X, we can derive a number that comes as close as possible in the sense that it minimizes the mean square error. We encountered the blind estimate in Section 3.8 where Theorem 3.13 shows that $\hat{X}_B = \mathrm{E}[X]$ is the minimum mean square error estimate in the absence of observations. The minimum error is $e_B^* = \mathrm{Var}[X]$. In introducing the idea of expected value, Chapter 3 describes $\mathrm{E}[X]$ as a "typical value" of X. Theorem 3.13 gives this description a mathematical meaning.

Example 12.1
Before a six-sided die is rolled, what is the minimum mean square error estimate of the number of spots X that will appear?
....
The probability model is $P_X(x) = 1/6$, $x = 1, 2, \ldots, 6$, otherwise $P_X(x) = 0$. For this model, $\mathrm{E}[X] = 3.5$. Even though $\hat{x}_B = 3.5$ is not in the range of X, it is the estimate that minimizes the mean square estimation error.

Estimation of X Given an Event

Suppose that in performing an experiment, instead of observing X directly, we learn only that $X \in A$. Given this information, what is the minimum mean square error estimate of X? Given A, X has a conditional PDF $f_{X|A}(x)$ or a conditional PMF $P_{X|A}(x)$. Our task is to minimize the *conditional mean square error* $e_{X|A} = \mathrm{E}[(X - \hat{x})^2|A]$. We see that this is essentially the same as the blind estimation problem with the conditional PDF $f_{X|A}(x|A)$ or the conditional PMF $P_{X|A}(x)$ replacing $f_X(x)$ or $P_X(x)$. Therefore, we have the following:

Theorem 12.1
Given the information $X \in A$, the minimum mean square error estimate of X is

$$\hat{x}_A = \mathrm{E}\left[X|A\right].$$

Example 12.2
The duration T minutes of a phone call is an exponential random variable with expected value $\mathrm{E}[T] = 3$ minutes. If we observe that a call has already lasted 2 minutes, what is the minimum mean square error estimate of the call duration?
....
The PDF of T is

$$f_T(t) = \begin{cases} \frac{1}{3}e^{-t/3} & t \geq 0, \\ 0 & \text{otherwise.} \end{cases} \qquad (12.2)$$

If the call is still in progress after 2 minutes, we have $t \in A = \{T > 2\}$. Therefore, the minimum mean square error estimate of T is $\hat{t}_A = \mathrm{E}[T|T > 2]$. In this case, the conditioning event is $T > 2$. The probability of the event is

$$\mathrm{P}[T > 2] = \int_2^\infty f_T(t)\, dt = e^{-2/3}. \qquad (12.3)$$

The conditional PDF of T given $T > 2$ is

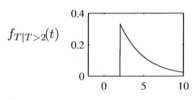

$$f_{T|T>2}(t) = \begin{cases} \frac{f_T(t)}{\mathrm{P}[T>2]} & t > 2, \\ 0 & \text{otherwise,} \end{cases}$$

$$= \begin{cases} \frac{1}{3}e^{-(t-2)/3} & t > 2, \\ 0 & \text{otherwise.} \end{cases}$$

Therefore,

$$\mathrm{E}[T|T > 2] = \int_2^\infty t\frac{1}{3}e^{-(t-2)/3}\, dt = 2 + 3 = 5 \text{ minutes.} \qquad (12.4)$$

Prior to the phone call, the minimum mean square error (blind) estimate of T is $\mathrm{E}[T] = 3$ minutes. After the call is in progress 2 minutes, the best estimate of the duration becomes $\mathrm{E}[T|T > 2] = 5$ minutes. This result is an example of the *memoryless property* of an exponential random variable. At any time during a call, the expected time remaining is the expected value of the call duration, $\mathrm{E}[T]$.

Minimum Mean Square Estimation of X Given Y

Consider an experiment that produces two random variables, X and Y. We can observe Y but we really want to know X. Therefore, the estimation task is to assign to every $y \in S_Y$ a number, $\hat{x}$, that is near X. As in the other techniques presented in this section, our accuracy measure is the mean square error

$$e_M = \mathrm{E}\left[(X - \hat{x}_M(y))^2|Y = y\right]. \qquad (12.5)$$

Because each $y \in S_Y$ produces a specific $\hat{x}_M(y)$, $\hat{x}_M(y)$ is a sample value of a random variable $\hat{X}_M(Y)$. The fact that $\hat{x}_M(y)$ is a sample value of a random variable is in contrast to blind estimation and estimation given an event. In those situations, $\hat{x}_B$ and $\hat{x}_A$ are parameters of the probability model of X.

In common with $\hat{x}_B$ in Theorem 3.13 and $\hat{x}_A$ in Theorem 12.1, the estimate of X given Y is an expected value of X based on available information. In this case, the available information is the value of Y.

Theorem 12.2

The minimum mean square error estimate of X given the observation $Y = y$ is

$$\hat{x}_M(y) = \mathrm{E}[X|Y = y].$$

════════**Example 12.3**════════

Suppose X and Y are independent random variables with PDFs $f_X(x)$ and $f_Y(y)$. What is the minimum mean square error estimate of X given Y?

· ·

In this case, $f_{X|Y}(x|y) = f_X(x)$ and the minimum mean square error estimate is

$$\hat{x}_M(y) = \int_{-\infty}^{\infty} x f_{X|Y}(x|y)\, dx = \int_{-\infty}^{\infty} x f_X(x)\, dx = \mathrm{E}[X] = \hat{x}_B. \tag{12.6}$$

That is, when X and Y are independent, the observation Y provides no information about X, and the best estimate of X is the blind estimate.

════════**Example 12.4**════════

Suppose that R has a uniform $(0,1)$ PDF and that given $R = r$, X is a uniform $(0,r)$ random variable. Find $\hat{x}_M(r)$, the minimum mean square error estimate of X given R.

· ·

From Theorem 12.2, we know $\hat{x}_M(r) = \mathrm{E}[X|R = r]$. To calculate the estimator, we need the conditional PDF $f_{X|R}(x|r)$. The problem statement implies that

$$f_{X|R}(x|r) = \begin{cases} 1/r & 0 \le x \le r, \\ 0 & \text{otherwise,} \end{cases} \tag{12.7}$$

permitting us to write

$$\hat{x}_M(r) = \int_0^r \frac{1}{r}\, dx = \frac{r}{2}. \tag{12.8}$$

════════

Although the estimate of X given $R = r$ is simply $r/2$, the estimate of R given $X = x$ for the same probability model is more complicated.

════════**Example 12.5**════════

Suppose that R has a uniform $(0,1)$ PDF and that given $R = r$, X is a uniform $(0,r)$ random variable. Find $\hat{r}_M(x)$, the minimum mean square error estimate of R given $X = x$.

· ·

From Theorem 12.2, we know $\hat{r}_M(x) = \mathrm{E}[R|X = x]$. To perform this calculation, we need to find the conditional PDF $f_{R|X}(r|x)$. The derivation of $f_{R|X}(r|x)$ appears in Example 7.14:

$$f_{R|X}(r|x) = \begin{cases} \frac{1}{-r \ln x} & 0 \le x \le r \le 1, \\ 0 & \text{otherwise.} \end{cases} \tag{12.9}$$

The corresponding estimator is, therefore,

$$\hat{r}_M(x) = \int_x^1 r \frac{1}{-r \ln x}\, dr = \frac{x - 1}{\ln x}. \tag{12.10}$$

The graph of this function appears at the end of Example 12.6.

While the solution of Example 12.4 is a simple function of r that can easily be obtained with a microprocessor or an analog electronic circuit, the solution of Example 12.5 is considerably more complex. In many applications, the cost of calculating this estimate could be significant. In these applications, engineers would look for a simpler estimate. Even though the simpler estimate produces a higher mean square error than the estimate in Example 12.5, the complexity savings might justify the simpler approach. For this reason, there are many applications of estimation theory that employ linear estimates, the subject of Section 12.2.

Quiz 12.1

The random variables X and Y have the joint probability density function

$$f_{X,Y}(x,y) = \begin{cases} 2(y+x) & 0 \le x \le y \le 1, \\ 0 & \text{otherwise.} \end{cases} \tag{12.11}$$

(a) What is $f_{X|Y}(x|y)$, the conditional PDF of X given $Y = y$?
(b) What is $\hat{x}_M(y)$, the MMSE estimate of X given $Y = y$?
(c) What is $f_{Y|X}(y|x)$, the conditional PDF of Y given $X = x$?
(d) What is $\hat{y}_M(x)$, the MMSE estimate of Y given $X = x$?

12.2   Linear Estimation of $\mathbf{X}$ given $\mathbf{Y}$

> The linear mean square error (LMSE) estimate of X given Y has the form $aY + b$. The optimum values of a and b depend on the expected values and variances of X and Y and the covariance of X and Y.

In this section we again use an observation, y, of random variable Y to produce an estimate, $\hat{x}$, of random variable X. Again, our accuracy measure is the mean square error in Equation (12.1). Section 12.1 derives $\hat{x}_M(y)$, the optimum estimate for each possible observation $Y = y$. By contrast, in this section the estimate is a single function that applies for all Y. The notation for this function is

$$\hat{x}_L(y) = ay + b \tag{12.12}$$

where a and b are constants for all $y \in S_Y$. Because $\hat{x}_L(y)$ is a linear function of y, the procedure is referred to as *linear estimation*. Linear estimation appears in many electrical engineering applications of statistical inference for several reasons:

- Linear estimates are easy to compute. Analog filters using resistors, capacitors, and inductors, and digital signal processing microcomputers perform linear operations efficiently.

- For some probability models, the optimum estimator $\hat{x}_M(y)$ described in Section 12.1 is a linear function of y. (See Example 12.4.) In other probability models, the error produced by the optimum linear estimator is not much higher than the error produced by the optimum estimator.

- The values of a, b of the optimum linear estimator and the corresponding value of the error depend only on $E[X]$, $E[Y]$, $Var[X]$, $Var[Y]$, and $Cov[X, Y]$. Therefore, it is not necessary to know the complete probability model of X and Y in order to design and evaluate an optimum linear estimator.

To present the mathematics of minimum mean square error linear estimation, we introduce the subscript L to denote the mean square error of a linear estimate:

$$e_L = E\left[\left(X - \hat{X}_L(Y)\right)^2\right]. \tag{12.13}$$

In this formula, we use $\hat{X}_L(Y)$ and not $\hat{x}_L(y)$ because the expected value in the formula is an unconditional expected value in contrast to the conditional expected value (Equation (12.5)) that is the quality measure for $\hat{x}_M(y)$. Minimum mean square error estimation in principle uses a different calculation for each $y \in S_Y$. By contrast, a linear estimator uses the same coefficients a and b for all y. The following theorem presents the important properties of optimum linear estimates in terms of the correlation coefficient $\rho_{X,Y}$ of X and Y introduced in Definition 5.6.

Theorem 12.3

Random variables X and Y have expected values μ_X and μ_Y, standard deviations σ_X and σ_Y, and correlation coefficient $\rho_{X,Y}$, The optimum linear mean square error (LMSE) estimator of X given Y is

$$\hat{X}_L(Y) = \rho_{X,Y}\frac{\sigma_X}{\sigma_Y}(Y - \mu_Y) + \mu_X.$$

This linear estimator has the following properties:

(a) The minimum mean square estimation error for a linear estimate is

$$e_L^* = E\left[(X - \hat{X}_L(Y))^2\right] = \sigma_X^2(1 - \rho_{X,Y}^2).$$

(b) The estimation error $X - \hat{X}_L(Y)$ is uncorrelated with Y.

Proof Replacing $\hat{X}_L(Y)$ by $aY + b$ and expanding the square, we have

$$e_L^* = E\left[X^2\right] - 2a\,E[XY] - 2b\,E[X] + a^2\,E\left[Y^2\right] + 2ab\,E[Y] + b^2. \tag{12.14}$$

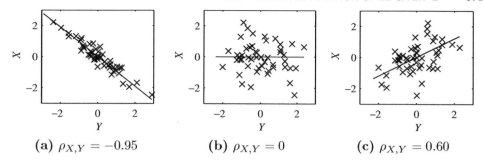

(a) $\rho_{X,Y} = -0.95$　　　　**(b)** $\rho_{X,Y} = 0$　　　　**(c)** $\rho_{X,Y} = 0.60$

Figure 12.1　Each graph contains 50 sample values of the random variable pair (X, Y), each marked by the symbol $\times$. In each graph, $E[X] = E[Y] = 0$, $\text{Var}[X] = \text{Var}[Y] = 1$. The solid line is the optimal linear estimator $\hat{X}_L(Y) = \rho_{X,Y} Y$.

The values of a and b that produce the minimum e_L are found by computing the partial derivatives of e_L with respect to a and b and setting the derivatives to zero, yielding

$$\frac{\partial e_L}{\partial a} = -2\,E\,[XY] + 2a\,E\,[Y^2] + 2b\,E\,[Y] = 0, \tag{12.15}$$

$$\frac{\partial e_L}{\partial b} = -2\,E\,[X] + 2a\,E\,[Y] + 2b = 0. \tag{12.16}$$

Solving the two equations for a and b, we find

$$a^* = \frac{\text{Cov}\,[X,Y]}{\text{Var}[Y]} = \rho_{X,Y}\frac{\sigma_X}{\sigma_Y}, \qquad\qquad b^* = E\,[X] - a^*\,E\,[Y]. \tag{12.17}$$

Some algebra will verify that $a^*Y + b^*$ is the optimum linear estimate $\hat{X}_L(Y)$. We confirm Theorem 12.3(a) by using $\hat{X}_L(Y)$ in Equation (12.13). To prove part (b) of the theorem, observe that the correlation of Y and the estimation error is

$$E\left[Y[X - \hat{X}_L(Y)]\right] = E\,[XY] - E\,[Y\,E\,[X]] - \frac{\text{Cov}\,[X,Y]}{\text{Var}[Y]}\left(E\,[Y^2] - E\,[Y\,E\,[Y]]\right)$$

$$= \text{Cov}\,[X,Y] - \frac{\text{Cov}\,[X,Y]}{\text{Var}[Y]}\,\text{Var}[Y] = 0. \tag{12.18}$$

Theorem 12.3(b) is referred to as the *orthogonality principle* of the LMSE. It states that the estimation error is orthogonal to the data used in the estimate. A geometric explanation of linear estimation is that the optimum estimate of X is the *projection* of X into the plane of linear functions of Y.

The correlation coefficient $\rho_{X,Y}$ plays a key role in the optimum linear estimator. Recall from Section 5.7 that $|\rho_{X,Y}| \leq 1$ and that $\rho_{X,Y} = \pm 1$ corresponds to a deterministic linear relationship between X and Y. This property is reflected in the fact that when $\rho_{X,Y} = \pm 1$, $e_L^* = 0$. At the other extreme, when X and Y are uncorrelated, $\rho_{X,Y} = 0$ and $\hat{X}_L(Y) = E[X]$, the blind estimate. With X and

Y uncorrelated, there is no linear function of Y that provides useful information about the value of X.

The magnitude of the correlation coefficient indicates the extent to which observing Y improves our knowledge of X, and the sign of $\rho_{X,Y}$ indicates whether the slope of the estimate is positive, negative, or zero. Figure 12.1 contains three different pairs of random variables X and Y. In each graph, the crosses are 50 outcomes x, y of the underlying experiment, and the line is the optimum linear estimate of X. In all three graphs, $\mathrm{E}[X] = \mathrm{E}[Y] = 0$ and $\mathrm{Var}[X] = \mathrm{Var}[Y] = 1$. From Theorem 12.3, we know that the optimum linear estimator of X given Y is the line $\hat{X}_L(Y) = \rho_{X,Y}Y$. For each pair (x, y), the estimation error equals the vertical distance to the estimator line. In the graph of Figure 12.1(a), $\rho_{X,Y} = -0.95$. Therefore, $e_L^* = 0.0975$, and all the observations are close to the estimate, which has a slope of -0.95. By contrast, in graph (b), with X and Y uncorrelated, the points are scattered randomly in the x, y plane and $e_L^* = \mathrm{Var}[X] = 1$. Lastly, in graph (c), $\rho_{X,Y} = 0.6$, and the observations, on average, follow the estimator $\hat{X}_L(Y) = 0.6Y$, although the estimates are less accurate than those in graph (a).

At the beginning of this section, we state that for some probability models, the optimum estimator of X given Y is a linear estimator. The following theorem shows that this is always the case when X and Y are jointly Gaussian random variables, described in Section 5.9.

Theorem 12.4

If X and Y are the bivariate Gaussian random variables in Definition 5.10, the optimum estimator of X given Y is the optimum linear estimator in Theorem 12.3.

Proof From Theorem 12.3, applying a^* and b^* to the optimal linear estimator $\hat{X}_L(Y) = a^*Y + b^*$ yields

$$\hat{X}_L(Y) = \rho_{X,Y}\frac{\sigma_X}{\sigma_Y}(Y - \mu_Y) + \mu_X. \tag{12.19}$$

From Theorem 7.16, we observe that when X and Y are jointly Gaussian, $\hat{X}_M(Y) = \mathrm{E}[X|Y]$ is identical to $\hat{X}_L(Y)$.

In the case of jointly Gaussian random variables, the optimum estimate of X given Y and the optimum estimate of Y given X are both linear. However, there are also probability models in which one of the optimum estimates is linear and the other one is not linear. This occurs in the probability model of Examples 12.4 and 12.5. Here $\hat{x}_M(r)$ (Example 12.4) is linear, and $\hat{r}_M(x)$ (Example 12.5) is nonlinear. In the following example, we derive the linear estimator $\hat{r}_L(x)$ for this probability model and compare it with the optimum estimator in Example 12.5.

Example 12.6

As in Examples 12.4 and 12.5, R is a uniform $(0, 1)$ random variable and given $R = r$, X is a uniform $(0, r)$ random variable. Derive the optimum linear estimator of R given X.

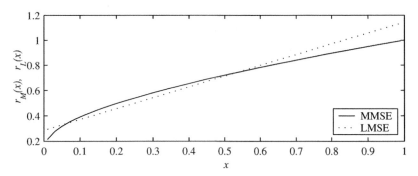

Figure 12.2   The minimum mean square error (MMSE) estimate $\hat{r}_M(x)$ in Example 12.5 and the optimum linear (LMSE) estimate $\hat{r}_L(x)$ in Example 12.6 of X given R.

From the problem statement, we know $f_{X|R}(x|r)$ and $f_R(r)$, implying that the joint PDF of X and R is

$$f_{X,R}(x,r) = f_{X|R}(x|r)\, f_R(r) = \begin{cases} 1/r & 0 \le x \le r \le 1, \\ 0 & \text{otherwise.} \end{cases} \tag{12.20}$$

The estimate we have to derive is given by Theorem 12.3:

$$\hat{r}_L(x) = \rho_{R,X}\frac{\sigma_R}{\sigma_X}\left(x - \mathrm{E}\left[X\right]\right) + \mathrm{E}\left[R\right]. \tag{12.21}$$

Since R is uniform on $[0,1]$, $\mathrm{E}[R] = 1/2$ and $\sigma_R = 1/\sqrt{12}$. Using the formula for $f_{X|R}(x|r)$ in Equation (12.7), we have

$$f_X(x) = \int_{-\infty}^{\infty} f_{X,R}(x,r)\, dr = \begin{cases} \int_x^1 (1/r)\, dr = -\ln x & 0 \le x \le 1, \\ 0 & \text{otherwise.} \end{cases} \tag{12.22}$$

From this marginal PDF, we can calculate $\mathrm{E}[X] = 1/4$ and $\sigma_X = \sqrt{7}/12$. Using the joint PDF, we obtain $\mathrm{E}[XR] = 1/6$, so that $\mathrm{Cov}[X,R] = \mathrm{E}[XR] - \mathrm{E}[X]\,\mathrm{E}[R] = 1/24$. Thus $\rho_{R,X} = \sqrt{3/7}$. Putting these values into Equation (12.21), the optimum linear estimator is

$$\hat{r}_L(x) = \frac{6}{7}x + \frac{2}{7}. \tag{12.23}$$

Figure 12.2 compares the optimum (MMSE) estimator and the optimum linear (LMSE) estimator. We see that the two estimators are reasonably close for all but extreme values of x (near 0 and 1). Note that for $x > 5/6$, the linear estimate is greater than 1, the largest possible value of R. By contrast, the optimum estimate $\hat{r}_M(x)$ is confined to the range of R for all x.

In this section, the examples apply to continuous random variables. For discrete random variables, the linear estimator is also described by Theorem 12.3. When X and Y are discrete, the parameters (expected value, variance, covariance) are sums containing the joint PMF $P_{X,Y}(x, y)$.

In Section 12.4, we use a linear combination of the random variables in a random vector to estimate another random variable.

──────Quiz 12.2──────

A telemetry signal, T, transmitted from a temperature sensor on a communications satellite is a Gaussian random variable with $E[T] = 0$ and $Var[T] = 9$. The receiver at mission control receives $R = T + X$, where X is a noise voltage independent of T with PDF

$$f_X(x) = \begin{cases} 1/6 & -3 \le x \le 3, \\ 0 & \text{otherwise.} \end{cases} \qquad (12.24)$$

The receiver uses R to calculate a linear estimate of the telemetry voltage:

$$\hat{t}_L(r) = ar + b. \qquad (12.25)$$

(a) What is $E[R]$, the expected value of the received voltage?

(b) What is $Var[R]$, the variance of the received voltage?

(c) What is $Cov[T, R]$, the covariance of the transmitted voltage and the received voltage?

(d) What is the correlation coefficient $\rho_{T,R}$ of T and R?

(e) What are a^* and b^*, the optimum mean square values of a and b in the linear estimator?

(f) What is e_L^*, the minimum mean square error of the linear estimate?

12.3   MAP and ML Estimation

The maximum a posteriori probability (MAP) estimate of X given $Y = y$ is the value of x that maximizes the conditional PDF $f_{X|Y}(x|y)$. The maximum likelihood (ML) estimate is the value of x that maximizes the conditional PDF $f_{Y|X}(y|x)$. The ML estimate is identical to the MAP estimate when X is a uniform random variable.

Sections 12.1 and 12.2 describe methods for minimizing the mean square error in estimating a random variable X given a sample value of another random variable Y.

In this section, we present the maximum a posteriori probability (MAP) estimator and the maximum likelihood (ML) estimator. Although neither of these estimates produces the minimum mean square error, they are convenient to obtain in some applications, and they often produce estimates with errors that are not much higher than the minimum mean square error.

As you might expect, MAP and ML estimation are closely related to MAP and ML hypothesis testing.

Definition 12.1 MAP Estimate

*The **maximum a posteriori probability (MAP) estimate** of X given an observation of Y is*

$$\text{Discrete:} \quad \hat{x}_{\text{MAP}}(y_j) = \arg \max_{x \in S_X} P_{X|Y}(x|y_j);$$

$$\text{Continuous:} \quad \hat{x}_{\text{MAP}}(y) = \arg \max_{x} f_{X|Y}(x|y).$$

The notation $\arg \max_x g(x)$ denotes a value of x that maximizes $g(x)$, where $g(x)$ is any function of a variable x. The properties of the conditional PMF and the conditional PDF lead to formulas calculating the MAP estimator that are used in applications. Recall from Theorem 7.10 that

$$f_{X|Y}(x|y) = \frac{f_{Y|X}(y|x) f_X(x)}{f_Y(y)} = \frac{f_{X,Y}(x,y)}{f_Y(y)}. \tag{12.26}$$

Because the denominator $f_Y(y)$ does not depend on x, maximizing $f_{X|Y}(x|y)$ over all x is equivalent to maximizing the numerator $f_{Y|X}(y|x)f_X(x)$. Similarly, maximizing $P_{X|Y}(x|y_j)$ is equivalent to finding x_i that corresponds to the maximum value of $P_{Y|X}(y|x)y_j|xP_X(x)x$. This implies the MAP estimation procedure can be written in the following way.

Theorem 12.5

The MAP estimate of X given $Y = y$ is

$$\text{Discrete:} \quad \hat{x}_{\text{MAP}}(y_j) = \arg \max_{x \in S_X} P_{X|Y}(x|y_j);$$

$$\text{Continuous:} \quad \hat{x}_{\text{MAP}}(y) = \arg \max_{x} f_{Y|X}(y|x) f_X(x).$$

When X and Y are discrete random variables, the MAP estimate is similar to the result of a multiple hypothesis test in Chapter 11, where each outcome x_i in the sample space of X corresponds to a hypothesis H_i. The MAP estimate maximizes the probability of choosing the correct x_i.

When X and Y are continuous random variables and we observe the event $Y = y$, we let H_x denote the hypothesis that $x \leq X \leq x + dx$. Since x is a continuous parameter, we have a continuum of hypotheses H_x. Deciding hypothesis $H_{\hat{x}}$ corresponds to choosing $\hat{x}$ as an estimate for X. The MAP estimator $\hat{x}_{\mathrm{MAP}}(y) = x$ maximizes the probability of H_x given the observation $Y = y$.

Theorem 12.5 indicates that the MAP estimation procedure uses the PMF $P_X(x)$ or the PDF $f_X(x)$, the a priori probability model for random variable X. This is analogous to the requirement of the MAP hypothesis test that we know the a priori probabilities $P[H_i]$. In the absence of this a priori information, we can instead implement a maximum likelihood estimator.

━━━━━━Definition 12.2━━━━━━Maximum Likelihood (ML) Estimate

*The **maximum likelihood (ML) estimate** of X given the observation $Y = y$ is*

$$\text{Discrete:} \quad \hat{x}_{\mathrm{MAP}}(y_j) = \arg\max_{x \in S_X} P_{Y|X}(y_j|x);$$

$$\text{Continuous:} \quad \hat{x}_{\mathrm{MAP}}(y) = \arg\max_x f_{Y|X}(y|x).$$

The primary difference between the MAP and ML estimates is that the maximum likelihood estimate does not use information about the a priori probability model of X. This is analogous to the situation in hypothesis testing in which the ML hypothesis-testing rule does not use information about the a priori probabilities of the hypotheses. The ML estimate is the same as the MAP estimate when all possible values of X are equally likely.

The following example, we observe relationships among five estimates studied in this chapter.

━━━━━━**Example 12.7**━━━━━━

Consider a collection of old coins. Each coin has random probability, q of landing with heads up when it is flipped. The probability of heads, q, is a sample value of a beta $(2, 2)$ random variable, Q, with PDF

$$f_Q(q) = \begin{cases} 6q(1 - q) & 0 \le q \le 1, \\ 0 & \text{otherwise.} \end{cases} \tag{12.27}$$

To estimate q for a coin we flip the coin n times and count the number of heads, k. Because each flip is a Bernoulli trial with probability of success q, k is a sample value of the binomial(n,q) random variable K. Given $K = k$, derive the following estimates of Q:

(a) The blind estimate $\hat{q}_B$,

(b) The maximum likelihood estimate $\hat{q}_{\mathrm{ML}}(k)$,

(c) The maximum a posteriori probability estimate $\hat{q}_{\mathrm{MAP}}(k)$,

(d) The minimum mean square error estimate $\hat{q}_M(k)$,

(e) The optimum linear estimate $\hat{q}_L(k)$.

(a) To derive the blind estimate, we refer to Appendix A for the properties of the beta $(i = 2, j = 2)$ random variable and find

$$\hat{q}_B = \mathrm{E}\,[Q] = \frac{i}{i+j} = 1/2. \tag{12.28}$$

(b) To find the other estimates, refer to the conditional PMF of the binomial(n,q) random variable K:

$$P_{K|Q}(k|q) = \binom{n}{k} q^k (1 - q)^{n-k}. \tag{12.29}$$

The ML estimate is the value of q that maximizes $P_{K|Q}(k|q)$. The derivative of $P_{K|Q}(k|q)$ with respect to q is

$$\frac{dP_{K|Q}(k|q)}{dq} = \binom{n}{k} q^{k-1}(1 - q)^{n-k-1}\left[k(1 - q) - (n - k)q\right]. \tag{12.30}$$

Setting $dP_{K|Q}(k|q)/dq = 0$, and solving for q yields

$$\hat{q}_{\mathrm{ML}}(k) = \frac{k}{n}, \tag{12.31}$$

the relative frequency of heads in n coin flips.

(c) The MAP estimator is the value of q that maximizes

$$f_{Q|K}(q|k) = \frac{P_{K|Q}(k|q)\,f_Q(q)}{P_K(k)}. \tag{12.32}$$

Since the denominator of Equation (12.32) is a constant with respect to q, we can obtain the maximum value by setting the derivative of the numerator to zero:

$$\frac{d\left[P_{K|Q}(k|q)\,f_Q(q)\right]}{dq} = 6\binom{n}{k} q^k (1 - q)^{n-k}\left[(k + 1)(1 - q) - (n - k + 1)q\right]$$

$$= 0. \tag{12.33}$$

Solving for q yields

$$\hat{q}_{\mathrm{MAP}}(k) = \frac{k + 1}{n + 2}. \tag{12.34}$$

(d) To compute the MMSE estimate $\hat{q}_M(k) = \mathrm{E}[Q|K = k]$, we have to analyze $f_{Q|K}(q|k)$ in Equation (12.32). The numerator terms, $f_Q(q)$ and $P_{K|Q}(k|q)$ appear in Equation (12.27) and Equation (12.29), respectively. To analyze $P_K(k)$ in the denominator of Equation (12.32), we refer to the properties of beta random variables in Appendix A:

$$P_K(k) = \int_{-\infty}^{\infty} P_{K|Q}(k|q)\,f_Q(q)\,dq. \tag{12.35}$$

Substituting $f_Q(q)$ and $P_{K|Q}(k|q)$ from Equations (12.27) and (12.29), we obtain

$$P_K(k) = 6\binom{n}{k} \int_0^1 q^{k+1}(1-q)^{n-k+1} \, dq. \tag{12.36}$$

The function of q in the integrand appears in a beta $(k+2, n-k+2)$ PDF. If we multiply the integrand by the constant $\beta(k+2, n-k+2)$, the resulting integral is 1. That is,

$$\int_0^1 \beta(k+2, n-k+2)q^{k+1}(1-q)^{n-k+1} \, dq = 1. \tag{12.37}$$

It follows from Equations (12.36) and (12.37) that

$$P_K(k) = \frac{6\binom{n}{k}}{\beta(k+2, n-k+2)} \tag{12.38}$$

for $k = 0, 1, \ldots, n$ and $P_K(k) = 0$ otherwise. From Equation (12.32),

$$f_{Q|K}(q|k) = \begin{cases} \beta(k+2, n-k+2)q^{k+1}(1-q)^{n-k+1} & 0 \le q \le 1, \\ 0 & \text{otherwise.} \end{cases}$$

That is, given $K = k$, Q is a beta $(i = k+2, j = n-k+2)$ random variable. Thus, from Appendix A,

$$\hat{q}_M(k) = \mathrm{E}[Q|K = k] = \frac{i}{i+j} = \frac{k+2}{n+4}. \tag{12.39}$$

(e) In Equation (12.39), the minimum mean square error estimator $\hat{q}_M(k)$ is the linear function of k: $\hat{q}_M(k) = a^*k + b^*$ where $a^* = 1/(n+4)$ and $b^* = 2/(n+4)$. Therefore, $\hat{q}_L(k) = \hat{q}_M(k)$.

It is instructive to compare the different estimates. The blind estimate, using only prior information, is simply $\mathrm{E}[Q] = 1/2$, regardless of the results of the Bernoulli trials. By contrast, the maximum likelihood estimate makes no use of prior information. Therefore, it estimates Q as k/n, the relative frequency of heads in n coin flips. When $n = 0$, there are no observations, and there is no maximum likelihood estimate. The other estimates use both prior information and data from the coin flips. In the absence of data $(n = 0)$, they produce $\hat{q}_{\mathrm{MAP}}(k) = \hat{q}_M(k) = \hat{q}_L(k) = 1/2 = \mathrm{E}[Q] = \hat{q}_B$. As n grows large, they all approach $k/n = \hat{q}_{\mathrm{ML}}(k)$, the relative frequency of heads. For low values of $n > 0$, $\hat{q}_M(k) = \hat{q}_L(k)$ is a little farther from $1/2$ relative to $\hat{q}_{\mathrm{MAP}}(k)$. This reduces the probability of high errors that occur when n is small and q is near 0 or 1.

Quiz 12.3

A receiver at a radial distance R from a radio beacon measures the beacon power to be

$$X = Y - 40 - 40\log_{10} R \text{ dB}, \tag{12.40}$$

where Y, called the shadow fading factor, is the Gaussian $(0, 8)$ random variable that is independent of R. When the receiver is equally likely to be at any point within a 1000 m radius circle around the beacon, the distance R has PDF

$$f_R(r) = \begin{cases} 2r/10^6 & 0 \le r \le 1000, \\ 0 & \text{otherwise.} \end{cases} \tag{12.41}$$

Find the ML and MAP estimates of R given the observation $X = x$.

12.4   Linear Estimation of Random Variables from Random Vectors

> Given an observation of a random vector, the coefficients of the optimum linear estimator of a random variable is the solution to a set of linear equations. The coefficients in the equations are elements of the autocorrelation matrix of the observed random vector. The right side is the cross-correlation matrix of the estimated random variable and the observed random vector. The estimation error of the optimum linear estimator is uncorrelated with the observed random variables.

There are many practical applications that use sample values of n random variables $Y_0, \dots Y_{n-1}$ to calculate a linear estimates of sample values of other random variables $X_0, \dots, X_{m-1}$. This section represents the random variables Y_i and X_j as elements of the random vectors $\mathbf{Y}$ and $\mathbf{X}$. We start with Theorem 12.6, a vector version of Theorem 12.3 in which we form a linear estimate of a random variable X based on the observation of a random vector $\mathbf{Y}$. Theorem 12.6 applies to the special case in which X and all of the elements of $\mathbf{Y}$ have zero expected value. This is followed by Theorem 12.7, which applies to the general case including X and $\mathbf{Y}$ with nonzero expected value. Finally, Theorem 12.8 provides the vector version of Theorem 12.7, in which the random vector $\mathbf{Y}$ is used to form a linear estimate of the sample value of random vector $\mathbf{X}$.

Theorem 12.6

X is a random variable with $\mathrm{E}[X] = 0$, and $\mathbf{Y}$ is an n-dimensional random vector with $\mathrm{E}[\mathbf{Y}] = \mathbf{0}$. The minimum mean square error linear estimator is

$$\hat{X}_L(\mathbf{Y}) = \mathbf{R}_{X\mathbf{Y}} \mathbf{R}_{\mathbf{Y}}^{-1} \mathbf{Y},$$

where $\mathbf{R}_{\mathbf{Y}}$ is the $n \times n$ correlation matrix of $\mathbf{Y}$ (Definition 8.8) and $\mathbf{R}_{X\mathbf{Y}}$ is the $1 \times n$ cross-correlation matrix of X and $\mathbf{Y}$ (Definition 8.10). This estimator has the following properties:

(a) The estimation error $X - \hat{X}_L(\mathbf{Y})$ is uncorrelated with the elements of $\mathbf{Y}$.

(b) The minimum mean square estimation error is

$$e_L^* = \mathrm{E}\left[(X - \mathbf{R}_{X\mathbf{Y}} \mathbf{R}_{\mathbf{Y}}^{-1} \mathbf{Y})^2\right] = \mathrm{Var}[X] - \mathbf{R}_{X\mathbf{Y}} \mathbf{R}_{\mathbf{Y}}^{-1} \mathbf{R}_{X\mathbf{Y}}'.$$

Proof In terms of $\mathbf{Y} = \begin{bmatrix} Y_0 & \cdots & Y_{n-1} \end{bmatrix}'$ and $\mathbf{a} = \begin{bmatrix} a_0 & \cdots & a_{n-1} \end{bmatrix}'$, we represent the linear estimator as $\hat{X}_L(\mathbf{Y}) = \mathbf{a}'\mathbf{Y}$. To derive the optimal $\mathbf{a}$, we write the mean square estimation error as

$$e_L = \mathrm{E}\left[(X - \hat{X}_L(\mathbf{Y}))^2\right] = \mathrm{E}\left[(X - a_0 Y_0 - a_1 Y_1 - \ldots - a_{n-1} Y_{n-1})^2\right]. \qquad (12.42)$$

The partial derivative of e_L with respect to a_i is

$$\frac{\partial e_L}{\partial a_i} = -2\,\mathrm{E}\left[Y_i(X - \hat{X}_L(\mathbf{Y}))\right]$$
$$= -2\,\mathrm{E}\left[Y_i(X - a_0 Y_0 - a_1 Y_1 - \ldots - a_{n-1} Y_{n-1})\right]. \qquad (12.43)$$

To minimize the error, we set $\partial e_L / \partial a_i = 0$ for all i. We recognize the first expected value in Equation (12.43) as the correlation of Y_i and the estimation error. Setting this correlation to zero for all Y_i establishes Theorem 12.6(a). Expanding the second expected value on the right side and setting it to zero, we obtain

$$a_0\,\mathrm{E}\left[Y_i Y_0\right] + a_1\,\mathrm{E}\left[Y_i Y_1\right] + \cdots + a_{n-1}\,\mathrm{E}\left[Y_i Y_{n-1}\right] = \mathrm{E}\left[Y_i X\right]. \qquad (12.44)$$

Recognizing that all the expected values are correlations, we write

$$a_0 r_{Y_i, Y_0} + a_1 r_{Y_i, Y_1} + \cdots + a_{n-1} r_{Y_i, Y_{n-1}} = r_{Y_i, X}. \qquad (12.45)$$

Setting the n partial derivatives to zero, we obtain a set of n linear equations in the n unknown elements of $\hat{\mathbf{a}}$. In matrix form, the equations are $\mathbf{R_Y a} = \mathbf{R}_{YX}$. Solving for $\mathbf{a} = \mathbf{R_Y}^{-1} \mathbf{R}_{YX}$ completes the proof of the first part of the theorem. To verify the minimum mean square error, we write

$$e_L^* = \mathrm{E}\left[(X - \hat{\mathbf{a}}'\mathbf{Y})^2\right] = \mathrm{E}\left[(X^2 - \hat{\mathbf{a}}'\mathbf{Y}X)\right] - \mathrm{E}\left[(X - \hat{\mathbf{a}}'\mathbf{Y})\hat{\mathbf{a}}'\mathbf{Y}\right]. \qquad (12.46)$$

The second term on the right side is zero because $\mathrm{E}[(X - \hat{\mathbf{a}}'\mathbf{Y})Y_j] = 0$ for $j = 0, 1, \ldots, n-1$. The first term is identical to the error expression of Theorem 12.6(b).

═══**Example 12.8**═══

Observe the random vector $\mathbf{Y} = \mathbf{X} + \mathbf{W}$, where $\mathbf{X}$ and $\mathbf{W}$ are independent random vectors with expected values $\mathrm{E}[\mathbf{X}] = \mathrm{E}[\mathbf{W}] = \mathbf{0}$ and correlation matrices

$$\mathbf{R_X} = \begin{bmatrix} 1 & 0.75 \\ 0.75 & 1 \end{bmatrix}, \qquad \mathbf{R_W} = \begin{bmatrix} 0.1 & 0 \\ 0 & 0.1 \end{bmatrix}. \qquad (12.47)$$

Find the coefficients $\hat{a}_1$ and $\hat{a}_2$ of the optimum linear estimator of the random variable $X = X_1$ given Y_1 and Y_2. Find the mean square error, e_L^*, of the optimum estimator.

In terms of Theorem 12.6, $n = 2$, and we wish to estimate X given the observation vector $\mathbf{Y} = \begin{bmatrix} Y_1 & Y_2 \end{bmatrix}'$. To apply Theorem 12.6, we need to find $\mathbf{R_Y}$ and $\mathbf{R}_{XY}$.

$$\mathbf{R_Y} = \mathrm{E}\left[\mathbf{Y}\mathbf{Y}'\right] = \mathrm{E}\left[(\mathbf{X} + \mathbf{W})(\mathbf{X}' + \mathbf{W}')\right]$$
$$= \mathrm{E}\left[\mathbf{X}\mathbf{X}' + \mathbf{X}\mathbf{W}' + \mathbf{W}\mathbf{X}' + \mathbf{W}\mathbf{W}'\right]. \qquad (12.48)$$

Because $\mathbf{X}$ and $\mathbf{W}$ are independent, $\mathrm{E}[\mathbf{X}\mathbf{W}'] = \mathrm{E}[\mathbf{X}]\,\mathrm{E}[\mathbf{W}'] = \mathbf{0}$. Similarly, $\mathrm{E}[\mathbf{W}\mathbf{X}'] = \mathbf{0}$. This implies

$$\mathbf{R_Y} = \mathrm{E}\left[\mathbf{X}\mathbf{X}'\right] + \mathrm{E}\left[\mathbf{W}\mathbf{W}'\right] = \mathbf{R_X} + \mathbf{R_W} = \begin{bmatrix} 1.1 & 0.75 \\ 0.75 & 1.1 \end{bmatrix}. \tag{12.49}$$

To find $\mathbf{R}_{XY}$, it is convenient to solve for the transpose $\mathbf{R}'_{XY} = \mathbf{R}_{YX}$.

$$\mathbf{R}_{YX} = \mathrm{E}\left[\mathbf{Y}X\right] = \begin{bmatrix} \mathrm{E}\left[Y_1 X\right] \\ \mathrm{E}\left[Y_2 X\right] \end{bmatrix} = \begin{bmatrix} \mathrm{E}\left[(X_1 + W_1)X_1\right] \\ \mathrm{E}\left[(X_2 + W_2)X_1\right] \end{bmatrix}. \tag{12.50}$$

Since $\mathbf{X}$ and $\mathbf{W}$ are independent vectors, $\mathrm{E}[W_1 X_1] = \mathrm{E}[W_1]\,\mathrm{E}[X_1] = 0$. For the same reason, $\mathrm{E}[W_2 X_1] = 0$. Thus

$$\mathbf{R}_{YX} = \begin{bmatrix} \mathrm{E}[X_1^2] \\ \mathrm{E}\left[X_2 X_1\right] \end{bmatrix} = \begin{bmatrix} 1 \\ 0.75 \end{bmatrix}. \tag{12.51}$$

Therefore, $\mathbf{R}_{XY} = \begin{bmatrix} 1 & 0.75 \end{bmatrix}$, and by Theorem 12.6, the optimum linear estimator of X given Y_1 and Y_2 is

$$\hat{X}_L(\mathbf{Y}) = \mathbf{R}_{XY}\mathbf{R}_\mathbf{Y}^{-1}\mathbf{Y}$$

$$= \begin{bmatrix} 1 & 0.75 \end{bmatrix} \begin{bmatrix} 1.1 & 0.75 \\ 0.75 & 1.1 \end{bmatrix}^{-1} \begin{bmatrix} Y_1 \\ Y_2 \end{bmatrix} = 0.830 Y_1 + 0.116 Y_2. \tag{12.52}$$

The mean square error is

$$\mathrm{Var}[X] - \mathbf{R}_{XY}\mathbf{R}_\mathbf{Y}^{-1}\mathbf{R}'_{XY} = 1 - \begin{bmatrix} 0.830 & 0.116 \end{bmatrix} \begin{bmatrix} 1 \\ 0.75 \end{bmatrix} = 0.0830. \tag{12.53}$$

The next theorem generalizes Theorem 12.6 to random variables with nonzero expected values. In this case the optimum estimate contains a constant term b, and the coefficients of the linear equations are covariances.

Theorem 12.7

X is a random variable with expected value $\mathrm{E}[X]$. $\mathbf{Y}$ is an n-dimensional random vector with expected value $\mathrm{E}[\mathbf{Y}]$ and $n \times n$ covariance matrix $\mathbf{C_Y}$. $\mathbf{C}_{XY}$ is the $1 \times n$ cross-covariance of X and $\mathbf{Y}$. The minimum mean square error (MMSE) linear estimator of X given $\mathbf{Y}$ is

$$\hat{X}_L(\mathbf{Y}) = \mathbf{C}_{XY}\mathbf{C}_\mathbf{Y}^{-1}(\mathbf{Y} - \mathrm{E}\left[\mathbf{Y}\right]) + \mathrm{E}\left[X\right].$$

This estimator has the following properties:

(a) The estimation error $X - \hat{X}_L(\mathbf{Y})$ is uncorrelated with the elements of $\mathbf{Y}$.

(b) The minimum mean square estimation error is

$$e_L^* = \mathrm{E}\left[(X - \hat{X}_L(\mathbf{Y}))^2\right] = \mathrm{Var}[X] - \mathbf{C}_{XY}\mathbf{C}_\mathbf{Y}^{-1}\mathbf{C}'_{XY}.$$

Proof We represent the optimum linear estimator as

$$\hat{X}_L(\mathbf{Y}) = \mathbf{a}'\mathbf{Y} + b. \tag{12.54}$$

For any $\mathbf{a}$, $\partial e_L / \partial b = 0$, implying $2\,\mathrm{E}[X - \mathbf{a}'\mathbf{Y} - b] = 0$. Hence $b = \mathrm{E}[X] - \mathbf{a}'\,\mathrm{E}[\mathbf{Y}]$. It follows from Equation (12.54) that

$$\hat{X}_L(\mathbf{Y}) - \mathrm{E}[X] = \mathbf{a}'(\mathbf{Y} - \mathrm{E}[\mathbf{Y}]). \tag{12.55}$$

Defining $U = X - \mathrm{E}[X]$ and $\mathbf{V} = \mathbf{Y} - \mathrm{E}[\mathbf{Y}]$, we can write Equation (12.55) as $\hat{U}_L(\mathbf{V}) = \mathbf{a}'\mathbf{V}$ where $\mathrm{E}[U] = 0$ and $\mathrm{E}[\mathbf{V}] = \mathbf{0}$. Theorem 12.6 implies that the optimum linear estimator of U given $\mathbf{V}$ is $\hat{U}_L(\mathbf{V}) = \mathbf{R}_{U\mathbf{V}}\mathbf{R}_{\mathbf{V}}^{-1}\mathbf{V}$. We next observe that Definition 8.11 implies that $\mathbf{R}_{\mathbf{V}} = \mathbf{C}_{\mathbf{Y}}$. Similarly $\mathbf{R}_{U\mathbf{V}} = \mathbf{C}_{X\mathbf{Y}}$. Therefore, $\mathbf{C}_{X\mathbf{Y}}\mathbf{C}_{\mathbf{Y}}^{-1}\mathbf{V}$ is the optimum estimator of U given $\mathbf{V}$. That is, over all choices of $\mathbf{a}$,

$$\mathrm{E}\left[(X - \mathrm{E}[X] - \mathbf{a}'(\mathbf{Y} - \mathrm{E}[\mathbf{Y}]))^2\right] = \mathrm{E}\left[(X - \mathbf{a}'\mathbf{Y} - \hat{b})^2\right] = \mathrm{E}\left[(X - \hat{X}_L(\mathbf{Y}))^2\right] \tag{12.56}$$

is minimized by $\mathbf{a}' = \mathbf{C}_{X\mathbf{Y}}\mathbf{C}_{\mathbf{Y}}^{-1}$. Thus $\hat{X}_L(\mathbf{Y}) = \mathbf{a}'\mathbf{Y} + b$ is the minimum mean square error estimate of X given $\mathbf{Y}$. The proofs of Theorem 12.7(a) and Theorem 12.7(b) use the same logic as the corresponding proofs in Theorem 12.6.

It is often convenient to represent the optimum linear estimator of Theorem 12.7 in the form

$$\hat{\mathbf{X}}_L(\mathbf{Y}) = \mathbf{a}'\mathbf{Y} + b, \tag{12.57}$$

with

$$\mathbf{a}' = \mathbf{C}_{X\mathbf{Y}}\mathbf{C}_{\mathbf{Y}}^{-1}, \qquad\qquad b = \mathrm{E}[X] - \mathbf{a}'\,\mathrm{E}[\mathbf{Y}]. \tag{12.58}$$

This form reminds us that $\mathbf{a}'$ is a row vector that is the solution to the set of linear equations

$$\mathbf{a}'\mathbf{C}_{\mathbf{Y}}^{-1} = \mathbf{C}_{X\mathbf{Y}}. \tag{12.59}$$

In many signal-processing applications, the vector $\mathbf{Y}$ is a collection of samples $Y(t_0), Y(t_1), \ldots, Y(t_{n-1})$ of a signal $Y(t)$. In this setting, $\mathbf{a}'$ is a vector representation of a linear filter.

Example 12.9

As in Example 8.10, consider the outdoor temperature at a certain weather station. On May 5, the temperature measurements in degrees Fahrenheit taken at 6 AM, 12 noon, and 6 PM are elements of the three-dimensional random vector $\mathbf{X}$ with $\mathrm{E}[\mathbf{X}] = \begin{bmatrix} 50 & 62 & 58 \end{bmatrix}'$. The covariance matrix of the three measurements is

$$\mathbf{C}_{\mathbf{X}} = \begin{bmatrix} 16.0 & 12.8 & 11.2 \\ 12.8 & 16.0 & 12.8 \\ 11.2 & 12.8 & 16.0 \end{bmatrix}. \tag{12.60}$$

Use the temperatures at 6 AM and 12 noon to predict the temperature at 6 PM: $\hat{X}_3 = \mathbf{a}'\mathbf{Y} + b$, where $\mathbf{Y} = \begin{bmatrix} X_1 & X_2 \end{bmatrix}'$.

(a) What are the coefficients of the optimum estimator $\hat{\mathbf{a}}$ and $\hat{b}$?

(b) What is the mean square estimation error?

(c) What are the coefficients a^* and b^* of the optimum estimator of X_3 given X_2?

(d) What is the mean square estimation error based on the observation X_2?

. . . .

(a) Let $X = X_3$. From Theorem 12.7, we know that

$$\mathbf{a}' = \mathbf{C}_{\mathbf{XY}}\mathbf{C}_{\mathbf{Y}}^{-1}, \tag{12.61}$$

$$b = \mathrm{E}\left[X\right] - \mathbf{C}_{\mathbf{XY}}\mathbf{C}_{\mathbf{Y}}^{-1}\mathrm{E}\left[\mathbf{Y}\right] = \mathrm{E}\left[X\right] - \mathbf{a}'\mathrm{E}\left[\mathbf{Y}\right]. \tag{12.62}$$

Thus we need to find the expected value $\mathrm{E}[\mathbf{Y}]$, the covariance matrix $\mathbf{C}_{\mathbf{Y}}$, and the cross-covariance matrix $\mathbf{C}_{\mathbf{XY}}$. Since $\mathbf{Y} = \begin{bmatrix} X_1 & X_2 \end{bmatrix}'$,

$$\mathrm{E}\left[\mathbf{Y}\right] = \begin{bmatrix} \mathrm{E}\left[X_1\right] & \mathrm{E}\left[X_2\right] \end{bmatrix}' = \begin{bmatrix} 50 & 62 \end{bmatrix}', \tag{12.63}$$

and we can find the covariance matrix of $\mathbf{Y}$ in $\mathbf{C}_{\mathbf{X}}$:

$$\mathbf{C}_{\mathbf{Y}} = \begin{bmatrix} C_X(1,1) & C_X(1,2) \\ C_X(2,1) & C_X(2,2) \end{bmatrix} = \begin{bmatrix} 16.0 & 12.8 \\ 12.8 & 16.0 \end{bmatrix}. \tag{12.64}$$

Since $X = X_3$, the elements of $\mathbf{C}_{\mathbf{XY}}$ are also in $\mathbf{C}_{\mathbf{X}}$. In particular, $\mathbf{C}_{\mathbf{XY}} = \mathbf{C}'_{\mathbf{YX}}$, where

$$\mathbf{C}_{YX} = \begin{bmatrix} \mathrm{Cov}\left[X_1, X_3\right] \\ \mathrm{Cov}\left[X_2, X_3\right] \end{bmatrix} = \begin{bmatrix} C_X(1,3) \\ C_X(2,3) \end{bmatrix} = \begin{bmatrix} 11.2 \\ 12.8 \end{bmatrix}. \tag{12.65}$$

Since $\mathbf{a}' = \mathbf{C}_{\mathbf{XY}}\mathbf{C}_{\mathbf{Y}}^{-1}$, $\mathbf{a}'$ solves $\mathbf{a}'\mathbf{C}_{\mathbf{Y}} = \mathbf{C}_{\mathbf{XY}}$, implying

$$\mathbf{a}' = \begin{bmatrix} 0.2745 & 0.6078 \end{bmatrix}. \tag{12.66}$$

Furthermore, $b = \mathrm{E}[X_3] - \mathbf{a}'\mathrm{E}[\mathbf{Y}] = 58 - 50a_1 - 62a_2 = 6.591$.

(b) The mean square estimation error is

$$e_L^* = \mathrm{Var}[X] - \mathbf{a}'\mathbf{C}'_{XY} = 16 - 11.2a_1 - 12.8a_2 = 5.145 \text{ degrees}^2.$$

Here, we have found $\mathrm{Var}[X] = \mathrm{Var}[X_3]$ in $\mathbf{C}_{\mathbf{X}}$: $\mathrm{Var}[X_3] = \mathrm{Cov}[X_3, X_3] = C_{\mathbf{X}}(3,3)$.

(c) Using only the observation $Y = X_2$, we apply Theorem 12.3 and find

$$a^* = \frac{\mathrm{Cov}\left[X_2, X_3\right]}{\mathrm{Var}[X_2]} = \frac{12.8}{16} = 0.8, \tag{12.67}$$

$$b^* = \mathrm{E}\left[X\right] - a^*\mathrm{E}\left[Y\right] = 58 - 0.8(62) = 8.4. \tag{12.68}$$

(d) The mean square error of the estimate based on $Y = X_2$ is

$$e_L^* = \text{Var}[X] - a^* \text{Cov}[Y, X] = 16 - 0.8(12.8) = 5.76 \text{ degrees}^2. \qquad (12.69)$$

In Example 12.9, we see that the estimator employing both X_1 and X_2 can exploit the correlation of X_1 and X_3 to offer a reduced mean square error compared to the estimator that uses just X_2.

If you go to `weather.com`, you will receive a comprehensive prediction of the future weather. If X_i is the temperature i hours from now, the website will make predictions $\hat{\mathbf{X}} = \begin{bmatrix} \hat{X}_1 & \cdots & \hat{X}_n \end{bmatrix}'$ of the vector $\mathbf{X} = \begin{bmatrix} X_1 & \cdots & X_n \end{bmatrix}'$ of future temperatures. These predictions are based on a vector $\mathbf{Y}$ of available observations. That is, a `weather.com` prediction is the vector function $\hat{\mathbf{X}} = \hat{\mathbf{X}}(\mathbf{Y})$ of observation $\mathbf{Y}$. When using vector $\mathbf{Y}$ to estimate vector $\mathbf{X}$, the MSE becomes

$$e = \text{E}\left[\left\|\hat{\mathbf{X}}(\mathbf{Y}) - \mathbf{X}\right\|^2\right] = \text{E}\left[\sum_{i=1}^{n}(\hat{X}_i(\mathbf{Y}) - X_i)^2\right] = \sum_{i=1}^{n} \text{MSE}_i. \qquad (12.70)$$

We see in Equation (12.70) that the MSE reduces to the sum of the expected square errors in estimating each component X_i. The MMSE solution is to use the observation $\mathbf{Y}$ to make an MMSE estimate of each component X_i of vector $\mathbf{X}$. In the context of linear estimation, the optimum linear estimate of each component X_i is $\hat{X}_i(\mathbf{Y}) = \mathbf{a}_i'\mathbf{Y} + b_i$, with $\mathbf{a}_i'$ and b_i as specified by Theorem 12.7 with $X = X_i$. The optimum linear vector estimate is

$$\hat{\mathbf{X}}_L(\mathbf{Y}) = \begin{bmatrix} \hat{X}_1(\mathbf{Y}) & \hat{X}_2(\mathbf{Y}) & \cdots & \hat{X}_m(\mathbf{Y}) \end{bmatrix}'. \qquad (12.71)$$

Writing $\hat{\mathbf{X}}_L(\mathbf{Y})$ in matrix form yields the vector generalization of Theorem 12.7.

Theorem 12.8

$\mathbf{X}$ *is an m-dimensional random vector with expected value* $\text{E}[\mathbf{X}]$. $\mathbf{Y}$ *is an n-dimensional random vector with expected value* $\text{E}[\mathbf{Y}]$ *and* $n \times n$ *covariance matrix* $\mathbf{C_Y}$. $\mathbf{X}$ *and* $\mathbf{Y}$ *have* $m \times n$ *cross-covariance matrix* $\mathbf{C_{XY}}$. *The minimum mean square error linear estimator of* $\mathbf{X}$ *given the observation* $\mathbf{Y}$ *is*

$$\hat{\mathbf{X}}_L(\mathbf{Y}) = \mathbf{C_{XY}}\mathbf{C_Y}^{-1}(\mathbf{Y} - \text{E}[\mathbf{Y}]) + \text{E}[\mathbf{X}].$$

Proof From Theorem 12.7,

$$\hat{X}_i(\mathbf{Y}) = \left(\mathbf{C_Y}^{-1}\mathbf{C}_{\mathbf{Y}X_i}\right)'(\mathbf{Y} - \text{E}[\mathbf{Y}]) + \text{E}[X_i]. \qquad (12.72)$$

Note that $(\mathbf{C_Y^{-1}C_{YX_i}})' = \mathbf{C'_{YX_i}(C_Y^{-1})'} = \mathbf{C_{X_iY}C_Y^{-1}}$. Thus (12.72) implies

$$\hat{\mathbf{X}}_L(\mathbf{Y}) = \begin{bmatrix} \hat{X}_1(\mathbf{Y}) \\ \vdots \\ \hat{X}_m(\mathbf{Y}) \end{bmatrix} = \begin{bmatrix} \mathbf{C}_{X_1\mathbf{Y}}\mathbf{C_Y^{-1}} \\ \vdots \\ \mathbf{C}_{X_m\mathbf{Y}}\mathbf{C_Y^{-1}} \end{bmatrix} (\mathbf{Y} - \mathrm{E}\,[\mathbf{Y}]) + \begin{bmatrix} \mathrm{E}\,[X_1] \\ \vdots \\ \mathrm{E}\,[X_m] \end{bmatrix}$$

$$= \mathbf{C_{XY}C_Y^{-1}}\,(\mathbf{Y} - \mathrm{E}\,[\mathbf{Y}]) + \mathrm{E}\,[\mathbf{X}]. \tag{12.73}$$

It is often convenient to represent the optimum linear estimator of Theorem 12.8 in the form

$$\hat{\mathbf{X}}_L(\mathbf{Y}) = \hat{\mathbf{A}}\mathbf{Y} + \hat{\mathbf{b}}, \tag{12.74}$$

with

$$\hat{\mathbf{A}} = \mathbf{C_{XY}C_Y^{-1}}, \qquad\qquad \hat{\mathbf{b}} = \mathrm{E}\,[\mathbf{X}] - \hat{\mathbf{A}}\,\mathrm{E}\,[\mathbf{Y}]. \tag{12.75}$$

When $\mathrm{E}[\mathbf{X}] = \mathbf{0}$ and $\mathrm{E}[\mathbf{Y}] = \mathbf{0}$, $\mathbf{C_Y} = \mathbf{R_Y}$ and $\mathbf{C_{XY}} = \mathbf{R_{XY}}$, and Theorem 12.8 reduces to

$$\hat{\mathbf{X}}_L(\mathbf{Y}) = \mathbf{R_{XY}R_Y^{-1}Y}, \tag{12.76}$$

the generalization of Theorem 12.6 to the estimation of the vector $\mathbf{X}$. In addition, because each component of $\hat{\mathbf{X}}_L(\mathbf{Y})$ is the optimum linear estimate of $\hat{X}_i$ from $\mathbf{Y}$ as given by Theorem 12.7, the MSE and orthogonality properties of $\hat{X}_i(\mathbf{Y})$ given in Theorem 12.7 remain the same.

The experiment in Example 12.9 consists of a sequence of $n+1$ subexperiments that produce random variables $X_1, X_2, \ldots X_{n+1}$. The estimator uses the outcomes of the first n experiments to form a linear estimate of the outcome of experiment $n+1$. We refer to this estimation procedure as *linear prediction* because it uses observations of earlier experiments to predict the outcome of a subsequent experiment. When the correlations of the random variables X_i have the property that r_{X_i,X_j} depends only on the difference $|i - j|$, the estimation equations in Theorem 12.8 have a structure that is exploited in many practical applications. To examine the implications of this property, we adopt the notation

$$R_X(i, j) = r_{|i-j|}. \tag{12.77}$$

In Chapter 13 we observe that this property is characteristic of random vectors derived from a *wide sense stationary random sequence*.

In the notation of the linear estimation model developed in Section 12.4, $X = X_{n+1}$ and $\mathbf{Y} = \begin{bmatrix} X_1 & X_2 & \cdots & X_n \end{bmatrix}'$. The elements of the correlation matrix $\mathbf{R_Y}$ and the cross-correlation matrix $\mathbf{R}_{\mathbf{Y}X}$ all have the form

$$\mathbf{R_Y} = \begin{bmatrix} r_0 & r_1 & \cdots & r_{n-1} \\ r_1 & r_0 & \cdots & r_{n-2} \\ \vdots & \vdots & \ddots & \vdots \\ r_{n-1} & \cdots & r_1 & r_0 \end{bmatrix}, \qquad \mathbf{R}_{\mathbf{Y}X} = \begin{bmatrix} r_n \\ r_{n-1} \\ \vdots \\ r_1 \end{bmatrix}. \tag{12.78}$$

Here $\mathbf{R_Y}$ and $\mathbf{R_{YX}}$ together have a special structure. There are only $n+1$ different numbers among the $n^2 + n$ elements of the two matrices, and each diagonal of $\mathbf{R_Y}$ consists of identical elements. This matrix is in a category referred to as *Toeplitz forms*. The properties of $\mathbf{R_Y}$ and $\mathbf{R_{YX}}$ make it possible to solve for $\mathbf{a}'$ in Equation (12.59) with far fewer computations than are required in solving an arbitrary set of n linear equations. Many audio compression techniques use algorithms for solving linear equations based on the properties of Toeplitz forms.

Quiz 12.4

$\mathbf{X} = \begin{bmatrix} X_1 & X_2 \end{bmatrix}'$ is a random vector with $\mathrm{E}[\mathbf{X}] = \mathbf{0}$ and autocorrelation matrix $\mathbf{R_X}$ with elements $R_X(i,j) = (-0.9)^{|i-j|}$. Observe the vector $\mathbf{Y} = \mathbf{X} + \mathbf{W}$, where $\mathrm{E}[\mathbf{W}] = \mathbf{0}$, $\mathrm{E}[W_1^2] = \mathrm{E}[W_2^2] = 0.1$, and $\mathrm{E}[W_1 W_2] = 0$. $\mathbf{W}$ and $\mathbf{X}$ are independent.

(a) Find a^*, the coefficient of the optimum linear estimator of X_2 given Y_2 and the mean square error of this estimator.

(b) Find the coefficients $\mathbf{a}' = \begin{bmatrix} a_1 & a_2 \end{bmatrix}$ of the optimum linear estimator of X_2 given Y_1 and Y_2, and the mean square error of this estimator.

12.5   MATLAB

> The matrix orientation of MATLAB makes it possible to write concise programs for generating the coefficients of a linear estimator and calculating the estimation error.

The following example explores the relationship of the mean square error to the number of observations used in a linear predictor of a random variable.

Example 12.10

The correlation matrix $\mathbf{R_X}$ of a 21-dimensional random vector $\mathbf{X}$ has i, jth element

$$R_X(i,j) = r_{|i-j|}, \qquad i, j = 1, 2, \ldots, 21. \qquad (12.79)$$

$\mathbf{W}$ is a random vector, independent of $\mathbf{X}$, with expected value $\mathrm{E}[\mathbf{W}] = \mathbf{0}$ and diagonal correlation matrix $\mathbf{R_W} = (0.1)\mathbf{I}$. Use the first n elements of $\mathbf{Y} = \mathbf{X} + \mathbf{W}$ to form a linear estimate of X_{21} and plot the mean square error of the optimum linear estimate as a function of n for

$$\text{(a)} \quad r_{|i-j|} = \frac{\sin(0.1\pi|i-j|)}{0.1\pi|i-j|}, \qquad \text{(b)} \quad r_{|i-j|} = \cos(0.5\pi|i-j|).$$

. . . .

In this problem, let $\mathbf{W}_{(n)}$, $\mathbf{X}_{(n)}$, and $\mathbf{Y}_{(n)}$ denote the vectors, consisting of the first

n components of $\mathbf{W}$, $\mathbf{X}$, and $\mathbf{Y}$. Similar to Example 12.8, independence of $\mathbf{X}_{(n)}$ and $\mathbf{W}_{(n)}$ implies that the correlation matrix of $\mathbf{Y}_{(n)}$ is

$$\mathbf{R}_{\mathbf{Y}_{(n)}} = \mathrm{E}\left[(\mathbf{X}_{(n)} + \mathbf{W}_{(n)})(\mathbf{X}_{(n)} + \mathbf{W}_{(n)})'\right] = \mathbf{R}_{\mathbf{X}_{(n)}} + \mathbf{R}_{\mathbf{W}_{(n)}}. \tag{12.80}$$

Note that $\mathbf{R}_{\mathbf{X}_{(n)}}$ and $\mathbf{R}_{\mathbf{W}_{(n)}}$ are the $n \times n$ upper-left submatrices of $\mathbf{R}_{\mathbf{X}}$ and $\mathbf{R}_{\mathbf{W}}$. In addition,

$$\mathbf{R}'_{X\mathbf{Y}_{(n)}} = \mathbf{R}_{\mathbf{Y}_{(n)}X} = \mathrm{E}\left[\begin{bmatrix} X_1 + W_1 \\ \vdots \\ X_n + W_n \end{bmatrix} X_{21}\right] = \begin{bmatrix} r_{20} \\ \vdots \\ r_{21-n} \end{bmatrix}. \tag{12.81}$$

Thus the optimum linear estimator based on the first n observations is

$$\mathbf{a}'_{(n)} = \mathbf{R}_{X\mathbf{Y}_{(n)}}\mathbf{R}_{\mathbf{Y}_{(n)}}^{-1}, \tag{12.82}$$

and the mean square error is

$$e_L^* = \mathrm{Var}[X_{21}] - \mathbf{a}'_{(n)}\mathbf{R}'_{X\mathbf{Y}_{(n)}}. \tag{12.83}$$

```
function e=mse(r)
N=length(r); e=[];
rr=fliplr(r(:)');
for n=1:N,
    RYX=rr(1:n)';
    RY=toeplitz(r(1:n))+0.1*eye(n);
    a=RY\RYX;
    en=r(1)-(a')*RYX;
    e=[e;en];
end
plot(1:N,e);
```

mse.m calculates the mean square error using Equation (12.83). The input r corresponds to the vector $\begin{bmatrix} r_0 & \cdots & r_{20} \end{bmatrix}$, which is the first row of the Toeplitz correlation matrix $\mathbf{R}_{\mathbf{X}}$. Note that $\mathbf{R}_{\mathbf{X}_{(n)}}$ is a Toeplitz matrix whose first row contains the first n elements of r. To plot the mean square error as a function of the number of observations, n, we generate the vector r and then run mse(r). For the correlation functions (a) and (b) in the problem statement, the necessary MATLAB commands and corresponding mean square estimation error output as a function of n are shown in Figure 12.3.

In comparing the results of cases (a) and (b) in Example 12.10, we see that the mean square estimation error depends strongly on the correlation structure given by $r_{|i-j|}$. For case (a), samples X_n for $n < 10$ have very little correlation with X_{21}. Thus for $n < 10$, the estimates of X_{21} are only slightly better than the blind estimate. On the other hand, for case (b), X_1 and X_{21} are completely correlated; $\rho_{X_1,X_{21}} = 1$. For $n = 1$, $Y_1 = X_1 + W_1$ is simply a noisy copy of X_{21}, and the estimation error is due to the variance of W_1. In this case, as n increases, the optimal linear estimator is able to combine additional noisy copies of X_{21}, yielding further reductions in the mean square estimation error.

Quiz 12.5

Estimate the Gaussian $(0,1)$ random variable X using the observation vector $\mathbf{Y} = \mathbf{1}X + \mathbf{W}$, where $\mathbf{1}$ is the vector of 20 1's. The noise vector $\mathbf{W} = \begin{bmatrix} W_0 & \cdots & W_{19} \end{bmatrix}'$ is

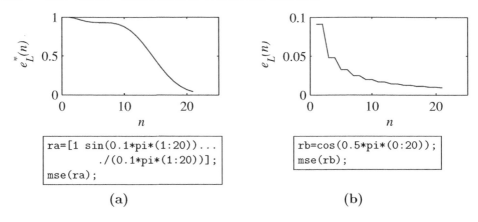

Figure 12.3   Two Runs of `mse.m`

independent of X, has zero expected value, and has a correlation matrix with i, jth entry $R_\mathbf{W}(i, j) = c^{|i-j|-1}$. Find $\hat{X}_L(\mathbf{Y})$, the linear MMSE estimate of X given $\mathbf{Y}$. For c in the range $0 < c < 1$, what value of c minimizes the mean square error?

Further Reading: The final chapter of [WS01] presents the basic theory of estimation of random variables as well as extensions to stochastic process estimation in the time domain and frequency domain.

Problems

Difficulty: ● Easy   ■ Moderate   ◆ Difficult   ◆◆ Experts Only

12.1.1● Generalizing the solution of Example 12.2, let the call duration T be an exponential (λ) random variable. For $t_0 > 0$, show that the minimum mean square error estimate of T, given that $T > t_0$ is

$$\hat{T} = t_0 + \mathrm{E}[T].$$

12.1.2● X and Y have the joint PDF

$$f_{X,Y}(x, y) = \begin{cases} 6(y - x) & 0 \le x \le y \le 1, \\ 0 & \text{otherwise.} \end{cases}$$

(a) What is $f_X(x)$?

(b) What is the blind estimate $\hat{x}_B$?

(c) What is the minimum mean square error estimate of X given $X < 0.5$?

(d) What is $f_Y(y)$?

(e) What is the blind estimate $\hat{y}_B$?

(f) What is the minimum mean square error estimate of Y given $Y > 0.5$?

12.1.3■ X and Y have the joint PDF

$$f_{X,Y}(x, y) = \begin{cases} 2 & 0 \le x \le y \le 1, \\ 0 & \text{otherwise.} \end{cases}$$

(a) What is $f_X(x)$?

(b) What is the blind estimate $\hat{x}_B$?

(c) What is the minimum mean square error estimate of X given $X > 1/2$?

(d) What is $f_Y(y)$?

(e) What is the blind estimate $\hat{y}_B$?

(f) What is the minimum mean square error estimate of Y given $X > 1/2$?

12.1.4● X and Y have the joint PDF

$$f_{X,Y}(x,y) = \begin{cases} 6(y-x) & 0 \le x \le y \le 1, \\ 0 & \text{otherwise.} \end{cases}$$

(a) What is $f_{X|Y}(x|y)$?

(b) What is $\hat{x}_M(y)$, the MMSE estimate of X given $Y = y$?

(c) What is $f_{Y|X}(y|x)$?

(d) What is $\hat{y}_M(x)$, the MMSE estimate of Y given $X = x$?

12.1.5■ X and Y have the joint PDF

$$f_{X,Y}(x,y) = \begin{cases} 2 & 0 \le x \le y \le 1, \\ 0 & \text{otherwise.} \end{cases}$$

(a) What is $f_{X|Y}(x|y)$?

(b) What is $\hat{x}_M(y)$, the MMSE estimate of X given $Y = y$?

(c) What is

$$e^*(0.5) = \mathrm{E}\left[(X - \hat{x}_M(0.5))^2 \,|Y=0.5\right],$$

the minimum mean square error of the estimate of X given $Y = 0.5$?

12.1.6■ A signal X and noise Z are independent Gaussian $(0,1)$ random variables, and $Y = X + Z$ is a noisy observation of the signal X. Usually, we want to use Y to estimate of X; however, in this problem we will use Y to estimate the noise Z.

(a) Find $\hat{Z}(Y)$, the MMSE estimator of Z given Y.

(b) Find the mean squared error $e = \mathrm{E}[(Z - \hat{Z}(Y))^2]$.

12.1.7◆ Random variable $Y = X - Z$ is a noisy observation of the continuous random variable X. The noise Z has zero expected value and unit variance and is independent of X. Find the conditional expectation $\mathrm{E}[X|Y]$.

12.1.8◆ In a BPSK communication system, a source wishes to communicate a random bit X to a receiver. The possible inputs $X = 1$ and $X = -1$ are equally likely. In this system, the source transmits X multiple times. In the ith transmission, the receiver observes $Y_i = X + W_i$. After n transmissions of X, the receiver has observed $\mathbf{Y} = \mathbf{y} = \begin{bmatrix} y_1 & \cdots & y_n \end{bmatrix}'$.

(a) Find $\hat{X}_n(\mathbf{y})$, the MMSE estimate of X given the observation $\mathbf{Y} = \mathbf{y}$. Express your answer in terms of the likelihood ratio

$$L(\mathbf{y}) = \frac{f_{\mathbf{Y}|X}(\mathbf{y}|-1)}{f_{\mathbf{Y}|X}(\mathbf{y}|1)}.$$

(b) Simplify your answer when the W_i are iid Gaussian $(0,1)$ random variables, independent of X.

12.2.1■ Random variables X and Y have joint PMF

$P_{X,Y}(x,y)$	$y = -3$	$y = -1$	$y = 1$	$y = 3$
$x = -1$	1/6	1/8	1/24	0
$x = 0$	1/12	1/12	1/12	1/12
$x = 1$	0	1/24	1/8	1/6

(a) Find the marginal probability mass functions $P_X(x)$ and $P_Y(y)$.

(b) Are X and Y independent?

(c) Find $\mathrm{E}[X]$, $\mathrm{Var}[X]$, $\mathrm{E}[Y]$, $\mathrm{Var}[Y]$, and $\mathrm{Cov}[X,Y]$.

(d) Let $\hat{X}(Y) = aY + b$ be a linear estimator of X. Find a^* and b^*, the values of a and b that minimize the mean square error e_L.

(e) What is e_L^*, the minimum mean square error of the optimum linear estimate?

(f) Find $P_{X|Y}(x|-3)$, the conditional PMF of X given $Y = -3$.

(g) Find $\hat{x}_M(-3)$, the optimum (nonlinear) mean square estimator of X given $Y = -3$.

(h) Find the mean square error

$$e^*(-3) = \mathrm{E}\left[(X - \hat{x}_M(-3))^2 \,|Y=-3\right].$$

12.2.2● A telemetry voltage V, transmitted from a position sensor on a ship's rudder, is a random variable with PDF

$$f_V(v) = \begin{cases} 1/12 & -6 \leq v \leq 6, \\ 0 & \text{otherwise.} \end{cases}$$

A receiver in the ship's control room receives $R = V + X$, The random variable X is a Gaussian $(0, \sqrt{3})$ noise voltage that is independent of V. The receiver uses R to calculate a linear estimate of the telemetry voltage: $\hat{V} = aR + b$. Find

(a) the expected received voltage $\mathrm{E}[R]$,

(b) the variance $\mathrm{Var}[R]$ of the received voltage,

(c) the covariance $\mathrm{Cov}[V, R]$ of the transmitted and received voltages,

(d) a^* and b^*, the optimum coefficients in the linear estimate,

(e) e_L^*, the minimum mean square error of the estimate.

12.2.3● Random variables X and Y have joint PMF given by the following table:

$P_{X,Y}(x,y)$	$y = -1$	$y = 0$	$y = 1$
$x = -1$	3/16	1/16	0
$x = 0$	1/6	1/6	1/6
$x = 1$	0	1/8	1/8

We estimate Y by $\hat{Y}_L(X) = aX + b$.

(a) Find a and b to minimize the mean square estimation error.

(b) Find the MMSE e_L^*.

12.2.4■ The random variables X and Y have the joint probability density function

$$f_{X,Y}(x,y) = \begin{cases} 2(y+x) & 0 \leq x \leq y \leq 1, \\ 0 & \text{otherwise.} \end{cases}$$

What is $\hat{X}_L(Y)$, the linear minimum mean square error estimate of X given Y?

12.2.5■ For random variables X and Y from Problem 12.1.4, find $\hat{X}_L(Y)$, the linear minimum mean square error estimator of X given Y.

12.2.6■ Random variable X has a second-order Erlang PDF

$$f_X(x) = \begin{cases} \lambda x e^{-\lambda x} & x \geq 0, \\ 0 & \text{otherwise.} \end{cases}$$

Given $X = x$, Y is a uniform $(0, x)$ random variable. Find

(a) the MMSE estimate of Y given $X = x$, $\hat{y}_M(x)$,

(b) the MMSE estimate of X given $Y = y$, $\hat{x}_M(y)$,

(c) the LMSE estimate of Y given X, $\hat{Y}_L(X)$,

(d) the LMSE estimate of X given Y, $\hat{X}_L(Y)$.

12.2.7■ Random variable R has an exponential PDF with expected value 1. Given $R = r$, X has an exponential PDF with expected value $1/r$. Find

(a) the MMSE estimate of R given $X = x$, $\hat{r}_M(x)$,

(b) the MMSE estimate of X given $R = r$, $\hat{x}_M(r)$,

(c) the LMSE estimate of R given X, $\hat{R}_L(X)$,

(d) the LMSE estimate of X given R, $\hat{X}_L(R)$.

12.2.8■ For random variables X and Y, we wish to use Y to estimate X. However, our estimate must be of the form $\hat{X} = aY$.

(a) Find a^*, the value of a that minimizes the mean square error

$$e = \mathrm{E}\left[(X - aY)^2\right].$$

(b) For $a = a^*$, what is the minimum mean square error e^*?

(c) Under what conditions is $\hat{X}$ the LMSE estimate of X?

12.2.9◆ Random variable $Y = X - Z$ is a noisy observation of the continuous random variable X. The noise Z has zero expected value and unit variance and is independent of X. Consider the following argument:

Since $X = Y + Z$, we see that if $Y = y$, then $X = y + Z$. Thus, by Theorem 6.4, the conditional PDF of X given $Y = y$ is $f_{X|Y}(x|y) = f_Z(x - y)$. It follows that

$$E[X|Y = y] = \int_{-\infty}^{\infty} x f_{X|Y}(x|y)\, dx$$

$$= \int_{-\infty}^{\infty} x f_Z(x - y)\, dx.$$

With the variable substitution, $z = x - y$,

$$E[X|Y = y] = \int_{-\infty}^{\infty} (z + y) f_Z(z)\, dz$$

$$= E[Z] + y = y.$$

We conclude that $E[X|Y] = Y$. Since $E[X|Y]$ is optimal in the mean square sense, we conclude that the optimal linear estimator $\hat{X}(Y) = \hat{a}Y$ must satisfy $\hat{a} = 1$.

Prove that this conclusion is wrong. What is the error in the above argument? Hint: Find the LMSE estimator $\hat{X}_L(Y) = aY$.

12.2.10◆ Here are four joint PMFs:

$P_{X,Y}(x,y)$	$x = -1$	$x = 0$	$x = 1$
$y = -1$	1/9	1/9	1/9
$y = 0$	1/9	1/9	1/9
$y = 1$	1/9	1/9	1/9

$P_{U,V}(u,v)$	$u = -1$	$u = 0$	$u = 1$
$v = -1$	0	0	1/3
$v = 0$	0	1/3	0
$v = 1$	1/3	0	0

$P_{S,T}(s,t)$	$s = -1$	$s = 0$	$s = 1$
$t = -1$	1/6	0	1/6
$t = 0$	0	1/3	0
$t = 1$	1/6	0	1/6

$P_{Q,R}(q,r)$	$q = -1$	$q = 0$	$q = 1$
$r = -1$	1/12	1/12	1/6
$r = 0$	1/12	1/6	1/12
$r = 1$	1/6	1/12	1/12

(a) For each pair of random variables, indicate whether the two random variables are independent, and compute the correlation coefficient ρ.

(b) Compute the least mean square linear estimator $\hat{U}_L(V)$ of U given V. What is the mean square error? Do the same for the pairs X, Y, Q, R, and S, T.

12.3.1■ Suppose in Quiz 12.3 that R, measured in meters, has a uniform $(0, 1000)$ PDF. Find the MAP estimate of R given $X = x$. Are the MAP and ML estimators the same?

12.3.2◆ Let R be an exponential random variable with expected value $1/\mu$. If $R = r$, then over an interval of length T, the number of phone calls N that arrive at a telephone switch has a Poisson PMF with expected value rT. Find:

(a) The MMSE estimate of N given R.

(b) The MAP estimate of N given R.

(c) The ML estimate of N given R.

12.3.3◆ Let R be an exponential random variable with expected value $1/\mu$. If $R = r$, then over an interval of length T, the number of phone calls N that arrive at a telephone switch has a Poisson PMF with expected value rT.

(a) Find the MMSE estimate of R given N.

(b) Find the MAP estimate of R given N.

(c) Find the ML estimate of R given N.

12.3.4◆ Flip a coin n times. For each flip, the probability of heads is $Q = q$ independent of all other flips. Q is a uniform $(0, 1)$ random variable. K is the number of heads in n flips.

(a) Find the ML estimator of Q given K.

(b) What is the PMF of K? What is $E[K]$?

(c) Find the conditional PDF $f_{Q|K}(q|k)$.

(d) Find the MMSE estimator of Q given $K = k$.

12.4.1● You would like to know a sample value of X, a Gaussian $(0, 4)$ random variable. However, you only can observe noisy observations of the form $Y_i = X + N_i$. In terms of a vector of noisy observations, you observe

$$\mathbf{Y} = \begin{bmatrix} Y_1 \\ Y_2 \end{bmatrix} = \begin{bmatrix} 1 \\ 1 \end{bmatrix} X + \begin{bmatrix} N_1 \\ N_2 \end{bmatrix},$$

where N_1 is a Gaussian $(0, 1)$ random variable and N_2 is a Gaussian $(0, 2)$ random variable. Under the assumption that X, N_1, and N_2 are mutually independent, answer the following questions:

(a) Suppose you use Y_1 as an estimate of X. The error in the estimate is $D_1 = Y_1 - X$. Find the expected error $E[D_1]$ and the expected squared error $E[D_1^2]$.

(b) Suppose we use $Y_3 = (Y_1 + Y_2)/2$ as an estimate of X. The error for this estimate is $D_3 = Y_3 - X$. Find the expected squared error $E[D_3^2]$. Is Y_3 or Y_1 a better estimate of X?

(c) Let $Y_4 = \mathbf{A}\mathbf{Y}$ where $\mathbf{A} = \begin{bmatrix} a & 1-a \end{bmatrix}$ is a 1×2 matrix. Let $D_4 = Y_4 - X$ denote the error in using Y_4 as an estimate for X. In terms of a, what is the expected squared error $E[D_4^2]$? What value of a minimizes $E[D_4^2]$?

12.4.2● X is a three-dimensional random vector with $E[\mathbf{X}] = \mathbf{0}$ and autocorrelation matrix $\mathbf{R_X}$ with elements $r_{ij} = (-0.80)^{|i-j|}$. Use X_1 and X_2 to form a linear estimate of X_3: $\hat{X}_3 = a_1 X_2 + a_2 X_1$.

(a) What are the optimum coefficients $\hat{a}_1$ and $\hat{a}_2$ and corresponding minimum mean square error e_L^*?

(b) Use X_2 to form a linear estimate of X_3: $\hat{X}_3 = aX_2 + b$. What are the optimum coefficients a^* and b^* and corresponding minimum mean square error e_L^*?

12.4.3■ X is a 3-dimensional random vector with $E[\mathbf{X}] = \mathbf{0}$ and autocorrelation matrix $\mathbf{R_X}$ with elements

$$R_{\mathbf{X}}(i,j) = 1 - 0.25|i - j|.$$

$\mathbf{Y}$ is a two-dimensional random vector with

$$Y_1 = X_1 + X_2, \qquad Y_2 = X_2 + X_3.$$

Use $\mathbf{Y}$ to form $\hat{X}_1 = \begin{bmatrix} a_1 & a_2 \end{bmatrix} \mathbf{Y}$, a linear estimate of X_1.

(a) Find the optimum coefficients $\hat{a}_1$ and $\hat{a}_2$ and the linear MMSE e_L^*.

(b) Use Y_1 to form a linear estimate of X_1: $\hat{X}_1 = aY_1 + b$. Find the optimum coefficients a^* and b^* and the MMSE e_L^*.

12.4.4■ X is a three-dimensional random vector with $E[\mathbf{X}] = \mathbf{0}$ and correlation matrix $\mathbf{R_X}$ with elements

$$R_{\mathbf{X}}(i,j) = 1 - 0.25|i - j|.$$

$\mathbf{W}$ is a two-dimensional random vector, independent of $\mathbf{X}$, with $E[\mathbf{W}] = \mathbf{0}$, $E[W_1 W_2] = 0$, and $E[W_1^2] = E[W_2^2] = 0.1$. $\mathbf{Y}$ is a two-dimensional random vector with

$$Y_1 = X_1 + X_2 + W_1,$$
$$Y_2 = X_2 + X_3 + W_2.$$

Use $\mathbf{Y}$ to form $\hat{X}_1 = \begin{bmatrix} a_1 & a_2 \end{bmatrix} \mathbf{Y}$, a linear estimate of X_1.

(a) Find the optimum coefficients $\hat{a}_1$ and $\hat{a}_2$ and minimum mean square error e_L^*.

(b) Use Y_1 to form a linear estimate of X_1: $\hat{X}_1 = aY_1 + b$. Find the optimum coefficients a^* and b^* and the MMSE e_L^*.

12.4.5■ Suppose

$$Y_k = q_0 + q_1 k + q_2 k^2 + Z_k,$$

where $q_0 + q_1 k + q_2 k^2$ is an unknown quadratic function of k and Z_k is a sequence of iid Gaussian $(0, 1)$ noise random variables. We wish to estimate the unknown parameters q_0, q_1, and q_2 of the quadratic function. Suppose we assume q_0, q_1, and q_2 are samples of iid Gaussian $(0, 1)$ random variables. Find the optimum linear estimator $\hat{\mathbf{Q}}(\mathbf{Y})$ of $\mathbf{Q} = \begin{bmatrix} q_0 & q_1 & q_2 \end{bmatrix}'$ given the observation $\mathbf{Y} = \begin{bmatrix} Y_1 & \cdots & Y_n \end{bmatrix}'$.

12.4.6■ X is a three-dimensional random vector with $E[\mathbf{X}] = \begin{bmatrix} -1 & 0 & 1 \end{bmatrix}'$ and correlation matrix $\mathbf{R_X}$ with elements

$$R_{\mathbf{X}}(i,j) = 1 - 0.25|i - j|.$$

$\mathbf{W}$ is a two-dimensional random vector, independent of $\mathbf{X}$, with $E[\mathbf{W}] = \mathbf{0}$, $E[W_1 W_2] = 0$, and $E[W_1^2] = E[W_2^2] = 0.1$. $\mathbf{Y}$ is a two-dimensional random vector with

$$Y_1 = X_1 + X_2 + W_1,$$
$$Y_2 = X_2 + X_3 + W_2.$$

Use $\mathbf{Y}$ to form the LMSE estimate of X_1:

$$\hat{X}_1 = \begin{bmatrix} \hat{a}_1 & \hat{a}_2 \end{bmatrix} \mathbf{Y} + \hat{b}.$$

(a) Find the coefficients $\hat{a}_1$, $\hat{a}_2$, and $\hat{b}$.

(b) Find the MMSE e_L^*.

(c) Use Y_1 to form a linear estimate of X_1: $\hat{X}_1 = aY_1 + b$. Find the optimum coefficients a^* and b^* and the MMSE e_L^*.

12.4.7■ When X and Y have expected values $\mu_X = \mu_Y = 0$, Theorem 12.3 says that $\hat{X}_L(Y) = \rho_{X,Y} \frac{\sigma_X}{\sigma_Y} Y$. Show that this result is a special case of Theorem 12.8 when random vector $\mathbf{Y}$ is the random variable Y.

12.4.8■ Prove the following theorem: $\mathbf{X}$ is an n-dimensional random vector with $E[\mathbf{X}] = \mathbf{0}$ and autocorrelation matrix $\mathbf{R_X}$ with elements $r_{ij} = c^{|i-j|}$, where $|c| < 1$. The optimum linear estimator of X_n,

$$\hat{X}_n = a_1 X_{n-1} + a_2 X_{n-2} + \cdots + a_{n-1} X_1,$$

is $\hat{X}_n = cX_{n-1}$. The minimum mean square estimation error is $e_L^* = 1 - c^2$. Hint: Consider the $n - 1$ equations $\partial e_L / \partial a_i = 0$.

12.4.9◆ In the CDMA communications system of Problem 11.2.8, each user i transmits an independent data bit X_i such that the vector $\mathbf{X} = \begin{bmatrix} X_1 & \cdots & X_n \end{bmatrix}'$ has iid components with $P_{X_i}(1) = P_{X_i}(-1) = 1/2$. The received signal is

$$\mathbf{Y} = \sum_{i=1}^{k} X_i \sqrt{p_i} \mathbf{S}_i + \mathbf{N},$$

where $\mathbf{N}$ is a Gaussian $(\mathbf{0}, \sigma^2 \mathbf{I})$ noise.

(a) Based on the observation $\mathbf{Y}$, find the LMSE estimate $\hat{X}_i(\mathbf{Y}) = \hat{\mathbf{a}}_i' \mathbf{Y}$ of X_i.

(b) Let $\hat{\mathbf{X}} = \begin{bmatrix} \hat{X}_1 & \cdots & \hat{X}_k \end{bmatrix}'$ denote the vector of LMSE estimates of bits trans-

mitted by users $1, \ldots, k$. Show that

$$\hat{\mathbf{X}} = \mathbf{P}^{1/2} \mathbf{S}' (\mathbf{S P S}' + \sigma^2 \mathbf{I})^{-1} \mathbf{Y}.$$

12.5.1● Continuing Example 12.10, the 21-dimensional vector $\mathbf{X}$ has correlation matrix $\mathbf{R_X}$ with i, jth element

$$R_{\mathbf{X}}(i, j) = \frac{\sin(\phi_0 \pi |i - j|)}{\phi_0 \pi |i - j|}.$$

We use the observation vector $\mathbf{Y} = \mathbf{Y}_{(n)} = \begin{bmatrix} Y_1 & \cdots & Y_n \end{bmatrix}'$ to estimate $X = X_{21}$. Find the LMSE estimate $\hat{X}_L(\mathbf{Y}_{(n)}) = \hat{\mathbf{a}}_{(n)} \mathbf{Y}_{(n)}$. Graph the mean square error $e_L^*(n)$ as a function of the number of observations n for $\phi_0 \in \{0.1, 0.5, 0.9\}$. Interpret your results. Does smaller ϕ_0 or larger ϕ_0 yield better estimates?

12.5.2● Repeat Problem 12.5.1 when

$$R_{\mathbf{X}}(i, j) = \cos(\phi_0 \pi |i - j|).$$

12.5.3■ In a variation on Example 12.10, we use the observation vector $\mathbf{Y} = \mathbf{Y}_{(n)} = \begin{bmatrix} Y_1 & \cdots & Y_n \end{bmatrix}'$ to estimate $X = X_1$. The 21-dimensional vector $\mathbf{X}$ has correlation matrix $\mathbf{R_X}$ with i, jth element

$$R_{\mathbf{X}}(i, j) = r_{|i-j|}.$$

Graph the MSE $e_L^*(n)$ of the LMSE estimate $\hat{X}_L(\mathbf{Y}_{(n)}) = \hat{\mathbf{a}}_{(n)} \mathbf{Y}_{(n)}$ as a function of n and interpret your results for

(a) $r_{|i-j|} = \dfrac{\sin(0.1\pi|i - j|)}{0.1\pi|i - j|}$,

(b) $r_{|i-j|} = \cos(0.5\pi|i - j|)$.

12.5.4◆◆ In the k user CDMA system in Problem 12.4.9, the receiver employs the LMSE bit estimate $\hat{X}_i$ in the bit decision rule $\tilde{X}_i = \text{sgn}\left(\hat{X}_i\right)$ for user i. Using the approach in Problem 11.3.6, construct a simulation to estimate the BER for a system with processing gain $n = 32$, with each user operating at 6 dB SNR. Graph your results as a function of k for $k = 1, 2, 4, 8, 16, 32$. Make sure to average your results over the choice of code vectors $\mathbf{S}_i$.

13

Stochastic Processes

Our study of probability refers to an experiment consisting of a procedure and observations. When we study random variables, each observation corresponds to one or more numbers. When we study stochastic processes, each observation corresponds to a function of time. The word *stochastic* means random. The word *process* in this context means function of time. Therefore, when we study stochastic processes, we study random functions of time. Almost all practical applications of probability involve multiple observations taken over a period of time. For example, our earliest discussion of probability in this book refers to the notion of the relative frequency of an outcome when an experiment is performed a large number of times. In that discussion and subsequent analyses of random variables, we have been concerned only with *how frequently* an event occurs. When we study stochastic processes, we also pay attention to the *time sequence* of the events.

In this chapter, we apply and extend the tools we have developed for random variables to introduce stochastic processes. We present a model for the randomness of a stochastic process that is analogous to the model of a random variable, and we describe some families of stochastic processes (Poisson, Brownian, Gaussian) that arise in practical applications. We then define the *autocorrelation function* and *autocovariance function* of a stochastic process. These time functions are useful summaries of the time structure of a process, just as the expected value and variance are useful summaries of the amplitude structure of a random variable. *Wide sense stationary processes* appear in many electrical and computer engineering applications of stochastic processes. In addition to descriptions of a single random process, we define the *cross-correlation* to describe the relationship between two wide sense stationary processes.

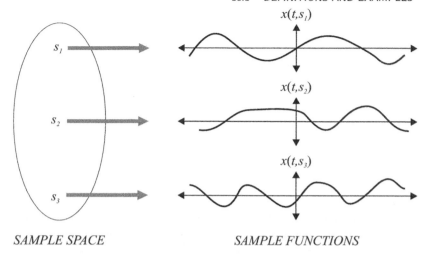

Figure 13.1 Conceptual representation of a random process.

13.1 Definitions and Examples

> The stochastic process $X(t)$ is a mapping of outcomes of an experiment to functions of time. $X(t)$ is both the name of the process and the name of the random variable observed at time t.

The definition of a stochastic process resembles Definition 3.1 of a random variable.

Definition 13.1 Stochastic Process

*A **stochastic process** $X(t)$ consists of an experiment with a probability measure P[·] defined on a sample space S and a function that assigns a time function $x(t, s)$ to each outcome s in the sample space of the experiment.*

Essentially, the definition says that the outcomes of the experiment are all functions of time. Just as a random variable assigns a number to each outcome s in a sample space S, a stochastic process assigns a *sample function* to each outcome s.

Definition 13.2 Sample Function

*A **sample function** $x(t, s)$ is the time function associated with outcome s of an experiment.*

A sample function corresponds to an outcome of a stochastic process experiment. It is one of the possible time functions that can result from the experiment. Figure 13.1 shows the correspondence between the sample space of an experiment and the ensemble of sample functions of a stochastic process. It also displays the two-dimensional notation for sample functions $x(t, s)$. In this notation, $X(t)$ is the name of the stochastic process, s indicates the particular outcome of the experiment, and t indicates the time dependence. Corresponding to the sample space of an experiment and to the range of a random variable, the *ensemble* of a stochastic process is defined as follows.

====Definition 13.3====Ensemble

*The **ensemble** of a stochastic process is the set of all possible time functions that can result from an experiment.*

====Example 13.1====

Starting at launch time $t = 0$, let $X(t)$ denote the temperature in Kelvins on the surface of a space shuttle. With each launch s, we record a temperature sequence $x(t, s)$. The ensemble of the experiment can be viewed as a catalog of the possible temperature sequences that we may record. For example,

$$x(8073.68, 175) = 207 \tag{13.1}$$

indicates that in the 175th entry in the catalog of possible temperature sequences, the temperature at $t = 8073.68$ seconds after the launch is 207 K.

Just as with random variables, one of the main benefits of the stochastic process model is that it lends itself to calculating averages. Corresponding to the two-dimensional nature of a stochastic process, there are two kinds of averages. With t fixed at $t = t_0$, $X(t_0)$ is a random variable, and we have the averages (for example, the expected value and the variance) that we have studied already. In the terminology of stochastic processes, we refer to these averages as *ensemble averages*. The other type of average applies to a specific sample function, $x(t, s_0)$, and produces a typical number for this sample function. This is a *time average* of the sample function.

====Example 13.2====

In Example 13.1 of the space shuttle, over all possible launches, the average temperature after 8073.68 seconds is $E[X(8073.68)] = 217$ K. This is an ensemble average taken over all possible temperature sequences. In the 175th entry in the catalog of possible temperature sequences, the average temperature over that space shuttle mission is

$$\frac{1}{671,208.3} \int_0^{671,208.3} x(t, 175) \, dt = 187.43 \text{ K}, \tag{13.2}$$

where the integral limit $671,208.3$ is the duration in seconds of the shuttle mission.

Before delving into the mathematics of stochastic processes, it is instructive to examine the following examples of processes that arise when we observe time functions.

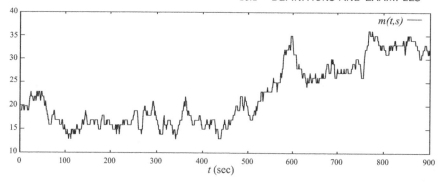

Figure 13.2   A sample function $m(t, s)$ of the random process $M(t)$ described in Example 13.4.

Example 13.3

Starting on January 1, we measure the noontime temperature (in degrees Celsius) at Newark Airport every day for one year. This experiment generates a sequence, $C(1), C(2), \ldots, C(365)$, of temperature measurements. With respect to the two kinds of averages of stochastic processes, people make frequent reference to both ensemble averages, such as "the average noontime temperature for February 19," and time averages, such as the "average noontime temperature for 1986."

Example 13.4

Consider an experiment in which we record $M(t)$, the number of active calls at a telephone switch at time t, at each second over an interval of 15 minutes. One trial of the experiment might yield the sample function $m(t, s)$ shown in Figure 13.2. Each time we perform the experiment, we would observe some other function $m(t, s)$. The exact $m(t, s)$ that we do observe will depend on many random variables including the number of calls at the start of the observation period, the arrival times of the new calls, and the duration of each call. An ensemble average is the average number of calls in progress at $t = 403$ seconds. A time average is the average number of calls in progress during a specific 15-minute interval.

The fundamental difference between Examples 13.3 and 13.4 and experiments from earlier chapters is that the randomness of the experiment depends explicitly on time. Moreover, the conclusions that we draw from our observations will depend on time. For example, in the Newark temperature measurements, we would expect the temperatures $C(1), \ldots, C(30)$ during the month of January to be low in comparison to the temperatures $C(181), \ldots, C(210)$ in the middle of summer. In this case, the randomness we observe will depend on the absolute time of our observation. We might also expect that for a day t that is within a few days of t', the temperatures $C(t)$ and $C(t')$ are likely to be similar. In this case, the randomness we observe may depend on the time difference between observations. We will see that characterizing the effects of the absolute time of an observation and the relative time between observations will be a significant step toward understanding stochastic processes.

━━━Example 13.5━━━

Suppose that at time instants $T = 0, 1, 2, \ldots$, we roll a die and record the outcome N_T where $1 \leq N_T \leq 6$. We then define the random process $X(t)$ such that for $T \leq t < T + 1$, $X(t) = N_T$. In this case, the experiment consists of an infinite sequence of rolls and a sample function is just the waveform corresponding to the particular sequence of rolls. This mapping is depicted on the right.

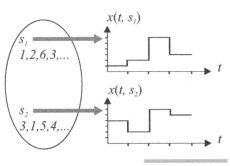

Although the stochastic process model in Figure 13.1 and Definition 13.1 refers to one experiment producing an observation s, associated with a sample function $x(t, s)$, our experience with practical applications of stochastic processes can better be described in terms of an ongoing sequence of observations of random events. In the experiment of Example 13.4, if we observe $m(17, s) = 22$ calls in progress after 17 seconds, then we know that unless in the next second at least one of the 22 calls ends or one or more new calls begin, $m(18, s)$ would remain at 22. We could say that each second we perform an experiment to observe the number of calls beginning and the number of calls ending. In this sense, the sample function $m(t, s)$ is the result of a sequence of experiments, with a new experiment performed every second. The observations of each experiment produce several random variables related to the sample functions of the stochastic process.

━━━Example 13.6━━━

The observations related to the waveform $m(t, s)$ in Example 13.4 could be

- $m(0, s)$, the number of ongoing calls at the start of the experiment,
- $X_1, \ldots, X_{m(0,s)}$, the remaining time in seconds of each of the $m(0, s)$ ongoing calls,
- N, the number of new calls that arrive during the experiment,
- $S_1, \ldots, S_N$, the arrival times in seconds of the N new calls,
- $Y_1, \ldots, Y_N$, the call durations in seconds of each of the N new calls.

Some thought will show that samples of each of these random variables, by indicating when every call starts and ends, correspond to one sample function $m(t, s)$. Keep in mind that although these random variables completely specify $m(t, s)$, there are other sets of random variables that also specify $m(t, s)$. For example, instead of referring to the duration of each call, we could instead refer to the time at which each call ends. This yields a different but equivalent set of random variables corresponding to the sample function $m(t, s)$. This example emphasizes that stochastic processes can be quite complex in that each sample function $m(t, s)$ is related to a large number of

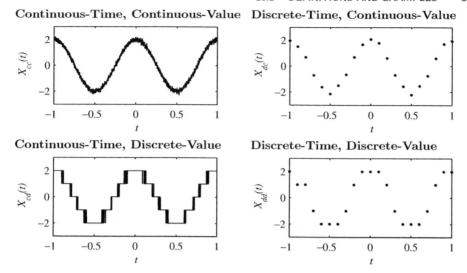

Figure 13.3 Sample functions of four kinds of stochastic processes. $X_{cc}(t)$ is a continuous-time, continuous-value process. $X_{dc}(t)$ is discrete-time, continuous-value process obtained by sampling X_{cc}) every 0.1 seconds. Rounding $X_{cc}(t)$ to the nearest integer yields $X_{cd}(t)$, a continuous-time, discrete-value process. Lastly, $X_{dd}(t)$, a discrete-time, discrete-value process, can be obtained either by sampling $X_{cd}(t)$ or by rounding $X_{dc}(t)$.

random variables, each with its own probability model. A complete model of the entire process, $M(t)$, is the model (joint probability mass function or joint probability density function) of all of the individual random variables.

Just as we developed different ways of analyzing discrete and continuous random variables, we can define categories of stochastic processes that can be analyzed using different mathematical techniques. To establish these categories, we characterize both the range of possible values at any instant t as well as the time instants at which changes in the random process can occur.

Definition 13.4————**Discrete-Value and Continuous-Value Processes**
$X(t)$ *is a **discrete-value process** if the set of all possible values of $X(t)$ at all times t is a countable set S_X; otherwise $X(t)$ is a **continuous-value process**.*

Definition 13.5————**Discrete-Time and Continuous-Time Processes**
*The stochastic process $X(t)$ is a **discrete-time process** if $X(t)$ is defined only for a set of time instants, $t_n = nT$, where T is a constant and n is an integer; otherwise $X(t)$ is a **continuous-time process**.*

In Figure 13.3, we see that the combinations of continuous/discrete time and continuous/discrete value result in four categories. For a discrete-time process, the sample function is completely described by the ordered sequence of random variables $X_n = X(nT)$.

════════**Definition 13.6**════════**Random Sequence**

A random sequence X_n is an ordered sequence of random variables $X_0, X_1, \ldots$

════════**Quiz 13.1**════════

For the temperature measurements of Example 13.3, construct examples of the measurement process such that the process is

(a) discrete-time, discrete-value,   (b) discrete-time, continuous-value,

(c) continuous-time, discrete-value,   (d) continuous-time, continuous-value.

13.2   Random Variables from Random Processes

> The probability model for the random process $X(t)$ specifies for all possible $\{t_1, \ldots, t_k\}$ the joint PMF $P_{X(t_1), \ldots, X(t_k)}(x_1, \ldots, x_k)$ or the joint PDF $f_{X(t_1), \ldots, X(t_k)}(x_1, \ldots, x_k)$.

Suppose we observe a stochastic process at a particular time instant t_1. In this case, each time we perform the experiment, we observe a sample function $x(t, s)$ and that sample function specifies the value of $x(t_1, s)$. Each time we perform the experiment, we have a new s and we observe a new $x(t_1, s)$. Therefore, each $x(t_1, s)$ is a sample value of a random variable. We use the notation $X(t_1)$ for this random variable. Like any other random variable, it has either a PDF $f_{X(t_1)}(x)$ or a PMF $P_{X(t_1)}(x)$. Note that the notation $X(t)$ can refer to either the random process or the random variable that corresponds to the value of the random process at time t. As our analysis progresses, when we write $X(t)$, it will be clear from the context whether we are referring to the entire process or to one random variable.

════════**Example 13.7**════════

In Example 13.5 of repeatedly rolling a die, what is the PMF of $X(3.5)$?

The random variable $X(3.5)$ is the value of the die roll at time 3. In this case,

$$P_{X(3.5)}(x) = \begin{cases} 1/6 & x = 1, \ldots, 6, \\ 0 & \text{otherwise.} \end{cases} \tag{13.3}$$

════════**Example 13.8**════════

Let $X(t) = R|\cos 2\pi ft|$ be a rectified cosine signal having a random amplitude R with the exponential PDF

$$f_R(r) = \begin{cases} \frac{1}{10} e^{-r/10} & r \geq 0, \\ 0 & \text{otherwise.} \end{cases} \tag{13.4}$$

What is the PDF $f_{X(t)}(x)$?

. . . .

Since $X(t) \geq 0$ for all t, $P[X(t) \leq x] = 0$ for $x < 0$. If $x \geq 0$, and $\cos 2\pi ft > 0$,

$$P[X(t) \leq x] = P[R \leq x/|\cos 2\pi ft|]$$

$$= \int_0^{x/|\cos 2\pi ft|} f_R(r) \, dr = 1 - e^{-x/10|\cos 2\pi ft|}. \tag{13.5}$$

When $\cos 2\pi ft \neq 0$, the complete CDF of $X(t)$ is

$$F_{X(t)}(x) = \begin{cases} 0 & x < 0, \\ 1 - e^{-x/10|\cos 2\pi ft|} & x \geq 0. \end{cases} \tag{13.6}$$

When $\cos 2\pi ft \neq 0$, the PDF of $X(t)$ is

$$f_{X(t)}(x) = \frac{dF_{X(t)}(x)}{dx} = \begin{cases} \frac{1}{10|\cos 2\pi ft|} e^{-x/10|\cos 2\pi ft|} & x \geq 0, \\ 0 & \text{otherwise.} \end{cases} \tag{13.7}$$

When $\cos 2\pi ft = 0$ corresponding to $t = \pi/2 + k\pi$, $X(t) = 0$ no matter how large R may be. In this case, $f_{X(t)}(x) = \delta(x)$. In this example, there is a different random variable for each value of t.

With respect to a single random variable X, we found that all the properties of X are determined from the PDF $f_X(x)$. Similarly, for a pair of random variables X_1, X_2, we needed the joint PDF $f_{X_1,X_2}(x_1, x_2)$. In particular, for the pair of random variables, we found that the marginal PDF's $f_{X_1}(x_1)$ and $f_{X_2}(x_2)$ were not enough to describe the pair of random variables. A similar situation exists for random processes. If we sample a process $X(t)$ at k time instants $t_1, \ldots, t_k$, we obtain the k-dimensional random vector $\mathbf{X} = \begin{bmatrix} X(t_1) & \cdots & X(t_k) \end{bmatrix}'$.

To answer questions about the random process $X(t)$, we must be able to answer questions about any random vector $\mathbf{X} = \begin{bmatrix} X(t_1) & \cdots & X(t_k) \end{bmatrix}'$ *for any value of* k *and any set of time instants* $t_1, \ldots, t_k$. In Section 8.1, the random vector is described by the joint PMF $P_{\mathbf{X}}(\mathbf{x})$ for a discrete-value process $X(t)$ or by the joint PDF $f_{\mathbf{X}}(\mathbf{x})$ for a continuous-value process.

For a random variable X, we could describe X by its PDF $f_X(x)$, without specifying the exact underlying experiment. In the same way, knowledge of the joint PDF $f_{X(t_1),\ldots,X(t_k)}(x_1, \ldots, x_k)$ for all k will allow us to describe a random process without reference to an underlying experiment. This is convenient because many experiments lead to the same stochastic process. This is analogous to the situation we described earlier in which more than one experiment (for example, flipping a coin or transmitting one bit) produces the same random variable.

In Section 13.1, there are two examples of random processes based on measurements. The real-world factors that influence these measurements can be very complicated. For example, the sequence of daily temperatures of Example 13.3 is the result of a very large dynamic weather system that is only partially understood. Just as we developed random variables from idealized models of experiments, we will construct random processes that are idealized models of real phenomena. The next three sections examine the probability models of specific types of stochastic processes.

Quiz 13.2

In a production line for 1000 Ω resistors, the actual resistance in ohms of each resistor is a uniform $(950, 1050)$ random variable R. The resistances of different resistors are independent. The resistor company has an order for 1% resistors with a resistance between 990 Ω and 1010 Ω. An automatic tester takes one resistor per second and measures its exact resistance. (This test takes one second.) The random process $N(t)$ denotes the number of 1% resistors found in t seconds. The random variable T_r seconds is the elapsed time at which r 1% resistors are found.

(a) What is p, the probability that any single resistor is a 1% resistor?
(b) What is the PMF of $N(t)$?
(c) What is $E[T_1]$ seconds, the expected time to find the first 1% resistor?
(d) What is the probability that the first 1% resistor is found in exactly 5 seconds?
(e) If the automatic tester finds the first 1% resistor in 10 seconds, find the conditional expected value $E[T_2|T_1 = 10]$ of the time of finding the second 1% resistor?

13.3   Independent, Identically Distributed Random Sequences

> The iid random sequence $X_1, X_2, \ldots$ is a discrete-time stochastic process consisting of a sequence of independent, identically distributed random variables.

An independent identically distributed (iid) random sequence is a random sequence X_n in which $\ldots, X_{-2}, X_{-1}, X_0, X_1, X_2, \ldots$ are iid random variables. An iid random sequence occurs whenever we perform independent trials of an experiment at a constant rate. An iid random sequence can be either discrete-value or continuous-value. In the discrete case, each random variable X_i has PMF $P_{X_i}(x) = P_X(x)$, while in the continuous case, each X_i has PDF $f_{X_i}(x) = f_X(x)$.

Example 13.9

In Quiz 13.2, each independent resistor test required exactly 1 second. Let R_n equal the number of 1% resistors found during minute n. The random variable R_n has the binomial PMF

$$P_{R_n}(r) = \binom{60}{r} p^r (1-p)^{60-r}. \tag{13.8}$$

Since each resistor is a 1% resistor independent of all other resistors, the number of 1% resistors found in each minute is independent of the number found in other minutes. Thus $R_1, R_2, \ldots$ is an iid random sequence.

For an iid random sequence, the probability model of $\mathbf{X} = \begin{bmatrix} X_1 & \cdots & X_n \end{bmatrix}'$ is easy to write since it is the product of the individual PMFs or PDFs.

=====**Theorem 13.1**=====

Let X_n denote an iid random sequence. For a discrete-value process, the sample vector $\mathbf{X} = \begin{bmatrix} X_{n_1} & \cdots & X_{n_k} \end{bmatrix}'$ has joint PMF

$$P_{\mathbf{X}}(\mathbf{x}) = P_X(x_1) P_X(x_2) \cdots P_X(x_k) = \prod_{i=1}^{k} P_X(x_i).$$

. . . .

For a continuous-value process, the joint PDF of $\mathbf{X} = \begin{bmatrix} X_{n_1} & \cdots, X_{n_k} \end{bmatrix}'$ is

$$f_{\mathbf{X}}(\mathbf{x}) = f_X(x_1) f_X(x_2) \cdots f_X(x_k) = \prod_{i=1}^{k} f_X(x_i).$$

Of all iid random sequences, perhaps the Bernoulli random sequence is the simplest.

=====**Definition 13.7**=====**Bernoulli Process**

A Bernoulli (p) process X_n is an iid random sequence in which each X_n is a Bernoulli (p) random variable.

=====**Example 13.10**=====

In a common model for communications, the output $X_1, X_2, \ldots$ of a binary source is modeled as a Bernoulli $(p = 1/2)$ process.

=====**Example 13.11**=====

For a Bernoulli (p) process X_n, find the joint PMF of $\mathbf{X} = \begin{bmatrix} X_1 & \cdots & X_n \end{bmatrix}'$.

. . . .

For a single sample X_i, we can write the Bernoulli PMF in the following way:

$$P_{X_i}(x_i) = \begin{cases} p^{x_i}(1-p)^{1-x_i} & x_i \in \{0,1\}, \\ 0 & \text{otherwise.} \end{cases} \qquad (13.9)$$

When $x_i \in \{0,1\}$ for $i = 1, \ldots, n$, the joint PMF can be written as

$$P_{\mathbf{X}}(\mathbf{x}) = \prod_{i=1}^{n} p^{x_i}(1-p)^{1-x_i} = p^k(1-p)^{n-k}, \qquad (13.10)$$

where $k = x_1 + \cdots + x_n$. The complete expression for the joint PMF is

$$P_{\mathbf{X}}(\mathbf{x}) = \begin{cases} p^{x_1+\cdots+x_n}(1-p)^{n-(x_1+\cdots+x_n)} & x_i \in \{0,1\}, i = 1, \ldots, n, \\ 0 & \text{otherwise.} \end{cases} \qquad (13.11)$$

===Quiz 13.3===

For an iid random sequence X_n of Gaussian $(0,1)$ random variables, find the joint PDF of $\mathbf{X} = \begin{bmatrix} X_1 & \cdots & X_m \end{bmatrix}'$.

13.4   The Poisson Process

> The Poisson process is a memoryless counting process in which an arrival at a particular instant is independent of an arrival at any other instant.

A counting process $N(t)$ starts at time 0 and counts the occurrences of events. These events are generally called *arrivals* because counting processes are most often used to model the arrivals of customers at a service facility. However, since counting processes have many applications, we will speak about arrivals without saying what is arriving.

Since we start at time $t = 0$, $n(t,s) = 0$ for all $t \le 0$. Also, the number of arrivals up to any $t > 0$ is an integer that cannot decrease with time.

===Definition 13.8=======Counting Process

*A stochastic process $N(t)$ is a **counting process** if for every sample function, $n(t,s) = 0$ for $t < 0$ and $n(t,s)$ is integer-valued and nondecreasing with time.*

We can think of $N(t)$ as counting the number of customers that arrive at a system during the interval $(0,t]$. A typical sample path of $N(t)$ is sketched in Figure 13.4. The jumps in the sample function of a counting process mark the arrivals, and the number of arrivals in the interval $(t_0, t_1]$ is just $N(t_1) - N(t_0)$.

We can use a Bernoulli process $X_1, X_2, \ldots$ to derive a simple counting process. In particular, consider a small time step of size Δ seconds such that there is one

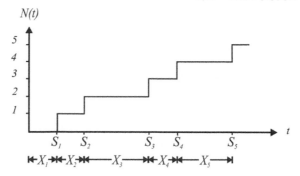

Figure 13.4 Sample path of a counting process.

arrival in the interval $(n\Delta, (n+1)\Delta]$ if and only if $X_n = 1$. For an average arrival rate $\lambda > 0$ arrivals/second, we can choose Δ such that $\lambda\Delta \ll 1$. In this case, we let the success probability of X_n be $\lambda\Delta$. This implies that the number of arrivals N_m before time $T = m\Delta$ has the binomial PMF

$$P_{N_m}(n) = \binom{m}{n}(\lambda T/m)^n (1 - \lambda T/m)^{m-n}. \tag{13.12}$$

In Theorem 3.8, we showed that as $m \to \infty$, or equivalently as $\Delta \to 0$, the PMF of N_m becomes a Poisson random variable $N(T)$ with PMF

$$P_{N(T)}(n) = \begin{cases} (\lambda T)^n e^{-\lambda T}/n! & n = 0, 1, 2, \dots, \\ 0 & \text{otherwise.} \end{cases} \tag{13.13}$$

We can generalize this argument to say that for any interval $(t_0, t_1]$, the number of arrivals would have a Poisson PMF with parameter λT where $T = t_1 - t_0$. Moreover, the number of arrivals in $(t_0, t_1]$ depends on the independent Bernoulli trials corresponding to that interval. Thus the number of arrivals in nonoverlapping intervals will be independent. In the limit as $\Delta \to 0$, we have obtained a counting process in which the number of arrivals in any interval is a Poisson random variable independent of the arrivals in any other nonoverlapping interval. We call this limiting process a *Poisson process*.

Definition 13.9 Poisson Process
*A counting process $N(t)$ is a **Poisson process** of rate λ if*

(a) *The number of arrivals in any interval $(t_0, t_1]$, $N(t_1) - N(t_0)$, is a Poisson random variable with expected value $\lambda(t_1 - t_0)$.*

(b) *For any pair of nonoverlapping intervals $(t_0, t_1]$ and $(t_0', t_1']$, the number of arrivals in each interval, $N(t_1) - N(t_0)$ and $N(t_1') - N(t_0')$, respectively, are independent random variables.*

We call λ the rate of the process because the expected number of arrivals per unit time is $E[N(t)]/t = \lambda$. By the definition of a Poisson random variable, $M = N(t_1) - N(t_0)$ has the PMF

$$P_M(m) = \begin{cases} \frac{[\lambda(t_1-t_0)]^m}{m!}e^{-\lambda(t_1-t_0)} & m = 0, 1, \ldots, \\ 0 & \text{otherwise.} \end{cases} \tag{13.14}$$

For a set of time instants $t_1 < t_2 < \cdots < t_k$, we can use the property that the number of arrivals in nonoverlapping intervals are independent to write the joint PMF of $N(t_1), \ldots, N(t_k)$ as a product of probabilities.

Theorem 13.2

For a Poisson process $N(t)$ of rate λ, the joint PMF of $\mathbf{N} = \left[N(t_1), \ldots, N(t_k)\right]'$ for ordered time instances $t_1 < \cdots < t_k$ is

$$P_{\mathbf{N}}(\mathbf{n}) = \begin{cases} \frac{\alpha_1^{n_1}e^{-\alpha_1}}{n_1!}\frac{\alpha_2^{n_2-n_1}e^{-\alpha_2}}{(n_2-n_1)!}\cdots\frac{\alpha_k^{n_k-n_{k-1}}e^{-\alpha_k}}{(n_k-n_{k-1})!} & 0 \le n_1 \le \cdots \le n_k, \\ 0 & \text{otherwise,} \end{cases}$$

where $\alpha_1 = \lambda t_1$, and for $i = 2, \ldots, k$, $\alpha_i = \lambda(t_i - t_{i-1})$.

Proof Let $M_1 = N(t_1)$ and for $i > 1$, let $M_i = N(t_i) - N(t_{i-1})$. By the definition of the Poisson process, $M_1, \ldots, M_k$ is a collection of independent Poisson random variables such that $E[M_i] = \alpha_i$.

$$\begin{aligned} P_{\mathbf{N}}(\mathbf{n}) &= P_{M_1, M_2, \ldots, M_k}(n_1, n_2 - n_1, \ldots, n_k - n_{k-1}) \\ &= P_{M_1}(n_1)P_{M_2}(n_2 - n_1)\cdots P_{M_k}(n_k - n_{k-1}). \end{aligned} \tag{13.15}$$

The theorem follows by substituting Equation (13.14) for $P_{M_i}(n_i - n_{i-1})$.

Keep in mind that the independent intervals property of the Poisson process must hold even for very small intervals. For example, the number of arrivals in $(t, t + \delta]$ must be independent of the arrival process over $[0, t]$ no matter how small we choose $\delta > 0$. Essentially, the probability of an arrival during any instant is independent of the past history of the process. In this sense, the Poisson process is *memoryless*.

This memoryless property can also be seen when we examine the times between arrivals. As depicted in Figure 13.4, the random time X_n between arrival $n-1$ and arrival n is called the nth *interarrival time*. In addition, we call the time X_1 of the first arrival the first interarrival time even though there is no previous arrival.

Theorem 13.3

For a Poisson process of rate λ, the interarrival times $X_1, X_2, \ldots$ are an iid random sequence with the exponential PDF

$$f_X(x) = \begin{cases} \lambda e^{-\lambda x} & x \ge 0, \\ 0 & \text{otherwise.} \end{cases}$$

Proof Given $X_1 = x_1, X_2 = x_2, \ldots, X_{n-1} = x_{n-1}$, arrival $n-1$ occurs at time

$$t_{n-1} = x_1 + \cdots + x_{n-1}. \tag{13.16}$$

For $x > 0$, $X_n > x$ if and only if there are no arrivals in the interval $(t_{n-1}, t_{n-1} + x]$. The number of arrivals in $(t_{n-1}, t_{n-1} + x]$ is independent of the past history described by $X_1, \ldots, X_{n-1}$. This implies

$$P[X_n > x | X_1 = x_1, \ldots, X_{n-1} = x_{n-1}] = P[N(t_{n-1} + x) - N(t_{n-1}) = 0]$$

$$= e^{-\lambda x}. \tag{13.17}$$

Thus X_n is independent of $X_1, \ldots, X_{n-1}$ and has the exponential CDF

$$F_{X_n}(x) = 1 - P[X_n > x] = \begin{cases} 1 - e^{-\lambda x} & x > 0, \\ 0 & \text{otherwise.} \end{cases} \tag{13.18}$$

From the derivative of the CDF, we see that X_n has the exponential PDF $f_{X_n}(x) = f_X(x)$ in the statement of the theorem.

From a sample function of $N(t)$, we can identify the interarrival times X_1, X_2 and so on. Similarly, from the interarrival times $X_1, X_2, \ldots$, we can construct the sample function of the Poisson process $N(t)$. This implies that an equivalent representation of the Poisson process is the iid random sequence $X_1, X_2, \ldots$ of exponentially distributed interarrival times.

Theorem 13.4

A counting process with independent exponential (λ) *interarrivals* $X_1, X_2, \ldots$ *is a Poisson process of rate* λ.

Quiz 13.4

Data packets transmitted by a modem over a phone line form a Poisson process of rate 10 packets/sec. Using M_k to denote the number of packets transmitted in the kth hour, find the joint PMF of M_1 and M_2.

13.5 Properties of the Poisson Process

> The sum $N(t) = N_1(t) + N_2(t)$ of independent Poisson processes $N_1(t)$ and $N_2(t)$ is a Poisson process. The Poisson process $N(t)$ can be decomposed into two independent Poisson processes $N_1(t)$ and $N_2(t)$.

The memoryless property of the Poisson process can also be seen in the exponential interarrival times. Since $P[X_n > x] = e^{-\lambda x}$, the conditional probability that $X_n >$

$t + x$, given $X_n > t$, is

$$P[X_n > t + x | X_n > t] = \frac{P[X_n > t + x, X_n > t]}{P[X_n > t]} = e^{-\lambda x}. \qquad (13.19)$$

The interpretation of Equation (13.19) is that if the arrival has not occurred by time t, the additional time until the arrival, $X_n - t$, has the same exponential distribution as X_n. That is, no matter how long we have waited for the arrival, the remaining time until the arrival remains an exponential (λ) random variable. The consequence is that if we start to watch a Poisson process at any time t, we see a stochastic process that is indistinguishable from a Poisson process started at time 0.

   This interpretation is the basis for ways of composing and decomposing Poisson processes. First we consider the sum $N(t) = N_1(t) + N_2(t)$ of two independent Poisson processes $N_1(t)$ and $N_2(t)$. Clearly, $N(t)$ is a counting process since any sample function of $N(t)$ is nondecreasing. Since interarrival times of each $N_i(t)$ are continuous exponential random variables, the probability that both processes have arrivals at the same time is zero. Thus $N(t)$ increases by one arrival at a time. Further, Theorem 9.7 showed that the sum of independent Poisson random variables is also Poisson. Thus for any time t_0, $N(t_0) = N_1(t_0) + N_2(t_0)$ is a Poisson random variable. This suggests (but does not prove) that $N(t)$ is a Poisson process. In the following theorem and proof, we verify this conjecture by showing that $N(t)$ has independent exponential interarrival times.

Theorem 13.5

Let $N_1(t)$ and $N_2(t)$ be two independent Poisson processes of rates λ_1 and λ_2. The counting process $N(t) = N_1(t) + N_2(t)$ is a Poisson process of rate $\lambda_1 + \lambda_2$.

Proof We show that the interarrival times of the $N(t)$ process are iid exponential random variables. Suppose the $N(t)$ process just had an arrival. Whether that arrival was from $N_1(t)$ or $N_2(t)$, X_i, the residual time until the next arrival of $N_i(t)$, has an exponential PDF since $N_i(t)$ is a memoryless process. Further, X, the next interarrival time of the $N(t)$ process, can be written as $X = \min(X_1, X_2)$. Since X_1 and X_2 are independent of the past interarrival times, X must be independent of the past interarrival times. In addition, we observe that $X > x$ if and only if $X_1 > x$ and $X_2 > x$. This implies $P[X > x] = P[X_1 > x, X_2 > x]$. Since $N_1(t)$ and $N_2(t)$ are independent processes, X_1 and X_2 are independent random variables so that

$$P[X > x] = P[X_1 > x]\, P[X_2 > x] = \begin{cases} 1 & x < 0, \\ e^{-(\lambda_1 + \lambda_2)x} & x \geq 0. \end{cases} \qquad (13.20)$$

Thus X is an exponential ($\lambda_1 + \lambda_2$) random variable.

   We derived the Poisson process of rate λ as the limiting case (as $\Delta \to 0$) of a Bernoulli arrival process that has an arrival in an interval of length Δ with probability $\lambda\Delta$. When we consider the sum of two independent Poisson processes

$N_1(t) + N_2(t)$ over an interval of length Δ, each process $N_i(t)$ can have an arrival with probability $\lambda_i \Delta$. The probability that both processes have an arrival is $\lambda_1 \lambda_2 \Delta^2$. As $\Delta \to 0$, $\Delta^2 \ll \Delta$ and the probability of two arrivals becomes insignificant in comparison to the probability of a single arrival.

Example 13.12

Cars, trucks, and buses arrive at a toll booth as independent Poisson processes with rates $\lambda_c = 1.2$ cars/minute, $\lambda_t = 0.9$ trucks/minute, and $\lambda_b = 0.7$ buses/minute. In a 10-minute interval, what is the PMF of N, the number of vehicles (cars, trucks, or buses) that arrive?

. . . .

By Theorem 13.5, the arrival of vehicles is a Poisson process of rate $\lambda = 1.2+0.9+0.7 = 2.8$ vehicles per minute. In a 10-minute interval, $\lambda T = 28$ and N has PMF

$$P_N(n) = \begin{cases} 28^n e^{-28}/n! & n = 0, 1, 2, \ldots, \\ 0 & \text{otherwise.} \end{cases} \tag{13.21}$$

Theorem 13.5 describes the composition of a Poisson process. Now we examine the decomposition of a Poisson process into two separate processes. Suppose whenever a Poisson process $N(t)$ has an arrival, we flip a biased coin to decide whether to call this a type 1 or type 2 arrival. That is, each arrival of $N(t)$ is independently labeled either type 1 with probability p or type 2 with probability $1 - p$. This results in two counting processes, $N_1(t)$ and $N_2(t)$, where $N_i(t)$ denotes the number of type i arrivals by time t. We will call this procedure of breaking down the $N(t)$ processes into two counting processes a *Bernoulli decomposition*.

Theorem 13.6

The counting processes $N_1(t)$ and $N_2(t)$ derived from a Bernoulli decomposition of the Poisson process $N(t)$ are independent Poisson processes with rates λp and $\lambda(1 - p)$.

Proof Let $X_1^{(i)}, X_2^{(i)}, \ldots$ denote the interarrival times of the process $N_i(t)$. We will verify that $X_1^{(1)}, X_2^{(1)}, \ldots$ and $X_1^{(2)}, X_2^{(2)}, \ldots$ are independent random sequences, each with exponential CDFs. We first consider the interarrival times of the $N_1(t)$ process. Suppose time t marked arrival $n - 1$ of the $N_1(t)$ process. The next interarrival time $X_n^{(1)}$ depends only on future coin flips and future arrivals of the memoryless $N(t)$ process and thus is independent of all past interarrival times of either the $N_1(t)$ or $N_2(t)$ processes. This implies the $N_1(t)$ process is independent of the $N_2(t)$ process. All that remains is to show that $X_n^{(1)}$ is an exponential random variable. We observe that $X_n^{(1)} > x$ if there are no type 1 arrivals in the interval $[t, t + x]$. For the interval $[t, t + x]$, let N_1 and N denote the number of arrivals of the $N_1(t)$ and $N(t)$ processes. In terms of N_1 and N, we can write

$$\mathrm{P}\left[X_n^{(1)} > x\right] = P_{N_1}(0) = \sum_{n=0}^{\infty} P_{N_1|N}(0|n) P_N(n). \tag{13.22}$$

Given $N = n$ total arrivals, $N_1 = 0$ if each of these arrivals is labeled type 2. This will occur with probability $P_{N_1|N}(0|n) = (1-p)^n$. Thus

$$P\left[X_n^{(1)} > x\right] = \sum_{n=0}^{\infty}(1-p)^n \frac{(\lambda x)^n e^{-\lambda x}}{n!} = e^{-p\lambda x}\underbrace{\sum_{n=0}^{\infty}\frac{[(1-p)\lambda x]^n e^{-(1-p)\lambda x}}{n!}}_{1}. \tag{13.23}$$

Thus $P[X_n^{(1)} > x] = e^{-p\lambda x}$; each $X_n^{(1)}$ has an exponential PDF with mean $1/(p\lambda)$. It follows that $N_1(t)$ is a Poisson process of rate $\lambda_1 = p\lambda$. The same argument can be used to show that each $X_n^{(2)}$ has an exponential PDF with mean $1/[(1-p)\lambda]$, implying $N_2(t)$ is a Poisson process of rate $\lambda_2 = (1-p)\lambda$.

Example 13.13

A corporate Web server records hits (requests for HTML documents) as a Poisson process at a rate of 10 hits per second. Each page is either an internal request (with probability 0.7) from the corporate intranet or an external request (with probability 0.3) from the Internet. Over a 10-minute interval, what is the joint PMF of I, the number of internal requests, and X, the number of external requests?

....

By Theorem 13.6, the internal and external request arrivals are independent Poisson processes with rates of 7 and 3 hits per second. In a 10-minute (600-second) interval, I and X are independent Poisson random variables with parameters $\alpha_I = 7(600) = 4200$ and $\alpha_X = 3(600) = 1800$ hits. The joint PMF of I and X is

$$\begin{aligned} P_{I,X}(i,x) &= P_I(i)\, P_X(x) \\ &= \begin{cases} \frac{(4200)^i e^{-4200}}{i!}\frac{(1800)^x e^{-1800}}{x!} & i, x \in \{0,1,\ldots\}, \\ 0 & \text{otherwise.} \end{cases} \end{aligned} \tag{13.24}$$

The Bernoulli decomposition of two Poisson processes and the sum of two Poisson processes are closely related. Theorem 13.6 says two independent Poisson processes $N_1(t)$ and $N_2(t)$ with rates λ_1 and λ_2 can be constructed from a Bernoulli decomposition of a Poisson process $N(t)$ with rate $\lambda_1+\lambda_2$ by choosing the success probability to be $p = \lambda_1/(\lambda_1 + \lambda_2)$. Furthermore, given these two independent Poisson processes $N_1(t)$ and $N_2(t)$ derived from the Bernoulli decomposition, the original $N(t)$ process is the sum of the two processes. That is, $N(t) = N_1(t) + N_2(t)$. Thus whenever we observe two independent Poisson processes, we can think of those processes as being derived from a Bernoulli decomposition of a single process. This view leads to the following conclusion.

Theorem 13.7

Let $N(t) = N_1(t)+N_2(t)$ be the sum of two independent Poisson processes with rates λ_1 and λ_2. Given that the $N(t)$ process has an arrival, the conditional probability that the arrival is from $N_1(t)$ is $\lambda_1/(\lambda_1 + \lambda_2)$.

Proof We can view $N_1(t)$ and $N_2(t)$ as being derived from a Bernoulli decomposition of $N(t)$ in which an arrival of $N(t)$ is labeled a type 1 arrival with probability $\lambda_1/(\lambda_1 + \lambda_2)$. By Theorem 13.6, $N_1(t)$ and $N_2(t)$ are independent Poisson processes with rate λ_1 and λ_2, respectively. Moreover, given an arrival of the $N(t)$ process, the conditional probability that an arrival is an arrival of the $N_1(t)$ process is also $\lambda_1/(\lambda_1 + \lambda_2)$.

A second way to prove Theorem 13.7 is outlined in Problem 13.5.5.

Quiz 13.5

Let $N(t)$ be a Poisson process of rate λ. Let $N'(t)$ be a process in which we count only even-numbered arrivals; that is, arrivals $2, 4, 6, \ldots$, of the process $N(t)$. Is $N'(t)$ a Poisson process?

13.6 The Brownian Motion Process

> The Brownian motion process describes a one-dimensional random walk in which at every instant, the position changes by a small increment that is independent of the current position and past history of the process. The position change over any time interval is a Gaussian random variable with zero expected value and variance proportional to the time interval.

The Poisson process is an example of a continuous-time, discrete-value stochastic process. Now we will examine Brownian motion, a continuous-time, continuous-value stochastic process.

Definition 13.10 Brownian Motion Process
*A **Brownian motion process** $W(t)$ has the property that $W(0) = 0$, and for $\tau > 0$, $W(t+\tau) - W(t)$ is a Gaussian $(0, \sqrt{\alpha\tau})$ random variable that is independent of $W(t')$ for all $t' \le t$.*

For Brownian motion, we can view $W(t)$ as the position of a particle on a line. For a small time increment δ,

$$W(t + \delta) = W(t) + [W(t + \delta) - W(t)]. \tag{13.25}$$

Although this expansion may seem trivial, by the definition of Brownian motion, the increment $X = W(t + \delta) - W(t)$ is independent of $W(t)$ and is a Gaussian $(0, \sqrt{\alpha\delta})$ random variable. This property of the Brownian motion is called *independent increments*. Thus after a time step δ, the particle's position has moved by an amount X that is independent of the previous position $W(t)$. The position change X may be positive or negative.

Brownian motion was first described in 1827 by botanist Robert Brown when he was examining the movement of pollen grains in water. It was believed that the movement was the result of the internal processes of the living pollen. Brown found that the same movement could be observed for any finely ground mineral particles. In 1905, Albert Einstein identified the source of this movement as random collisions with water molecules in thermal motion. The Brownian motion process of Definition 13.10 describes this motion along one axis of motion.

Brownian motion is another process for which we can derive the PDF of the sample vector $\mathbf{W} = \left[W(t_1), \cdots, W(t_k)\right]'$.

==========Theorem 13.8==========

For the Brownian motion process $W(t)$, the PDF of $\mathbf{W} = \left[W(t_1), \ldots, W(t_k)\right]'$ is

$$f_{\mathbf{W}}(\mathbf{w}) = \prod_{n=1}^{k} \frac{1}{\sqrt{2\pi\alpha(t_n - t_{n-1})}} e^{-(w_n - w_{n-1})^2/[2\alpha(t_n - t_{n-1})]}.$$

Proof Since $W(0) = 0$, $W(t_1) = X(t_1) - W(0)$ is a Gaussian random variable. Given time instants $t_1, \ldots, t_k$, we define $t_0 = 0$ and, for $n = 1, \ldots, k$, we can define the increments $X_n = W(t_n) - W(t_{n-1})$. Note that $X_1, \ldots, X_k$ are independent random variables such that X_n is Gaussian $(0, \sqrt{\alpha(t_n - t_{n-1})})$.

$$f_{X_n}(x) = \frac{1}{\sqrt{2\pi\alpha(t_n - t_{n-1})}} e^{-x^2/[2\alpha(t_n - t_{n-1})]}. \tag{13.26}$$

Note that $\mathbf{W} = \mathbf{w}$ if and only if $W_1 = w_1$ and for $n = 2, \ldots, k$, $X_n = w_n - w_{n-1}$. Although we omit some significant steps that can be found in Problem 13.6.5, this does imply

$$f_{\mathbf{W}}(\mathbf{w}) = \prod_{n=1}^{k} f_{X_n}(w_n - w_{n-1}). \tag{13.27}$$

The theorem follows from substitution of Equation (13.26) into Equation (13.27).

==========Quiz 13.6==========

Let $W(t)$ be a Brownian motion process with variance $\mathrm{Var}[W(t)] = \alpha t$. Show that $X(t) = W(t)/\sqrt{\alpha}$ is a Brownian motion process with variance $\mathrm{Var}[X(t)] = t$.

13.7   Expected Value and Correlation

> The expected value of a stochastic process is a function of time. The autocovariance and autocorrelation are functions of two time variables. All three functions indicate the rate of change of the sample functions of a stochastic process.

In studying random variables, we often refer to properties of the probability model such as the expected value, the variance, the covariance, and the correlation. These parameters are a few numbers that summarize the complete probability model. In the case of stochastic processes, deterministic functions of time provide corresponding summaries of the properties of a complete model.

For a stochastic process $X(t)$, $X(t_1)$, the value of a sample function at time instant t_1, is a random variable. Hence it has a PDF $f_{X(t_1)}(x)$ and expected value $E[X(t_1)]$. Of course, once we know the PDF $f_{X(t_1)}(x)$, everything we have learned about random variables and expected values can be applied to $X(t_1)$ and $E[X(t_1)]$. Since $E[X(t)]$ is simply a number for each value of t, the expected value $E[X(t)]$ is a deterministic function of t. Since $E[X(t)]$ is a somewhat cumbersome notation, the next definition is just a new notation that emphasizes that the expected value is a function of time.

Definition 13.11 The Expected Value of a Process

*The **expected value** of a stochastic process $X(t)$ is the deterministic function*

$$\mu_X(t) = E\left[X(t)\right].$$

From the PDF $f_{X(t)}(x)$, we can also calculate the variance of $X(t)$. While the variance is of some interest, the covariance function of a stochastic process provides very important information about the time structure of the process. Recall that $\text{Cov}[X, Y]$ is an indication of how much information random variable X provides about random variable Y. When the magnitude of the covariance is high, an observation of X provides an accurate indication of the value of Y. If the two random variables are observations of $X(t)$ taken at two different times, t_1 seconds and $t_2 = t_1 + \tau$ seconds, the covariance indicates how much the process is likely to change in the τ seconds elapsed between t_1 and t_2. A high covariance indicates that the sample function is unlikely to change much in the τ-second interval. A covariance near zero suggests rapid change. This information is conveyed by the *autocovariance* function.

Definition 13.12 Autocovariance

*The **autocovariance** function of the stochastic process $X(t)$ is*

$$C_X(t, \tau) = \text{Cov}\left[X(t), X(t + \tau)\right].$$

. . . .

*The **autocovariance** function of the random sequence X_n is*

$$C_X[m, k] = \text{Cov}\left[X_m, X_{m+k}\right].$$

For random sequences, we have slightly modified the notation for autocovariance by placing the arguments in square brackets just as a reminder that the functions have integer arguments. For a continuous-time process $X(t)$, the autocovariance definition at $\tau = 0$ implies $C_X(t, t) = \text{Var}[X(t)]$. Equivalently, for $k = 0$, $C_X[n, n] = \text{Var}[X_n]$. The prefix *auto* of autocovariance emphasizes that $C_X(t, \tau)$ measures the covariance between two samples of the same process $X(t)$. (There is also a cross-covariance function that describes the relationship between two different random processes.)

The autocorrelation function of a stochastic process is closely related to the autocovariance function.

========Definition 13.13========Autocorrelation Function

*The **autocorrelation function** of the stochastic process $X(t)$ is*

$$R_X(t, \tau) = \mathrm{E}\left[X(t)X(t+\tau)\right].$$

. . . .

*The **autocorrelation** function of the random sequence X_n is*

$$R_X[m, k] = \mathrm{E}\left[X_m X_{m+k}\right].$$

From Theorem 5.16(a), we have the following result.

========Theorem 13.9========

The autocorrelation and autocovariance functions of a process $X(t)$ satisfy

$$C_X(t, \tau) = R_X(t, \tau) - \mu_X(t)\mu_X(t+\tau).$$

. . . .

The autocorrelation and autocovariance functions of a random sequence X_n satisfy

$$C_X[n, k] = R_X[n, k] - \mu_X(n)\mu_X(n+k).$$

Since the autocovariance and autocorrelation are so closely related, it is reasonable to ask why we need both of them. It would be possible to use only one or the other in conjunction with the expected value $\mu_X(t)$. The answer is that each function has its uses. In particular, the autocovariance is more useful when we want to use $X(t)$ to predict a future value $X(t + \tau)$. On the other hand, since $R_X(t, 0) = \mathrm{E}[X^2(t)]$, the autocorrelation describes the average power of a random signal.

========Example 13.14========

Find the autocovariance $C_X(t\,\tau)$ and autocorrelation $R_X(t, \tau)$ of the Brownian motion process $X(t)$.

. . . .

From the definition of the Brownian motion process, we know that $\mu_X(t) = 0$. Thus the autocorrelation and autocovariance are equal: $C_X(t, \tau) = R_X(t, \tau)$. To find the autocorrelation $R_X(t, \tau)$, we exploit the independent increments property of Brownian motion. For the moment, we assume $\tau \geq 0$ so we can write $R_X(t, \tau) = \mathrm{E}[X(t)X(t+\tau)]$. Because the definition of Brownian motion refers to $X(t+\tau) - X(t)$, we introduce this quantity by substituting $X(t + \tau) = X(t + \tau) - X(t) + X(t)$. The result is

$$R_X(t, \tau) = \mathrm{E}\left[X(t)[(X(t+\tau) - X(t)) + X(t)]\right]$$
$$= \mathrm{E}\left[X(t)[X(t+\tau) - X(t)]\right] + \mathrm{E}\left[X^2(t)\right]. \tag{13.28}$$

By the definition of Brownian motion, $X(t)$ and $X(t+\tau) - X(t)$ are independent, with zero expected value. This implies

$$\mathrm{E}\left[X(t)[X(t+\tau) - X(t)]\right] = \mathrm{E}\left[X(t)\right]\mathrm{E}\left[X(t+\tau) - X(t)\right] = 0. \qquad (13.29)$$

Furthermore, since $\mathrm{E}[X(t)] = 0$, $\mathrm{E}[X^2(t)] = \mathrm{Var}[X(t)]$. Therefore, Equation (13.28) implies

$$R_X(t,\tau) = \mathrm{E}\left[X^2(t)\right] = \alpha t, \qquad \tau \geq 0. \qquad (13.30)$$

When $\tau < 0$, we can reverse the labels in the preceding argument to show that $R_X(t,\tau) = \alpha(t + \tau)$. For arbitrary t and τ we can combine these statements to write

$$R_X(t,\tau) = \alpha \min\left\{t, t + \tau\right\}. \qquad (13.31)$$

Quiz 13.7

$X(t)$ has expected value $\mu_X(t)$ and autocorrelation $R_X(t,\tau)$. We make the noisy observation $Y(t) = X(t) + N(t)$, where $N(t)$ is a random noise process independent of $X(t)$ with $\mu_N(t) = 0$ and autocorrelation $R_N(t,\tau)$. Find the expected value and autocorrelation of $Y(t)$.

13.8 Stationary Processes

> A stochastic process is stationary if the probability model does not vary with time.

Recall that in a stochastic process, $X(t)$, there is a random variable $X(t_1)$ at every time instant t_1 with PDF $f_{X(t_1)}(x)$. For most random processes, the PDF $f_{X(t_1)}(x)$ depends on t_1. For example, when we make daily temperature readings, we expect that readings taken in the winter will be lower than temperatures recorded in the summer.

However, for a special class of random processes known as *stationary processes*, $f_{X(t_1)}(x)$ does not depend on t_1. That is, for any two time instants t_1 and $t_1 + \tau$,

$$f_{X(t_1)}(x) = f_{X(t_1+\tau)}(x) = f_X(x). \qquad (13.32)$$

Therefore, in a stationary process, we observe the same random variable at all time instants. The key idea of stationarity is that the statistical properties of the process do not change with time. Equation (13.32) is a necessary condition but not a sufficient condition for a stationary process. Since the statistical properties of a random process are described by PDFs of random vectors $[X(t_1), \ldots, X(t_m)]$, we have the following definition.

═══Definition 13.14═══   Stationary Process

*A stochastic process $X(t)$ is **stationary** if and only if for all sets of time instants $t_1, \ldots, t_m$, and any time difference τ,*

$$f_{X(t_1),\ldots,X(t_m)}(x_1, \ldots, x_m) = f_{X(t_1+\tau),\ldots,X(t_m+\tau)}(x_1, \ldots, x_m).$$

. . . .

*A random sequence X_n is **stationary** if and only if for any set of integer time instants $n_1, \ldots, n_m$, and integer time difference k,*

$$f_{X_{n_1},\ldots,X_{n_m}}(x_1, \ldots, x_m) = f_{X_{n_1+k},\ldots,X_{n_m+k}}(x_1, \ldots, x_m).$$

Generally it is not obvious whether a stochastic process is stationary. Usually a stochastic process is not stationary. However, proving or disproving stationarity can be tricky. Curious readers may wish to determine which of the processes in earlier examples are stationary.

═══Example 13.15═══

Is the Brownian motion process with parameter α introduced in Section 13.6 stationary?

. . . .

For Brownian motion, $X(t_1)$ is the Gaussian $(0, \sqrt{\alpha \tau_1})$ random variable. Similarly, $X(t_2)$ is Gaussian $(0, \sqrt{\alpha \tau_2})$. Since $X(t_1)$ and $X(t_2)$ do not have the same variance, $f_{X(t_1)}(x) \neq f_{X(t_2)}(x)$, and the Brownian motion process is not stationary.

The following theorem applies to applications in which we modify one stochastic process to produce a new process. If the original process is stationary and the transformation is a linear operation, the new process is also stationary.

═══Theorem 13.10═══

Let $X(t)$ be a stationary random process. For constants $a > 0$ and b, $Y(t) = aX(t) + b$ is also a stationary process.

Proof For an arbitrary set of time samples $t_1, \ldots, t_n$, we need to find the joint PDF of $Y(t_1), \ldots, Y(t_n)$. We have solved this problem in Theorem 8.5 where we found that

$$f_{Y(t_1),\ldots,Y(t_n)}(y_1, \ldots, y_n) = \frac{1}{|a|^n} f_{X(t_1),\ldots,X(t_n)}\left(\frac{y_1 - b}{a}, \ldots, \frac{y_n - b}{a}\right). \tag{13.33}$$

Since the process $X(t)$ is stationary, we can write

$$\begin{aligned}
f_{Y(t_1+\tau),\ldots,Y(t_n+\tau)}(y_1, \ldots, y_n) &= \frac{1}{a^n} f_{X(t_1+\tau),\ldots,X(t_n+\tau)}\left(\frac{y_1 - b}{a}, \ldots, \frac{y_n - b}{a}\right) \\
&= \frac{1}{a^n} f_{X(t_1),\ldots,X(t_n)}\left(\frac{y_1 - b}{a}, \ldots, \frac{y_n - b}{a}\right) \\
&= f_{Y(t_1),\ldots,Y(t_n)}(y_1, \ldots, y_n). \tag{13.34}
\end{aligned}$$

Thus $Y(t)$ is also a stationary random process.

There are many consequences of the time-invariant nature of a stationary random process. For example, setting $m = 1$ in Definition 13.14 leads immediately to Equation (13.32). Equation (13.32) implies, in turn, that the expected value function in Definition 13.11 is a constant. Furthermore, the autocovariance function and the autocorrelation function defined in Definition 13.12 and Definition 13.13 are independent of t and depend only on the time-difference variable τ. Therefore, we adopt the notation $C_X(\tau)$ and $R_X(\tau)$ for the autocovariance function and autocorrelation function of a stationary stochastic process.

Theorem 13.11

For a stationary process $X(t)$, the expected value, the autocorrelation, and the autocovariance have the following properties for all t:

(a) $\mu_X(t) = \mu_X$,

(b) $R_X(t, \tau) = R_X(0, \tau) = R_X(\tau)$,

(c) $C_X(t, \tau) = R_X(\tau) - \mu_X^2 = C_X(\tau)$.

. . . .

For a stationary random sequence X_n the expected value, the autocorrelation, and the autocovariance satisfy for all n

(a) $E[X_n] = \mu_X$,

(b) $R_X[n, k] = R_X[0, k] = R_X[k]$,

(c) $C_X[n, k] = R_X[k] - \mu_X^2 = C_X[k]$.

Proof By Definition 13.14, stationarity of $X(t)$ implies $f_{X(t)}(x) = f_{X(0)}(x)$, so that

$$\mu_X(t) = \int_{-\infty}^{\infty} x f_{X(t)}(x)\ dx = \int_{-\infty}^{\infty} x f_{X(0)}(x)\ dx = \mu_X(0). \tag{13.35}$$

Note that $\mu_X(0)$ is just a constant that we call μ_X. Also, by Definition 13.14,

$$f_{X(t), X(t+\tau)}(x_1, x_2) = f_{X(t-t), X(t+\tau-t)}(x_1, x_2), \tag{13.36}$$

so that

$$R_X(t, \tau) = E\left[X(t)X(t+\tau)\right] = \int_{-\infty}^{\infty}\int_{-\infty}^{\infty} x_1 x_2 f_{X(0), X(\tau)}(x_1, x_2)\ dx_1\ dx_2$$
$$= R_X(0, \tau) = R_X(\tau). \tag{13.37}$$

Lastly, by Theorem 13.9,

$$C_X(t, \tau) = R_X(t, \tau) - \mu_X^2 = R_X(\tau) - \mu_X^2 = C_X(\tau). \tag{13.38}$$

We obtain essentially the same relationships for random sequences by replacing $X(t)$ and $X(t + \tau)$ with X_n and X_{n+k}.

Example 13.16

At the receiver of an AM radio, the received signal contains a cosine carrier signal at the carrier frequency f_c with a random phase Θ that is a sample value of the uniform $(0, 2\pi)$ random variable. The received carrier signal is

$$X(t) = A\cos(2\pi f_c t + \Theta). \tag{13.39}$$

What are the expected value and autocorrelation of the process $X(t)$?

. . . .

The phase has PDF

$$f_\Theta(\theta) = \begin{cases} 1/(2\pi) & 0 \le \theta \le 2\pi, \\ 0 & \text{otherwise.} \end{cases} \tag{13.40}$$

For any fixed angle α and integer k,

$$E\left[\cos(\alpha + k\Theta)\right] = \int_0^{2\pi} \cos(\alpha + k\theta)\frac{1}{2\pi}\,d\theta$$

$$= \left.\frac{\sin(\alpha + k\theta)}{k}\right|_0^{2\pi} = \frac{\sin(\alpha + k2\pi) - \sin\alpha}{k} = 0. \tag{13.41}$$

Choosing $\alpha = 2\pi f_c t$, and $k = 1$, $E[X(t)]$ is

$$\mu_X(t) = E\left[A\cos(2\pi f_c t + \Theta)\right] = 0. \tag{13.42}$$

We will use the identity $\cos A \cos B = [\cos(A - B) + \cos(A + B)]/2$ to find the autocorrelation:

$$R_X(t, \tau) = E\left[A\cos(2\pi f_c t + \Theta)A\cos(2\pi f_c(t + \tau) + \Theta)\right]$$

$$= \frac{A^2}{2} E\left[\cos(2\pi f_c \tau) + \cos(2\pi f_c(2t + \tau) + 2\Theta)\right]. \tag{13.43}$$

For $\alpha = 2\pi f_c(t + \tau)$ and $k = 2$,

$$E\left[\cos(2\pi f_c(2t + \tau) + 2\Theta)\right] = E\left[\cos(\alpha + k\Theta)\right] = 0. \tag{13.44}$$

Thus

$$R_X(t, \tau) = \frac{A^2}{2}\cos(2\pi f_c \tau) = R_X(\tau). \tag{13.45}$$

Therefore, $X(t)$ has the properties of a stationary stochastic process listed in Theorem 13.11.

Quiz 13.8

Let $X_1, X_2, \ldots$ be an iid random sequence. Is $X_1, X_2, \ldots$ a stationary random sequence?

13.9 Wide Sense Stationary Stochastic Processes

A stochastic process is wide sense stationary if the expected value is constant with time and the autocorrelation depends only on the time difference between two random variables. A wide sense stationary process is ergodic if expected values such as $E[X(T)]$ and $E[X^2(t)]$ are equal to corresponding time averages.

There are many applications of probability theory in which investigators do not have a complete probability model of an experiment. Even so, much can be accomplished with partial information about the model. Often the partial information takes the form of expected values, variances, correlations, and covariances. In the context of stochastic processes, when these parameters satisfy the conditions of Theorem 13.11, we refer to the relevant process as *wide sense stationary*.

Definition 13.15 Wide Sense Stationary

$X(t)$ *is a **wide sense stationary stochastic process** if and only if for all t,*

$$E[X(t)] = \mu_X, \quad and \quad R_X(t, \tau) = R_X(0, \tau) = R_X(\tau).$$

. . . .

X_n *is a **wide sense stationary random sequence** if and only if for all n,*

$$E[X_n] = \mu_X, \quad and \quad R_X[n, k] = R_X[0, k] = R_X[k].$$

Theorem 13.11 implies that every stationary process or sequence is also wide sense stationary. However, if $X(t)$ or X_n is wide sense stationary, it may *or may not* be stationary. Thus wide sense stationary processes include stationary processes as a subset. Some texts use the term *strict sense stationary* for what we have simply called *stationary*.

Example 13.17

In Example 13.16, we observe that $\mu_X(t) = 0$ and $R_X(t, \tau) = (A^2/2) \cos 2\pi f_c \tau$. Thus the random phase carrier $X(t)$ is a wide sense stationary process.

The autocorrelation function of a wide sense stationary process has a number of important properties.

Theorem 13.12

For a wide sense stationary process $X(t)$, the autocorrelation function $R_X(\tau)$ has the following properties:

$$R_X(0) \geq 0, \quad R_X(\tau) = R_X(-\tau), \quad R_X(0) \geq |R_X(\tau)|.$$

. . . .

If X_n is a wide sense stationary random sequence:

$$R_X[0] \geq 0, \qquad R_X[k] = R_X[-k], \qquad R_X[0] \geq |R_X[k]|.$$

Proof For the first property, $R_X(0) = R_X(t,0) = \mathrm{E}[X^2(t)]$. Since $X^2(t) \geq 0$, we must have $\mathrm{E}[X^2(t)] \geq 0$. For the second property, we substitute $u = t + \tau$ in Definition 13.13 to obtain

$$R_X(t,\tau) = \mathrm{E}\left[X(u - \tau)X(u)\right] = R_X(u,-\tau). \tag{13.46}$$

Since $X(t)$ is wide sense stationary,

$$R_X(t,\tau) = R_X(\tau) = R_X(u,-\tau) = R_X(-\tau). \tag{13.47}$$

The proof of the third property is a little more complex. First, we note that when $X(t)$ is wide sense stationary, $\mathrm{Var}[X(t)] = C_X(0)$, a constant for all t. Second, Theorem 5.14 implies that

$$C_X(t,\tau) \leq \sigma_{X(t)}\sigma_{X(t+\tau)} = C_X(0). \tag{13.48}$$

Now, for any numbers a, b, and c, if $a \leq b$ and $c \geq 0$, then $(a + c)^2 \leq (b + c)^2$. Choosing $a = C_X(t,\tau)$, $b = C_X(0)$, and $c = \mu_X^2$ yields

$$\left(C_X(t,\tau) + \mu_X^2\right)^2 \leq \left(C_X(0) + \mu_X^2\right)^2. \tag{13.49}$$

In this expression, the left side equals $(R_X(\tau))^2$ and the right side is $(R_X(0))^2$, which proves the third part of the theorem. The proof for the random sequence X_n is essentially the same. Problem 13.9.10 asks the reader to confirm this fact.

$R_X(0)$ has an important physical interpretation for electrical engineers.

Definition 13.16 Average Power

*The **average power** of a wide sense stationary process $X(t)$ is $R_X(0) = \mathrm{E}[X^2(t)]$.*

*The **average power** of a wide sense stationary sequence X_n is $R_X[0] = \mathrm{E}[X_n^2]$.*

This definition relates to the fact that in an electrical circuit, a signal is measured as either a voltage $v(t)$ or a current $i(t)$. Across a resistor of $R\ \Omega$, the instantaneous power dissipated is $v^2(t)/R = i^2(t)R$. When the resistance is $R = 1\ \Omega$, the instantaneous power is $v^2(t)$ when we measure the voltage, or $i^2(t)$ when we measure the current. When we use $x(t)$, a sample function of a wide sense stationary stochastic process, to model a voltage or a current, the instantaneous power across a $1\ \Omega$ resistor is $x^2(t)$. We usually assume implicitly the presence of a $1\ \Omega$ resistor and refer to $x^2(t)$ as the instantaneous power of $x(t)$. By extension, we refer to the random variable $X^2(t)$ as the instantaneous power of the process $X(t)$. Definition 13.16 uses the terminology *average power* for the expected value of the instantaneous power of a process. Recall that Section 13.1 describes ensemble averages and time averages of stochastic processes. In our presentation of stationary processes, we have encountered only ensemble averages including the expected value, the auto-correlation, the autocovariance, and the average power. Engineers, on the other hand, are accustomed to observing time averages. For example, if $X(t)$ models a voltage, the time average of sample function $x(t)$ over an interval of duration $2T$ is

$$\overline{X}(T) = \frac{1}{2T} \int_{-T}^{T} x(t)\, dt. \tag{13.50}$$

This is the *DC voltage* of $x(t)$, which can be measured with a voltmeter. Similarly, a time average of the power of a sample function is

$$\overline{X^2}(T) = \frac{1}{2T} \int_{-T}^{T} x^2(t)\, dt. \tag{13.51}$$

The relationship of these time averages to the corresponding ensemble averages, μ_X and $\mathrm{E}[X^2(t)]$, is a fascinating topic in the study of stochastic processes. When $X(t)$ is a stationary process such that $\lim_{T \to \infty} \overline{X}(T) = \mu_X$, the process is referred to as *ergodic*. In words, for an ergodic process, the time average of the sample function of a wide sense stationary stochastic process is equal to the corresponding ensemble average. For an electrical signal modeled as a sample function of an ergodic process, μ_X and $\mathrm{E}[X^2(t)]$ and many other ensemble averages can be observed with familiar measuring equipment.

Although the precise definition and analysis of ergodic processes are beyond the scope of this introductory text, we can use the tools of Chapter 10 to make some additional observations. For a stationary process $X(t)$, we can view the time average $\overline{X}(T)$ as an estimate of the parameter μ_X, analogous to the sample mean $M_n(X)$. The difference, however, is that the sample mean is an average of independent random variables, whereas sample values of the random process $X(t)$ are correlated. However, if the autocovariance $C_X(\tau)$ approaches zero quickly, then as T becomes large, most of the sample values have little or no correlation, and we would expect the process $X(t)$ to be ergodic. This idea is made more precise in the following theorem.

═══════**Theorem 13.13**═══════

Let $X(t)$ be a stationary random process with expected value μ_X and autocovariance $C_X(\tau)$. If $\int_{-\infty}^{\infty} |C_X(\tau)|\, d\tau < \infty$, then $\overline{X}(T), \overline{X}(2T), \ldots$ is an unbiased, consistent sequence of estimates of μ_X.

Proof First we verify that $\overline{X}(T)$ is unbiased:

$$\mathrm{E}\left[\overline{X}(T)\right] = \frac{1}{2T}\mathrm{E}\left[\int_{-T}^{T} X(t)\, dt\right] = \frac{1}{2T}\int_{-T}^{T}\mathrm{E}\left[X(t)\right] dt = \frac{1}{2T}\int_{-T}^{T}\mu_X\, dt = \mu_X. \tag{13.52}$$

To show consistency, it is sufficient to show that $\lim_{T \to \infty} \text{Var}[\overline{X}(T)] = 0$. First, we observe that $\overline{X}(T) - \mu_X = \frac{1}{2T} \int_{-T}^{T} (X(t) - \mu_X)\, dt$. This implies

$$
\begin{aligned}
\text{Var}[\overline{X}(T)] &= \text{E}\left[\left(\frac{1}{2T} \int_{-T}^{T} (X(t) - \mu_X)\, dt\right)^2\right] \\
&= \text{E}\left[\frac{1}{(2T)^2} \left(\int_{-T}^{T} (X(t) - \mu_X)\, dt\right)\left(\int_{-T}^{T} (X(t') - \mu_X)\, dt'\right)\right] \\
&= \frac{1}{(2T)^2} \int_{-T}^{T} \int_{-T}^{T} \text{E}\left[(X(t) - \mu_X)(X(t') - \mu_X)\right]\, dt'\, dt \\
&= \frac{1}{(2T)^2} \int_{-T}^{T} \int_{-T}^{T} C_X(t' - t)\, dt'\, dt.
\end{aligned}
\tag{13.53}
$$

We note that

$$
\int_{-T}^{T} C_X(t' - t)\, dt' \leq \int_{-T}^{T} \left|C_X(t' - t)\right|\, dt'
$$

$$
\leq \int_{-\infty}^{\infty} \left|C_X(t' - t)\right|\, dt' = \int_{-\infty}^{\infty} \left|C_X(\tau)\right|\, d\tau < \infty.
\tag{13.54}
$$

Hence there exists a constant K such that

$$
\text{Var}[\overline{X}(T)] \leq \frac{1}{(2T)^2} \int_{-T}^{T} K\, dt = \frac{K}{2T}.
\tag{13.55}
$$

Thus $\lim_{T \to \infty} \text{Var}[\overline{X}(T)] \leq \lim_{T \to \infty} \frac{K}{2T} = 0$.

Quiz 13.9

Which of the following functions are valid autocorrelation functions?

(a) $R_1(\tau) = e^{-|\tau|}$

(b) $R_2(\tau) = e^{-\tau^2}$

(c) $R_3(\tau) = e^{-\tau} \cos \tau$

(d) $R_4(\tau) = e^{-\tau^2} \sin \tau$

13.10   Cross-Correlation

The cross-covariance and cross-correlation functions partially describe the probability model of two wide sense stationary processes.

In many applications, it is necessary to consider the relationship of two stochastic processes $X(t)$ and $Y(t)$, or two random sequences X_n and Y_n. For certain experiments, it is appropriate to model $X(t)$ and $Y(t)$ as independent processes. In this simple case, any set of random variables $X(t_1), \ldots, X(t_k)$ from the $X(t)$ process is independent of any set of random variables $Y(t_1'), \ldots, Y(t_j')$ from the $Y(t)$ process. In general, however, a complete probability model of two processes consists of a joint PMF or a joint PDF of all sets of random variables contained in the processes. Such a joint probability function completely expresses the relationship of the two processes. However, finding and working with such a joint probability function is usually prohibitively difficult.

To obtain useful tools for analyzing a pair of processes, we recall that the covariance and the correlation of a pair of random variables provide valuable information about the relationship between the random variables. To use this information to understand a pair of stochastic processes, we work with the correlation and covariance of the random variables $X(t)$ and $Y(t + \tau)$.

Definition 13.17 **Cross-Correlation Function**
*The **cross-correlation** of continuous-time random processes $X(t)$ and $Y(t)$ is*

$$R_{XY}(t, \tau) = \mathrm{E}\left[X(t)Y(t + \tau)\right].$$

*The **cross-correlation** of random sequences X_n and Y_n is*

$$R_{XY}[m, k] = \mathrm{E}\left[X_m Y_{m+k}\right].$$

Just as for the autocorrelation, there are many interesting practical applications in which the cross-correlation depends only on one time variable, the time difference τ or the index difference k.

Definition 13.18 **Jointly Wide Sense Stationary Processes**
*Continuous-time random processes $X(t)$ and $Y(t)$ are **jointly wide sense stationary** if $X(t)$ and $Y(t)$ are both wide sense stationary, and the cross-correlation depends only on the time difference between the two random variables:*

$$R_{XY}(t, \tau) = R_{XY}(\tau).$$

*Random sequences X_n and Y_n are **jointly wide sense stationary** if X_n and Y_n are both wide sense stationary and the cross-correlation depends only on the index difference between the two random variables:*

$$R_{XY}[m, k] = R_{XY}[k].$$

We encounter cross-correlations in experiments that involve noisy observations of a wide sense stationary random process $X(t)$.

Example 13.18
Suppose we are interested in $X(t)$ but we can observe only

$$Y(t) = X(t) + N(t), \tag{13.56}$$

where $N(t)$ is a noise process that interferes with our observation of $X(t)$. Assume $X(t)$ and $N(t)$ are independent wide sense stationary processes with $\mathrm{E}[X(t)] = \mu_X$

and $E[N(t)] = \mu_N = 0$. Is $Y(t)$ wide sense stationary? Are $X(t)$ and $Y(t)$ jointly wide sense stationary? Are $Y(t)$ and $N(t)$ jointly wide sense stationary?

....

Since the expected value of a sum equals the sum of the expected values,

$$E[Y(t)] = E[X(t)] + E[N(t)] = \mu_X. \tag{13.57}$$

Next, we must find the autocorrelation

$$\begin{aligned}
R_Y(t,\tau) &= E[Y(t)Y(t+\tau)] \\
&= E[(X(t) + N(t))(X(t+\tau) + N(t+\tau))] \\
&= R_X(\tau) + R_{XN}(t,\tau) + R_{NX}(t,\tau) + R_N(\tau). \tag{13.58}
\end{aligned}$$

Since $X(t)$ and $N(t)$ are independent, $R_{NX}(t,\tau) = E[N(t)]\,E[X(t+\tau)] = 0$. Similarly, $R_{XN}(t,\tau) = \mu_X \mu_N = 0$. This implies

$$R_Y(t,\tau) = R_X(\tau) + R_N(\tau). \tag{13.59}$$

The right side of this equation indicates that $R_Y(t,\tau)$ depends only on τ, which implies that $Y(t)$ is wide sense stationary. To determine whether $Y(t)$ and $X(t)$ are jointly wide sense stationary, we calculate the cross-correlation

$$\begin{aligned}
R_{YX}(t,\tau) = E[Y(t)X(t+\tau)] &= E[(X(t) + N(t))X(t+\tau)] \\
&= R_X(\tau) + R_{NX}(t,\tau) = R_X(\tau). \tag{13.60}
\end{aligned}$$

We can conclude that $X(t)$ and $Y(t)$ are jointly wide sense stationary. Similarly, we can verify that $Y(t)$ and $N(t)$ are jointly wide sense stationary by calculating

$$\begin{aligned}
R_{YN}(t,\tau) = E[Y(t)N(t+\tau)] &= E[(X(t) + N(t))N(t+\tau)] \\
&= R_{XN}(t,\tau) + R_N(\tau) = R_N(\tau). \tag{13.61}
\end{aligned}$$

In the following example, we observe that a random sequence Y_n derived from a wide sense stationary sequence X_n may also be wide sense stationary even though X_n and Y_n are not jointly wide sense stationary.

Example 13.19

X_n is a wide sense stationary random sequence with autocorrelation function $R_X[k]$. The random sequence Y_n is obtained from X_n by reversing the sign of every other random variable in X_n: $Y_n = -1^n X_n$.

(a) Express the autocorrelation function of Y_n in terms of $R_X[k]$.
(b) Express the cross-correlation function of X_n and Y_n in terms of $R_X[k]$.
(c) Is Y_n wide sense stationary?
(d) Are X_n and Y_n jointly wide sense stationary?

....

The autocorrelation function of Y_n is

$$R_Y[n, k] = \mathrm{E}[Y_n Y_{n+k}] = \mathrm{E}\left[(-1)^n X_n (-1)^{n+k} X_{n+k}\right]$$
$$= (-1)^{2n+k} \mathrm{E}[X_n X_{n+k}]$$
$$= (-1)^k R_X[k]. \tag{13.62}$$

Y_n is wide sense stationary because the autocorrelation depends only on the index difference k. The cross-correlation of X_n and Y_n is

$$R_{XY}[n, k] = \mathrm{E}[X_n Y_{n+k}] = \mathrm{E}\left[X_n (-1)^{n+k} X_{n+k}\right]$$
$$= (-1)^{n+k} \mathrm{E}[X_n X_{n+k}]$$
$$= (-1)^{n+k} R_X[k]. \tag{13.63}$$

X_n and Y_n are not jointly wide sense stationary because the cross-correlation depends on both n and k. When n and k are both even or when n and k are both odd, $R_{XY}[n, k] = R_X[k]$; otherwise $R_{XY}[n, k] = -R_X[k]$.

Theorem 13.12 indicates that the autocorrelation of a wide sense stationary process $X(t)$ is symmetric about $\tau = 0$ (continuous-time) or $k = 0$ (random sequence). The cross-correlation of jointly wide sense stationary processes has a corresponding symmetry.

==========Theorem 13.14==========

If $X(t)$ and $Y(t)$ are jointly wide sense stationary continuous-time processes, then

$$R_{XY}(\tau) = R_{YX}(-\tau).$$

. . . .

If X_n and Y_n are jointly wide sense stationary random sequences, then

$$R_{XY}[k] = R_{YX}[-k].$$

Proof From Definition 13.17, $R_{XY}(\tau) = \mathrm{E}[X(t)Y(t + \tau)]$. Making the substitution $u = t + \tau$ yields

$$R_{XY}(\tau) = \mathrm{E}[X(u - \tau)Y(u)] = \mathrm{E}[Y(u)X(u - \tau)] = R_{YX}(u, -\tau). \tag{13.64}$$

Since $X(t)$ and $Y(t)$ are jointly wide sense stationary, $R_{YX}(u, -\tau) = R_{YX}(-\tau)$. The proof is similar for random sequences.

==========Quiz 13.10==========

$X(t)$ is a wide sense stationary stochastic process with autocorrelation function $R_X(\tau)$. $Y(t)$ is identical to $X(t)$, except that time is reversed: $Y(t) = X(-t)$.

(a) Express the autocorrelation function of $Y(t)$ in terms of $R_X(\tau)$. Is $Y(t)$ wide sense stationary?

(b) Express the cross-correlation function of $X(t)$ and $Y(t)$ in terms of $R_X(\tau)$. Are $X(t)$ and $Y(t)$ jointly wide sense stationary?

13.11 Gaussian Processes

> For a Gaussian process $X(t)$, every vector of sample values $\mathbf{X} = \begin{bmatrix} X(t_1) & \cdots & X(t_k) \end{bmatrix}'$ is a Gaussian random vector.

The central limit theorem (Theorem 9.10) helps explain the proliferation of Gaussian random variables in nature. The same insight extends to Gaussian stochastic processes. For electrical and computer engineers, the noise voltage in a resistor is a pervasive example of a phenomenon that is accurately modeled as a Gaussian stochastic process. In a Gaussian process, every collection of sample values is a Gaussian random vector (Definition 8.12).

Definition 13.19   Gaussian Process

$X(t)$ is a Gaussian stochastic process if and only if $\mathbf{X} = \begin{bmatrix} X(t_1) & \cdots & X(t_k) \end{bmatrix}'$ is a Gaussian random vector for any integer $k > 0$ and any set of time instants $t_1, t_2, \ldots, t_k$.

X_n is a Gaussian random sequence if and only if $\mathbf{X} = \begin{bmatrix} X_{n_1} & \cdots & X_{n_k} \end{bmatrix}'$ is a Gaussian random vector for any integer $k > 0$ and any set of time instants $n_1, n_2, \ldots, n_k$.

In Problem 13.11.5, we ask you to show that the Brownian motion process in Section 13.6 is a special case of a Gaussian process. Although the Brownian motion process is not stationary (see Example 13.15), our primary interest will be in wide sense stationary Gaussian processes. In this case, the probability model for the process is completely specified by the expected value μ_X and the autocorrelation function $R_X(\tau)$ or $R_X[k]$. As a consequence, a wide sense stationary Gaussian process is stationary.

Theorem 13.15

If $X(t)$ is a wide sense stationary Gaussian process, then $X(t)$ is a stationary Gaussian process.

If X_n is a wide sense stationary Gaussian sequence, X_n is a stationary Gaussian sequence.

Proof Let $\boldsymbol{\mu}$ and $\mathbf{C}$ denote the expected value vector and the covariance matrix of the random vector $\mathbf{X} = \begin{bmatrix} X(t_1) & \ldots & X(t_k) \end{bmatrix}'$. Let $\bar{\boldsymbol{\mu}}$ and $\bar{\mathbf{C}}$ denote the same quantities for the time-shifted random vector $\bar{\mathbf{X}} = \begin{bmatrix} X(t_1 + T) & \cdots & X(t_k + T) \end{bmatrix}'$. Since $X(t)$ is wide sense stationary, $\mathrm{E}[X(t_i)] = \mathrm{E}[X(t_i + T)] = \mu_X$. The i, jth entry of $\mathbf{C}$ is

$$
\begin{aligned}
C_{ij} = C_X(t_i, t_j) &= C_X(t_j - t_i) \\
&= C_X(t_j + T - (t_i + T)) = C_X(t_i + T, t_j + T) = \bar{C}_{ij}.
\end{aligned} \tag{13.65}
$$

Thus $\mu = \bar{\mu}$ and $\mathbf{C} = \bar{\mathbf{C}}$, implying that $f_{\mathbf{X}}(\mathbf{x}) = f_{\bar{\mathbf{X}}}(\mathbf{x})$. Hence $X(t)$ is a stationary process. The same reasoning applies to a Gaussian random sequence X_n.

The *white Gaussian noise process* is a convenient starting point for many studies in electrical and computer engineering.

Definition 13.20 White Gaussian Noise

$W(t)$ *is a white Gaussian noise process if and only if $W(t)$ is a stationary Gaussian stochastic process with the properties $\mu_W = 0$ and $R_W(\tau) = \eta_0 \delta(\tau)$.*

A consequence of the definition is that for any collection of distinct time instants $t_1, \ldots, t_k$, $W(t_1), \ldots, W(t_k)$ is a set of independent Gaussian random variables. In this case, the value of the noise at time t_i tells nothing about the value of the noise at time t_j. While the white Gaussian noise process is a useful mathematical model, it does not conform to any signal that can be observed physically. Note that the average noise power is

$$\mathrm{E}\left[W^2(t)\right] = R_W(0) = \infty. \tag{13.66}$$

That is, white noise has infinite average power, which is physically impossible. The model is useful, however, because any Gaussian noise signal observed in practice can be interpreted as a filtered white Gaussian noise signal with finite power.

Quiz 13.11

$X(t)$ is a stationary Gaussian random process with $\mu_X(t) = 0$ and autocorrelation function $R_X(\tau) = 2^{-|\tau|}$. What is the joint PDF of $X(t)$ and $X(t+1)$?

13.12 MATLAB

> Stochastic processes appear in models of many phenomena studied by engineers. When the phenomena are complicated, MATLAB simulations are valuable analysis tools.

To produce MATLAB simulations we need to develop codes for stochastic processes. For example, to simulate the cellular telephone switch of Example 13.4, we need to model both the arrivals and departures of calls. A Poisson process $N(t)$ is a conventional model for arrivals.

Example 13.20

Use MATLAB to generate the arrival times $S_1, S_2, \ldots$ of a rate λ Poisson process over a time interval $[0, T]$.

```
t=0.01*(0:1000);
lambda=5;
N=poissonprocess(lambda,t);
plot(t,N)
xlabel('\it t');
ylabel('\it N(t)');
```

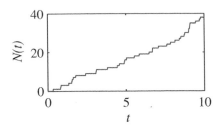

Figure 13.5   A Poisson process sample path $N(t)$ generated by `poissonprocess.m`.

```
function s=poissonarrivals(lam,T)
%arrival times s=[s(1) ... s(n)]
%  s(n)<= T < s(n+1)
n=ceil(1.1*lam*T);
s=cumsum(exponentialrv(lam,n));
while (s(length(s))< T),
  s_new=s(length(s))+ ...
    cumsum(exponentialrv(lam,n));
  s=[s; s_new];
end
s=s(s<=T);
```

To generate Poisson arrivals at rate λ, we employ Theorem 13.4, which says that the interarrival times are independent exponential (λ) random variables. Given interarrival times X_i, the ith arrival time is the cumulative sum

$$S_i = X_1 + X_2 + \cdots + X_i.$$

The function `poissonarrivals` outputs the cumulative sums of independent exponential random variables; it returns the vector s with

s(i) corresponding to S_i, the ith arrival time. Note that the length of s is a Poisson (λT) random variable because the number of arrivals in $[0, T]$ is random.

When we wish to examine a Poisson arrival process graphically, the vector of arrival times is not so convenient. A direct representation of the process $N(t)$ is often more useful.

═══Example 13.21═══

Generate a sample path of $N(t)$, a rate $\lambda = 5$ arrivals/min Poisson process. Plot $N(t)$ over a 10-minute interval.

. . . .

```
function N=poissonprocess(lambda,t)
%N(i) = no. of arrivals by t(i)
s=poissonarrivals(lambda,max(t));
N=count(s,t);
```

Given $\mathbf{t} = \begin{bmatrix} t_1 & \cdots & t_m \end{bmatrix}'$, the function `poissonprocess` generates the vector $\mathbf{N} = \begin{bmatrix} N_1 & \cdots & N_m \end{bmatrix}'$ where $N_i = N(t_i)$ for a rate λ Poisson process $N(t)$. The ba-

sic idea of `poissonprocess.m` is that given the arrival times $S_1, S_2, \ldots$, $N(t) = \max\{n | S_n \leq t\}$ is the number of arrivals that occur by time t. In particular, in N=count(s,t), N(i) is the number of elements of s that are less than or equal to t(i). A sample path generated by `poissonprocess.m` appears in Figure 13.5.

Note that the number of arrivals generated by `poissonprocess` depends only on $T = \max_i t_i$ but not on how finely we represent time. That is,

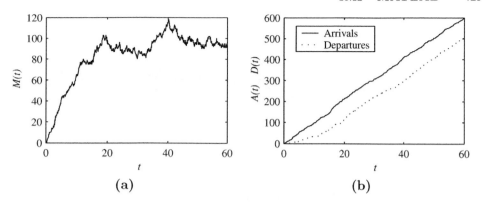

Figure 13.6 Output for `simswitch`: **(a)** The active-call process $M(t)$. **(b)** The arrival and departure processes $A(t)$ and $D(t)$ such that $M(t) = A(t) - D(t)$.

$$t=0.1*(0:10*T) \quad \text{or} \quad t=0.001*(0:1000*T)$$

both generate a Poisson number N, with $\mathrm{E}[N] = \lambda T$, of arrivals over the interval $[0, T]$. What changes is how finely we observe the output $N(t)$.

Now that MATLAB can generate a Poisson arrival process, we can simulate systems such as the telephone switch of Example 13.4.

Example 13.22

Simulate 60 minutes of activity of the telephone switch of Example 13.4 under the following assumptions.

(a) The switch starts with $M(0) = 0$ calls.

(b) Arrivals occur as a Poisson process of rate $\lambda = 10$ calls/min.

(c) The duration of each call (often called the holding time) in minutes is an exponential $(1/10)$ random variable independent of the number of calls in the system and the duration of any other call.

. . . .

```
function M=simswitch(lambda,mu,t)
%Poisson arrivals, rate lambda
%Exponential (mu) call duration
%For vector t of times
%M(i) = no. of calls at time t(i)
s=poissonarrivals(lambda,max(t));
y=s+exponentialrv(mu,length(s));
A=count(s,t);
D=count(y,t);
M=A-D;
```

In `simswitch.m`, the vectors s and x mark the arrival times and call durations. The ith call arrives at time s(i), stays for time x(i), and departs at time y(i)=s(i)+x(i). Thus the vector y=s+x denotes the call completion times, also known as *departures*. By counting the arrivals s and departures y, we produce the arrival and departure processes A and D. At any given time t, the number of calls in the system equals the number of arrivals minus the number of departures. Hence M=A-D is the number of calls in the system. One run of `simswitch.m` depicting sample functions of $A(t)$, $D(t)$, and $M(t) = A(t) - D(t)$ appears in Figure 13.6.

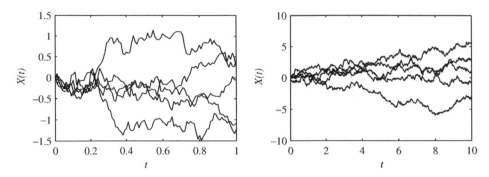

Figure 13.7 Sample paths of Brownian motion.

Similar techniques can be used to produce a Brownian motion process $Y(t)$.

Example 13.23

Generate a Brownian motion process $W(t)$ with parameter α.

....

```
function w=brownian(alpha,t)
%Brownian motion process
%sampled at t(1)<t(2)< ...
t=t(:);
n=length(t);
delta=t-[0;t(1:n-1)];
x=sqrt(alpha*delta).*gaussrv(0,1,n);
w=cumsum(x);
```

The function brownian outputs a Brownian motion process $W(t)$ sampled at times $t_i = $ t(i). The vector x consists of the independent increments x(i) scaled to have variance $\alpha(t_i - t_{i-1})$.

Each graph in Figure 13.7 shows five sample paths of a Brownian motion processes with $\alpha = 1$. For plot (a), $0 \le t \le$ 1, for plot (b), $0 \le t \le 10$. Note that the plots have different y-axis scaling because $\text{Var}[X(t)] = \alpha t$. Thus as time increases, the excursions from the origin tend to get larger.

For an arbitrary Gaussian process $X(t)$, we can use MATLAB to generate random sequences $X_n = X(nT)$ that represent sampled versions of $X(t)$. For the sampled process, the vector $\mathbf{X} = \begin{bmatrix} X_0 & \cdots & X_{n-1} \end{bmatrix}'$ is a Gaussian random vector with expected value $\boldsymbol{\mu}_\mathbf{X} = \begin{bmatrix} \text{E}[X(0)] & \cdots \text{E}[X((n-1)T)] \end{bmatrix}'$ and covariance matrix $\mathbf{C_X}$ with i,jth element $C_\mathbf{X}(i,j) = C_X(iT, jT)$. We can generate m samples of $\mathbf{X}$ using x=gaussvector(mu,C,m). As described in Section 8.6, mu is a length n vector and C is the $n \times n$ covariance matrix.

When $X(t)$ is wide sense stationary, the sampled sequence is wide sense stationary with autocovariance $C_X[k]$. In this case, the vector $\mathbf{X} = \begin{bmatrix} X_0 & \cdots & X_{n-1} \end{bmatrix}'$ has

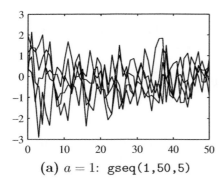
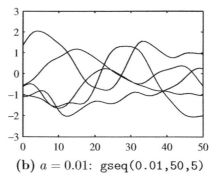

(a) $a = 1$: gseq(1,50,5) **(b)** $a = 0.01$: gseq(0.01,50,5)

Figure 13.8 Two sample outputs for Example 13.24.

covariance matrix $\mathbf{C_X}$ with $C_{\mathbf{X}}(i,j) = C_X[i-j]$. Since $C_X[k] = C_X[-k]$,

$$\mathbf{C_X} = \begin{bmatrix} C_X[0] & C_X[1] & \cdots & C_X[n-1] \\ C_X[1] & C_X[0] & \ddots & \vdots \\ \vdots & \ddots & \ddots & C_X[1] \\ C_X[n-1] & \cdots & C_X[1] & C_X[0] \end{bmatrix}. \tag{13.67}$$

We see that $\mathbf{C_X}$ is constant along each diagonal. A matrix with constant diagonals is called a Toeplitz matrix. When the covariance matrix $\mathbf{C_X}$ is Toeplitz, it is completely specified by the vector $\mathbf{c} = \begin{bmatrix} C_X[0] & C_X[1] & \cdots & C_X[n-1] \end{bmatrix}'$, whose elements are both the first column and first row of $\mathbf{C_X}$. Thus the PDF of $\mathbf{X}$ is completely described by the expected value $\mu_X = \mathrm{E}[X_i]$ and the vector $\mathbf{c}$. In this case, a function that generates sample vectors $\mathbf{X}$ needs only the scalar μ_X and vector $\mathbf{c}$ as inputs. Since generating sample vectors $\mathbf{X}$ corresponding to a stationary Gaussian sequence is quite common, we extend the function gaussvector(mu,C,m) introduced in Section 8.6 to make this as simple as possible.

```
function x=gaussvector(mu,C,m)
%output: m Gaussian vectors,
%each with mean mu
%and covariance matrix C
if (min(size(C))==1)
    C=toeplitz(C);
end
n=size(C,2);
if (length(mu)==1)
    mu=mu*ones(n,1);
end
[U,D,V]=svd(C);
x=V*(D^(0.5))*randn(n,m)...
    +(mu(:)*ones(1,m));
```

If C is a length n row or column vector, it is assumed to be the first row of an $n \times n$ Toeplitz covariance matrix that we create with the statement C=toeplitz(C). In addition, when mu is a scalar value, it is assumed to be the expected value $\mathrm{E}[X_n]$ of a stationary sequence. The program extends mu to a length n vector with identical elements. When mu is an n-element vector and C is an $n \times n$ covariance matrix, as was required in the original gaussvector.m, they are left unchanged. The real work of gaussvector still occurs in the last two lines, which are identical to the simpler version of gaussvector.m in Section 8.6.

=======Example 13.24======

Write a MATLAB function x=gseq(a,n,m) that generates m sample vectors $\mathbf{X} = \begin{bmatrix} X_0 & \cdots & X_n \end{bmatrix}'$ of a stationary Gaussian sequence with

$$\mu_X = 0, \qquad C_X[k] = \frac{1}{1+ak^2}. \tag{13.68}$$

. . . .

```
function x=gseq(a,n,m)
nn=0:n;
cx=1./(1+a*nn.^2);
x=gaussvector(0,cx,m);
plot(nn,x);
```

All we need to do is generate the vector cx corresponding to the covariance function. Figure 13.8 shows sample outputs for graphs

(a) $a = 1$: gseq(1,50,5),

(b) $a = 0.01$: gseq(0.01,50,5).

We observe in Figure 13.8 that each graph shows $m = 5$ sample paths even though graph (a) may appear to have many more. The graphs look very different because for $a = 1$, samples just a few steps apart are nearly uncorrelated and the sequence varies quickly with time. That is, the sample paths in graph (a) zig-zag around. By contrast, when $a = 0.01$, samples have significant correlation and the sequence varies slowly. that is, in graph (b), the sample paths look realtively smooth.

=======Quiz 13.12======

The switch simulation of Example 13.22 is unrealistic in the assumption that the switch can handle an arbitrarily large number of calls. Modify the simulation so that the switch blocks (i.e., discards) new calls when the switch has $c = 120$ calls in progress. Estimate $\mathrm{P}[B]$, the probability that a new call is blocked. Your simulation may need to be significantly longer than 60 minutes.

Further Reading: [Doo90] contains the original (1953) mathematical theory of stochastic processes. [HSP87] is a concise introduction to basic principles for readers familiar with probability and random variables. The second half of [PP02] is a comprehensive treatise on stochastic processes.

Problems

Difficulty: ● Easy   ■ Moderate   ◆ Difficult   ◆◆ Experts Only

13.1.1● For Example 13.4, define a set of random variables that could produce the sample function $m(t,s)$. Do not duplicate the set listed in Example 13.6.

13.1.2● For the random processes of Examples 13.3, 13.4 and 13.5, identify whether the process is discrete-time or continuous-time, discrete-value or continuous-value.

13.1.3■ In a binary phase shift keying (BPSK) communications system, one of two equally probable bits, 0 or 1, is transmitted every T seconds. If the kth bit is $j \in \{0,1\}$, the waveform $x_j(t) = \cos(2\pi f_0 t + j\pi)$ is transmitted over the interval $[(k-1)T, kT]$. Let $X(t)$ denote the random process in which three symbols are transmitted in the interval $[0, 3T]$. Assuming f_0 is an integer multiple of $1/T$, sketch the sample space and corresponding sample functions of the process $X(t)$.

13.1.4■ True or false: For a continuous-value random process $X(t)$, the random variable $X(t_0)$ is always a continuous random variable.

13.2.1■ Let W be an exponential random variable with PDF

$$f_W(w) = \begin{cases} e^{-w} & w \geq 0, \\ 0 & \text{otherwise.} \end{cases}$$

Find the CDF $F_{X(t)}(x)$ of the time-delayed ramp process $X(t) = t - W$.

13.2.2■ In a production line for 10 kHz oscillators, the output frequency of each oscillator is a random variable W uniformly distributed between 9980 Hz and 1020 Hz. The frequencies of different oscillators are independent. The oscillator company has an order for one part in 10^4 oscillators with frequency between 9999 Hz and 10,001 Hz. A technician takes one oscillator per minute from the production line and measures its exact frequency. (This test takes one minute.) The random variable T_r minutes is the elapsed time at which the technician finds r acceptable oscillators.

(a) Find p, the probability that an single oscillator has one-part-in-10^4 accuracy.

(b) What is $E[T_1]$ minutes, the expected time for the technician to find the first one-part-in-10^4 oscillator?

(c) What is the probability that the technician will find the first one-part-in-10^4 oscillator in exactly 20 minutes?

(d) What is $E[T_5]$, the expected time of finding the fifth one-part-in-10^4 oscillator?

13.2.3■ For the random process of Problem 13.2.2, what is the conditional PMF

of T_2 given T_1? If the technician finds the first oscillator in 3 minutes, what is $E[T_2|T_1 = 3]$, the conditional expected value of the time of finding the second one-part-in-10^4 oscillator?

13.2.4◆ Let $X(t) = e^{-(t-T)}u(t-T)$ be an exponential pulse with a random delay T. The delay T has a PDF $f_T(t)$. Find the PDF of $X(t)$.

13.3.1● Suppose that at the equator, we can model the noontime temperature in degrees Celsius, X_n, on day n by a sequence of iid Gaussian random variables with expected value 30 degrees and standard deviation of 5 degrees. A new random process $Y_k = [X_{2k-1} + X_{2k}]/2$ is obtained by averaging the temperature over two days. Is Y_k an iid random sequence?

13.3.2■ For the equatorial noontime temperature sequence X_n of Problem 13.3.1, a second sequence of averaged temperatures is $W_n = [X_n + X_{n-1}]/2$. Is W_n an iid random sequence?

13.3.3■ Let Y_k denote the number of failures between successes $k-1$ and k of a Bernoulli (p) random process. Also, let Y_1 denote the number of failures before the first success. What is the PMF $P_{Y_k}(y)$? Is Y_k an iid random sequence?

13.4.1● The arrivals of new telephone calls at a telephone switching office is a Poisson process $N(t)$ with an arrival rate of $\lambda = 4$ calls per second. An experiment consists of monitoring the switching office and recording $N(t)$ over a 10-second interval.

(a) What is $P_{N(1)}(0)$, the probability of no phone calls in the first second of observation?

(b) What is $P_{N(1)}(4)$, the probability of exactly four calls arriving in the first second of observation?

(c) What is $P_{N(2)}(2)$, the probability of exactly two calls arriving in the first two seconds?

13.4.2● Queries presented to a computer database are a Poisson process of rate $\lambda = 6$ queries per minute. An experiment consists of monitoring the database for m minutes and recording $N(m)$, the number of queries presented. The answer to each of the following questions can be expressed in terms of the PMF $P_{N(m)}(k) = P[N(m) = k]$.

(a) What is the probability of no queries in a one-minute interval?

(b) What is the probability of exactly six queries arriving in a one-minute interval?

(c) What is the probability of exactly three queries arriving in a one-half-minute interval?

13.4.3 At a successful garage, there is always a backlog of cars waiting to be serviced. The service times of cars are iid exponential random variables with a mean service time of 30 minutes. Find the PMF of $N(t)$, the number of cars serviced in the first t hours of the day.

13.4.4 The count of students dropping the course "Probability and Stochastic Processes" is known to be a Poisson process of rate 0.1 drops per day. Starting with day 0, the first day of the semester, let $D(t)$ denote the number of students that have dropped after t days. What is $P_{D(t)}(d)$?

13.4.5 Customers arrive at the Veryfast Bank as a Poisson process of rate λ customers per minute. Each arriving customer is immediately served by a teller. After being served, each customer immediately leaves the bank. The time a customer spends with a teller is called the service time. If the service time of a customer is exactly two minutes, what is the PMF of the number of customers $N(t)$ in service at the bank at time t?

13.4.6 Given a Poisson process $N(t)$, identify which of the following are Poisson processes..

(a) $N(2t)$, (b) $N(t/2)$,

(c) $2N(t)$, (d) $N(t)/2$,

(e) $N(t+2)$, (f) $N(t) - N(t-1)$.

13.4.7 Starting at any time t, the number N_τ of hamburgers sold at a White Castle in the τ minute interval $(t, t+\tau)$ has the Poisson PMF

$$P_{N_\tau}(n) = \begin{cases} (10\tau)^n e^{-10\tau}/n! & n = 0, 1, \ldots \\ 0 & \text{otherwise} \end{cases}$$

(a) Find the expected number $E[N_{60}]$ of hamburgers sold in one hour (60 minutes).

(b) What is the probability that no hamburgers are sold in the 10-minute interval starting at 12 noon?

(c) You arrive at the White Castle at 12 noon. You wait a random time W (minutes) until you see a hamburger sold. What is the PDF of W? Hint: Find $P[W > w]$.

13.4.8 A sequence of queries are made to a database system. The response time of the system, T seconds, is the exponential $(1/8)$ random variable. As soon as the system responds to a query, the next query is made. Assuming the first query is made at time zero, let $N(t)$ denote the number of queries made by time t.

(a) What is $P[T \geq 4]$, the probability that a single query will last at least four seconds?

(b) If the database user has been waiting five seconds for a response, what is $P[T \geq 13 | T \geq 5]$, the probability that the user will wait at least eight more seconds?

(c) What is the PMF of $N(t)$?

13.4.9 The proof of Theorem 13.3 neglected the first interarrival time X_1. Show that X_1 has an exponential (λ) PDF.

13.4.10 $U_1, U_2, \ldots$ are independent identically distributed uniform random variables with parameters 0 and 1.

(a) Let $X_i = -\ln U_i$. What is $P[X_i > x]$?

(b) What kind of random variable is X_i?

(c) Given a constant $t > 0$, let N denote the value of n, such that

$$\prod_{i=1}^{n} U_i \ge e^{-t} > \prod_{i=1}^{n+1} U_i.$$

Note that we define $\prod_{i=1}^{0} U_i = 1$. What is the PMF of N?

13.5.1● Customers arrive at a casino as a Poisson process of rate 100 customers per hour. Upon arriving, each customer must flip a coin, and only those customers who flip heads actually enter the casino. Let $N(t)$ denote the process of customers entering the casino. Find the PMF of N, the number of customers who arrive between 5 PM and 7 PM.

13.5.2● A subway station carries both blue (B) line and red (R) line trains. Red line trains and blue line trains arrive as independent Poisson processes with rates $\lambda_R = 0.05$ and $\lambda_B = 0.15$ trains/min, respectively. You arrive at a random time t and wait until a red train arrives. Let N denote the number of blue line trains that pass through the station while you are waiting. What is the PMF $P_N(n)$?

13.5.3● A subway station carries both blue (B) line and red (R) line trains. Red line trains and blue line trains arrive as independent Poisson processes with rates $\lambda_R = 0.15$ and $\lambda_B = 0.30$ trains/min respectively. You arrive at the station at random time t and watch the trains for one hour.

(a) What is the PMF of N, the number of trains that you count passing through the station?

(b) Given that you see $N = 30$ trains, what is the conditional PMF of R, the number of red trains that you see?

13.5.4■ Buses arrive at a bus stop as a Poisson process of rate $\lambda = 1$ bus/minute. After a very long time t, you show up at the bus stop.

(a) Let X denote the interarrival time between two bus arrivals. What is the PDF $f_X(x)$?

(b) Let W equal the time you wait after time t until the next bus arrival. What is the PDF $f_W(w)$?

(c) Let V equal the time (in minutes) that has passed since the most recent bus arrival. What is the PDF $f_V(v)$?

(d) Let U equal the time gap between the most recent bus arrival and the next bus arrival. What is the PDF of U?

13.5.5■ For a Poisson process of rate λ, the Bernoulli arrival approximation assumes that in any very small interval of length Δ, there is either 0 arrivals with probability $1 - \lambda\Delta$ or 1 arrival with probability $\lambda\Delta$. Use this approximation to prove Theorem 13.7.

13.5.6◆ Continuing Problem 13.4.5, suppose each service time is either one minute or two minutes equiprobably, independent of the arrival process or the other service times. What is the PMF of the number of customers $N(t)$ in service at the bank at time t?

13.5.7◆ Ten runners compete in a race starting at time $t = 0$. The runners' finishing times $R_1, \ldots, R_{10}$ are iid exponential random variables with expected value $1/\mu = 10$ minutes.

(a) What is the probability that the last runner will finish in less than 20 minutes?

(b) What is the PDF of X_1, the finishing time of the winning runner?

(c) Find the PDF of $Y = R_1 + \cdots + R_{10}$.

(d) Let $X_1, \ldots, X_{10}$ denote the runners' interarrival times at the finish line. Find the joint PDF $f_{X_1, \ldots, X_{10}}(x_1, \ldots, x_{10})$.

13.5.8◆◆ Let N denote the number of arrivals of a Poisson process of rate λ over the interval $(0, T)$. Given $N = n$, let $S_1, \ldots, S_n$ denote the corresponding arrival times. Prove that

$$f_{S_1, \ldots, S_n | N}(S_1, \ldots, S_n | n)$$
$$= \begin{cases} n!/T^n & 0 \le s_1 < \cdots < s_n \le T, \\ 0 & \text{otherwise.} \end{cases}$$

Conclude that, given $N(T) = n$, $S_1, \ldots, S_n$ are the order statistics of a collection of n uniform $(0, T)$ random variables. (See Problem 5.10.11.)

13.6.1● Over the course of a day, the stock price of a widely traded company can be modeled as a Brownian motion process where $X(0)$ is the opening price at the morning bell. Suppose the unit of time t is an hour, the exchange is open for eight hours, and the standard deviation of the daily price change (the difference between the opening bell and closing bell prices) is $1/2$ point. What is the Brownian motion parameter α?

13.6.2■ $X_0, X_1, \ldots$ is an iid Gaussian $(0, 1)$ random sequence. The random sequence Y_n is defined by $Y_0 = 0$ and $Y_{n+1} = X_{n+1} + Y_n$. Find the autocorrelation function $R_Y[n, k]$.

13.6.3■ Let $X(t)$ be a Brownian motion process with variance $\text{Var}[X(t)] = \alpha t$. For a constant $c > 0$, determine whether $Y(t) = X(ct)$ is a Brownian motion process.

13.6.4◆ For a Brownian motion process $X(t)$, let $X_0 = X(0), X_1 = X(1), \ldots$ represent samples of a Brownian motion process with variance αt. The discrete-time continuous-value process $Y_1, Y_2, \ldots$ defined by $Y_n = X_n - X_{n-1}$ is called an *increments process*. Show that Y_n is an iid random sequence.

13.6.5◆ This problem works out the missing steps in the proof of Theorem 13.8. For **W** and **X** as defined in the proof of the theorem, show that **W** = **AX**. What is the matrix **A**? Use Theorem 8.11 to find $f_{\mathbf{W}}(\mathbf{w})$.

13.7.1● X_n is an iid random sequence with expected value $\text{E}[X_n] = \mu_X$ and variance $\text{Var}[X_n] = \sigma_X^2$. What is the autocovariance $C_X[m, k]$?

13.7.2■ For the time-delayed ramp process $X(t)$ from Problem 13.2.1, find for any $t \geq 0$:

(a) The expected value function $\mu_X(t)$,

(b) The autocovariance function $C_X(t, \tau)$. Hint: $\text{E}[W] = 1$ and $\text{E}[W^2] = 2$.

13.7.3■ A simple model (in degrees Celsius) for the daily temperature process $C(t)$ of Example 13.3 is

$$C_n = 16 \left[1 - \cos \frac{2\pi n}{365} \right] + 4X_n,$$

where $X_1, X_2, \ldots$ is an iid random sequence of Gaussian $(0, 1)$ random variables.

(a) What is $\text{E}[C_n]$?

(b) Find the autocovariance function $C_C[m, k]$.

(c) Why is this model overly simple?

13.7.4■ The input to a digital filter is an iid random sequence $\ldots, X_{-1}, X_0, X_1, \ldots$ with $\text{E}[X_i] = 0$ and $\text{Var}[X_i] = 1$. The output $\ldots, Y_{-1}, Y_0, Y_1, \ldots$ is related to the input by the formula

$$Y_n = X_n + X_{n-1}$$

for all integers n. For the output Y_n, find the expected value $\text{E}[Y_n]$ and autocovariance function $C_Y[m, k]$.

13.7.5◆ A different model for the daily temperature process $C(n)$ of Example 13.3 is

$$C_n = \frac{1}{2} C_{n-1} + 4X_n,$$

where $C_0, X_1, X_2, \ldots$ is an iid random sequence of Gaussian $(0, 1)$ random variables.

(a) Find the mean and variance of C_n.

(b) Find the autocovariance $C_C[m, k]$.

(c) Is this a plausible model for the daily temperature over the course of a year?

(d) Would $C_1, \ldots, C_{31}$ constitute a plausible model for the daily temperature for the month of January?

13.7.6■ For a Poisson process $N(t)$ of rate λ, show that for $s < t$, the autocovariance is $C_N(s, t) = \lambda s$. If $s > t$, what is $C_N(s, t)$? Is there a general expression for $C_N(s, t)$?

13.7.7◆ $N(t)$ is a Poisson process of rate $\lambda = 1$ and $X_0, X_1, X_2, \ldots$ is an iid sequence of Gaussian $(0, \sigma)$ random variables that are independent of $N(t)$. Consider the process $\{Y(t)|t \geq 0\}$ defined by $Y(t) = X_{N(t)}$. Find the expected value $\mu_Y(t) = E[Y(t)]$ and the covariance function $C_Y(t, \tau)$. (Assume $|\tau| < t$.)

13.7.8◆ X_n is an iid random sequence with $E[X_n] = 0$ and $\text{Var}[X_n] = 3$. Find the autocorrelation function $C_Y[n, k]$ of the process $Y_n = X_{n-1} X_n$.

13.8.1● For an arbitrary constant a, let $Y(t) = X(t + a)$. If $X(t)$ is a stationary random process, is $Y(t)$ stationary?

13.8.2● $\mathbf{X} = \begin{bmatrix} X_1 & X_2 \end{bmatrix}'$ has expected value $E[\mathbf{X}] = \mathbf{0}$ and covariance matrix

$$\mathbf{C}_X = \begin{bmatrix} 2 & 1 \\ 1 & 1 \end{bmatrix}.$$

Does there exist a stationary process $X(t)$ and time instances t_1 and t_2 such that $\mathbf{X}$ is actually a pair of observations $\begin{bmatrix} X(t_1) & X(t_2) \end{bmatrix}'$ of the process $X(t)$?

13.8.3● For an arbitrary constant a, let $Y(t) = X(at)$. If $X(t)$ is a stationary random process, is $Y(t)$ stationary?

13.8.4● Let $X(t)$ be a stationary continuous-time random process. By sampling $X(t)$ every Δ seconds, we obtain the discrete-time random sequence $Y_n = X(n\Delta)$. Is Y_n a stationary sequence?

13.8.5● Given a stationary random sequence X_n, we can *subsample* X_n by extracting every kth sample: $Y_n = X_{kn}$. Is Y_n a stationary random sequence?

13.8.6■ Let A be a nonnegative random variable that is independent of any collection of samples $X(t_1), \ldots, X(t_k)$ of a stationary random process $X(t)$. Is $Y(t) = AX(t)$ a stationary random process?

13.8.7◆ Let $g(x)$ be deterministic function. If $X(t)$ is a stationary random process, is $Y(t) = g(X(t))$ a stationary process?

13.9.1● Which of the following are valid autocorrelation functions?

$R_1(\tau) = \delta(\tau)$ $\qquad$ $R_2(\tau) = \delta(\tau) + 10$

$R_3(\tau) = \delta(\tau - 10)$ $\quad$ $R_4(\tau) = \delta(\tau) - 10$

13.9.2● Let A be a nonnegative random variable that is independent of any collection of samples $X(t_1), \ldots, X(t_k)$ of a wide sense stationary random process $X(t)$. Is $Y(t) = A + X(t)$ a wide sense stationary process?

13.9.3● True or False: If X_n is a wide sense stationary random sequence with $E[X_n] = 0$, then $Y_n = X_n - X_{n-1}$ is a wide sense stationary random sequence.

13.9.4● Let X_n denote an iid sequence of Bernoulli $(p = 1/2)$ random variables. Find the autocorrelation function $R_X[n, k]$ and the autocovariance function $C_X[n, k]$.

13.9.5● X_n is an iid sequence with $E[X_n] = \mu$ and $\text{Var}[X_n] = \sigma^2$. Find the autocorrelation function $R_X[n, k]$.

13.9.6■ $X(t)$ and $Y(t)$ are independent wide sense stationary processes. Determine if these processes are wide-sense stationary:

(a) $V(t) = X(t) + Y(t)$,

(b) $W(t) = X(t)Y(t)$.

13.9.7■ True or False: If X_n is a wide sense stationary random sequence with $E[X_n] = 0$, then $Y_n = X_n + (-1)^{n-1} X_{n-1}$ is a wide sense stationary random sequence.

13.9.8■ Consider the random process

$$W(t) = X \cos(2\pi f_0 t) + Y \sin(2\pi f_0 t),$$

where X and Y are uncorrelated random variables, each with expected value 0 and variance σ^2. Find the autocorrelation $R_W(t, \tau)$. Is $W(t)$ wide sense stationary?

13.9.9■ $X(t)$ is a wide sense stationary random process with average power equal to 1. Let Θ denote a random variable with uniform distribution over $[0, 2\pi]$ such that $X(t)$ and Θ are independent.

(a) What is $E[X^2(t)]$?

(b) What is $E[\cos(2\pi f_c t + \Theta)]$?

(c) Let $Y(t) = X(t) \cos(2\pi f_c t + \Theta)$. What is $E[Y(t)]$?

(d) What is the average power of $Y(t)$?

13.9.10■ Prove the properties of $R_X[n]$ given in Theorem 13.12.

13.9.11◆ Let X_n be a wide sense stationary random sequence with expected value μ_X and autocovariance $C_X[k]$. For $m = 0, 1, \ldots$, we define

$$\overline{X}_m = \frac{1}{2m+1} \sum_{n=-m}^{m} X_n$$

as the sample mean process. Prove that if $\sum_{k=-\infty}^{\infty} C_X[k] < \infty$, then $\overline{X}_0, \overline{X}_1, \ldots$ is an unbiased consistent sequence of estimates of μ_X.

13.9.12◆ Determine whether each of these statements is true or false:

(a) If X_n and Y_n are independent stationary processes, then $V_n = X_n/Y_n$ is wide-sense stationary.

(b) If X_n and Y_n are independent wide sense stationary processes, then $W_n = X_n/Y_n$ is wide-sense stationary.

13.10.1◉ $X(t)$ and $Y(t)$ are independent wide sense stationary processes with expected values μ_X and μ_Y and autocorrelation functions $R_X(\tau)$ and $R_Y(\tau)$, respectively. Let $W(t) = X(t)Y(t)$.

(a) Find μ_W and $R_W(t, \tau)$ and show that $W(t)$ is wide sense stationary.

(b) Are $W(t)$ and $X(t)$ jointly wide sense stationary?

13.10.2■ $X(t)$ is a wide sense stationary random process. For each process $X_i(t)$ defined below, determine whether $X_i(t)$ and $X(t)$ are jointly wide sense stationary.

(a) $X_1(t) = X(t + a)$

(b) $X_2(t) = X(at)$

13.10.3■ $X(t)$ is a wide sense stationary stochastic process with autocorrelation function

$$R_X(\tau) = 10 \sin(2\pi 1000\tau)/(2\pi 1000\tau).$$

The process $Y(t)$ is a version of $X(t)$ delayed by 50 microseconds: $Y(t) = X(t - t_0)$ where $t_0 = 5 \times 10^{-5}$s.

(a) Derive the autocorrelation function of $Y(t)$.

(b) Derive the cross-correlation function of $X(t)$ and $Y(t)$.

(c) Is $Y(t)$ wide-sense stationary?

(d) Are $X(t)$ and $Y(t)$ jointly wide sense stationary?

13.11.1◉ A stationary Gaussian process $X(t)$ is observed at times t_1 and t_2 to form the random vector $\mathbf{X} = \begin{bmatrix} X(t_1) & X(t_2) \end{bmatrix}'$ with expected value $E[\mathbf{X}] = \mathbf{0}$ and covariance matrix $\mathbf{C}_X = \begin{bmatrix} \sigma_1^2 & 1 \\ 1 & \sigma_2^2 \end{bmatrix}$. What is the range of valid values (if any) of σ_1^2 and σ_2^2?

13.11.2■ Given a Gaussian process $X(t)$, identify which of the following, if any, are Gaussian processes.

(a) $2X(t)$,   (b) $X(t/2)$,

(c) $X(t)/2$,   (d) $X(t) - X(t-1)$,

(e) $X(2t)$.

13.11.3■ A white Gaussian noise process $N(t)$ with autocorrelation $R_N(\tau) = \alpha\delta(\tau)$ is passed through an integrator yielding the output

$$Y(t) = \int_0^t N(u)\, du.$$

Find $E[Y(t)]$ and the autocorrelation function $R_Y(t, \tau)$. Show that $Y(t)$ is a nonstationary process.

13.11.4■ Let $X(t)$ be a Gaussian process with mean $\mu_X(t)$ and autocovariance $C_X(t, \tau)$. In this problem, we verify that the for two samples $X(t_1), X(t_2)$, the multivariate Gaussian density reduces to the bivariate Gaussian PDF. In the following steps, let σ_i^2 denote the variance of $X(t_i)$ and let $\rho = C_X(t_1, t_2 - t_1)/(\sigma_1\sigma_2)$ equal the correlation coefficient of $X(t_1)$ and $X(t_2)$.

(a) Find the covariance matrix $\mathbf{C}$ and show that the determinant is

$$|\mathbf{C}| = \sigma_1^2 \sigma_2^2 (1 - \rho^2).$$

(b) Show that the inverse of the correlation matrix is

$$\mathbf{C}^{-1} = \frac{1}{1 - \rho^2} \begin{bmatrix} \frac{1}{\sigma_1^2} & \frac{-\rho}{\sigma_1 \sigma_2} \\ \frac{-\rho}{\sigma_1 \sigma_2} & \frac{1}{\sigma_1^2} \end{bmatrix}.$$

(c) Now show that the multivariate Gaussian density for $X(t_1), X(t_2)$ is the bivariate Gaussian density.

13.11.5▧ Show that the Brownian motion process is a Gaussian random process. Hint: For $\mathbf{W}$ and $\mathbf{X}$ as defined in the proof of the Theorem 13.8, find matrix $\mathbf{A}$ such that $\mathbf{W} = \mathbf{AX}$ and then apply Theorem 8.11.

13.11.6◆◆ Let $X_1, X_2, \ldots$ denote a sequence of iid Gaussian $(0, 1)$ random variables. Let $N(t)$ denote a Poisson process of rate λ that is independent of the sequence X_n. Consider the random process

$$Y(t) = \sum_{n=0}^{N(t)} X_n.$$

(a) Find the conditional CDF

$$F_{Y(t)|N(t)}(y|n) = \mathrm{P}\left[Y(t) \le y|N(t) = n\right].$$

Express your answer in terms of the $\Phi(\cdot)$ function.

(b) Is $Y(t)$ a Gaussian process?

(c) Is $Y(t)$ a stationary process?

(d) Is $Y(t)$ wide-sense stationary?

13.12.1◉ Write a MATLAB program that generates and graphs the noisy cosine sample paths $X_{cc}(t)$, $X_{dc}(t)$, $X_{cd}(t)$, and $X_{dd}(t)$ of Figure 13.3. Note that the mathematical definition of $X_{cc}(t)$ is

$$X_{cc}(t) = 2\cos(2\pi t) + N(t), \quad -1 \le t \le 1.$$

Note that $N(t)$ is a white noise process with autocorrelation $R_N(\tau) = 0.01\delta(\tau)$. Practically, the graph of $X_{cc}(t)$ in Figure 13.3

is a sampled version $X_{cc}[n] = X_{cc}(nT_s)$, where the sampling period is $T_s = 0.001$s. In addition, the discrete-time functions are obtained by subsampling $X_{cc}[n]$. In subsampling, we generate $X_{dc}[n]$ by extracting every kth sample of $X_{cc}[n]$; see Problem 13.8.5. In terms of MATLAB, which starts indexing a vector x with first element x(1),

$$\texttt{Xdc(n)=Xcc(1+(n-1)k).}$$

The discrete-time graphs of Figure 13.3 used $k = 100$.

13.12.2◉ For the telephone switch of Example 13.22, we can estimate the expected number of calls in the system, $\mathrm{E}[M(t)]$, after T minutes using the time average estimate

$$\overline{M}_T = \frac{1}{T} \sum_{k=1}^{T} M(k).$$

Perform a 600-minute switch simulation and graph the sequence $\overline{M}_1, \overline{M}_2, \ldots, \overline{M}_{600}$. Does it appear that your estimates are converging? Repeat your experiment ten times and interpret your results.

13.12.3◉ A particular telephone switch handles only automated junk voicemail calls that arrive as a Poisson process of rate $\lambda = 100$ calls per minute. Each automated voicemail call has duration of exactly one minute. Use the method of Problem 13.12.2 to estimate the expected number of calls $\mathrm{E}[M(t)]$. Do your results differ very much from those of Problem 13.12.2?

13.12.4▧ Recall that for a rate λ Poisson process, the expected number of arrivals in $[0, T]$ is λT. Inspection of the code for `poissonarrivals(lambda,T)` will show that initially $n = \lceil 1.1\lambda T \rceil$ arrivals are generated. If $S_n > T$, the program stops and returns $\{S_j | S_j \le T\}$. Otherwise, if $S_n < T$, then we generate an additional n arrivals and check if $S_{2n} > T$. This process may be repeated an arbitrary number of times k until $S_{kn} > T$. Let K equal the number of times this process is repeated. What is $\mathrm{P}[K = 1]$? What is the disadvantage of choosing larger n so as to increase $P[K = 1]$?

13.12.5■ In this problem, we employ the result of Problem 13.5.8 as the basis for a function `s=newarrivals(lambda,T)` that generates a Poisson arrival process. The program `newarrivals.m` should do the following:

- Generate a sample value of N, a Poisson (λT) random variable.

- Given $N = n$, generate $\{U_1, \ldots, U_n\}$, a set of n uniform $(0, T)$ random variables.

- Sort $\{U_1, \ldots, U_n\}$ from smallest to largest and return the vector of sorted elements.

Write the program `newarrivals.m` and experiment to find out whether this program is any faster than `poissonarrivals.m`.

13.12.6◆ Suppose the Brownian motion process with $\alpha = 1$ is constrained by barriers. That is, we wish to generate a process $Y(t)$ such that $-b \leq Y(t) \leq b$ for a constant $b > 0$. Build a simulation of this system. Estimate $P[Y(t) = b]$.

13.12.7◆ For the departure process $D(t)$ of Example 13.22, let D_n denote the time of the nth departure. The nth *interdeparture time* is then $V_n = D_n - D_{n-1}$. From a sample path containing 1000 departures, estimate the PDF of V_n. Is it reasonable to model V_n as an exponential random variable? What is the mean interdeparture time?

Appendix A
Families of Random Variables

A.1 Discrete Random Variables

Bernoulli (p)

For $0 \leq p \leq 1$,

$$P_X(x) = \begin{cases} 1-p & x = 0 \\ p & x = 1 \\ 0 & \text{otherwise} \end{cases} \qquad \phi_X(s) = 1 - p + pe^s$$

$$E[X] = p$$
$$\text{Var}[X] = p(1-p)$$

Binomial (n, p)

For a positive integer n and $0 \leq p \leq 1$,

$$P_X(x) = \binom{n}{x} p^x (1-p)^{n-x} \qquad \phi_X(s) = (1 - p + pe^s)^n$$

$$E[X] = np$$
$$\text{Var}[X] = np(1-p)$$

Discrete Uniform (k, l)

For integers k and l such that $k < l$,

$$P_X(x) = \begin{cases} 1/(l-k+1) & x = k, k+1, \ldots, l \\ 0 & \text{otherwise} \end{cases} \qquad \phi_X(s) = \frac{e^{sk} - e^{s(l+1)}}{(l-k+1)(1-e^s)}$$

$$E[X] = \frac{k+l}{2}$$
$$\text{Var}[X] = \frac{(l-k)(l-k+2)}{12}$$

Geometric (p)

For $0 < p \leq 1$,

$$P_X(x) = \begin{cases} p(1-p)^{x-1} & x = 1, 2, \ldots \\ 0 & \text{otherwise} \end{cases} \qquad \phi_X(s) = \frac{pe^s}{1 - (1-p)e^s}$$

$$E[X] = 1/p$$
$$\text{Var}[X] = (1-p)/p^2$$

Multinomial

For integer $n > 0$, $p_i \geq 0$ for $i = 1, \ldots, n$, and $p_1 + \cdots + p_n = 1$,

$$P_{X_1, \ldots, X_r}(x_1, \ldots, x_r) = \binom{n}{x_1, \ldots, x_r} p_1^{x_1} \cdots p_r^{x_r}$$

$$E[X_i] = np_i$$
$$\text{Var}[X_i] = np_i(1 - p_i)$$

Pascal (k, p)

For positive integer k, and $0 < p < 1$,

$$P_X(x) = \binom{x-1}{k-1} p^k (1-p)^{x-k} \qquad \phi_X(s) = \left(\frac{pe^s}{1 - (1-p)e^s} \right)^k$$

$$E[X] = k/p$$
$$\text{Var}[X] = k(1-p)/p^2$$

Poisson (α)

For $\alpha > 0$,

$$P_X(x) = \begin{cases} \dfrac{\alpha^x e^{-\alpha}}{x!} & x = 0, 1, 2, \ldots \\ 0 & \text{otherwise} \end{cases} \qquad \phi_X(s) = e^{\alpha(e^s - 1)}$$

$$E[X] = \alpha$$
$$\text{Var}[X] = \alpha$$

━━━**Zipf** (n, α)━━━

For positive integer $n > 0$ and constant $\alpha \geq 1$,

$$P_X(x) = \begin{cases} \dfrac{c(n, \alpha)}{x^\alpha} & x = 1, 2, \ldots, n \\ 0 & \text{otherwise} \end{cases}$$

where

$$c(n, \alpha) = \left(\sum_{k=1}^{n} \frac{1}{k^\alpha} \right)^{-1}$$

A.2 Continuous Random Variables

━━━**Beta** (i, j)━━━

For positive integers i and j, the beta function is defined as

$$\beta(i, j) = \frac{(i + j - 1)!}{(i - 1)!(j - 1)!}$$

For a $\beta(i, j)$ random variable X,

$$f_X(x) = \begin{cases} \beta(i, j)x^{i-1}(1 - x)^{j-1} & 0 < x < 1 \\ 0 & \text{otherwise} \end{cases}$$

$$\mathrm{E}[X] = \frac{i}{i + j}$$

$$\mathrm{Var}[X] = \frac{ij}{(i + j)^2(i + j + 1)}$$

━━━**Cauchy** (a, b)━━━

For constants $a > 0$ and $-\infty < b < \infty$,

$$f_X(x) = \frac{1}{\pi} \frac{a}{a^2 + (x - b)^2} \qquad\qquad \phi_X(s) = e^{bs - a|s|}$$

Note that $\mathrm{E}[X]$ is undefined since $\int_{-\infty}^{\infty} x f_X(x)\, dx$ is undefined. Since the PDF is symmetric about $x = b$, the mean can be defined, in the sense of a principal value, to be b.

$$\mathrm{E}[X] \equiv b$$

$$\mathrm{Var}[X] = \infty$$

Erlang (n, λ)

For $\lambda > 0$, and a positive integer n,

$$f_X(x) = \begin{cases} \dfrac{\lambda^n x^{n-1} e^{-\lambda x}}{(n-1)!} & x \geq 0 \\ 0 & \text{otherwise} \end{cases} \qquad \phi_X(s) = \left(\dfrac{\lambda}{\lambda - s}\right)^n$$

$$\mathrm{E}[X] = n/\lambda$$

$$\mathrm{Var}[X] = n/\lambda^2$$

Exponential (λ)

For $\lambda > 0$,

$$f_X(x) = \begin{cases} \lambda e^{-\lambda x} & x \geq 0 \\ 0 & \text{otherwise} \end{cases} \qquad \phi_X(s) = \dfrac{\lambda}{\lambda - s}$$

$$\mathrm{E}[X] = 1/\lambda$$

$$\mathrm{Var}[X] = 1/\lambda^2$$

Gamma (a, b)

For $a > -1$ and $b > 0$,

$$f_X(x) = \begin{cases} \dfrac{x^a e^{-x/b}}{a! b^{a+1}} & x > 0 \\ 0 & \text{otherwise} \end{cases} \qquad \phi_X(s) = \dfrac{1}{(1 - bs)^{a+1}}$$

$$\mathrm{E}[X] = (a + 1)b$$

$$\mathrm{Var}[X] = (a + 1)b^2$$

Gaussian (μ, σ)

For constants $\sigma > 0$, $-\infty < \mu < \infty$,

$$f_X(x) = \dfrac{e^{-(x-\mu)^2/2\sigma^2}}{\sigma\sqrt{2\pi}} \qquad \phi_X(s) = e^{s\mu + s^2\sigma^2/2}$$

$$\mathrm{E}[X] = \mu$$

$$\mathrm{Var}[X] = \sigma^2$$

━━━━**Laplace** (a, b)━━━━

For constants $a > 0$ and $-\infty < b < \infty$,

$$f_X(x) = \frac{a}{2} e^{-a|x-b|} \qquad\qquad \phi_X(s) = \frac{a^2 e^{bs}}{a^2 - s^2}$$

$$E[X] = b$$

$$\mathrm{Var}[X] = 2/a^2$$

━━━━**Log-normal** (a, b, σ)━━━━

For constants $-\infty < a < \infty$, $-\infty < b < \infty$, and $\sigma > 0$,

$$f_X(x) = \begin{cases} \dfrac{e^{-(\ln(x-a)-b)^2/2\sigma^2}}{\sqrt{2\pi}\sigma(x-a)} & x > a \\[2mm] 0 & \text{otherwise} \end{cases}$$

$$E[X] = a + e^{b+\sigma^2/2}$$

$$\mathrm{Var}[X] = e^{2b+\sigma^2}\left(e^{\sigma^2} - 1\right)$$

━━━━**Maxwell** (a)━━━━

For $a > 0$,

$$f_X(x) = \begin{cases} \sqrt{2/\pi}\, a^3 x^2 e^{-a^2 x^2/2} & x > 0 \\ 0 & \text{otherwise} \end{cases}$$

$$E[X] = \sqrt{\frac{8}{a^2 \pi}}$$

$$\mathrm{Var}[X] = \frac{3\pi - 8}{\pi a^2}$$

━━━━**Pareto** (α, μ)━━━━

For $\alpha > 0$ and $\mu > 0$,

$$f_X(x) = \begin{cases} (\alpha/\mu)(x/\mu)^{-(\alpha+1)} & x \geq \mu \\ 0 & \text{otherwise} \end{cases}$$

$$E[X] = \frac{\alpha\mu}{\alpha - 1} \qquad\qquad\qquad (\alpha > 1)$$

$$\mathrm{Var}[X] = \frac{\alpha\mu^2}{(\alpha - 2)(\alpha - 1)^2} \qquad\qquad (\alpha > 2)$$

Rayleigh (a)

For $a > 0$,

$$f_X(x) = \begin{cases} a^2 x e^{-a^2 x^2/2} & x > 0 \\ 0 & \text{otherwise} \end{cases}$$

$$\mathrm{E}\,[X] = \sqrt{\frac{\pi}{2a^2}}$$

$$\mathrm{Var}[X] = \frac{2 - \pi/2}{a^2}$$

Uniform (a, b)

For constants $a < b$,

$$f_X(x) = \begin{cases} \dfrac{1}{b - a} & a < x < b \\ 0 & \text{otherwise} \end{cases} \qquad\qquad \phi_X(s) = \frac{e^{bs} - e^{as}}{s(b - a)}$$

$$\mathrm{E}\,[X] = \frac{a + b}{2}$$

$$\mathrm{Var}[X] = \frac{(b - a)^2}{12}$$

Appendix B
A Few Math Facts

This text assumes that the reader knows a variety of mathematical facts. Often these facts go unstated. For example, we use many properties of limits, derivatives, and integrals. Generally, we have omitted comment or reference to mathematical techniques typically employed by engineering students.

However, when we employ math techniques that a student may have forgotten, the result can be confusion. It becomes difficult to separate the math facts from the probability facts. To decrease the likelihood of this event, we have summarized certain key mathematical facts. In the text, we have noted when we use these facts. If any of these facts are unfamiliar, we encourage the reader to consult with a textbook in that area.

Trigonometric Identities

Math Fact B.1 Half Angle Formulas

$$\cos(A + B) = \cos A \cos B - \sin A \sin B \qquad \sin(A + B) = \sin A \cos B + \cos A \sin B$$
$$\cos 2A = \cos^2 A - \sin^2 A \qquad \qquad \sin 2A = 2\sin A \cos A$$

Math Fact B.2 Products of Sinusoids

$$\sin A \sin B = \frac{1}{2}\left[\cos(A - B) - \cos(A + B)\right]$$
$$\cos A \cos B = \frac{1}{2}\left[\cos(A - B) + \cos(A + B)\right]$$
$$\sin A \cos B = \frac{1}{2}\left[\sin(A + B) + \sin(A - B)\right]$$

Math Fact B.3 The Euler Formula

The Euler formula $e^{j\theta} = \cos\theta + j\sin\theta$ is the source of the identities

$$\cos\theta = \frac{e^{j\theta} + e^{-j\theta}}{2} \qquad\qquad \sin\theta = \frac{e^{j\theta} - e^{-j\theta}}{2j}$$

Sequences and Series

=====Math Fact B.4=====Finite Geometric Series

The finite geometric series is

$$\sum_{i=0}^{n} q^i = 1 + q + q^2 + \cdots + q^n = \frac{1 - q^{n+1}}{1 - q}.$$

. . . .

To see this, multiply left and right sides by $(1 - q)$ to obtain

$$(1 - q) \sum_{i=0}^{n} q^i = (1 - q)(1 + q + q^2 + \cdots + q^n) = 1 - q^{n+1}.$$

=====Math Fact B.5=====Infinite Geometric Series

When $|q| < 1$,

$$\sum_{i=0}^{\infty} q^i = \lim_{n \to \infty} \sum_{i=0}^{n} q^i = \lim_{n \to \infty} \frac{1 - q^{n+1}}{1 - q} = \frac{1}{1 - q}.$$

=====Math Fact B.6=====

$$\sum_{i=1}^{n} i q^i = \frac{q \left(1 - q^n [1 + n(1 - q)]\right)}{(1 - q)^2}.$$

=====Math Fact B.7=====

If $|q| < 1$,

$$\sum_{i=1}^{\infty} i q^i = \frac{q}{(1 - q)^2}.$$

=====**Math Fact B.8**=====

$$\sum_{j=1}^{n} j = \frac{n(n+1)}{2}$$

=====**Math Fact B.9**=====

$$\sum_{j=1}^{n} j^2 = \frac{n(n+1)(2n+1)}{6}$$

Calculus

=====**Math Fact B.10**=====**Integration by Parts**

The integration by parts formula is

$$\int_{a}^{b} u\, dv = uv\Big|_{a}^{b} - \int_{a}^{b} v\, du.$$

=====**Math Fact B.11**=====**Gamma Function**

The gamma function is defined as

$$\Gamma(z) = \int_{0}^{\infty} t^{z-1} e^{-t}\, dt.$$

If $z = n$, a positive integer, then $\Gamma(n) = (n-1)!$. Also note that $\Gamma(1/2) = \sqrt{\pi}$, $\Gamma(3/2) = \sqrt{\pi}/2$, and $\Gamma(5/2) = 3\sqrt{\pi}/4$.

════Math Fact B.12════Leibniz's Rule

The function

$$R(\alpha) = \int_{a(\alpha)}^{b(\alpha)} r(\alpha, x)\, dx$$

has derivative

$$\frac{dR(\alpha)}{d\alpha} = -r\big(\alpha, a(\alpha)\big)\frac{da(\alpha)}{d\alpha} + r\big(\alpha, b(\alpha)\big)\frac{db(\alpha)}{d\alpha} + \int_{a(\alpha)}^{b(\alpha)} \frac{\partial r(\alpha, x)}{\partial \alpha}\, dx.$$

In the special case when $a(\alpha) = a$ and $b(\alpha) = b$ are constants,

$$R(\alpha) = \int_a^b r(\alpha, x)\, dx,$$

and Leibniz's rule simplifies to

$$\frac{dR(\alpha)}{d\alpha} = \int_a^b \frac{\partial r(\alpha, x)}{\partial \alpha}\, dx.$$

════Math Fact B.13════Change-of-Variable Theorem

Let $\mathbf{x} = T(\mathbf{y})$ be a continuously differentiable transformation from $\mathcal{U}^n$ to $\mathcal{R}^n$. Let R be a set in $\mathcal{U}^n$ having a boundary consisting of finitely many smooth sets. Suppose that R and its boundary are contained in the interior of the domain of T, T is one-to-one of R, and $\det(()T')$, the Jacobian determinant of T, is nonzero on R. Then, if $f(\mathbf{x})$ is bounded and continuous on $T(R)$,

$$\int_{T(R)} f(\mathbf{x})dV_{\mathbf{x}} = \int_R f(T(\mathbf{y}))|\det(T)'|\, dV_{\mathbf{y}}.$$

Vectors and Matrices

════Math Fact B.14════Vector/Matrix Definitions

(a) Vectors $\mathbf{x}$ and $\mathbf{y}$ are *orthogonal* if $\mathbf{x}'\mathbf{y} = 0$.

(b) A number λ is an *eigenvalue* of a matrix $\mathbf{A}$ if there exists a vector $\mathbf{x}$ such that $\mathbf{A}\mathbf{x} = \lambda\mathbf{x}$. The vector $\mathbf{x}$ is an *eigenvector* of matrix $\mathbf{A}$.

(c) A matrix $\mathbf{A}$ is *symmetric* if $\mathbf{A} = \mathbf{A}'$.

(d) A square matrix $\mathbf{A}$ is *unitary* if $\mathbf{A}'\mathbf{A}$ equals the identity matrix $\mathbf{I}$.

(e) A real symmetric matrix $\mathbf{A}$ is *positive definite* if $\mathbf{x}'\mathbf{A}\mathbf{x} > 0$ for every nonzero vector $\mathbf{x}$.

(f) A real symmetric matrix $\mathbf{A}$ is *positive semidefinite* if $\mathbf{x}'\mathbf{A}\mathbf{x} \geq 0$ for every nonzero vector $\mathbf{x}$.

(g) A set of vectors $\{\mathbf{x}_1, \ldots, \mathbf{x}_n\}$ is *orthonormal* if $\mathbf{x}'_i\mathbf{x}_j = 1$ if $i = j$ and otherwise equals zero.

(h) A matrix $\mathbf{U}$ is *unitary* if its columns $\{\mathbf{u}_1, \ldots, \mathbf{u}_n\}$ are orthonormal.

════Math Fact B.15════Real Symmetric Matrices

A real symmetric matrix $\mathbf{A}$ has the following properties:

(a) All eigenvalues of $\mathbf{A}$ are real.

(b) If $\mathbf{x}_1$ and $\mathbf{x}_2$ are eigenvectors of $\mathbf{A}$ corresponding to eigenvalues $\lambda_1 \neq \lambda_2$, then $\mathbf{x}_1$ and $\mathbf{x}_2$ are orthogonal vectors.

(c) $\mathbf{A}$ can be written as $\mathbf{A} = \mathbf{U}\mathbf{D}\mathbf{U}'$ where $\mathbf{D}$ is a diagonal matrix and $\mathbf{U}$ is a unitary matrix with columns that are n orthonormal eigenvectors of $\mathbf{A}$.

════Math Fact B.16════Positive Definite Matrices

For a real symmetric matrix $\mathbf{A}$, the following statements are equivalent:

(a) $\mathbf{A}$ is a *positive definite* matrix.

(b) $\mathbf{x}'\mathbf{A}\mathbf{x} > 0$ for all nonzero vectors $\mathbf{x}$.

(c) Each eigenvalue λ of $\mathbf{A}$ satisfies $\lambda > 0$.

(d) There exists a nonsingular matrix $\mathbf{W}$ such that $\mathbf{A} = \mathbf{W}\mathbf{W}'$.

Math Fact B.17 Positive Semidefinite Matrices

For a real symmetric matrix $\mathbf{A}$, the following statements are equivalent:

(a) $\mathbf{A}$ is a positive semidefinite matrix.

(b) $\mathbf{x}'\mathbf{A}\mathbf{x} \geq 0$ for all vectors $\mathbf{x}$.

(c) Each eigenvalue λ of $\mathbf{A}$ satisfies $\lambda \geq 0$.

(d) There exists a matrix $\mathbf{W}$ such that $\mathbf{A} = \mathbf{W}\mathbf{W}'$.

References

Ber98.   P. L. Bernstein. *Against the Gods: The Remarkable Story of Risk*. John Wiley, 1998.

Bil12.   P. Billingsley. *Probability and Measure*. John Wiley & Sons, anniversary edition, 2012.

BT08.   D.P. Bertsekas and J.N. Tsitsiklis. *Introduction to Probability*. Athena Scientific, 2nd edition, 2008.

Dav10.   T. A. Davis. *MATLAB Primer*. CRC Press, 8th edition, 2010.

Doo90.   J. L. Doob. *Stochastic Processes*. Wiley Reprint, 1990.

Dra67.   A. W. Drake. *Fundamentals of Applied Probability Theory*. McGraw-Hill, New York, 1967.

Dur94.   R. Durrett. *The Essentials of Probability*. Duxbury, 1994.

Gal13.   R. G. Gallager. *Stochastic Processes: Theory for Applications*. Cambridge University Press, 2013.

GS93.   L. Gonick and W. Smith. *The Cartoon Guide to Statistics*. Harper Perennial, 1993.

Gub06.   J. Gubner. *Probability and Random Processes for Electrical and Computer Engineers*. Cambridge University Press, 2006.

Hay01.   Simon Haykin. *Communication Systems*. John Wiley, 4th edition, 2001.

HL11.   D. Hanselman and B. Littlefield. *Mastering MATLAB*. Prentice Hall, 2011.

HSP87.   P. G. Hoel, C. J. Stone, and S. C. Port. *Introduction to Stochastic Processes*. Waveland Press, 1987.

Kay98.   S. M. Kay. *Fundamentals of Statistical Signal Processing Volume II: Detection theory*. Prentice Hall, 1998.

KMT12.   H. Kobayashi, B. Mark, and W. Turin. *Probability, Random Processes, and Statistical Analysis: Applications to Communications, Signal Processing, Queueing Theory and Mathematical Finance*. Cambridge University Press, 2012.

LG11.   A. Leon-Garcia. *Probability, Statistics, and Random Processes for Electrical Engineering*. Prentice Hall, third edition, 2011.

MR10.   D. C. Montgomery and G. C. Runger. *Applied Statistics and Probability for Engineers*. John Wiley & Sons, fifth edition, 2010.

Pos01.   K. Poskitt. *Do You Feel Lucky? The Secrets of Probability*. Scholastic, 2001.

PP02.   A. Papoulis and S. U. Pillai. *Probability, Random Variables and Stochastic Processes.* McGraw Hill, 4th edition, 2002.

Ros12.   S. M. Ross. *A First Course in Probability.* Pearson, ninth edition, 2012.

SMM10.   R. L. Scheaffer, M. Mulekar, and J. T. McClave. *Probability and Statistics for Engineers.* Cengage Learning, 5th edition, 2010.

Str98.   G. Strang. *Introduction to Linear Algebra.* Wellesley Cambridge Press, second edition, 1998.

Ver98.   S. Verdú. *Multiuser Detection.* Cambridge University Press, New York, 1998.

WS01.   J. W. Woods and H. Stark. *Probability and Random Processes with Applications to Signal Processing.* Prentice Hall, 3rd edition, 2001.

Index